Reliability and Risk Analysis

Completely updated for a new edition, this book introduces reliability and risk analysis for both practicing engineers and engineering students at the undergraduate and graduate levels. Since reliability analysis is a multidisciplinary subject, this book draws together a wide range of topics and presents them in a way that applies to most engineering disciplines.

Reliability and Risk Analysis, Second Edition, emphasizes an introduction and explanation of the practical methods used in reliability and risk studies, with a discussion of their uses and limitations. It offers basic and advanced methods in reliability analysis that are commonly used in daily practice and provides methods that address unique topics such as dependent failure analysis, importance analysis, and analysis of repairable systems. This book goes on to present a comprehensive overview of modern probabilistic life assessment methods such as Bayesian estimation, system reliability analysis, and human reliability. End-of-chapter problems and a solutions manual are available to support any course adoptions.

This book is refined, simple, and focused on fundamentals. The audience is the beginner with no background in reliability engineering and rudimentary knowledge of probability and statistics. It can be used by new practitioners, undergraduates, and first-year graduate students.

What Every Engineer Should Know

Series Editor:
Phillip A. Laplante
Pennsylvania State University

What Every Engineer Should Know about MATLAB® and Simulink®
Adrian B. Biran

Green Entrepreneur Handbook: The Guide to Building and Growing a Green and Clean Business
Eric Koester

What Every Engineer Should Know about Cyber Security and Digital Forensics
Joanna F. DeFranco

What Every Engineer Should Know about Modeling and Simulation
Raymond J. Madachy and Daniel Houston

What Every Engineer Should Know about Excel, Second Edition
J.P. Holman and Blake K. Holman

Technical Writing: A Practical Guide for Engineers, Scientists, and Nontechnical Professionals, Second Edition
Phillip A. Laplante

What Every Engineer Should Know About the Internet of Things
Joanna F. DeFranco and Mohamad Kassab

What Every Engineer Should Know about Software Engineering
Phillip A. Laplante and Mohamad Kassab

What Every Engineer Should Know About Cyber Security and Digital Forensics
Joanna F. DeFranco and Bob Maley

Ethical Engineering: A Practical Guide with Case Studies
Eugene Schlossberger

What Every Engineer Should Know About Data-Driven Analytics
Phillip A. Laplante and Satish Mahadevan Srinivasan

Reliability and Risk Analysis
Mohammad Modarres and Katrina Groth

For more information about this series, please visit:
www.routledge.com/What-Every-Engineer-Should-Know/book-series/CRCWEESK

Reliability and Risk Analysis

Second Edition

Mohammad Modarres
Katrina Groth

CRC Press is an imprint of the
Taylor & Francis Group, an **informa** business

Designed cover image: Vincent Paglioni and Katrina Groth

Second edition published 2023
by CRC Press
6000 Broken Sound Parkway NW, Suite 300, Boca Raton, FL 33487-2742

and by CRC Press
4 Park Square, Milton Park, Abingdon, Oxon, OX14 4RN

CRC Press is an imprint of Taylor & Francis Group, LLC

First edition published by CRC Press 1992

Library of Congress Cataloging-in-Publication Data
Names: Modarres, M. (Mohammad), author. | Groth, Katrina, author.
Title: What every engineer should know about reliability and risk analysis /
Mohammad Modarres, Katrina Groth.
Description: Second edition. | Boca Raton : CRC Press, [2023] |
Includes index. Identifiers: LCCN 2022050183 (print) | LCCN 2022050184 (ebook) |
ISBN 9781032309729 (pbk) | ISBN 9781032309736 (hbk) | ISBN 9781003307495 (ebk)
Subjects: LCSH: Reliability (Engineering) | Risk assessment.
Classification: LCC TA169 .M63 2023 (print) | LCC TA169 (ebook) |
DDC 620/.00452—dc23/eng/20221018
LC record available at https://lccn.loc.gov/2022050183
LC ebook record available at https://lccn.loc.gov/2022050184

ISBN: 978-1-032-30973-6 (hbk)
ISBN: 978-1-032-30972-9 (pbk)
ISBN: 978-1-003-30749-5 (ebk)

DOI: 10.1201/9781003307495

Typeset in Times
by codeMantra

Dedication

Dedicated to all engineers who understand the importance of the field and continue to advance it in research and practice.

Contents

Preface to the Second Edition

Reliability engineering and risk analysis are essential branches of science and engineering that enable us to uphold the integrity of engineered systems and the engineering profession. These capabilities are relevant to various engineering disciplines, including mechanical, chemical, aerospace, nuclear, oil and gas, transportation, electrical, and more. Since engineering designs and processes impose various hazards, risk and reliability analyses are universally important. Interest in reliability and risk analysis has increased due to accidents that have recently resulted in significant public attention, such as the nuclear accidents at Three Mile Island and Fukushima Daiichi, the Deepwater Horizon oil spill in the Gulf of Mexico, the Space Shuttle Challenger disintegration, the Bhopal chemical disaster in India, the Concord crash in Paris, the 2016 and 2021 Tesla's self-driving cars crashes, and the San Bruno pipeline explosion. The nature of reliability and risk make this an inherently multidisciplinary field because systems and their failures involve the interplay between machines, software, humans, organizations, policy, economics, and the physical world. As systems become more complex and interconnected, learning the methods used to prevent their failures becomes more essential.

This book provides a comprehensive introduction to reliability and an overview of risk analysis for engineering students and practicing engineers. The focus is on the fundamental methods commonly used in practice, along with numerous examples. The book's emphasis is on the introduction and explanation of the practical concepts and techniques used in reliability and risk studies and discussion of their use. These concepts form the foundation of a degree focused on reliability and prepare the student for coursework on advanced reliability methods. This book corresponds to the first course a graduate student takes in reliability engineering and is also suitable for a senior-level undergraduate engineering course or self-study by a practitioner. This book assumes that the readers have taken multivariate calculus and several years of engineering coursework and have some basic familiarity with probability and statistics.

The first edition was developed by Mohammad Modarres over 30 years ago from material he developed presented over 10 years in undergraduate and graduate-level courses in Reliability Analysis and Risk Assessment at the University of Maryland. This new version of the book restructures and updates the content based on the further development of the course by Katrina Groth over the past 5 years.

Information has been streamlined and simplified to enhance understanding of the foundational elements of the discipline. The notation has been standardized across chapters, and some outdated methods and examples have been removed and replaced with a discussion of more current techniques. Chapters have been condensed and reorganized to enhance readability and understanding. Some aspects of the book have been changed considerably.

We structured the book to provide a series of fundamental elements that build upon each other, and thus each chapter builds upon the chapters that come before it. The introduction chapter (Chapter 1) defines reliability, availability, and risk analysis.

Chapter 2 introduces the probability mathematics necessary to conduct reliability engineering, while Chapter 3 explains how the basic probabilistic reliability methods characterize the reliability of a basic engineering unit (i.e., a component). Chapter 4 introduces the statistical techniques used to conduct reliability data analysis. In Chapter 5, we discuss the uses of probability, statistics, and data analysis techniques to assess and predict component reliability.

In Chapter 6, we present the analytical methods used to assess the reliability of systems (i.e., an engineering unit consisting of many interacting components). The availability and reliability considerations for repairable systems are discussed in Chapter 7. This chapter also explains the corresponding use of the analytical and statistical methods discussed in the earlier chapters and connects them to performing availability analysis of repairable components and engineering systems. In Chapter 8, we introduce four important advanced topics sometimes overlooked in introductory texts but which we believe are foundational to the field and essential for dealing with modern systems. The topics covered are uncertainty analysis, dependent failures, importance measures, and human reliability analysis. While Chapter 8 does not cover the topics in depth, the description should whet the reader's appetite and motivate further study. This book concludes with Chapter 9, which brings these pieces together under the umbrella of risk analysis, a process that integrates nearly all the analytical methods explained in the preceding eight chapters.

For a 16-week undergraduate course, we recommend covering the material in Chapters 1–6 and concluding with Chapter 9. A 16-week graduate course in reliability engineering should be able to cover the entire book. Once these fundamentals are mastered, a student is prepared to take subsequent coursework which delves into more advanced topics in reliability engineering, such as comprehensive probabilistic risk assessment, quantifying and propagating parameter uncertainties, physics of failure, human reliability analysis, dynamic risk assessment, machine learning, prognosis and health management.

This book could not have been materialized without the help of numerous colleagues and graduate students since the publication of its first edition. Of particular note are Vincent Paglioni and Yu-Shu Hu, who were instrumental in compiling and proofreading the manuscript and drawing the figures. Numerous other students and colleagues contributed to varying degrees; it is not easy to name all, but we are grateful for their contributions. In addition, we are grateful to V. Paglioni, T. Williams and S. Karunakaran for drawing figures for the second edition. Finally, we are grateful to our spouses for their endless patience with the long hours and disruption that came with writing this book and for reminding us why it all matters.

Mohammad Modarres and Katrina Groth

Authors

Mohammad Modarres is a scientist and educator in risk and reliability engineering. He is the Nicole Y. Kim Eminent Professor at the University of Maryland. Dr. Modarres co-founded the world's first degree-granting graduate curriculum in Reliability Engineering. As the University of Maryland Center for Risk and Reliability (CRR) director, Professor Modarres serves as an international expert on reliability and risk analysis. He has authored numerous textbooks, book chapters, and hundreds of scholarly papers. His areas of research interests include probabilistic risk assessment, degradation and physics of failure of materials, nuclear safety, and fracture mechanics. His works in structural integrity and prognosis and health management include both experimental and probabilistic model development. He is a University of Maryland Distinguished Scholar-Teacher, a fellow of the Institute of Electrical and Electronics (IEEE), a fellow of the American Nuclear Society (ANS), a fellow of the Asia-Pacific AI Association (AAIA), and a recipient of the prestigious ANS Tommy Thompson Award for nuclear safety. Dr. Modarres received his BS in Mechanical Engineering from Tehran Polytechnic, MS in Mechanical Engineering, and MS and Ph.D. in Nuclear Engineering from Massachusetts Institute of Technology.

Katrina Groth is an associate professor of Mechanical Engineering and the associate director of the Center for Risk and Reliability at the University of Maryland. Her research focuses on engineering safer systems through risk assessment and reliability engineering, with direct impact on hydrogen fueling stations, hydrogen storage, gas pipelines, aviation, and nuclear power plants. She has published over 130 archival papers and reports, authored multiple software packages, and holds 2 patents. Prior to joining UMD, Groth was a principal R&D engineer at Sandia National Laboratories, where she invented the HyRAM (Hydrogen Risk Assessment Models) toolkit used in dozens of countries to establish hydrogen safety standards. Groth holds a B.S. in Engineering and M.S. and Ph.D. in Reliability Engineering from the University of Maryland. She is the recipient of an NSF CAREER award, a technical leadership award from the U.S. Department of Energy's Hydrogen and Fuel Cell Technology Office, and the 2021 David Okrent Award for Nuclear Safety from the American Nuclear Society. Groth believes that risk assessment should be in every engineer's toolbox.

1 Reliability Engineering Perspective and Fundamentals

1.1 WHY STUDY RELIABILITY?

Structures, systems, and components designs are not perfect. While a naïve view would insist that it is technologically possible to engineer a perfect system and design out all failures, unfortunately, this view is idealistic, impractical, and economically infeasible. It is an essential job of engineers to prevent, mitigate, respond to, and recover from failures. Reliability engineering helps ensure robust design, product reliability, and economic competitiveness. Beyond this, reliability engineering helps engineers achieve their duty to hold paramount the safety, health, and welfare of the public, workers, and the environment. Reliability engineering and risk analysis are essential capabilities that enable us to uphold the integrity of the systems we design, build, operate, maintain, and ultimately attain the high integrity of the engineering profession.

Myriad examples of engineering failures motivate the study of reliability. For instance, on March 11, 2011, an earthquake and tsunami induced a loss of reactor cooling at Japan's Fukushima Daiichi nuclear power plant. This resulted in meltdown, evacuations, radiation release, and environmental contamination. The cleanup continued a decade later with a cost estimated to exceed $200 billion.

Another example comes from the space sector: on February 1, 2003, the Space Shuttle Columbia disintegrated on reentry. The failure scenario was initiated when a small piece of foam insulation broke off during takeoff, causing damage that allowed hot gases to penetrate and destroy the wing. The failure led to the loss of all seven crew members, and the U.S. space shuttle program was halted for over 2 years while NASA investigated the accident and implemented new safety measures. While the Space Shuttle Columbia incident resulted in disaster, NASA was aware that problems with foam had been happening for years. More minor failures happen—and careful study of these is an opportunity to prevent catastrophic failure.

An example from the automotive sector demonstrates additional reasons for studying reliability. Widespread Firestone/Ford tire failures in the 1990s led to a recall in 2000. It is estimated that these tire failures cost Firestone's parent company, Bridgestone, $1.6+ billion and Ford Motor Company $500+ million. In addition, severe financial impacts extended beyond the immediate losses: subsequent corporate restructuring cost Bridgestone $2 billion. The failures also resulted in new regulatory oversight with the passage of the Transportation Recall Enhancement, Accountability and Documentation (TREAD Act), which requires vehicle manufacturers to report safety recalls and defects to the National Highway & Transportation Safety Administration and holds vehicle manufacturers criminally liable if these reporting requirements are violated.

DOI: 10.1201/9781003307495-1

These are a few examples among many that motivate our studies. Other incidents and near-misses have happened in many industries. In the nuclear industry, the Three Mile Island core damage in 1979 was a well-known disaster and regulatory turning point for the U.S. nuclear industry. For NASA, the 1986 accident of the Space Shuttle Challenger preceded the Columbia disaster. In the chemical, oil, and gas industry, Bhopal, Deepwater Horizon, Texas City, and Piper Alpha accidents have led to significant economic, environmental, and regulatory consequences. Building fires such as the Grenfell Tower fire and the Station nightclub fire led to significant loss of life. On the aviation side: Tenerife, Air France 447, and numerous other crashes have led to the loss of human life and changes to the aviation system.

Reliability consideration is not limited only to technologies with catastrophic failure potential. Designers, manufacturers, and end-users strive to minimize the occurrence and recurrence of failures in the components, devices, products, and systems we encounter daily. However, even if technical knowledge is not a limiting factor in designing, manufacturing, constructing, and operating a failure-free design, the cost of development, testing, materials, and engineering analysis may far exceed economic prospects for such a design. Therefore, practical and financial limitations dictate the use of less than perfect designs.

A reliable design remains operational and attains its objective without failure during its mission time. However, achieving a reliable design first requires us to understand what an item is intended to do and why and how failures occur. Further, to maximize system performance and efficiently use resources, it is also critical to know *how often* such failures occur. Finally, it is important to understand various strategies to reduce failures. When failure is understood and appropriately considered in the life cycle of a design, the impact or occurrence rate can be minimized. Further, the system can be protected against failure impacts through careful engineering and economic analysis.

1.2 HISTORY AND EVOLUTION OF THE FIELD

Interest in establishing a quantitative measure for design quality began during World War II, as did the design concept that a chain is only as strong as its weakest link. However, the ideas were primitive and ad hoc. After the war, between 1945 and 1950, the U.S. Air Force became concerned about the quality of electronic products. The starting point of a formal reliability engineering practice is traced back to the Ad Hoc Group on Reliability of Electronic Equipment, established in December 1950. However, the formation and active involvement of the Advisory Group on the Reliability of Electronic Equipment (AGREE) by the U.S. Department of Defense from 1956 to 1958 is considered the turning point in modern reliability engineering.

Around 1953, the applications of the probabilistic notions of reliability represented by the exponential distribution became systematically and widely used. One of the main driving forces for this popularity was the simplicity of the corresponding reliability functions. This simplicity accelerated many improvements in traditional statistical and probabilistic approaches to measuring, predicting, and testing item reliability in the 1950s. The reliability block diagram (RBD) concept was adopted from other engineering applications and used to assess system reliability.

By the 1960s, the exponential distribution proved impractical for many applications and because it is sensitive to departure from the initial assumptions of constant failure rate. Applying this model for components with high reliability targets could result in unrealistic mean time to failure (MTTF) estimates. This was because the exponential model ignores any aging and degradation accumulated in the item. Distributions that considered degradation and wear-out were used, such as the Weibull distribution. The physics of failure (PoF) concept was introduced in the early 1960s as part of a series of symposia at the Rome Air Development Center (which later became the U.S. Air Force's Rome Laboratory). However, because of its complexity, it was not until the late 1980s that PoF became a serious alternative to statistical reliability prediction methods. The concept of failure modes and effects analysis (FMEA), although introduced by the U.S. military in the late 1940s, was revived by NASA in the early 1960s for its Apollo and other missions as an inductive approach to the analysis of complex systems for which the RBD approach was inadequate.

The 1970s experienced more complex system analysis methods, the most important of which was the deductive fault tree analysis approach. This approach was motivated by the aerospace industry's need for risk and safety assessment and later applied to assess nuclear power plant risks. In this decade, there was intense interest in system-level safety, risk, and reliability in different applications such as the gas, oil, and chemical industries, and above all, in nuclear power applications. As a result, more sophisticated statistical and probabilistic techniques for reliability life model development were developed on the computational side. These include maximum likelihood estimation (MLE) and Bayesian estimation techniques in reliability model developments.

The explosive growth of the integrated circuits (IC) in the 1980s renewed interest in the PoF modeling. As a result, accelerated life testing became critical to reliability analysis as a means of understanding the underlying processes and failure mechanisms that cause the failure of ICs. Further, the use of environmental screening methods improved the IC and other complex device mass production. The 1980s also marked the development of initiatives for modeling dependencies at the system level. Most of these efforts tackled common cause failures as frequent dependency problems in systems. The common cause failure (CCF), which is the failure of more than one component due to a shared root cause, is classified as a dependent failure. Finally, during the 1980s, applications of reliability growth became helpful during the design of complicated systems, particularly in defense systems.

Limited resources in conducting PoF resulted in limited test data. To account for the uncertainties in the PoF model, the 1990s experienced the development of the probabilistic physics of failure (PPoF) that would account for PoF model uncertainties, thus allowing for assignment measures of confidence over estimated reliability metrics (see Chatterjee and Modarres 2012). The 1990s also gave rise to the uses of time-varying accelerated tests (e.g., step-stress tests) and highly accelerated life testing (HALT) that introduced qualitative reliability information into the design process.

The decade of the 2000s proved to be significant to reliability assessments that required powerful simulation tools. With the appearance of practical Monte Carlo simulation algorithms such as Markov chain Monte Carlo (MCMC) simulation, recursive Bayes, and particle filtering, condition-based monitoring (CBM) and prognosis and health management (PHM) became dominant topics in the field. Additionally, integrated logic and probabilistic models such as combinations of the traditional logic trees coupled with PoF, Bayesian networks or Bayesian belief networks (BN or BBN), and other methods from machine learning were introduced.

Further extensions in big data, PHM, systems modeling, machine learning (ML), and related ML application fields had a commanding influence in the 2010s and beyond. Notably, complex regression models and formal accounting of uncertainties in reliability models, extensive causal Bayesian networks for modeling failure, semi-supervised and unsupervised deep learning from failure data and sensors monitoring performance systems, application of reliability modeling to highly complex systems such as the cyber-physical systems, PoF-informed deep learning models of reliability have been examples advances in this decade. A significant thrust of research in PoF modeling was to rely on thermodynamics and information theory to explain failure mechanisms and move toward developing science-based PoF models instead of empirical models that have formed the foundation of these physical reliability prediction models. Bayesian networks have been used widely for modeling failure under changing conditions. A significant body of work on causal modeling has also emerged.

1.3 RELIABILITY MODELING APPROACHES

The most widely accepted definition of ***reliability*** is *the ability of an item (e.g., a product, component, or a system) to operate without failure under specified operating conditions to attain a mission having the desired length of time or number of cycles*

By dissecting this definition, we can see that to assess reliability correctly, we must do the following:

1. Define the item (i.e., part, component, subsystem, system, or structure)
2. Define the mission and what constitutes success (or failure)
3. Define the designated conditions and environments of use
4. Specify a mission variable (e.g., time, cycles, stress level, and cumulative damage)
5. Assess ability (e.g., through testing, modeling, data collection, or analysis)

This process involves considerable engineering skills.

Reliability engineering can be approached from a deterministic or probabilistic perspective. This book deals with the reliability analyses involving predicting failures using probabilistic notions.

In a deterministic perspective, the primary view is that understanding failure mechanisms and the approaches to model them, such as physics of failure, and the resulting failure modes that may be associated with the design or that can be introduced from outside of the system (e.g., by users, maintainers, or the external operating environment) is sufficient. The probabilistic view builds on this by adding information about likelihood and uncertainty of events.

Most failure mechanisms, interactions, and degradation processes in a particular design are generally hard to predict and sometimes not well understood. Accordingly, predicting specific failure modes or any functional failure involves *uncertainty,* and thus is inherently a probabilistic problem. Therefore, reliability prediction, whether using physics of failure or relying on the historical occurrences of failures, is a probabilistic domain. It is the strength of the evidence and knowledge about the failure events and processes that describe failure mechanisms and associated failure modes and allow prediction of the reliability of the design and the corresponding confidence over that prediction.

This book considers how to probabilistically assess the reliability and risk of multiple classes of items (units, systems, subsystems, structures, components, software, human operators, etc.). We will refer to two primary layers as components and systems. First, we discuss how things fail in the remainder of this chapter. Then, in Chapters 2–5, we will elaborate on how to conduct the reliability analysis of a component (i.e., an item for which enough information (data) is available to predict its reliability without deconstructing it further). Then, in Chapter 6, we will discuss methods to model the reliability of a system, which is a collection of multiple components whose proper, coordinated operation leads to the appropriate functioning of the system. Assessment of the reliability of a system based on its basic elements is one of the most important aspects of reliability analysis. In system reliability analysis, it is therefore imperative to model the relationship between various items (and their failures) and the reliability of the individual items themselves to determine the overall system's reliability.

The domain of coverage in reliability engineering is extensive. Engineering systems are composed of interacting hardware, software, humans, operational environments, and more (see Figure 1.1). The conditions and environment cannot be neglected, nor can the role of humans. When considering an engineering system, the same hardware design operating in different locations, under varied organizations, or cultures can have vastly different reliability outcomes. Imagine a vehicle operating in the desert vs. one operating in the arctic. Heat and dust cause different types of failures than cold and ice. Likewise, systems operated by experts will have lower failure rates than those operated by novices—one reason we enforce training requirements. Furthermore, organizational policies that enforce maintenance practices, monitor

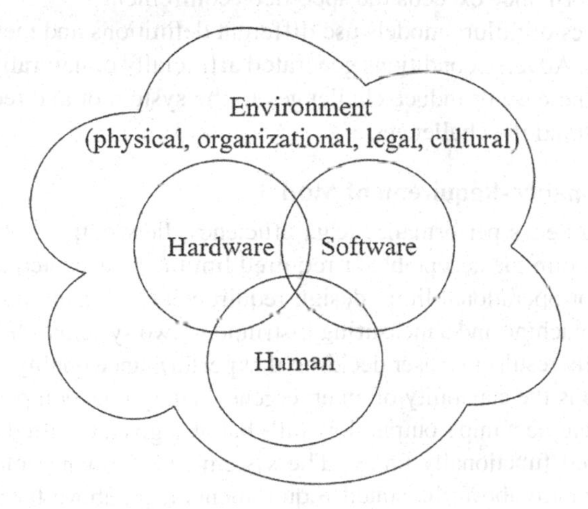

FIGURE 1.1 Domains of coverage in reliability engineering.

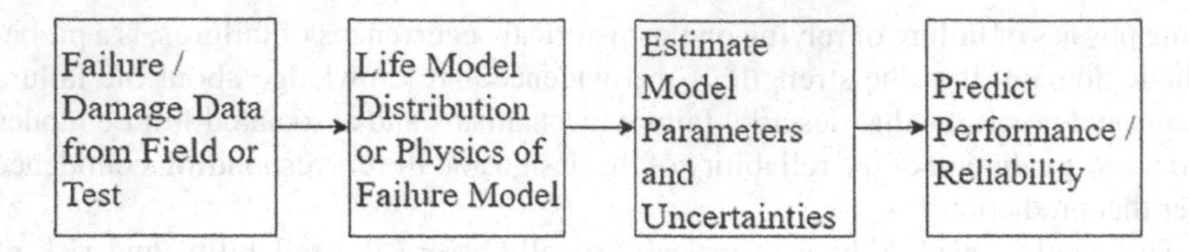

FIGURE 1.2 Framework for modeling failure.

failures, and invest in training the humans and software will inherently see high hardware reliability than those that neglect to monitor, maintain, and train.

Because this book takes a probabilistic modeling approach to reliability, multiple facets of probability modeling are relevant. First, the data-driven statistical models are concerned with when and how often failures occur. Statistical models are used to predict static conditions where appropriate data exist. A second facet is causal modeling, concerned with *why* failures occur. These causal models enable us to analyze and predict changing (uncertain) conditions.

Figure 1.2 depicts elements of a framework to construct failure models using capacity and challenge notions. Several simple failure models discussed by Dasgupta and Pecht (1991) follow the framework in Figure 1.2. A summary of these models has been provided below.

1.3.1 Physics of Failure Approach

Physics of failure views failures as occurring when *challenges* (e.g., operating stress, environmental stress, accumulated damage) exceed the *capacity* of the item to withstand the challenges. Both the challenges and the capacity may be affected by specific internal or external conditions. When challenges surpass the capacity of the system, a failure occurs. Or, viewing this in success space, reliability occurs when the system's performance exceeds the specified requirement.

Specific physics of failure models use different definitions and metrics for capacity and challenge. Adverse conditions generated artificially or naturally, internally, or externally, may increase or induce challenges to the system or/and reduce the item's capacity to withstand the challenges.

1.3.1.1 Performance-Requirement Model

In this model, an item's performance (e.g., efficiency, flow output, reliability) is adequate if it falls within the acceptable or required limit (e.g., accepted margin, such as a safety margin or operational limit, design requirement, warranty life). Examples of this are a copier machine and a measuring instrument, two systems where gradual degradation eventually results in a user deciding that performance quality is unacceptable. Another example is the capability of an emergency safety injection pump at a nuclear power plant. When the pump's output flow falls below a given required level, the pump may be considered functionally failed. The system's performance may decline over time but must remain above the stated requirements (e.g., above the minimum flow output) to achieve its function (e.g., to safely cool a reactor if needed).

1.3.1.2 Stress-Strength Model

In the stress-strength model, an item fails if the challenge (described by stress in the form of mechanical, thermal, etc.) exceeds the capacity (characterized by strength, such as yielding point, melting point, etc.). The *stress* represents an aggregate of the challenges and external conditions, often considered a sudden burst or overstressed, such as an item dropped on a hard surface. This failure model may depend on environmental conditions, applied loads, and the occurrence of critical events rather than the mere passage of time or cycles. Strength is treated either as a random variable (r.v.) showing stochastic variability or lack of knowledge about the item's strength or as a known deterministic value. Stress may be represented as an r.v. Two examples of this model are (a) a steel bar in tension and (b) a transistor with a voltage applied across the emitter-collector. In this model, the item fails only when stress exceeds strength, and the memory of past stresses applied to the item is rarely considered.

1.3.1.3 Damage-Endurance Model

This model is similar to the stress-strength model, but the scenario of interest is that *stress* (or load) causes damage over time that accumulates irreversibly, as in corrosion, wear, embrittlement, creep, and fatigue. The aggregate of these applied stresses leads to the metric represented as cumulative damage (e.g., total entropy, amount of wear, and crack length in case of fatigue-based fracture). The cumulative damage may not necessarily degrade performance; the item fails only when the cumulative damage exceeds the item's endurance to damage (i.e., the damage accumulates until the endurance of the item is reached). As such, an item's capacity is measured by its tolerance of damage or its endurance (such as fracture toughness). Unlike the stress-strength model, the accumulated damage does not disappear when the stresses causing the damage are removed; the stress leaves a memory of the past. However, sometimes treatments such as annealing reduce the amount of damage. In this modeling technique, damage and endurance may be treated as r.v.s.

Like the stress-strength model, endurance is an aggregate measure for the effects of challenges and external conditions on the item's ability to withstand cumulative stresses.

1.3.2 FAILURE AGENTS

In the physics of failure modeling approaches discussed in the previous three subsections, challenges are caused by failure-inducing agents (sometimes called *failure agents*). Examples of two of the most important failure-inducing agents are high stress and passage of time. High stress can be in mechanical (e.g., cyclic load), thermal (e.g., high heat, thermal cycling), electrical (e.g., voltage), chemical (e.g., salt), and radiation (e.g., neutron). For example, turning on and off a standby component may cause a thermal or mechanical load cycling to occur. The passage of time gives more opportunity for the normal environmental and operating stresses to cause small cumulative damage.

Careful consideration of reliability requires analyzing the two failure-inducing agents (high stresses and simple passage time under normal stresses). To properly depict performance and requirements, it is necessary to understand *why and how* such conditions occur that reduce performance.

The main body of this book addresses the probabilistic treatment of time or cycles to failure as agents of failure. Equally important, however, is understanding the failure mechanisms and their related models of physics of failure. It is also important to recognize that models representing failure mechanisms (related stress to damage or life) are also uncertain (i.e., their parameters are uncertain, and the model is associated with stochastic error due to failure data scatter). This means that the more appropriate model of failure mechanisms is the probabilistic physics of failure. For further readings on the physics of failure, see Modarres et al. (2017), Dasgupta and Pecht (1991), Pecht and Kang (2018), and Collins (1993).

1.4 DEFINITIONS

Now that we understand the role that a well-established reliability analysis and engineering program can play in influencing the performance of items, let us define the important elements of the performance.

The performance of an item is composed of four elements:

- *Capability*: the item's ability to attain the functional requirements.
- *Efficiency*: the item's ability to attain its functions economically and quickly with minimum waste.
- *Reliability*: the item's ability to start and continue to operate to fulfill a mission.
- *Availability*: the item's ability to become operational following a failure and remain operational during a mission.

All these measures are influenced by the item's design, construction, production, or manufacturing. Capability and efficiency reflect the levels to which the item is designed and built. For example, the designer ensures that the design meets the functional requirements. While reliability is influenced by design, reliability is also an operation-related performance metric influenced by the factors that enhance or degrade the ability to remain operational without failure. For a repairable item, the ease with which the item is maintained, repaired, and returned to operation is measured by its maintainability, which affects its availability. Capability, efficiency, reliability, and availability may be measured deterministically or probabilistically. For more discussions on this subject, see Modarres (2005). It is possible to have a highly reliable item that does not achieve high performance or capability. Examples include items that do not fully meet their stated design requirements. Humans play a significant role in the design, construction, production, operation, and maintenance of the item. These roles can significantly influence the values of the four performance measures. The role of the human is often determined by various programs and activities that support the four elements of performance, proper implementation of which leads to a *quality* item.

Reliability and availability play a vital role in the overall performance. Therefore, of the four elements of performance discussed above, we are mainly interested in reliability and availability in this book. Reliability is a critical element of achieving high performance since it directly and significantly influences the item's performance

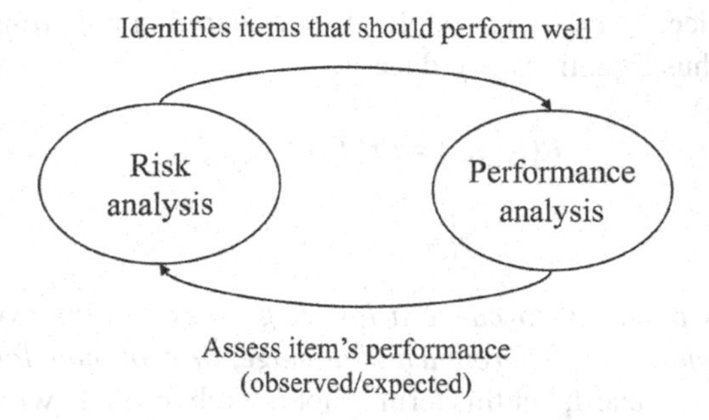

FIGURE 1.3 Synergistic effects between risk and reliability of an item.

and, ultimately, its life cycle cost. Conversely, poor reliability causes increased warranty costs, liabilities, recalls, and repair costs.

In this book, we are also interested in risk analysis. While the risk associated with an item is not a direct indicator of performance, a quantitative measure of risk can be an essential metric for identifying and highlighting items that contribute significantly to the risk and will be required to set and meet adequate performance levels for risk-significant items. Reliability and availability highly influence an item's risk. For example, a highly reliable item is expected to fail less frequently, resulting in lower risk. On the other hand, the risk assessment of an item may be used to identify items that should attain high reliability or availability. Accordingly, the risk and reliability of an item synergistically influence each other. This concept is depicted in Figure 1.3.

1.4.1 Reliability

The most widely accepted definition of reliability is the one in Section 1.3 repeated here for convenience: *reliability is the ability of an item to operate without failure under specified operating conditions to attain a mission having the desired length of time or number of cycles.* We typically model reliability as a function of use time or cycles of use.

Now, let us expand this definition to include a mathematical description of reliability. Probabilistically, an item's reliability can be expressed by

$$R(t_{\text{interest}}) = Pr(T_{\text{fail}} > t_{\text{interest}} | c_1, c_2, \ldots), \tag{1.1}$$

where $R(t_{\text{interest}})$ is the reliability of the item at the time of interest t_{interest}, given c_1, c_2,... the designated operating conditions, such as the environmental conditions. T_{fail} denotes the r.v. of time to failure. The time of interest t_{interest} is the designated mission time (i.e., the length of time or number of cycles desired for the item's operation). T could also represent the strength, endurance limit, or other quantities of interest. Clearly T_{fail} is most often an r.v., but rarely is a constant. t_{interest} is typically constant, but rarely an r.v.

Often, in practice, c_1, c_2, … are implicitly considered in the probabilistic reliability analysis, and thus Equation 1.1 reduces to

$$R(t_{\text{interest}}) = Pr(T_{\text{fail}} > t_{\text{interest}}). \tag{1.2}$$

1.4.2 Risk

Risk is the item's potential to cause a loss (e.g., loss to other systems, harm to individuals or populations, environmental damage, or economic loss).

A related term is hazard, but this term is not interchangeable with risk. We define *hazard* as *a source of damage, harm, or loss*. Inherently, risk involves both uncertainty and loss/damage. Risk can be addressed both qualitatively and quantitatively. Qualitatively speaking, when there is a source of danger (e.g., a hazard) and imperfect safeguards to prevent the danger (e.g., prevent or mitigate exposure to the hazard), there is a possibility of loss or injury. Sometimes this possibility is quantified and represented probabilistically as a metric for risk. *Risk can be more formally defined as the potential of loss (e.g., material, human, or environmental losses) resulting from exposure to a hazard.*

In complex engineering systems, there are often safeguards against exposure to hazards; safeguards may include physical barriers or preventive systems. The higher the level of safeguards, the lower the risk is. This also underscores the importance of highly reliable safeguard systems to prevent or mitigate losses and shows the roles and relationship between reliability analysis and risk analysis. An example of a risk metric could be the expected loss caused by an item during a mission or over a period. The expected loss as a measure of risk can also be viewed and assessed probabilistically.

This book is concerned with quantitative risk assessment (QRA). Since quantitative risk assessment involves an estimation of the degree or probability of an uncertain loss, risk analysis is fundamentally intertwined with the concept of probability of occurrence of hazards. Formal risk assessment consists of answers to three questions (Kaplan and Garrick, 1981):

1. What can go wrong [that could lead to a hazard exposure]?
2. How likely is it to happen?
3. If it happens, what are the consequences are expected?

A list of scenarios of events leading to the hazard exposures should be defined to answer question one. Then, the probability of these scenarios should be estimated (answer to question two), and the consequence of each scenario should be described (answer to question three). Therefore, the risk is defined, quantitatively, as the following triplet:

$$R = S_i, P_i, C_i, \quad i = 1,2,\ldots, n, \tag{1.3}$$

where S_i is a scenario of events that lead to hazard exposure, P_i is the probability or frequency of scenario i, and C_i is the consequence of scenario i, for example, a measure of the degree of damage or loss.

Since Equation 1.3 involves estimating the probability or frequency of occurrence of events (e.g., failure of protective, preventive, or mitigative safeguards), most of the methods described in Chapters 2–8 become relevant and useful. We have specifically devoted Chapter 9 to a more detailed, quantitative description of risk analysis. For more discussions on risk analysis, the reader is referred to Modarres (2006), Bedford and Cooke (2001), and Kelly and Smith (2011).

1.4.3 Availability and Maintainability

Availability analysis is performed on repairable systems to ensure that an item has a satisfactory probability of being operational to achieve its intended objectives. An item's availability can be thought of as a combination of its reliability and maintainability. *Maintainability is an item's ability to be quickly restored following a failure.*

Accordingly, when no maintenance or repair is performed (e.g., in nonrepairable items), reliability can be considered as instantaneous availability.

Mathematically, the availability of an item is a measure of the fraction of time that the item is in operating condition as a function of either total system time or calendar time. There are several availability measures, namely, inherent availability, achieved availability, and operational availability. Here, we describe inherent availability, defined most commonly in the literature.

A more formal definition of *availability is the probability that an item, when used under stated conditions and support environment (i.e., perfect spare parts, personnel, diagnosis equipment, procedures, etc.), will be operational at a given time.* Based on this definition, the mean availability of an item during an interval of time T can be expressed by

$$A = \frac{U}{U + D}, \tag{1.4}$$

where U is the uptime during time T, D is the downtime during time T, and total time $T = U + D$.

The mathematics and methods for reliability analysis discussed in this book are also equally applicable to availability analysis. Equation 1.4 can also be extended to represent a time-dependent availability. Time-dependent expressions of availability and availability measures for different types of equipment are discussed in more detail in Chapter 7. For further definition of these availability measures, see Ireson and Coombs (1988).

1.4.4 Failure Modes and Failure Mechanisms

Failure modes, failure mechanisms, and functions are building blocks of risk and reliability scenarios. The manner or way of the final failure occurrence is called the *failure mode.* Stated another way, a failure mode is a functional manner of item failure. An example of a failure mode for a valve would be *external leakage.* Failure modes answer the questions "How does the item fail?" or "What does the

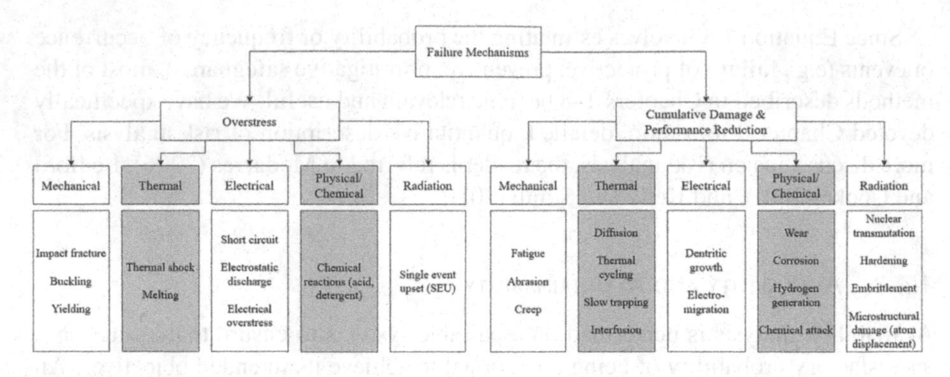

FIGURE 1.4 Relationship between environmental conditions and failure mechanisms in components.

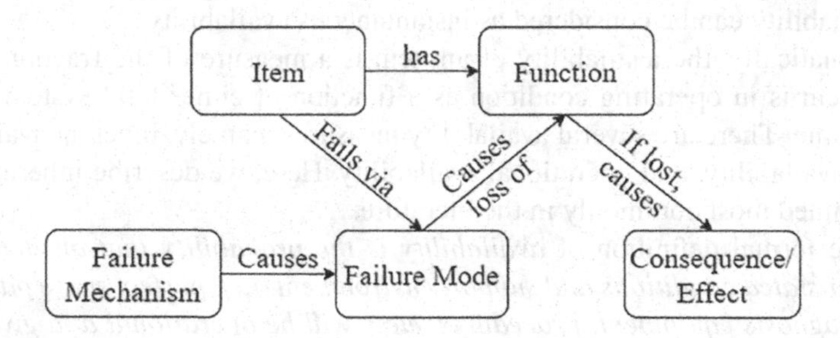

FIGURE 1.5 Relationship between item, functions, failure modes, and failure mechanisms.

item fail to do?" Typically, we assume that all item failure modes are mutually exclusive. From this discussion, we can conclude that the failure mode refers to a failed state of the item.

Failure mechanisms are physical processes through which damage occurs. Damage can occur rapidly (abruptly) or slowly (cumulatively). Examples of failure mechanisms leading to rapid damage include brittle facture, melting and yielding. Examples of failure mechanisms causing cumulative damage are fatigue, wear and corrosion. In both cases failure occurs when the resulting damage exceeds the item's capacity to withstand damage (e.g., strength, endurance, or required performance). An example of failure mechanism leading to external valve leakage (i.e., failure mode) would be corrosion. From this discussion, we can conclude that the failure mechanism refers to the process leading to the item's failure mode (i.e., failed state). Figure 1.4 shows the relationship between environmental conditions and common failure mechanisms. The relationship between failure modes, failure mechanisms, and functions is illustrated in Figure 1.5.

1.5 SYSTEMS, FUNCTIONS, AND FAILURE

Assessment of the reliability of a system based on its basic elements is one of the most important aspects of reliability analysis. A *system* is a collection of items whose coordinated operation achieves a specific function (or functions). These items may be subsystems, structures, components, software, algorithms, human operators, users, maintenance programs, and more. In reliability analysis, it is therefore imperative to model both the reliability of the individual items and the relationship between various items as they affect the system's reliability.

Several aspects must be considered in system reliability analysis. One is the reliability of the components that comprise the system. A *component* is a basic physical entity of a system analyzed from a reliability perspective (i.e., not further divided into more abstract entities). In Chapters 2–5, we elaborate on conducting the reliability analysis at a basic component level. A second consideration is the manner of the item's failure (the item failure mode) and how it affects the system's failure mode(s). Finally, another important consideration is the physical configuration of the system and its components. Beginning in Chapter 6, we discuss methods to model the relationship between components and systems, which allow us to determine overall system reliability. We will elaborate more on this in Chapter 6 when we explain methods and functions for modeling the reliability of a system with n components as a function of the reliability of those components.

For example, consider a pumping system that draws water from two independent storage tanks—one primary and one backup tank, each equipped with a supply valve. If the supply valve on one tank fails closed, the pumping system can still operate because of the second tank. Thus, both the reliability of each component and the system configuration affects system reliability. Now, consider the effect of failure modes. If the tank's supply valves fail in an open position, the pump can still draw water and successfully operate. However, if the supply valves fail in a closed position, the pump cannot work, and the pumping system fails. So, the failure mode of the valves is highly relevant for system reliability analysis.

The line between component and system is arbitrary and varies depending on the objectives, scope, resources of modeling and analysis, and the state of the art and conventions. This relationship is conceptually shown in Figure 1.6. It is necessary to define and clearly articulate the elements and boundaries of the system in any system reliability analysis or data source. One example of a system definition is shown in Figure 1.7.

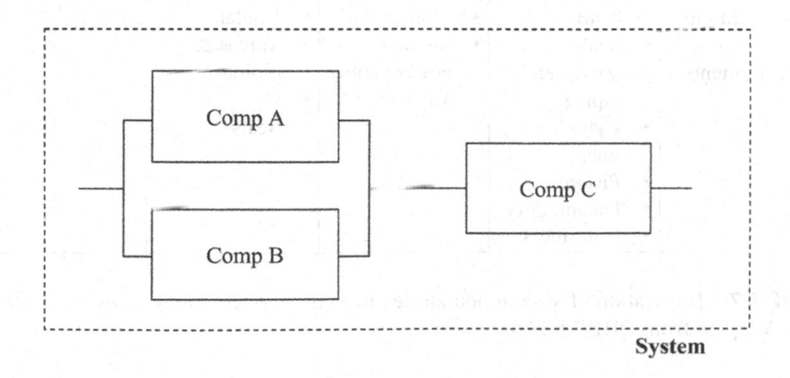

FIGURE 1.6 Notional relationship between components and systems.

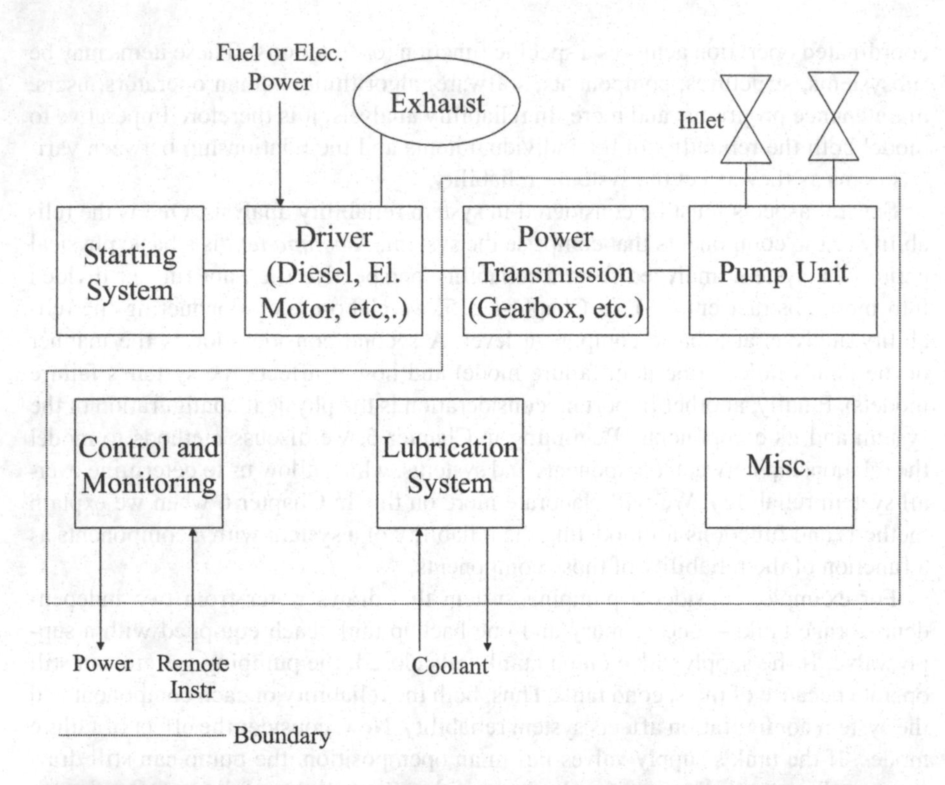

Pump System				
Power transmission	**Pump**	**Control and monitoring**	**Lubrication system**	**Miscellaneous**
• Gearbox/var. drive • Bearing • Seals • Lubrication • Coupling to driver • Coupling to driven unit • Instruments	• Support • Casing • Impeller • Shaft • Radial bearing • Thrust bearing • Seals • Valves & piping • Cylinder liner • Piston • Diaphragm • Instruments	• Instruments • Cabling & junction boxes • Control unit • Actuating device • Monitoring • Internal power supply • Valves	• Instruments • Reservoir w/ heating system • Pump w/ motor • Filter • Cooler • Valves & piping • Oil • Seals	• Purge air • Cooling/heat ing system • Filter, cyclone • Pulsation damper

FIGURE 1.7 Illustration of system boundaries (top) and system breakdown (bottom) for a pump. (Adapted from OREDA 2002.)

1.5.1 Failure Modes and Mechanisms for Mechanical Equipment

Some typical mechanical failure modes for active hardware components are:

1. Premature operation,
2. Failure to start operation when needed,
3. Failure to continue operation after the start,
4. Failure to stop operation at the specified time,
5. Degraded operation.

A more detailed example list can be generated by expanding upon these failure modes. Many reliability data collection systems provide extensive taxonomies of failure modes. One example of failure modes for a pump can be found in Table 1.1. (Source: OREDA 2002):

Now let's turn to failure mechanisms. Mechanical failure mechanisms can be divided into *damage-inducing*, *capacity-reducing,* and a combination of both. Damage-inducing mechanisms refer to mechanisms that cause or result from localized damage, such as cracking. The damage may be temporary or permanent (cumulative). For example, elastic deformation may result from a force applied on the item that causes damage that disappears when the applied force is removed. However, cumulative damage caused by a failure mechanism is permanent. For example, fatigue is a mechanical failure mechanism whose direct effect is irreversible cumulative damage in the form of crack initiation and growth.

TABLE 1.1
Detailed List of the Failure Modes of a Pump (OREDA, 2002)

AIR	Abnormal Instruments Reading
BRD	Breakdown
ERO	Erratic output
ELP	External leakage—Process medium
ELU	External leakage—Utility medium
FTS	Fail to start on demand
STP	Fail to stop on demand
HIC	High output
INL	Internal leakage
LOO	Low output
SER	Minor in service problems
NOI	Noise
OTH	Other
OHE	Overheating
PDE	Parameter deviation
UST	Spurious stop
STD	Structural deficiency
UNK	Unknown
VIB	Vibration

TABLE 1.2
Examples of Failure Mechanisms

Damage-Inducing	Capacity-Reducing
Wear	Fatigue
Corrosion	Embrittlement
Cracking	Thermal shock
Diffusion	Diffusion
Creep	Grain boundary migration
Fretting	Grain growth
Fatigue	Precipitation hardening

Capacity-reducing mechanisms are those that lead (indirectly) to a reduction of the item's strength or endurance to withstand applied stress or cumulative damage. For example, radiation may cause material embrittlement, thus reducing the material's capacity to withstand cracks or other damage. Table 1.2 shows examples of failure mechanisms in each class of mechanisms. Table 1.3 summarizes the cause, effect, and physical processes involving common mechanical failure mechanisms.

1.5.2 Failure Modes and Mechanisms for Electrical Equipment

Electrical failure mechanisms tend to be more complicated than those of purely mechanical failure mechanisms. This is caused by the complexity of the electrical items (e.g., devices) themselves. In integrated circuits, a typical electrical device, such as a resistor, capacitor, or transistor, is manufactured on a single crystalline chip of silicon, with multiple layers of various metals, oxides, nitrides, and organics on the surface, deposited in a controlled manner. Often a single electrical device comprises several million elements, compounding any reliability problem present at the single element level. Furthermore, once the electrical device is manufactured, it must be packaged with electrical connections to the outside world. These connections and the packaging are as vital to the proper operation of the device as the electrical elements themselves.

Electrical device failure mechanisms are usually divided into three types: *electrical stress failure, intrinsic* and *extrinsic failure mechanisms.* These are discussed below.

- *Electrical stress failure* occurs when an electrical device is subjected to voltage levels higher than design constraints, damaging the device and degrading electrical characteristics enough that the device may effectively fail. This failure mechanism is often a result of human error. Also known as electrical overstress (EOS), uncontrolled currents in the electrical device can cause resistive heating or localized melting at critical circuit points, which usually results in catastrophic failure but can also cause latent damage. Electrostatic discharge (ESD) is one common way of imparting large, undesirable currents into an electrical device.

TABLE 1.3
Examples of Leading Mechanical Failure Mechanisms

Mechanism	Examples of Causes	Examples of Effects	Description
Buckling	1. Compressive load exceeding Euler's critical level 2. Incorrect dimensions or material properties	1. Item deflects 2. Complete loss of load-carrying ability	When load applied to items such as struts, columns, plates, or thin-walled cylinders reaches a critical value, a sudden significant change in geometry, such as bowing, winking, or bending, occurs
Corrosion	1. Electrochemical oxidation on the surface of the item 2. High concentrations of species such as chloride 3. Contact between two dissimilar metals (galvanic corrosion) 4. Improper welding of copper, chromium, nickel, aluminum, magnesium, and zinc alloys 5. Presence of abrasive or viscid flow of chemicals over the surface 6. Living organisms in contact with the item 7. High stress in a chemically active environment (stress-corrosion-cracking)	1. Pitting 2. Cracking 3. Fracture 4. Geometry changes	Gradual damage of materials (primarily metals) by chemical and/or electrochemical reactions with the environment. Corrosion also interacts with other mechanisms such as cracking, wear, and fatigue
Impact	1. Sudden load from dropping an item or having been struck 2. An item coming forcibly into contact with an object	1. Localized stresses 2. Deformation 3. Fracture	Failure or damage by the interaction of generated dynamic or abrupt loads that result in significant local stresses and strains
Fatigue	1. Fluctuating force (loads) 2. Thermal cycling 3. Repeated bending 4. Random vibration 5. Surface finish 6. Stress raisers (e.g., notches)	1. Cracking initiation 2. Crack growth leading to deformation and fracture	Application of fluctuating loads (even far below the yield point) leading to a progressive failure phenomenon that initiates and propagates cracks

(*Continued*)

TABLE 1.3 (*Continued*)
Examples of Leading Mechanical Failure Mechanisms

Mechanism	Examples of Causes	Examples of Effects	Description
Wear	1. Solid surfaces in rubbing contact 2. Particles entrapped between rubbing surfaces 3. The corrosive environment near rubbing contacts	1. Progressive loss of material 2. Cumulative change in dimensions 3. Deformation	Wear is not a single process. It can be a complex combination of local shearing, plowing, welding, and tearing, causing the gradual removal of discrete particles from contacting surfaces in motion. Particles entrapped between mating surfaces. Corrosion often interacts with wear processes and changes the character of the surfaces
Embrittlement	1. Hydrogen absorption 2. Diffusion of metal atoms 3. Low temperatures (environmental or from cold treatments) 4. *In plastics*: Loss of plasticizers from overheating or aging 5. Neutron absorption	1. Loss of ductility (reducing in a brittle material) 2. Brittle fracture	Failure or damage from loss of material ductility (i.e., becoming brittle), often resulting from a combination of processing factors (e.g., cold or heat treatment), stress concentration points, and presence of hydrogen.
Creep	1. Loading (tensile or compressive) applied at very high temperature	1. High strain 2. Deformation of item 3. Rupture	Plastic deformation in an item accrues over some time under tensile, compressive, or bending stress until the accumulated dimensional changes interfere with the item's ability to function properly.
Thermal shock	1. Rapid cooling 2. Rapid heating 3. Sudden large differential temperature	1. Yield fracture 2. Embrittlement	Thermal gradients in an item causing major differential thermal strains that, if exceeding the ability of the material to withstand, lead to failure and fracture.
Yield	1. Large static force 2. Operational load or random load exceeding the yield strength	1. Permanent geometry changes 2. Irreversible deformation	High enough tensile, compressive, or bending stress leads to plastic deformation in an item by operational loads or motion.
Radiation damage	1. Neutron radiation 2. Ionizing radiation 3. Electromagnetic radiation 4. Cosmic rays	1. Changes in material property 2. Loss of ductility 3. Microstructural changes such as displacement of atoms 4. Formation of reactive compounds	Radiation causes rigidity and loss of ductility. Polymers are more susceptible than metals.

- *Intrinsic failure mechanisms* are related to the electrical element itself. Most failure mechanisms related to the semiconductor chip and electrically active layers grown on its surface are in this category. Intrinsic failures are associated with the basic electrical activity of the device and usually result from poor manufacturing or design procedures. Thus, intrinsic failures cause both reliability and manufacturing yield problems. Common intrinsic failure mechanisms are gate oxide breakdown, ionic contamination, surface charge spreading, and hot electrons. In recent years semiconductor technology has reached a high level of maturity, with a correspondingly high level of control over intrinsic failure mechanisms.
- *Extrinsic failure mechanisms* are external mechanisms for electrical devices that stem from device packaging and interconnections problems. For example, failures caused by an error during the design, layout, fabrication, or assembly process or by a defect in the fabrication or assembly materials are considered extrinsic failures. Also, a failure caused by a defect created during manufacturing is classified as extrinsic. Most extrinsic failure mechanisms are mechanical. However, embedded deficiencies or errors in the electronic device and packaging manufacturing process often cause these extrinsic mechanisms to occur, though the operating environment strongly affects the damage caused by these failure mechanisms. Because of recent reductions in damage accumulation through intrinsic failure mechanisms, extrinsic failures have become more critical to the reliability of the latest generation of electronic devices.

Many electrical failure mechanisms are interrelated. A partial failure due to one mechanism can often evolve into another. For example, oxide breakdown may be caused by poor oxide processing during manufacturing, but it may also be exasperated by ESD, damaging an otherwise intact oxide layer. Likewise, corrosion and ionic contamination may be initiated when a packaging failure allows unwanted chemical species to contact the electronic devices. Then failure can occur through trapping, piping, or surface charge spreading. Also, intrinsic failure mechanisms may be initiated by an extrinsic problem once the package of an electrical device is damaged. There are a variety of intrinsic failure mechanisms that may manifest themselves in the device.

Tables 1.4–1.6 summarize the cause, effect, and physical processes involving common electrical stress and intrinsic and extrinsic failure mechanisms.

TABLE 1.4
Electrical Stress Failure Mechanisms

Mechanism	Causes	Effect	Description
EOS	Improper application of handling	Localized melting Gate oxide breakdown	The device is subjected to voltages higher than design constraints.
ESD	Common static charge buildup	Localized melting Gate oxide breakdown	Contact with static charge buildup during device fabrication or later handling results in high voltage discharge into the device.

TABLE 1.5
Intrinsic Failure Mechanisms

Mechanism	Causes	Effect	Description
Gate oxide breakdown	1) EOS 2) ESD 3) Poor gate oxide processing	1) Degradation in current-voltage (I–V) characteristics	The oxide layer that separates gate metal from the semiconductor is damaged or degrades with time.
Ionic contamination	1) Introduction of undesired ionic species into the semiconductor	1) Degradation in I–V characteristics 2) Increase in threshold voltage	Undesired chemical species can be introduced to the device through human contact, processing materials, improper packaging, and so on.
Surface charge spreading	1) Ionic contamination 2) Excess surface moisture	1) Short-circuiting between devices 2) Threshold voltage shifts or parasitic formation	Undesired formation of conductive pathways on surfaces alters the electrical characteristic of the device.
Slow trapping	1) Poor interface quality	1) Shifting Threshold voltage	Defects at gate oxide interface trap electrons, producing undesired electric fields.
Hot electrons	1) High electric fields in the conduction channel	1) Threshold voltage shifts	High electric fields create electrons with sufficient energy to enter the oxide.
Piping	1) Crystal defects 2) Phosphorus or gold diffusion	1) Electrical shorts in emitter or collector	Diffusion along crystal defects in the silicon during device fabrication causes electrical shorts.

1.5.3 Human Functions, Failure, and Reliability

Humans play many roles in engineering systems. Since the focus of modern reliability engineering is on the engineered system, not only the hardware, analyzing the reliability of humans (or, more appropriately, the human-machine teams (Groth et al., 2019)) that contribute to functioning of an engineered system is an essential part of reliability analysis. *Human reliability analysis* (HRA) is the aspect of reliability engineering that provides the methodologies to model these human failures.

HRA is an embedded part of the reliability engineering process, not a separate process to be conducted independently. While in some aspects of reliability and risk analyses, human failures have been treated as a separate analysis, this unnecessarily fragments the reliability modeling of complex engineering systems. Reliability and risk analysis should treat human functions and failures as embedded in reliability analysis.

The human-machine team is inherently another item considered as part of system reliability similar to the hardware. However, unlike many types of hardware components, humans can play multiple roles in a system—so they can be viewed as components achieving various functions at several levels of abstraction. Addressing human reliability within an engineered system requires techniques built upon those

TABLE 1.6
Extrinsic Failure Mechanisms

Mechanism	Causes	Effect	Description
Packaging failures	1. Most mechanical failure mechanisms can cause electrical device packaging failures	1. Increased resistance 2. Open circuits	See section on mechanical failure mechanisms
Corrosion	1. Moisture 2. DC operating voltages 3. Na or Cl ionic species	1. Open circuits	The combination of moisture, DC operating voltages, and ionic catalysts causes electrochemical movement of material, usually the metallization
Electromigration	1. High current densities 2. Poor device processing	1. Open circuits	High electron velocities impact and move atoms, resulting in altered metallization geometry and, eventually, open circuits
Contact migration	1. Uncontrolled material diffusion	1. Open or short circuits	Poor interface control causes metallization to diffuse into a semiconductor. Often this occurs as metallic "spikes"
Microcracks	1. Poorly processed oxide steps	1. Open circuits	The formation of a metallization path on top of a sharp oxide step results in a break in the metal or a weakened area prone to further damage
Stress migration	1. High mechanical stress in electrical device	1. Short circuits	Metal migration occurs to relieve high mechanical stress in the device
Bonding failures	1. Poor bond control	1. Open circuits	Electrical contact to device package (bonds) are areas of high mechanical instability and can separate if the processing is not strictly controlled
Die attachment failures	1. Poor die-attach integrity 2. Corrosion	1. Hot spots 2. Parametric shifts in-circuit 3. Mechanical failures	Corrosion or poor fabrication causes voids in die-attach or partial or complete deadhesion
Particulate contamination	1. Poor manufacturing and chip breakage	1. Short circuits	Conductive particles may be sealed in a hermetic package or may be generated through chip breakage
Radiation	1. Trace radioactive elements in a device 2. External radiation source	1. Various degrading effects	High energy radiation (ionizing or cosmic) can create hot electron-hole pairs that can interfere with and degrade device performance

used in component and systems analysis. Thus, HRA is one of the special topics covered in Chapter 8.

1.6 PUTTING IT ALL TOGETHER: RISK ASSESSMENT MODELING

As we conclude this chapter, let's briefly discuss the general steps in risk assessment, which provides an integrated framework for explaining how the book fits together (Figure 1.8) as it builds toward Chapter 9. The key elements of a risk assessment include:

1. Define system and assemble information.
   a. Define the system to be analyzed and its components, interfaces, and boundaries,
   b. Collect, analyze, and model component failure data.
2. Define the problem scope objectives and select the appropriate methodology.
   a. Identify hazards, operating modes, and/or initiating events of interest.
3. Construct causal models and conduct causal analysis.
   a. Scenario development and modeling,
   b. Identify root causes (failure modes and failure mechanisms if needed),
   c. Identify failure effects,
   d. Construct logic models.
4. Qualitatively evaluate those models.
   a. System failure logic.
5. Integrate logic models,
   a. For example, many nuclear probabilistic risk assessments (PRAs) use a combination of event trees and fault trees. For complex systems, it is common to build separate models for different functions or subsystems or separate parts of the analysis (e.g., hardware vs. human reliability).
6. Evaluate consequences,
   a. Use a variety of models to assess effects and outcomes of the scenarios.

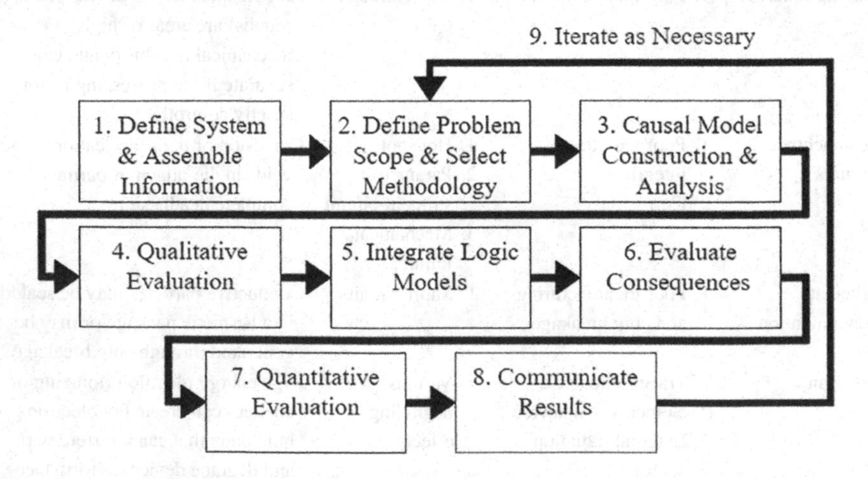

FIGURE 1.8 A simplified view of the process of risk assessment forms the framework of this book.

7. Perform quantitative evaluation of the models,
   a. Quantify (parameterize) all basic events by using appropriate data sources and/or models for each,
   b. Evaluate logic models using quantified elements to obtain event and scenario probabilities,
   c. Dependency (common cause failure) analysis,
   d. Uncertainty analysis,
   e. Sensitivity analysis,
   f. Importance ranking.
8. Communicate results (as part of risk management).
9. Iterate and repeat as appropriate.

For Step (1), we discussed how to define (and document) the system to be analyzed in Chapter 1, (or we can either add content about or find references to techniques such as functional breakdown and structural breakdown, structural and functional block diagrams, and more). Next, data collection and analysis will be discussed in Chapters 4 and 5.

For Step (2), several objectives are possible, and knowing your objectives will help you select the appropriate methodology. Objectives can range from design insight, demonstration of risk acceptability or tolerability, regulatory or code compliance, operational design support, system health management, and more. In Chapter 1, we define critical terms used establishing objectives and methods. The remainder of this textbook discusses a range of methodologies to support this step.

For Step (3), in this chapter, we discuss failure mode and mechanism taxonomies. Chapter 6 will expand upon this and connect to identifying failure causes and effects through our discussion of several qualitative and quantitative models. Chapter 6 also discusses the construction of the qualitative logic models.

For Steps (4) and (5), we will discuss the integration and evaluation of the logic models within Chapter 6 and Chapter 9. See Moradi and Groth (2020) for discussion of integrating additional types of information.

Step (6) typically falls into the domain of deterministic modeling or loss calculations. It may include use of multiphysics models or detailed fire behavior models. While full coverage of this falls outside the scope of this book, Chapter 9 illustrates the use of fire modeling within this process via the HyRAM methodology (Groth and Hecht, 2017).

For Step (7), we discuss Step 7a extensively in Chapters 3 and 5 (for nonrepairable items), Chapter 7 (for repairable items) and Chapter 8 (for common cause failure analysis, importance ranking, uncertainty and sensitivity analysis and the human reliability). Step 7b is explained in Chapter 6. The remaining aspects of Step (7) are advanced topics beyond the scope of this first course in reliability engineering—each of these nuanced topics is covered in more depth in advanced books such as Modarres (2006).

For Step (8), see Modarres (2006).

Finally, Step (9) is an important reminder that this is an iterative process, and not a one-time activity that we place on the shelf after the system is built. Some aspects of this step are presented in Chapter 9. This step is described in more detail in Modarres (2006).

This book culminates in Chapter 9, which presents the integration of this process and presents several examples.

1.7 EXERCISES

1.1 Explain the difference between a failure mechanism and a failure mode.

1.2 Discuss the relationship between reliability and failure, both in words and mathematically.

1.3 Select a failure mechanism and find a news article where it has been relevant to an engineering system.

1.4 Select a failure mechanism and explain what activates or causes it to happen, what is the process through which damage occurs, what is the nature of the damage, what is the capacity (strength or endurance) to the damage, and what failure mode(s) can result at the end.

1.5 Describe the relationship between reliability and availability.

1.6 Select an accident or failure that you are familiar with. Write a failure scenario narrative containing the following:

a. A brief description of the system and the accident or failure and what initiates the accident.

b. The system failure mode, operating environment, and health, safety or environmental consequences.

c. At least one of each of the following: a hardware failure mode, a failure mechanism and influencing factor contributed to this system failure.

d. A human activity that contributed to the failure.

e. A relevant image.

1.7 Read about the accident at a chemical plant in Bhopal, India. Describe what are the three fundamental elements of the risk triplet.

1.8 An item is down an average of 16 hours a year for repair and maintenance. What is the item's mean availability?

1.9 Described five critical information that reliability analysis provides as input to a risk assessment.

1.10 Graphically and conceptually show the deterministic stress/damage vs. capacity (e.g., strength and endurance) for the three PoF models. Repeat the graphs to conceptually depict the three PoF models probabilistically.

REFERENCES

Bedford, T. and R. Cooke, *Probabilistic Risk Analysis: Foundations and Methods*. Cambridge University Press, London, 2001.

Chatterjee, K. and M. Modarres. "A Probabilistic Physics-of-Failure Approach to Prediction of Steam Generator Tube Rupture Frequency." *Nuclear Science and Engineering*, 170(2), 136–150, 2012.

Collins, J. A., *Failure of Materials in Mechanical Design, Analysis, Prediction, and Prevention*. 2nd edition, Wiley, New York, 1993.

Dasgupta, A. and M. Pecht, "Materials Failure Mechanisms and Damage Models." *IEEE Transactions on Reliability*, 40 (5), 531–536, 1991.

Groth, K. M. and E. S. Hecht, HyRAM: "A Methodology and Toolkit for Quantitative Risk Assessment of Hydrogen Systems." *International Journal of Hydrogen Energy*, 42, 7485–749, 2017.

Groth, K. M., R. Smith, and R. Moradi. "A Hybrid Algorithm for Developing Third Generation HRA Methods Using Simulator Data, Causal Models, and Cognitive Science." *RWeliability Engineering & System Safety*, 191, 106507, 2019.

Ireson, W. G. and C. F. Coombs, eds., *Handbook of Reliability Engineering and Management*. McGraw-Hill, New York, 1988.

Kaplan, S. and J. Garrick, "On the Quantitative Definition of Risk." *Risk Analysis*, 1(1), 11–27, 1981.

Kelly, D. L. and C. Smith, *Bayesian Inference for Probabilistic Risk Assessment: A Practitioner's Guidebook*. Springer -Verlag, London, 2011.

Modarres, M., "Technology-Neutral Nuclear Power Plant Regulation: Implications of a Safety Goals-Driven Performance-Based Regulation." *Nuclear Engineering and Technology*, 37(3), 221–230, 2005.

Modarres, M., M. Amiri, and C. R. Jackson, *Probabilistic Physics of Failure Approach to Reliability: Modeling, Accelerated Testing, Prognosis and Reliability Assessment*. John Wiley, New York, 2017.

Modarres, M., *Risk Analysis in Engineering*. CRC Press, Boca Raton, FL, 2006.

Moradi, R. and K. M. Groth, "Modernizing Risk Assessment: A Systematic Integration of PRA and PHM Techniques." *Reliability Engineering & System Safety*, 204, 107194, 2020.

OREDA. *Offshore Reliability Data Handbook*, 4th edition, OREDA Participants, Høvik, Norway 2002.

Pecht, M. and M. Kang, eds., *Prognostics and Health Management of Electronics: Fundamentals, Machine Learning, and the Internet of Things*. John Wiley & Sons, New York, 2018.

2 Basic Reliability Mathematics *Probability*

2.1 INTRODUCTION

In this chapter, we discuss the elements of probability theory relevant to the study of reliability of engineered systems. Probability is a concept that people use formally and casually every day. Weather forecasts are probabilistic in nature. People use probability in their casual conversations to show their perception of the likely occurrence or nonoccurrence of particular events. Odds are given for the outcomes of sport events and are used in gambling. The formal use of probability concepts is widespread in physics, astronomy, biology, finance, and engineering.

We begin with a presentation of basic concepts of probability and Boolean algebra used in reliability analysis. The emphasis in this chapter is on presenting these fundamental mathematical concepts which are used and expanded upon in the remaining chapters in this book.

For some readers, this chapter is a review of mathematics that they have learned in previous courses. Some details may be omitted. It is worth reviewing this chapter briefly to see the established terminology, notation, and key assumptions used throughout the book.

2.2 EVENTS AND RANDOM VARIABLES USED IN RELIABILITY

There are two basic units of analysis involved in the probabilistic approach to reliability: events and random variables. An *event*, E, is an outcome or set of outcomes of a process. An event occurs with some probability. The *sample space*, S, is a mutually exclusive and collectively exhaustive list of all possible outcomes of a process. For example, examining the outcomes of rolling a die, the sample space contains events $S = \{1,2,3,4,5,6\}$. Thus, an event is a combination of one or more sample points of interest. For example, the event of "an odd outcome when rolling a die" represents the event space containing outcome points 1, 3, and 5. In reliability engineering, E could be event of a pump failure, or that a pump fails to start on demand, or that multiple pumps fail simultaneously. One use of probability is to communicate the chance of various combinations of events, such as multiple components failing simultaneously in a specific period of time.

We may also be concerned with characterizing *random variables* (r.v.) associated with events. For instance, for an event of component failure, the r.v. X could be an r.v. representing time to failure of a component, the number of failures that occur by a given time, the number of cycles until a first failure, the number of failure

DOI: 10.1201/9781003307495-2

components drawn from a population, the failure rate, the rate of occurrence of failures, and more. The lowercase x could be the probability of the event of interest.

Random variables can be divided into two classes, namely *discrete* and *continuous*. An r.v. is said to be discrete if the values it can take are finite and countable, such as the number of system breakdowns in 1 year. An r.v. is said to be continuous if it can take on a continuum of values. It takes on a range of values from an interval(s) as opposed to a specific countable number. Continuous r.v.s result from measured variables as opposed to counted data. For example, the failure time of a light bulb can be modeled by a continuous r.v. time, T, which can take on a value within a continuous range. For any specific light bulb, we can observe its failure time t_i.

In this book, we will use an uppercase letter (e.g., X, Y) to represent an event or an r.v. We will use a lowercase letter to denote the value that the r.v. can take. For example, if X represents the number of system breakdowns in a process plant, then x_i is the actual number of observed breakdowns during a given time interval.

2.3 SETS AND BOOLEAN ALGEBRA

In reliability engineering, we are analyzing specific events, data, or combinations. A *set* is a collection of items or elements, each with some specific characteristics. The elements of the set are contained in brackets {}. We use sets to talk about combinations or items of interest, e.g., data or events. For example, we may have a population of light bulb failure times representing early failures (e.g., those light bulbs with failure times less than 100 hours). Or a set containing two pumps, A and B. Suppose event A is the event that pump A fails to start, and event B is the event that pump B fails to start. For the event that both pumps fail to start—this event would be the set consisting of A and B.

A set that includes all possible items of interest is called a *universal set*, denoted by Ω (which sometimes is replaced by 1 in engineering notation). A *null set* or *empty set*, ϕ (or 0 in some engineering notations), refers to a set that contains no items. A *subset*, denoted by $\subset$ or $\subseteq$, refers to a collection of items that belong to another set. For example, if set Ω represents the collection of all pumps in a power plant, then the collection of auxiliary feedwater pumps, A, is a subset of Ω, denoted as $A \subset \Omega$.

Graphically, the relationship between subsets and sets can be illustrated through Venn diagrams. The Venn diagram in Figure 2.1 shows the universal set Ω by a rectangle and subsets A, B, and C by circles. It can also be seen that B is a subset of A. The relationship between subsets A and B and the universal set can be symbolized by $B \subset A \subset \Omega$. In this diagram, A is a *superset* of B, denoted as $A \supset B$. Subsets A and C

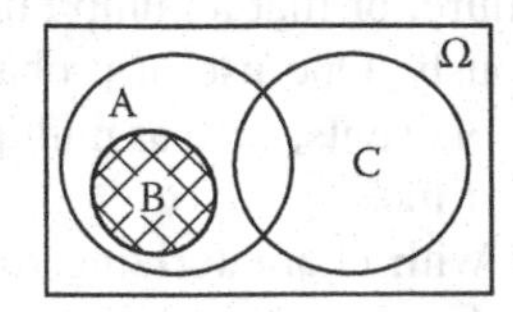

FIGURE 2.1 Venn diagram showing the relationship between subsets A, B, C and universal set Ω.

have elements in common, but B and C do not. In this case B and C are called *disjoint* or *mutually exclusive* events.

The *complement* of a set A, denoted by $\bar{A}$ and called *A not,* represents negation. It is the set of all events in the universal set that do not belong to subset A. In Figure 2.1, the non-shaded area outside of the set B bounded by the rectangle represents $\bar{B}$. The sets B and $\bar{B}$ together comprise Ω. One can easily see that the complement of a universal set is a null set, and vice versa. That is, $\bar{\Omega} = \varnothing$ and $\Omega = \bar{\varnothing}$.

The *union* of two sets, A with B, is a set that contains all items that belong to A or *B or both.* The union is symbolized either by $A \cup B$ or by $A + B$ and is read "*A or B*." The shaded area in Figure 2.2 shows the union of sets A or B.

Suppose A and B represent odd and even numbers between 1 and 10, respectively. Then $A = \{1,3,5,7,9\}$, and $B = \{2, 4, 6, 8, 10\}$. Therefore, the union of these two sets is $A \cup B = \{1, 2, 3, 4, 5, 6, 7, 8, 9, 10\}$. If $A = \{x, y, z\}$ and $B = \{t, x, z\}$, then $A \cup B = \{t, x, y, z\}$. Note that elements x and z are in both sets A and B but appear only once in their union.

The *intersection* of two sets, A and B, is the set of items that are contained in both A and B. The intersection is symbolized by $A \cap B$ or $A \cdot B$, which is usually simplified to AB. The intersection is read "*A and B*." In Figure 2.3, the shaded area represents the intersection of A and B. If we have discrete sets $A = \{x, y, z\}$ and $B = \{t, x, z\}$, then $A \cap B = \{x, z\}$.

Now, consider a set representing continuous items. Suppose A is a set of manufactured devices that operate for $t > 0$ but failed before 1,000 hours of operation. If set B represents a set of the devices that operated between 500 and 2,000 hours, then $A \cap B$ represents devices that have worked between 500 and 1,000 hours and belong to both sets and can be expressed as follows:

$$A = \{t \mid 0 < t < 1{,}000\},\ B = \{t \mid 500 < t < 2{,}000\}, A \cap B = \{t \mid 500 < t < 1{,}000\}.$$

In two *mutually exclusive* or *disjoint* sets, A and B, $A \cap B = \varnothing$. In this case, A and B have no elements in common. Two mutually exclusive sets are illustrated in Figure 2.4.

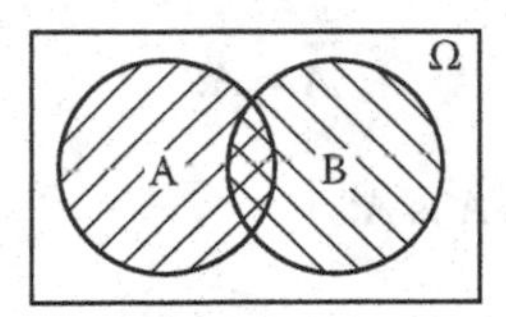

FIGURE 2.2 Union of two sets, denoted *A or B*, $(A \cup B)$.

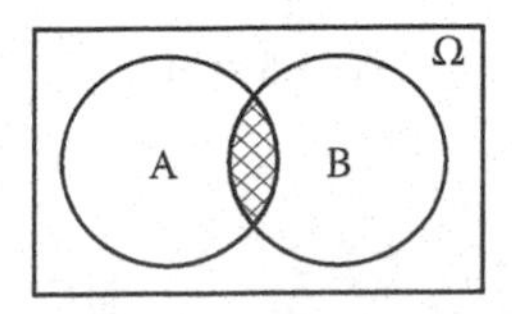

FIGURE 2.3 Intersection of two sets, denoted *A and B*, $(A \cap B)$.

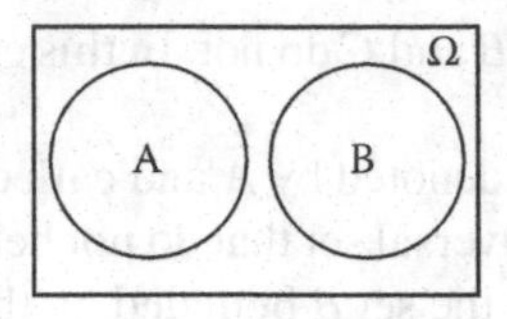

FIGURE 2.4 Mutually exclusive sets A and B, $(A \cap B) = \varnothing$.

From the discussions thus far, and from the examination of the Venn diagram, these conclusions can be drawn:
The intersection of set A and a null set is a null set:

$$A \cap \varnothing = \varnothing. \tag{2.1}$$

The union of set A and a null set is A:

$$A \cup \varnothing = A. \tag{2.2}$$

The intersection of set A and its complement A is a null set:

$$A \cap \bar{A} = \varnothing. \tag{2.3}$$

The intersection of set A and a universal set is A:

$$A \cap \Omega = A. \tag{2.4}$$

The union of set A and a universal set is the universal set:

$$A \cup \Omega = \Omega. \tag{2.5}$$

The complement of the complement of set A is A:

$$\bar{\bar{A}} = A. \tag{2.6}$$

The union of two identical sets A is A:

$$A \cup A = A. \tag{2.7}$$

The intersection of two identical sets A is A:

$$A \cap A = A. \tag{2.8}$$

TABLE 2.1
Boolean Algebra Laws

Designation	Mathematical Notation	Engineering Notation[a]
Identity laws	$A \cup \varnothing = A$	$A + 0 = A$
	$A \cup \Omega = \Omega$	$A + 1 = 1$
	$A \cap \varnothing = \varnothing$	$A \cdot 0 = 0$
	$A \cap \Omega = A$	$A \cdot 1 = A$
Idempotent laws	$A \cap A = A$	$A \cdot A = A$
	$A \cup A = A$	$A + A = A$
Complement laws	$A \cap \bar{A} = \varnothing$	$A \cdot \bar{A} = 0$
	$A \cup \bar{A} = \Omega$	$A \cdot \bar{A} = 1$
Law of absorption	$A \cup (A \cap B) = A$	$A + (A \cdot B) = A$
	$A \cap (A \cup B) = A$	$A \cdot (A + B) = A$
de Morgan's theorem	$\overline{(A \cap B)} = \bar{A} \cup \bar{B}$	$\overline{(A \cdot B)} = \bar{A} + \bar{B}$
	$\overline{(A \cup B)} = \bar{A} \cap \bar{B}$	$\overline{(A + B)} = \bar{A} \cdot \bar{B}$
Commutative laws	$A \cap B = B \cap A$	$A \cdot B = B \cdot A$
	$A \cup B = B \cup A$	$A + B = B + A$
Associative laws	$A \cup (B \cup C) = (A \cup B) \cup C = A \cup B \cup C$	$A + (B + C) = (A + B) + C = A + B + C$
	$A \cap (B \cap C) = (A \cap B) \cap C = A \cap B \cap C$	$A \cdot (B \cdot C) = (A \cdot B) \cdot C = A \cdot B \cdot C$
Distributive laws	$A \cap (B \cup C) = (A \cap B) \cup (A \cap C)$	$A \cdot (B + C) = (A \cdot B) + (A \cdot C)$
	$A \cup (B \cap C) = (A \cup B) \cap (A \cup C)$	$A + (B \cdot C) = (A + B) \cdot (A + C)$

[a] In this table, we use $A \cdot B$ to denote Boolean intersection. In the remainder of the book, we simplify this to AB.

These identity laws and idempotent laws are a few axioms of *Boolean algebra*. Boolean algebra provides a means of evaluating sets. The rules are fairly simple. The axioms in Table 2.1 provide the major relations of interest in Boolean algebra.

Example 2.1

Simplify the expression: $E \cap (E \cap \Omega)$.

Solution:

Since $E \cap \Omega = E$ and $E \cap E = E$, then the expression reduces to E.

Example 2.2

Simplify the following Boolean expression:

$\overline{\left[(A\cap B)\cup\left(A\cap\bar{B}\right)\cup\left(\bar{A}\cap\bar{B}\right)\right]}$	de Morgan's theorem
$=\overline{(A\cap B)}\cap\overline{\left(A\cap\bar{B}\right)}\cap\overline{\left(\bar{A}\cap\bar{B}\right)}$	de Morgan's theorem
$=\left(\bar{A}\cup\bar{B}\right)\cap\left(\bar{A}\cup\bar{\bar{B}}\right)\cap\left(\bar{\bar{A}}\cup\bar{\bar{B}}\right)$	Complementation law
$=\left\{\left[\bar{A}\cap\left(\bar{A}\cup B\right)\right]\cup\left[\bar{B}\cap\left(\bar{A}\cup B\right)\right]\right\}\cap(A\cup B)$	Distributive law
$=\left\{\left[\bar{A}\cup\left(B\cap\bar{A}\right)\right]\cup\left[\left(\bar{B}\cap\bar{A}\right)\cup\left(\bar{B}\cap B\right)\right]\right\}\cap(A\cup B)$	Distributive law (twice)
$=\left[\bar{A}\cup\left(B\cap\bar{A}\right)\right]\cap(A\cup B)$	Absorption law
$=\bar{A}\cap(A\cup B)$	Absorption law
$=\bar{A}\cap B$	

Note that other solutions are possible. The reader is encouraged to explore additional solutions to this example.

2.4 PROBABILITY TERMINOLOGY AND INTERPRETATIONS

In using probability theory, we are concerned with assigning a probability to an event. So, what is probability? While this question seems simple, it is one with multiple answers, owing to different schools of thought regarding how to interpret probabilities and what data underpin them. See Cox (1946) for further discussion. Generally, we define *probability* as *a numerical measure of the chance that an event occurs or more generally that a hypothesis is true. It is used to quantitatively express the uncertainty about the occurrence of (or outcome of) an event.*

A closely related term is *frequency*, which is defined as *the rate of occurrence of events*, i.e., the number of times an event occurs over a given period of time (or space or number of trials). To differentiate the concepts of *probability* and *frequency*, consider a cloud server with an average frequency of failure of 5×10^{-2} failures per year. This means that the *average time between* failures will be 20 years, not that a failure will occur exactly at 20 years operation. Frequency can also be used to determine the probability of an event per unit of time, and/or to compute the frequency of multiple, jointly occurring events (e.g., multiple system failures).

Likelihood as a quantitative measure should be differently interpreted than probability. Generally, whereas probability is about an event, or more generally, a proposition given available evidence or observed data, likelihood describes the probability of observing the evidence or data given an event or a proposition. For example, consider if one flips a coin three times and gets the data {Head, Tail, Head}. Given this observation, you may ask, "what is the *likelihood* that the proposition that the coin is fair is true?" Reversely, by assuming the truth of the proposition that the coin is fair, you may ask, "what is the probability that in three flips of a fair coin we get {Head, Tail, Head}?" Note that these questions yield the same numerical answer, but the interpretation of that number is quite different between the likelihood and probability. We will further elaborate on the more specific interpretations of probability and likelihood from two schools of thought—namely frequentist and Bayesian interpretations.

While probability theory is concerned with determining the probability of events, *statistics* is the field of study concerned with collecting, organizing, analyzing, presenting, and interpreting data. Statistics applies probability theory to draw conclusions from data. For example, to draw conclusions about the likelihood that the coin is fair in the example above, we use statistics to evaluate the data and reach the conclusion.

The probability will be associated with an event of interest or a discrete random variable E. We denote this as $Pr(E=e)$ or simply $Pr(e)$, both of which indicate the probability that random variable E takes on value *e*. We may also refer to this generally as $Pr(E)$ denoting some event of interest which may take on states $E_1, \ldots, E_n$.

It is important that the reader appreciates the intuitive differences between the three major conceptual interpretations of probability, all of which are used in different aspects of reliability engineering.

2.4.1 Classical Interpretation of Probability (Equally Likely Concept, or Sample Space Partitioning)

In this interpretation the probability of an event, *E*, can be obtained from partitioning the sample space, *S*, provided that the same space contains *n* equally likely and different outcomes. This interpretation gives rise to the classical interpretation of probability. However, this is rarely the case in engineering.

The probability of any event *E* describes a comparison of the relative size of the subset represented by the event *E* to the sample space *S*. It can be obtained from the following equation:

$$\Pr(E) = \frac{n_e}{n}, \tag{2.9}$$

where *n* denotes the number of elements in the sample space *S*, of which n_e have an outcome (event) *E*. For example, the probability of the event "rolling an odd number on a die" is determined by using: sample space $n = 6$, and $n_e(odd\ outcomes) = 3$. Here, $Pr(odd\ outcomes) = \frac{3}{6} = 0.5$.

This definition is often inadequate for engineering applications. For example, if failures of a pump in a process plant are observed, it is unknown whether all failure modes are equally likely to occur. Nor is it clear if the whole spectrum of possible events is observed. When all sample points are not equally likely to be the outcome, the sample points may be weighted according to their relative frequency of occurrence over many trials or according to subjective judgment.

2.4.2 Frequency Interpretation of Probability

In this interpretation, the limitation of the lack of knowledge about the overall sample space is remedied by defining the probability as the relative number of occurrences of an event in a large number of identical and independent trials—that is the limit of n_e/n as n becomes large. Therefore,

$$Pr(E=e)= \lim_{n \to \infty} \left(\frac{n_e}{n}\right). \tag{2.10}$$

This interpretation provides an empirical method to estimate probabilities by counting over the ensemble of many trials. Thus, if we have observed $n=2{,}000$ starts of a pump in which $n_e = 20$ failed, and if we assume that 2,000 is a large number and these pumps are identical, independent and exchangeable, then the probability of the pump's failure to start is $\frac{20}{2{,}000} = 0.01$.

In the frequentist interpretation the idea of a large ensemble of trials of identical, independently distributed, and exchangeable events is vital. In some cases, there is a natural ensemble such as tossing a coin or repeated failures of the same equipment over many identical and independent trials reflecting identical use cases. However, this interpretation does not cover cases in which little or no data is available, nor cases where estimates are intuitive. Thus, a broader definition of probability is required, which has led to the subjective interpretation of probability.

2.4.3 Subjective Interpretation of Probability (Bayesian Probability; Evidential Probability)

In many problems involving one-of-a-kind events (e.g., the weather at a specific location on a specific date), or when limited data and information are available (e.g., for new technologies), definition and creation of a large set of independent, identical, and interchangeable events is impossible. For example, consider the probability of the existence of intelligent life on another planet. To implement the frequentist interpretation requires creating an ensemble of identical, interchangeable planets that would be purely hypothetical. Another example might be the performance of a particular student on an upcoming exam in this class; while we can assess evidence-based probabilities of past performance of a population of students on the exam, any particular student would balk at considering their exam grade as identical and interchangeable with another student. When there is no historical data, a small population, or other limitations the frequency interpretation becomes inappropriate and we instead use the subjective, Bayesian, or evidential interpretation of probability.

In this interpretation, which can be traced back to Bernoulli and Laplace, a probability can be assigned to any statement as a way of representing possibility or the degree to which the statement is supported by the available evidence. That is, the probability $Pr(E)$ is interpreted as the *degree of belief* (or confidence) of an individual in the truth of a proposition or event E. It is a representation of an individual's state of knowledge, and when using it we generally assume that any two rational individuals with the same knowledge, information, and beliefs will assign the same probability.

These probabilities are referred to as Bayesian because of the central role they play in the uses of Bayes' Theorem. Different individuals may have dissimilar degrees of belief about the truth of a proposition, but they base those beliefs and can change their individual beliefs, based on introspection or revelation and new data and information.

To better understand this interpretation, consider the probability of improving a system by making a design change. The designer believes that such a change will result in a performance improvement in one out of three missions in which the system is used. It would be difficult to describe this problem through the classical or frequentist interpretations. The classical interpretation is inadequate since there is no reason to believe that performance is equally likely to improve or not improve. The frequency interpretation is not applicable because no historical data exist to show how often a design change resulted in improving the system. Thus, the subjective interpretation provides a broad definition of the probability concept.

Clearly, the subjective interpretation provides a consistent and rational method for reasoning about unique and singular events; it is not limited to repeatable and identical events (but can also be applied to those events). While the subjective component of interpretation cannot be entirely eliminated, the calibration reduces this subjectivity. The Bayesian rules for processing information and updating probabilities (as discussed in the next section) are themselves objective. Accordingly, new data and information can be objectively combined into posterior probabilities that allow one to continuously suppress the subjectivity in the probabilities elements as we consider increasingly more information.

Note that the way a frequentist view looks at the data and proposition differs from a subjectivist view. Whereas a frequentist view treats the proposition as fixed and the data as random, the subjectivist view treats the truth of a proposition as random and places a degree of belief (or certainty) on its truth considering the fixed observed data or evidence. We will further discuss this in Chapter 4.

2.5 LAWS AND MATHEMATICS OF PROBABILITY

The basic mathematics underpinning probability are the same, regardless of the interpretation of probability.

2.5.1 Definitions

The *marginal (or unconditional) probability, Pr(A)* is the probability of event *A* occurring. The *joint probability* of two events, *A* and *B*, is the probability that both events occur and is denoted as $Pr(A \cap B)$, $Pr(A, B)$, or $Pr(A \cdot B)$. The *conditional probability* $Pr(A|B)$ is the probability of event *A* occurring given that B has occurred. This is read as "the probability of *A*, given that *B* has occurred."

The conditional probability is a function of the joint and marginal probabilities:

$$Pr(A|B) = \frac{Pr(A \cap B)}{Pr(B)}. \tag{2.11}$$

Note that conditional probability is not commutative, i.e., $Pr(A|B) \neq Pr(B|A)$ except under the special condition that $Pr(A) = Pr(B)$.

It is important to understand two additional terms that are sometimes confused: mutually exclusive and independent. Recall the concept of *mutually exclusive* events from Section 2.3. Two events are mutually exclusive when $A \cap B = \varnothing$, i.e., they have no elements in common. This means that A and B cannot happen simultaneously. For example, if A = rolling an even number on a die, and B = rolling an odd number on a die. Another example of mutually exclusive events is: event E_1 = engine is in a failed state and event E_2 = engine is in an operational state. By definition, the probability of occurrence of two mutually exclusive states is zero:

$$Pr(A \cap B) = 0. \tag{2.12}$$

By contrast, two events are *independent*, which is denoted as $A \perp B$, if the occurrence or nonoccurrence of one does not depend on or change the probability of the occurrence of the other. For example, if event A is rolling the number one on the first roll of a die, and event B is rolling a one on the second roll of a die. Another example could be event A being a valve failing to open and event B being an unrelated engine failing to start. This is expressed as

$$Pr(A|B) = Pr(A). \tag{2.13}$$

It is important to emphasize the difference between independent events and mutually exclusive events, since these two concepts are sometimes confused. In fact, *two events that are mutually exclusive are not independent.* Since two mutually exclusive events A and B have no intersection, that is, $A \cap B = \varnothing$, then $Pr(A \cap B) = Pr(A) \cdot Pr(B|A) = 0$. This means that $Pr(B|A) = 0$, since $Pr(A) \neq 0$. For two independent events, we expect to have $Pr(B|A) = Pr(B)$, which is not zero except for the trivial case of $Pr(B) = 0$. This indicates that two mutually exclusive events are indeed dependent. An example to illustrate this would be flipping a coin. In this scenario there are only two possible outcomes, a head or a tail, but not both, hence the events are mutually exclusive. However, if a head is obtained, then the probability of the tail is zero. Hence, these two events are not independent.

Example 2.3

Let's consider the result of a test on 200 manufactured identical parts. Let L represent the event that a part does not meet the specified length and H represent the event that the part does not meet the specified height. It is observed that 23 parts fail to meet the length limitation imposed by the designer and 18 fail to meet the height limitation. Additionally, seven parts fail to meet both length and height limitations. Therefore, 152 parts meet both specified requirements. Are events L and H dependent?

Solution:

According to Equation 2.9, $Pr(L) = (23+7)/200 = 0.15$ and $Pr(H) = (18+7)/200 = 0.125$. Furthermore, among 25 parts (18+7) that have at least event H, seven parts also have event L. Thus, $Pr(L|H)=7/25=0.28$. Since $Pr(L|H)\neq Pr(L)$, events L and H are dependent.

2.5.2 Axioms of Probability and Their Implications

The axioms of probability, or Kolmogorov axioms, are defined as:

1. $Pr(E_i)\geq 0$, for every event E_i. (2.14)
2. $Pr(\Omega)=1$. (2.15)
3. $Pr(E_1\cup E_2\cup\ldots\cup E_n)=Pr(E_1)+Pr(E_2)+\ldots+Pr(E_n)$,
   when the events $E_1, \ldots, E_n$ are mutually exclusive. (2.16)

By examination of these axioms, there are several implications that become evident. First, that probabilities are numbers that range from 0 to 1:

$$0\leq Pr(E_i)\leq 1. \quad (2.17)$$

The probability of the null set is 0,

$$Pr(\varnothing)=0. \quad (2.18)$$

We can also state that if $A\subset B$ then

$$Pr(A)\leq Pr(B). \quad (2.19)$$

Finally, the probability of a complement of an event is 1 minus the probability of that event:

$$Pr(\overline{E_i})=1-Pr(E_i). \quad (2.20)$$

2.5.3 Mathematics of Probability

A few rules are necessary to enable manipulation of probabilities. By rearranging the definition of conditional probability (Equation 2.11), we can see that the joint probability of two events can be obtained from the following expression:

$$Pr(A\cap B)=Pr(B|A)Pr(A). \quad (2.21)$$

The *chain rule of probability* or the *multiplication rule* is the generalized form of this relationship. It gives the joint occurrence of n events $E_1, \ldots, E_n$ as a product of conditionals:

$$Pr(E_1\cap E_2\cap\ldots\cap E_n)=Pr(E_1)\cdot Pr(E_2|E_1)\cdot Pr(E_3|E_1\cap E_2)\ldots$$
$$Pr(E_n|E_1\cap E_2\cap\ldots\cap E_{n-1}) \quad (2.22)$$

where, $Pr(E_3 \mid E_1 \cap E_2 \cap \ldots)$ denotes the conditional probability of E_3, given the occurrence of both E_1 and E_2, and so on.

It is easy to see that when A and B are independent and Equation 2.13 is applied, Equation 2.21 reduces to:

$$Pr(A \cap B) = Pr(A)Pr(B). \tag{2.23}$$

And thus, if all events are independent (i.e., $E_1 \perp E_2 \perp \ldots \perp \ldots E_n$) the chain rule of probability simplifies to:

$$Pr(E_1 \cap E_2 \cap \ldots \cap E_n) = Pr(E_1) \cdot Pr(E_2) \ldots \cdot Pr(E_n) = \prod_{i=1}^{n} Pr(E_i). \tag{2.24}$$

Example 2.4

Suppose that Vendor 1 provides 40% and Vendor 2 provides 60% of chips used in a computer. It is further known that 2.5% of Vendor 1's supplies are defective and 1% of Vendor 2's supplies are defective. What is the probability that a randomly selected chip is both defective and supplied by Vendor 1? What is the same probability for Vendor 2?

Solution:

E_1 = the event that a chip is from Vendor 1
E_2 = the event that a chip is from Vendor 2
D = the event that a chip is defective
$D|E_1$ = the event that a defective chip is supplied by Vendor 1
$D|E_2$ = the event that a defective chip is supplied by Vendor 2

Then $Pr(E_1) = 0.40$, $Pr(E_2) = 0.60$, $Pr(D \mid E_1) = 0.025$, and $Pr(D \mid E_2) = 0.01$.

From Equation 2.22, the probability that a randomly selected chip is defective and from Vendor 1 is:

$$Pr(E_1 \cap D) = Pr(E_1)Pr(D \mid E_1) = (0.4)(0.025) = 0.01.$$

Similarly, $Pr(E_2 \cap D) = 0.006$.

Another probability rule, the *inclusion-exclusion principle* or *addition law of probability*, deals with the union of inclusive events. Recall that the union of mutually exclusive events, $A \cup B$, is given by one of the axioms of probability to be: $Pr(A \cup B) = Pr(A) + Pr(B)$. But what if A and B are not mutually exclusive? The inclusion-exclusion principle applies:

$$Pr(A \cup B) = Pr(A) + Pr(B) - Pr(A \cap B). \tag{2.25}$$

To illustrate the origin of this law, let's consider the 200 electronic parts that we discussed in Example 2.3. The union of two events L and H includes those parts that do not meet the length requirement, or the height requirement, or both, that is, a total of $23+18+7=48$. Thus, $Pr(L \cup H) = \frac{48}{200} = 0.24$. In other words, 24% of the parts do not meet one or both requirements. We can easily see that $Pr(L \cup H) \neq Pr(L) + Pr(H)$, since $0.24 \neq 0.125+0.15$. The reason for this inequality is that the two events L and H are not mutually exclusive. In turn, *Pr(L)* will include the probability of inclusive events where both L and H fail, (i.e., $L \cap H$). *Pr(H)* will also include the events where both fail, ($L \cap H$). Thus, joint events are counted twice in the expression $Pr(L) + Pr(H)$. Therefore, $Pr(L \cap H)$ must be subtracted from this expression. This overlap, which can also be seen in a Venn diagram, leads to the need to correct for this double counting, as shown in Equation 2.25. Since $Pr(L \cap H) = \frac{7}{200} = 0.035$, then $Pr(L \cup H) = 0.125 + 0.15 - 0.035 = 0.24$, which is what we expect to get. From Equation 2.25 one can see that if A and B are mutually exclusive, then $Pr(A \cup B) = Pr(A) + Pr(B)$, again matching the axiom given in Equation 2.14.

The addition law of probability for two events can be logically extended to n events:

$$\begin{aligned} Pr(E_1 \cup E_2 \cup \ldots \cup E_n) = {} & Pr(E_1) + Pr(E_2) + \ldots + Pr(E_n) \\ & - \left[Pr(E_1 \cap E_2) + Pr(E_1 \cap E_3) + \ldots + Pr(E_{n-1} \cap E_n) \right] \\ & + \left[Pr(E_1 \cap E_2 \cap E_3) + Pr(E_1 \cap E_2 \cap E_4) + \ldots \right] \ldots \\ & + (-1)^{n-1} Pr(E_1 \cap E_2 \cap \ldots \cap E_n). \end{aligned} \quad (2.26)$$

Note that Equation 2.25 can be expanded into the form involving the conditional probability by using Equation 2.21:

$$Pr(A \cup B) = Pr(A) + Pr(B) - Pr(B|A) Pr(A). \quad (2.27)$$

Equation 2.26 can be expanded likewise.

If all events are independent, then Equation 2.27 simplifies further to

$$Pr(A \cup B) = Pr(A) + Pr(B) - Pr(A) Pr(B), \quad (2.28)$$

which can be reformatted into the more compact form of $Pr(A \cup B) = 1 - (1 - Pr(A))(1 - Pr(B))$. Generalizing this expression results in the addition law of probability for n independent events:

$$Pr(E_1 \cup E_2 \cup \ldots \cup E_n) = 1 - \prod_{i=1}^{n} [1 - Pr(E_i)]. \quad (2.29)$$

Example 2.5

A particular type of valve is manufactured by three suppliers. It is known that 5% of valves from Supplier 1, 3% from Supplier 2, and 8% from Supplier 3 are defective. Assume the supplies are independent. If one valve is selected from each supplier, what is the probability that at least one valve is defective?

Solution:

$D1$ = the event that a valve from Supplier 1 is defective
$D2$ = the event that a valve from Supplier 2 is defective
$D3$ = the event that a valve from Supplier 3 is defective

$D1 \cup D2 \cup D3$ is the event that at least one valve from Suppliers 1, 2, or 3 is defective. Since the occurrence of events $D1$, $D2$, and $D3$ is independent, we can use Equation 2.29 to determine the probability of $D1 \cup D2 \cup D3$. Thus,

$$Pr(D_1 \cup D_2 \cup D_3) = 1 - (1-0.05)(1-0.03)(1-0.08) = 0.152.$$

For a large number of events, the addition law of probability can result in combinatorial explosion. Equation 2.26 consists of $2^n - 1$ terms. A useful method for dealing with this is the *rare event approximation*. In this approximation, we ignore the joint probability terms (i.e., we treat the n events as if they were mutually exclusive), and use Equation 2.14 (rewritten below as Equation 2.30 for convenience) for the union of events $E_1, \ldots, E_n$. This rare event approximation is used if all probabilities of interest, $Pr(E_i)$, are small, for example, $Pr(E_i) < \frac{1}{50n}$.

$$Pr(E_1 \cup E_2 \cup \ldots \cup E_n) = Pr(E_1) + Pr(E_2) + \ldots + Pr(E_n). \tag{2.30}$$

Example 2.6

Determine the maximum error in applying the rare event approximation. If Equation 2.30 is used instead of Equation 2.26, and $Pr(E_i) < \frac{1}{50n}$, what is the maximum error in Equation 2.26? Find this error for $n = 2, 3, 4$.

Solution:

For $n = 2$, using Equation 2.26,

$$Pr(E_1 \cup E_2) = \frac{2}{50(2)} - \left[\frac{1}{50(2)}\right]^2 = 0.01990.$$

Using Equation 2.30,

$$Pr(E_1 \cup E_2) = \frac{2}{50(2)} = 0.02000,$$

$$|\max \%\,\text{Error}| = \left|\frac{0.1990 - 0.02000}{0.1990} \times 100\right| = 0.50\%.$$

For $n=3$, using Equation 2.26,

$$Pr(E_1 \cup E_2 \cup E_3) = \frac{3}{50(3)} - 3\left[\frac{1}{50(3)}\right]^2 + \left[\frac{1}{50(3)}\right]^3 = 0.01987.$$

For $n=3$, using Equation 2.30, $Pr(E_1 \cup E_2) = \frac{3}{50(3)} = 0.02000$, $|\max\%\text{Error}| = 0.65\%$.

Similarly, for $n=4$, $|\max\%\text{Error}| = 0.76\%$.

The *law of total probability* defines the relationship between joint and marginal distributions; we use it to conduct *marginalization* over all of the states of a variable.

$$Pr(A) = \sum_{i=1}^{n} Pr(A \cap B_i) = \sum_{i=1}^{n} Pr(A|B_i)Pr(B_i). \tag{2.31}$$

For example, to marginalize out a binary variable, B, with states B_1 and B_2 (or B and $\bar{B}$)

$$Pr(A) = Pr(A \cap B_1) + Pr(A \cap B_2), \tag{2.32}$$

$$Pr(A) = Pr(A|B_1)Pr(B_1) + Pr(A|B_2)Pr(B_2). \tag{2.33}$$

If event A is conditional on another event C, the law of total probability may be extended to

$$Pr(A|C) = \sum_{i=1}^{n} Pr(A|B_i \cap C|)Pr(B_i|C). \tag{2.34}$$

Bayes' Theorem follows directly from the concepts of joint and conditional probability, a form of which is described using Equation 2.21. By solving for $Pr(A|B)$, one can quickly derive Bayes' Theorem:

$$Pr(A|B) = \frac{Pr(A)Pr(B|A)}{Pr(B)}. \tag{2.35}$$

It is easy to see that the denominator term in Bayes' Theorem (i.e., $Pr(B)$) can be replaced with the law of total probability (Equation 2.31) to provide a generalized expression for the Bayes' Theorem:

$$Pr(A|B) = \frac{Pr(A)Pr(B|A)}{\sum_{i=1}^{n} Pr(B|A_i)Pr(A_i)}. \tag{2.36}$$

The right-hand side of Bayes' Theorem consists of two terms: $Pr(A)$ is called the *prior probability* or *a priori probability of A*. The second term,

$$\frac{Pr(B|A)}{\sum_{i=1}^{n} Pr(B|A_i)Pr(A_i)}, \tag{2.37}$$

is the *relative likelihood;* this is the model by which the prior probability is revised based on evidence or observations (e.g., limited failure observations).

$Pr(A|B)$ is called the *posterior probability;* that is, given evidence or information B, the probability of event A can be updated from prior probability $Pr(A)$ by way of the likelihood function to create a posterior probability of A.

Clearly, when more evidence (in the form of events B) becomes available, $Pr(A|B)$ can be further updated. Bayes' Theorem provides a means of changing one's knowledge about an event considering new evidence or data related to the event. This has powerful implications. We return to this topic and its application in failure data evaluation in Chapter 5.

Example 2.7

Suppose that 70% of an inventory of the memory chips used by a computer manufacturer comes from Vendor 1 and 30% from Vendor 2, and that 99% of the chips from Vendor 1 and 88% of the chips from Vendor 2 are not defective. If a chip from the manufacturer's inventory is selected and is defective, what is the probability that the chip was made by Vendor 1? What is the probability of selecting a defective chip (irrespective of the vendor)?

Solution:

Let

A_1 = event that a chip is supplied by Vendor 1,
A_2 = event that a chip is supplied by Vendor 2,
D = event that a chip is defective,
$D|A_1$ = event that a chip known to be made by Vendor 1 is defective,
$A_1|D$ = event that a chip known to be defective is made by Vendor 1.

Thus,

$Pr(A_1) = 0.7$,
$Pr(A_2) = 0.3$,
$Pr(D|A_1) = 1 - 0.99 = 0.01$,
$Pr(D|A_2) = 1 - 0.88 = 0.12$.

Using Bayes' Theorem

$$Pr(A_1|D) = \frac{0.7(0.01)}{0.7(0.01) + 0.3(0.12)} = 0.163.$$

Thus, given the new evidence that the chosen unit is defective, the prior probability that Vendor 1 was the supplier ($Pr(A_1) = 0.7$) is changed to a posterior probability of $Pr(A_1|D) = 0.163$.

From the law of total probability, we find the probability of selecting a defective chip:

$$Pr(D)=\sum_{i=1}^{n}Pr(A_i)Pr(D|A_i)=0.043.$$

Example 2.8

A passenger air bag disable switch is used to deactivate it when the passenger seat of a car is not occupied. Data show that a commercial van driver usually has a passenger 30% of the time. In addition, the probability of driver having a collision when passenger seat is occupied is 40% of the probability of collision when passenger seat is not occupied, because drivers are believed to be more careful when passengers are in the van. Given a collision has occurred, what is the probability of the passenger seat is occupied?

Solution:

Denote

$Pr(P)$ = probability of a passenger seat being occupied
$Pr(\bar{P})$ = probability of a passenger seat not being occupied
$Pr(C)$ = probability of getting into a collision
$Pr(C|P)$ = probability of getting into a collision, given the passenger seat is occupied
$Pr(P|C)$ = probability of the passenger seat being occupied when a collision happens.

Using Bayes' Theorem, this probability can be found as

$$Pr(P|C)=\frac{Pr(P)Pr(C|P)}{Pr(P)Pr(C|P)+Pr(\bar{P})Pr(C|\bar{P})}.$$

According to the statement of the problem, the probabilities of getting into a collision with and without a passenger are related to each other in the following manner:

$$Pr(C|P)=0.4\cdot Pr(C|\bar{P}).$$

Remembering this, and that $Pr(P)=1-Pr(\bar{P})$, one obtains

$$Pr(P|C)=\frac{Pr(P)}{Pr(P)+(1/0.4)[1-Pr(P)]}=\frac{0.3}{0.3+(1/0.4)(1-0.3)}=0.146.$$

Example 2.9

In prognosis and health monitoring we often need to interpret an event of interest (e.g., failure of a part) using some sensor information as indirect evidence (truth) about the event. Suppose a test using an acoustic emission sensor to detect the failure event turns out to be positive. One may at this point jump to the conclusion that the part has most likely failed. However, tests can have false positives. Further examination of the part's failure history shows that the probability of the part failing at its current age is 0.01. Calculate the probability of failure given the sensor results using these assumptions about the test:

a. This test is known be 98% accurate at detecting the true state of the part (both true positives (failures) and true negatives (non-failures)).
b. This test is 98% accurate at detecting true failures and 85% accurate at detecting true non-failures.

Solution:

$Pr(F)$ = Prior probability of failure of the part at this age = 0.01.
$Pr(\bar{F})$ = Prior probability of no failure of the part at this age $= 1 - Pr(F) = 0.99$.
$Pr(S|F)$ = Probability of sensor signaling failure given failure of the part = 0.98.
$Pr(S|\bar{F})$ = Probability of sensor signaling failure given no failure of the part $= 1 - 0.98 = 0.02$.
$Pr(F|S)$ = Probability of failure of the part at this age given sensor results.

$$Pr(F|S) = \frac{Pr(S|F)Pr(F)}{Pr(S|F)Pr(F) + Pr(S|\bar{F})Pr(\bar{F})} = \frac{0.98 \cdot 0.01}{0.98 \cdot 0.01 + 0.02 \cdot 0.99} = 0.331.$$

For assumption a, despite the high probability of correct detection, the result remains highly uncertain due to the low probability of part failure at this age.

For assumption b, $Pr(F)$, $Pr(\bar{F})$, and $Pr(S|F)$ are the same as for assumption a. However, $Pr(S|\bar{F})$ = Probability of sensor signaling failure given no failure of the part $= 1 - 0.85 = 0.15$

$$Pr(F|S) = \frac{Pr(S|F)Pr(F)}{Pr(S|F)Pr(F) + Pr(S|\bar{F})Pr(\bar{F})} = \frac{0.98 \cdot 0.01}{0.98 \cdot 0.01 + 0.15 \cdot 0.99} = 0.0619.$$

In this case, the probability of a failed device is even lower, reflecting the high probability of a false positive from this test.

2.6 PROBABILITY DISTRIBUTION BASICS

In this section, we provide information on a variety of *parametric probability distributions* from a mathematical perspective. We introduce several commonly used distributions in reliability engineering. We provide worked examples for the first several distributions and introduce several additional distributions which will be discussed further in Chapter 3. In Chapter 3, we also discuss applications of those most used in reliability analysis and provide further examples illustrating their use.

Sometimes parametric distributions are not a good fit for a model or set of data. In these cases, *empirical distributions* can be created directly from data. If a parametric distribution cannot be used, other modeling options exist too, such as those created through combining probability distributions, using logic-based models, or using Bayesian networks. These will be discussed later in this book.

2.6.1 Probability Distributions Defined

The probability of an event is not directly observable, and models are necessary to help turn data and information into probabilities. There are various types of observable data we may see, e.g., we can observe if events (e.g., failures) occur, we can observe how many events occur in a certain number of trials, and we can observe the times at which things fail. We can count combinations of items or options. We can collect measurements and run large sets of simulation models. We can ask experts to assign probabilities using their implicit models.

A *probability distribution* is a function that assigns a probability to each possible value that a random variable can take (i.e., each point in the sample space $S = \{x_1, \dots, x_n\}$ where n is either finite or infinite), such that the total probability assigned is 1.0. If n is finite the probability distribution is discrete, otherwise it is continuous.

The function that assigns probability is called the *probability mass function* (pmf) for discrete variables or the *probability density function* (pdf) for continuous variables—in both cases, we denote this as $f(x)$. This function describes the general form of a probability distribution. For brevity, we may use the term "pdf" generically to refer to the pdf or the pmf when referring to an unspecified random variable.

The *probability mass function* (pmf) is a function that assigns the probability of each event in a discrete sample space. That is, $f(x) = Pr(X = x)$, if it meets the following properties:

$$f(x_i) \geq 0 \text{ for all } x_i;\ i = 1, \dots, n, \tag{2.38}$$

and

$$\sum_{i=1}^{k} f(x_i) = 1. \tag{2.39}$$

The *probability density function* (pdf) assigns a probability to an interval of values, and has these properties:

$$f(x) \geq 0,$$

and

$$\int_{-\infty}^{\infty} f(x)\,dx = 1. \tag{2.40}$$

In both cases, $f(x) = 0$ for values outside of the range or sample space of the variable. The value of $f(x)$ can be used to determine a relative likelihood of the value of the random variable. Note that a continuous variable has a probability of zero for the r.v. taking exactly any of its possible values, because there are infinite possible values it could take. In this situation, it is appropriate to introduce the probability associated with a small interval of values. For example, one can determine $Pr(x_1 < X < x_2)$, that is, the probability that the pump would fail sometime between 50 and 60 hours.

A related function is called the *cumulative distribution function* (cdf), $F(x)$, which describes the probability that the random variable X is less than or equal to a given value, x, and is defined as:

$$F(x) = Pr(X \leq x) = \begin{cases} \sum_{X \leq x} f(x_i) & \text{for a discrete r.v.} \\ \int_{-\infty}^{x} f(t)dt & \text{for a continuous r.v.} \end{cases} \tag{2.41}$$

The cdf can be used to determine the probability that a random variable falls into an interval:

$$Pr(a \leq X \leq b) = \int_a^b f(x)dx = F(b) - F(a). \tag{2.42}$$

For a discrete variable, the cdf is defined similarly, recognizing that integration is the continuous analog of summation:

$$Pr(a \leq X \leq b) = \sum_{i=1}^{b} f(x_i) - \sum_{i=1}^{a-1} f(x_i) = F(b) - F(a-1), \tag{2.43}$$

$$Pr(a < X \leq b) = \sum_{i=1}^{b} f(x_i) - \sum_{i=1}^{a} f(x_i) = F(b) - F(a). \tag{2.44}$$

Note the use of a–1 in in Equation 2.43 vs Equation 2.44. For continuous distributions $Pr(a < X)$ is the same as $Pr(a \leq X)$, but for discrete distributions the boundary values matter.

The cdf must be monotonically increasing and must satisfy conditions derived from the axioms of probability:

$$F(-\infty) = 0, \tag{2.45}$$

$$0 \leq F(x) \leq 1, \tag{2.46}$$

$$F(x_1) \leq F(x_2) \text{ if } x_1 \leq x_2, \tag{2.47}$$

$$F(\infty) = 1. \tag{2.48}$$

Further, additional relationships hold:

$$f(x) = \frac{dF(x)}{dt}. \tag{2.49}$$

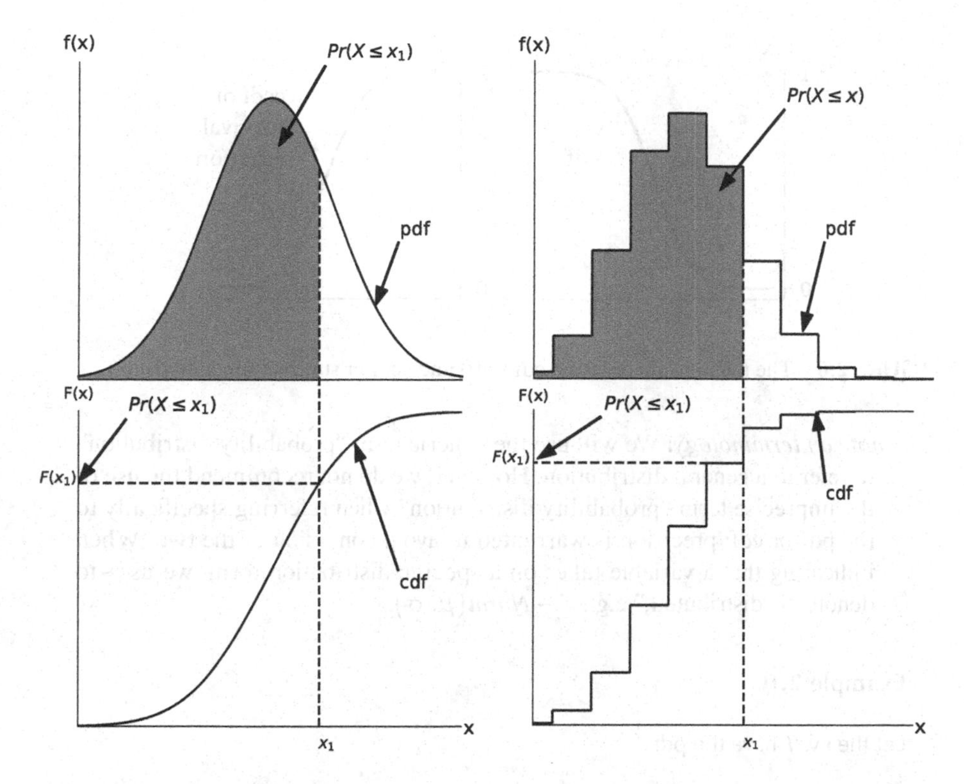

FIGURE 2.5 Top left: pdf for a continuous r.v., top right: pdf for a discrete r.v., bottom left: cdf for a continuous r.v., bottom right: cdf for a discrete r.v.

Figure 2.5 illustrates some basic features of pdfs and cdfs and illustrates the relationship between the two.

In some applications, the *complementary cumulative distribution function* (ccdf) is used. This is expressed as $R(x)$ to denote reliability (or sometimes $S(x)$ to denote survival as done in survival analysis). Most often, the cdfs $F(x)$ we encounter in reliability engineering refer to failure, and thus $R(x)$ is a suitable abbreviation for the ccdf. However, this is not a strict rule and care should be taken to ensure that the function naming does not cause confusion.

$$R(x)=1-F(x)=Pr(X>x)=\begin{cases} 1-\sum_{X\le x} f(x_i) \text{ for a discrete r.v.} \\ \int_x^{\infty} f(t)dt \text{ for a continuous r.v.} \end{cases} \tag{2.50}$$

The relationship between the ccdf and cdf is shown in Figure 2.6.

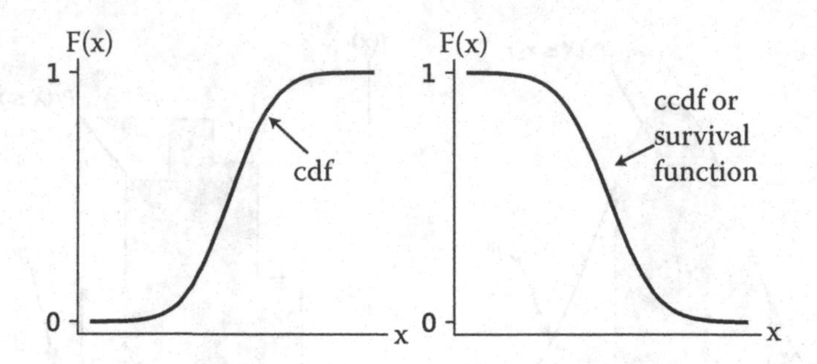

FIGURE 2.6 The relationship between cdf (left) and ccdf or survival function (right).

A note on terminology: We will use the generic term "probability distribution" to refer to a general distribution. However, we do not recommend the use of the imprecise term "probability distribution" when referring specifically to the pdf or cdf; precision is warranted to avoid confusion of the two. When indicating that a variable takes on a specific distribution form, we use ~ to denote "is distributed," e.g., $X \sim Norm(\mu, \sigma)$.

Example 2.10

Let the r.v. T have the pdf

$$f(T) = \begin{cases} \dfrac{t^2}{c}, & 0 < t < 6; \\ 0 & \text{otherwise.} \end{cases}$$

Find the value of constant c. Then find $Pr(1 < T < 3)$

Solution:

$$\int_{-\infty}^{\infty} f(t)dt = \int_{t_1}^{t_2} f(t)dt = 1,$$

$$\int_{t_1}^{t_2} f(t)dt = \int_0^6 \frac{t^2}{c}\, dt = \frac{t^3}{3c}\Bigg|_0^6 = \frac{216}{3c} - 0 = 1.$$

Then,

$$c = \frac{216}{3} = 72,$$

$$Pr(1 < T < 3) = \int_1^3 \frac{t^2}{72}\, dt = \frac{t^3}{216}\Bigg|_1^3 = \frac{27}{216} - \frac{1}{216} = 0.12.$$

It is common to use *parametric probability distributions,* wherein the distributions share the same functional form (i.e., the same pdf and cdf), but differ based on a set of parameters. A common example is the normal distribution, which is has two parameters μ and σ, its mean and standard deviation. In this book, we use Greek letters to denote parameters in most cases. Parameters are generally described by their effect on the distribution, including:

- *Location* parameters which shift the location of the distribution on the principal axis
- *Shape* parameters define the distribution shape
- *Scale* parameters define the spread of the distribution
- *Rate* parameters are the reciprocal of the scale

Note that many commonly used parametric distributions have several possible parameterizations. It is prudent to confirm the parameterization with software and transform parameters as necessary. We also recommend providing the pdf equation when writing about a distribution that has multiple parameterizations; this helps your reader and eliminates the need for them to track down your parameterization or guess which transformation is necessary.

2.6.2 Distribution Properties

In addition to the distribution parameters, there are other useful quantities that can be calculated to characterize and describe the distribution. These include measures of central tendency, moments, and expectations.

2.6.2.1 Central Tendency

Three measures of *central tendency* are used often in reliability engineering.

The *median* of a distribution, denoted as $x_{0.5}$ is the point where the cdf is equal to 0.5,

$$F(x_{0.5}) = 0.5, \tag{2.51}$$

$$x_{0.5} = F^{-1}(0.5). \tag{2.52}$$

The *mode* of a distribution is the point of the highest density of the pdf, x_{mode}. For a discrete variable, the mode is the value of the r.v. that occurs most often or has the greater probability of occurring. For a continuous r.v., the mode occurs at the maximum relative density of the pdf.

The *expected value* or *mean* of a variable is the weighted average or first moment. This is denoted as $E(X)$ or μ_X. For a discrete r.v. X, $E(X)$ is defined as:

$$E(X) = \sum_i x_i f(x_i). \tag{2.53}$$

And for a continuous r.v. with a pdf $f(t)$, then the expectation of T is defined as:

$$E(X) = \int_{-\infty}^{\infty} x f(x)\,dx. \tag{2.54}$$

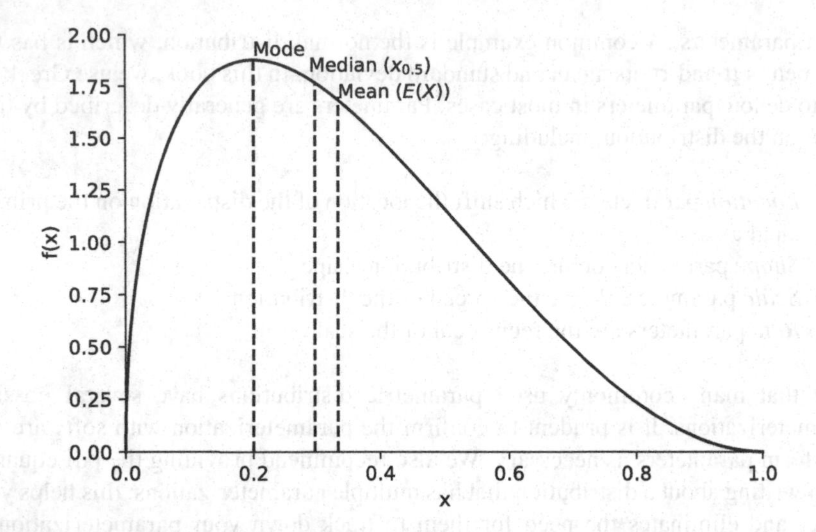

FIGURE 2.7 Measures of central tendency.

$E(X)$ is also called the *first moment about the origin*. The kth moment about the origin (the kth raw moment) is defined for all integer $k \geq 1$ as:

$$E\left(X^k\right) = \mu_k(X) = \int_{-\infty}^{\infty} x^k f(x)dx. \tag{2.55}$$

See Figure 2.7 for a conceptual depiction the three critical central tendency measures discussed above.

The expected value is also generalizable to other functions of X. For a discrete r.v. X with sample space $\{x_1, \ldots, x_k\}$, if $g(x)$ is a real-valued function of X then expected value of $g(x)$ is defined as:

$$E\left[g(X)\right] = \sum_{i=1}^{k} g(x_i)f(x_i). \tag{2.56}$$

Similarly, for continuous distribution with a pdf of $f(x)$,

$$E\left[g(X)\right] = \int_{-\infty}^{\infty} g(x)f(x)dx. \tag{2.57}$$

Note that in Equations 2.56 and 2.57, $g(x)$ is not the pdf.

The extension of expectation to joint distribution $f(x,y)$ is straightforward:

$$E(X) = \mu_X = \int_{-\infty}^{\infty}\int_{-\infty}^{\infty} xf(x, y)dxdy \text{ and } E(Y) = \mu_Y = \int_{-\infty}^{\infty}\int_{-\infty}^{\infty} yf(x, y)dxdy. \tag{2.58}$$

Equations 2.56 and 2.57 extend similarly.

2.6.2.2 Dispersion, Shape, and Spread

The *variance, var(X) or* σ_X^2, or *second central moment*, is used as a measure of dispersion or spread.

$$Var(X) = \sigma^2 = E\left[\left(X - E(X)\right)^2\right] \tag{2.59}$$

The *standard deviation, stdev(X) or* σ_X, is the square root of the variance. The standard deviation is more commonly used as a measure of dispersion than variance, because the units on standard deviation are the same as the units of the random variable, which aids interpretability.

Another measure used to characterize variability is the ratio of the standard deviation to the mean, known as the *coefficient of variation*:

$$c_v = \frac{\sigma_X}{\mu_X}. \tag{2.60}$$

There are two additional standardized moments of interest to us. The *skewness*, also known as third standardized moment, is used to describe the horizontal asymmetry or a distribution about its mean. For a unimodal distribution, positive skewness indicates that the long tail is on the right side of the distribution, negative skewness indicates that the long tail is on the left side of the distribution. A skewness close to zero indicates symmetry or balanced tails.

$$Skew(X) = \tilde{\mu}_3 = E\left[\left(\frac{X - \mu_1}{\sigma}\right)^3\right] = \frac{E(X - \mu_1)^3}{\sigma^3} = \frac{\mu_3}{\sigma^3}. \tag{2.61}$$

See Figure 2.8 to visualize skewness.

Kurtosis is a measure that refers to the combined size of the tails, it is referring to the extremity of the tails. It is sometimes described as peakedness, although this is not fully correct.

$$Kurt(X) = \tilde{\mu}_4 = E\left[\left(\frac{X - \mu_1}{\sigma}\right)^4\right] = \frac{\mu_4}{\sigma^4}. \tag{2.62}$$

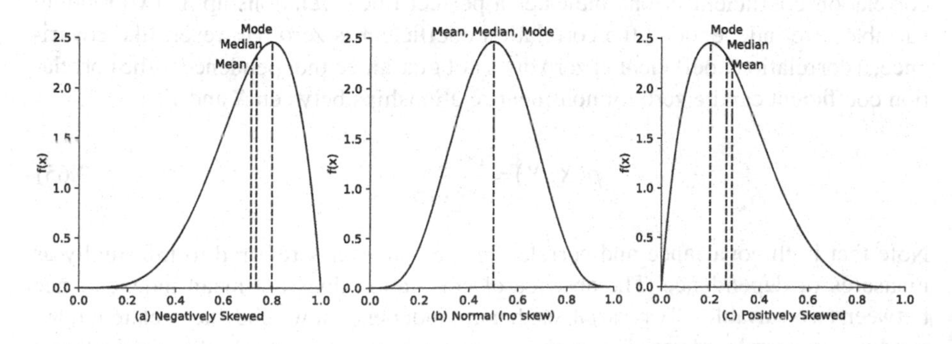

FIGURE 2.8 Types of skewness.

One common interpretation of kurtosis is by comparison to the normal distribution using the excess kurtosis. The kurtosis of any normal distribution is 3. The excess kurtosis is $\tilde{\mu}_4 - 3$. If the excess kurtosis is equal to zero, the distribution has the same peakedness as the normal distribution, if it is positive the distribution has more peakedness than normal, and if negative the distribution is flatter than normal distribution. Distributions with positive excess kurtosis have fatter tails than a normal distribution, while distributions with negative excess kurtosis have thinner tails.

Beyond these, additional moments can be described. These fall into the realm of higher-order statistics. Using Laplace transforms can simplify determination of some additional moments.

2.6.2.3 Covariance and Correlation

Covariance, Cov(X, Y), is a measure of the *linear* relationship between two random variables. Covariance is positive when high values of one variable correspond to high values of the other variable, and negative when high values of one variable correspond to low values of the other variable. If two random variables are independent, the covariance is equal to zero. However, if the covariance is zero it does not guarantee independence—the covariance can be zero for nonlinear relationships between X and Y.

$$Cov(X, Y) = E\{[X - E(X)][Y - E(Y)]\} = E[(X - \mu_x)(Y - \mu_Y)], \quad (2.63)$$

$$= E(XY) - E(X)E(Y). \quad (2.64)$$

The units of covariance are those of X times Y, so if X and Y are distances in m, the covariance has units of m^2.

Covariance can be difficult to interpret, so we normalize it into a dimensionless measure called *correlation coefficient,* $\rho(X, Y)$ or $\rho_{X,Y}$. A linear correlation coefficient sometimes known as the Pearson correlation coefficient is a bivariate correlation between two random variables as shown by Equation 2.65. It takes values between –1 and 1. The sign indicates the direction of the relationship: positive values mean that as X increases, Y tends to increase too, whereas negative values indicate that as X increases, Y tends to decrease. The absolute numerical value indicates the strength of the relationship: the higher the number, the stronger the relationship. A correlation coefficient of one indicates a perfect linear relationship. If two random variables are independent, the correlation coefficient is zero. However, like covariance, a correlation coefficient of zero does not guarantee independence—the correlation coefficient can be zero for nonlinear relationships between X and Y.

$$\rho(X, Y) = \frac{cov(X, Y)}{\sigma(x)\sigma(y)}. \quad (2.65)$$

Note that both covariance and correlation are sometimes referred to informally as measures of dependence. The absence of correlation does not mean independence between two variables. Two variables that are independent will be uncorrelated; however, two uncorrelated variables may or may not be independent. In this application it

TABLE 2.2
Properties of Expectation and Variance of r.v. X and Y

$E(c)=c$, where c is a constant.

$E(cX)=cE(X)$, where c is a constant.

$$E[g(X)\pm h(X)]=E[g(X)]\pm E[h(X)].$$
$$E[X\pm Y]=E[X]\pm E[Y].$$

$E[XY]=E[X]E[Y]$, if X and Y are independent.

$E[XY]=E(X)E(Y)+Cov(X,Y)$ if X and Y are dependent.

$Var(X+c)=Var(X).$

$Var(cX)=c^2Var(X).$

$Var(X\pm Y)=Var(X)+Var(Y)\pm 2Cov(X,Y).$

For two dependent variables, the variance of their product is:

$$\begin{aligned}Var(XY)&=E(X^2Y^2)-[E(XY)]^2\\&=Cov(X^2,Y^2)+E(X^2)E(Y^2)-[E(XY)]^2\\&=Cov(X^2,Y^2)+\left(Var(X)+[E(X)]^2\right)\left(Var(Y)+[E(Y)]^2\right)-[Cov(X,Y)+E(X)E(Y)]^2.\end{aligned}$$

For two independent variables, $X\perp Y$, this becomes:

$$\begin{aligned}Var(XY)&=[E(X)]^2\,Var(Y)+[E(Y)]^2\,Var(X)+Var(X)Var(Y)\\&=E(X^2)E(Y^2)-[E(X)]^2[E(Y)]^2.\end{aligned}$$

is necessary to be precise. Dependence of two variables means they do not meet the criteria for probabilistic independence.

2.6.2.4 Algebra of Expectations

There are useful laws and theorems that permit us to simplify complex expectations. Some of these are shown in Table 2.2.

2.7 PROBABILITY DISTRIBUTIONS USED IN RELIABILITY

2.7.1 Discrete Distributions

In this section, we discuss several commonly used discrete distributions. In each section, we also present some descriptive characteristics of the distributions. There are many additional discrete distributions that exist for various cases. The books by Johnson et al. (1995), Hahn and Shapiro (1994), Nelson (2003), and O'Connor et al. (2019) are good references for additional probability distributions.

2.7.1.1 Binomial Distribution

Consider a random trial having two possible outcomes: for instance, failure with probability p and success with probability $1-p$. Consider a series of n independent trials with these outcomes and assume the probability p is the same for each trial. Let r.v. X denote the total number of failures in n trials. The probability distribution of X is given by the binomial distribution, which gives the probability that a known event or outcome occurs exactly x times out of n trials. A single trial is also called a *Bernoulli trial* or Bernoulli experiment, and the corresponding binomial distribution for a single trial is called Bernoulli distribution. A sequence of trial outcomes is called a *Bernoulli process*. The binomial distribution $X \sim binom(n, p)$ depends on the number of trials, n, and the parameter p which is the probability that a given outcome will occur in single trial. Table 2.3 shows important properties of a binomial distribution.

TABLE 2.3
Binomial Distribution Properties

Parameters & Description $X \sim binom(n, p)$		
p	$0 \le p \le 1$	Probability of outcome x in a single trial
n	$n \in \{0,1,2,\ldots\}$	Number of trials
Range	$x \in \{0,1,\ldots,n\}$	Number of outcomes
Pdf, $f(x)$		
$f(x \mid n, p) = \binom{n}{x} p^x (1-p)^{n-x}$ 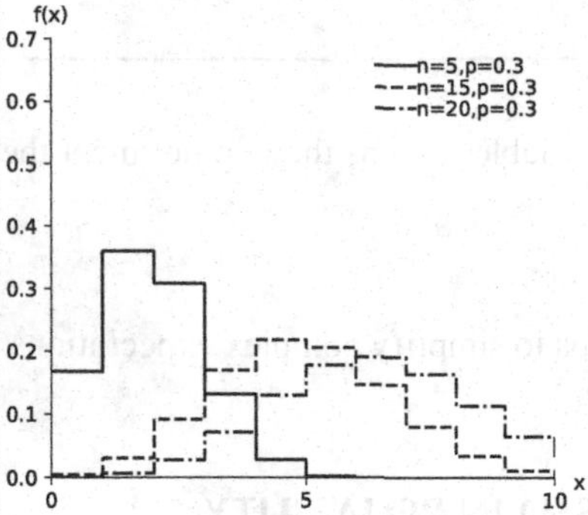 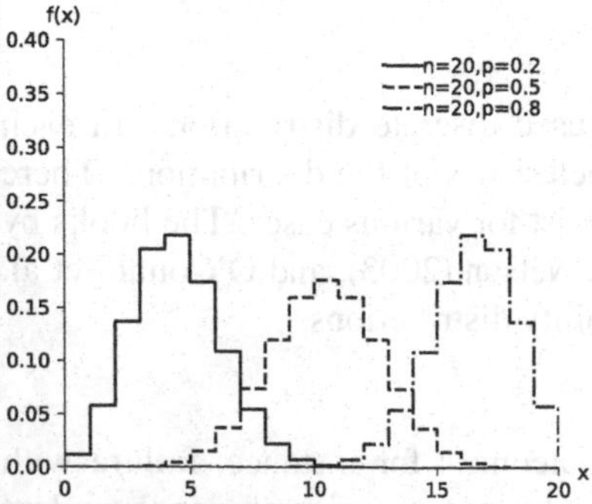		Excel: binom.dist (x, n, p, false) Matlab: binopdf (x, n, p) Python: scipy.stats.binom.pmf (x, n, p, 0)[a]

(*Continued*)

TABLE 2.3 (*Continued*)
Binomial Distribution Properties

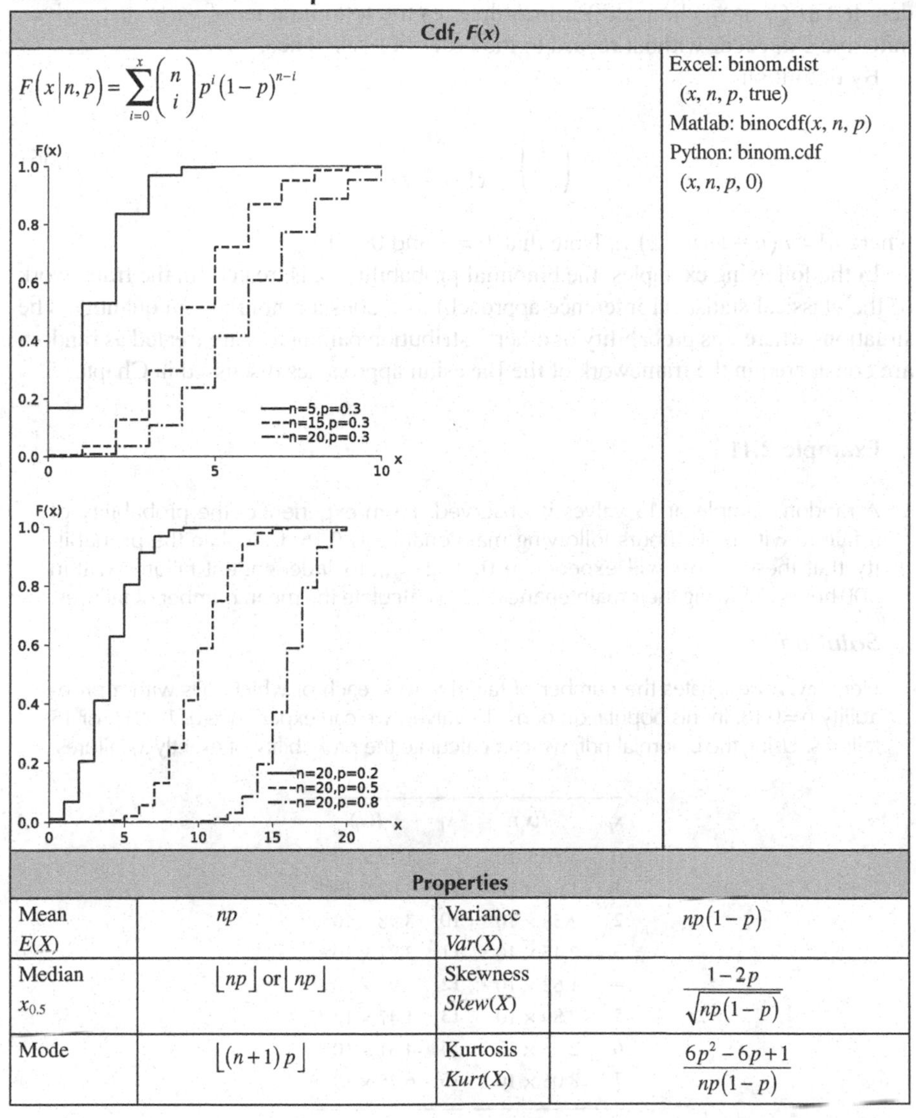

Cdf, *F*(*x*)	
$F\left(x\middle\|n,p\right)=\sum_{i=0}^{x}\binom{n}{i}p^{i}(1-p)^{n-i}$	Excel: binom.dist (*x*, *n*, *p*, true) Matlab: binocdf(*x*, *n*, *p*) Python: binom.cdf (*x*, *n*, *p*, 0)

Properties			
Mean $E(X)$	np	Variance $Var(X)$	$np(1-p)$
Median $x_{0.5}$	$\lfloor np \rfloor$ or $\lfloor np \rfloor$	Skewness $Skew(X)$	$\frac{1-2p}{\sqrt{np(1-p)}}$
Mode	$\lfloor (n+1)p \rfloor$	Kurtosis $Kurt(X)$	$\frac{6p^2-6p+1}{np(1-p)}$

[a] Unless otherwise stated, all Python commands in Section 2.7 use the "scipy.stats" package header, which is omitted for length. Additionally, the Python implementation of several distributions requires a location parameter which is not used in Section 2.7. In these cases, we set this parameter to 0.

The symbol $\binom{n}{x}$ is the binomial coefficient or combination operator (sometimes denoted as C_x^n or "n choose x"), which denotes the total number of ways that a given outcome can occur without regard to the order of occurrence.

By definition,

$$\binom{n}{x} = \frac{n!}{x!(n-x)!}, \tag{2.66}$$

where $n! = n(n-1)(n-2)\ldots$. Note that $1! = 1$ and $0! = 1$.

In the following examples, the binomial probability, p, is treated (in the framework of the classical statistical inference approach) as a constant nonrandom quantity. The situations where this probability or other distribution parameters are treated as random are considered in the framework of the Bayesian approaches discussed in Chapter 5.

Example 2.11

A random sample of 15 valves is observed. From experience, the probability of a failure within 500 hours following maintenance is 0.18. Calculate the probability that these valves will experience 0, 1, 2, ..., 15 independent failures within 500 hours following their maintenance. Also calculate the mean number of failures.

Solution:

Here, r.v. X designates the number of failed valves, each of which fails with a probability $p = 0.18$. In this population of $n = 15$ valves, we can experience 0, 1, 2, ... or 15 failures. Using the binomial pdf, we can calculate the probability of exactly x_i failures.

x_i	$f(x_i)$	x_i	$f(x_i)$
0	5.10×10^{-2}	8	1.77×10^{-3}
1	1.68×10^{-1}	9	3.02×10^{-4}
2	2.58×10^{-1}	10	3.98×10^{-5}
3	2.45×10^{-1}	11	3.97×10^{-6}
4	1.62×10^{-1}	12	2.90×10^{-7}
5	7.80×10^{-2}	13	1.47×10^{-8}
6	2.85×10^{-2}	14	4.61×10^{-10}
7	8.05×10^{-2}	15	6.75×10^{-12}

Notice the sum, $\sum f(x_i) = 1.0$, indicating that this is a true probability distribution.

The mean number of failures is E(X) = $np = 2.7$.

Example 2.12

In a process plant, there are two identical diesel generators for emergency power needs, units A and B. One of these diesels is sufficient to provide the needed emergency power. Operational history indicates that there is one failure in 100 demands for each diesel. Assume *A* and *B* are independent.[1]

a. What is the probability that at a given time of demand both diesel generators will fail?
b. If, on the average, there are 12 test-related demands per year for emergency power, what is the probability of at least one failure for diesel *A* in a year?
c. What is the probability that for the case described in (b), both diesels *A* and *B* will fail on demand simultaneously at least one time in a year?
d. What is the probability in (c) of exactly one simultaneous failure in a year?

Solution:

a. $p = \frac{1}{100} = 0.01$, $1-p=0.99$ where p denotes failure

$$Pr(A \cap B) = Pr(A) \cdot Pr(B) = (0.01)(0.01) = 0.0001.$$

That is, there is 1 in 10,000 chance that both *A* and *B* will fail on a given demand.

b. Using the binomial distribution, one can find

$$Pr(X=0) = \binom{12}{0}(0.01)^0(0.99)^{12} = 0.886,$$

which is the probability of observing no failure in 12 trials (demands). Therefore, the probability of at least one failure per diesel in a year is

$$Pr(X \geq 1) = 1 - Pr(X < 1) = 1 - Pr(X=0),$$

$$Pr(X \geq 1) = 1 - 0.886 = 0.114.$$

c. Using the results obtained in (a), the probability of simultaneous failure of generators *A* and *B* is $p=0.0001$. Then, similarly to the previous case,

$$Pr(X=12) = \binom{12}{12}(0.9999)^{12}(0.0001)^0 = 0.9988,$$

which is the probability of 12 successes for the system composed of both *A* and *B* on 12 demands (trials). Therefore, the probability of at least one failure of both *A* and *B* in a year is

$$Pr(X \geq 1) = 1 - Pr(X < 1) = 1 - Pr(X=0),$$

$$Pr(X \geq 1) = 1 - Pr(Y=12) = 1 - 0.9988 = 0.0012.$$

d. For exactly one simultaneous failure in a year,

$$Pr(X=1) = \binom{12}{1}(0.0001)^1(0.9999)^{11} = 0.00119.$$

---

[1] In this chapter, most examples will involve the assumption of independence. Methods for addressing dependent failures are discussed in Chapter 8.

2.7.1.2 Poisson Distribution

The Poisson distribution expresses the probability of an r.v. X representing the number of events occurring in a time (or over space) t for events that occur with a constant rate (or intensity) of occurrence λ. This model assumes that the occurrence of an event is independent of the time since the last event. The parameter of this distribution is $\mu = \lambda t$, which is also its mean. That is, $X \sim Poiss(\mu)$.

Essentially, the Poisson distribution assumes that objects or events of interest are evenly dispersed at random in a time or space domain at a constant rate. For example, X can represent the number of failures observed at a process plant per during a year (time domain) or the number of buses arriving at a station during 2-hour intervals, if they arrive randomly and independently in time. It can also represent the number of cracks or flaws per unit area over an area of a metal sheet (space domain). Table 2.4 shows important properties of a Poisson distribution.

TABLE 2.4
Poisson Distribution Properties

Parameters & Description $X \sim poiss(\mu)$		
μ	$\mu > 0$	Rate parameter $\mu = E(x) = \lambda t$ representing the mean number of events in a specified time or space, t, wherein events occur at a constant rate or intensity λ.
Range	$x \in \{0,1,2,\ldots\}$	Number of events that occur
Pdf, *f*(*x*)		
$f(x\|\mu) = \dfrac{\mu^x e^{-\mu}}{x!}$ 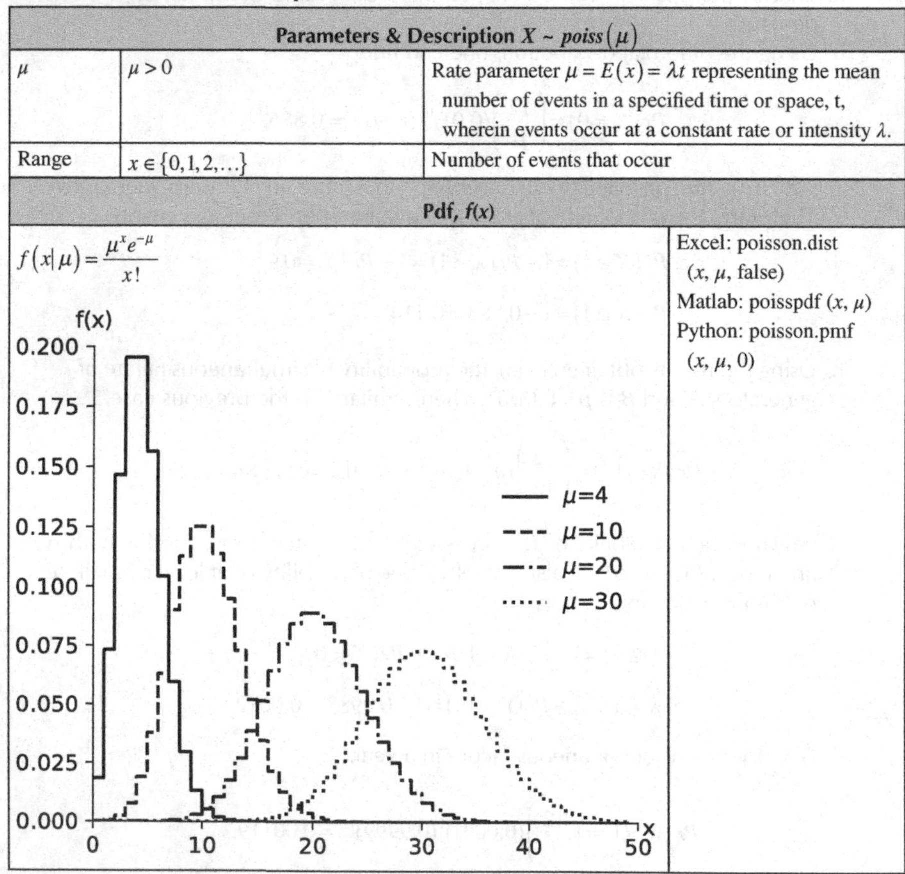		Excel: poisson.dist (x, μ, false) Matlab: poisspdf (x, μ) Python: poisson.pmf (x, μ, 0)

(Continued)

TABLE 2.4 (*Continued*)
Poisson Distribution Properties

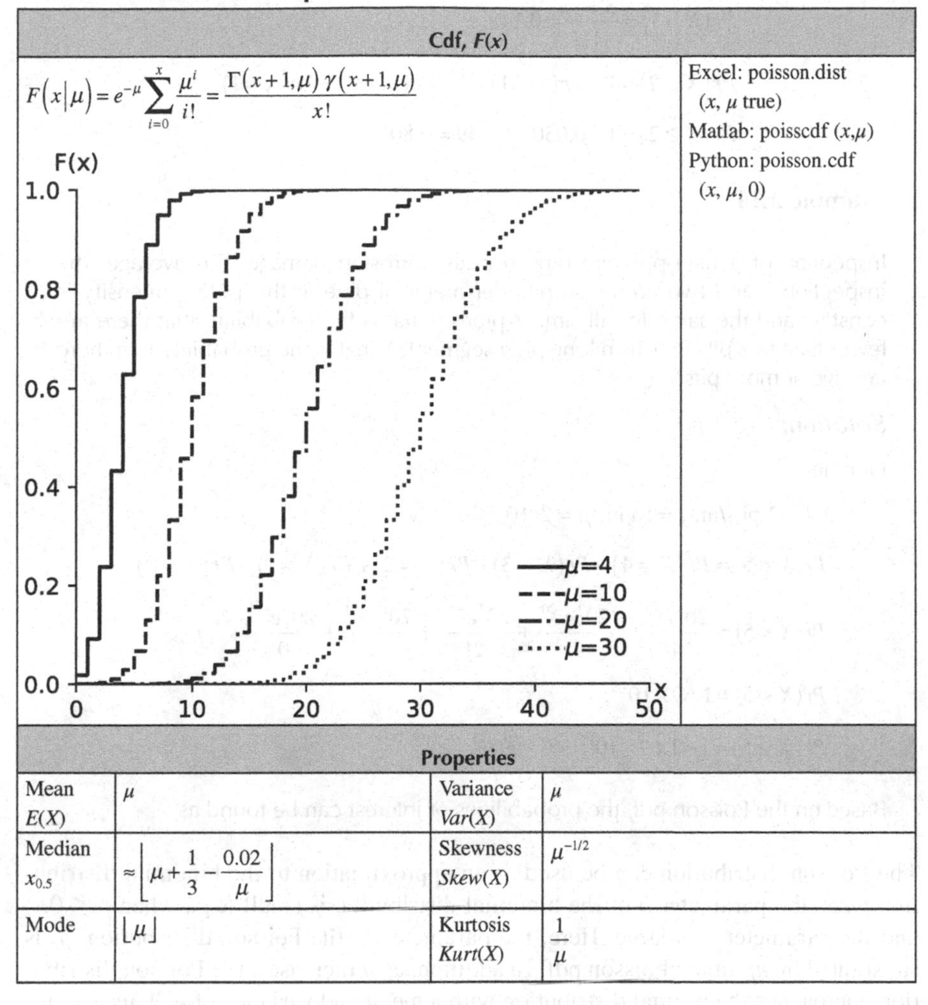

Cdf, $F(x)$	
$F(x\|\mu) = e^{-\mu}\sum_{i=0}^{x}\frac{\mu^i}{i!} = \frac{\Gamma(x+1,\mu)\,\gamma(x+1,\mu)}{x!}$	Excel: poisson.dist (x, μ true) Matlab: poisscdf (x,μ) Python: poisson.cdf (x, μ, 0)

Properties			
Mean $E(X)$	μ	Variance $Var(X)$	μ
Median $x_{0.5}$	$\approx\left\lfloor \mu+\frac{1}{3}-\frac{0.02}{\mu}\right\rfloor$	Skewness $Skew(X)$	$\mu^{-1/2}$
Mode	$\lfloor \mu \rfloor$	Kurtosis $Kurt(X)$	$\frac{1}{\mu}$

Example 2.13

A nuclear plant receives its electric power from a utility grid outside of the plant. From experience, loss of grid power occurs at an average rate of once a year. What is the probability that over 3 years no power outage will occur? That at least two power outages will occur?

Solution:

Denote $\lambda = 1$ outage/year, $t = 3$ years, $\mu = 1 \times 3 = 3$. Using the Poisson distribution, we find:

$$Pr(X=0)=\frac{3^0 e^{-3}}{0!}=0.050,$$

$$Pr(X=1)=\frac{3^1 e^{-3}}{1!}=0.149,$$

$$Pr(X\geq 2)=1-Pr(X\leq 1)=1-Pr(X=0)-Pr(X=1),$$

$$Pr(X\geq 2)=1-0.050-0.149=0.801.$$

Example 2.14

Inspection of a high-pressure pipe reveals corrosion damage. On average, the inspection found two corrosion pits per meter of pipe. If this pitting intensity is constant and the same for all similar pipes, what is the probability that there are fewer than five pits in a 10m long pipe segment? What is the probability that there are five or more pits?

Solution:

Denote

$$\lambda=2 \text{ pits/m},\ t=10 \text{ m},\ \mu=2\cdot 10=20.$$

$$Pr(X<5)=Pr(X=4)+Pr(X=3)+Pr(X=2)+Pr(X=1)+Pr(X=0),$$

$$Pr(X<5)=\frac{20^4 e^{-20}}{4!}+\frac{20^3 e^{-20}}{3!}+\frac{20^2 e^{-20}}{2!}+\frac{20^1 e^{-20}}{1!}+\frac{20^0 e^{-20}}{0!},$$

$$Pr(X<5)=1.69\times 10^{-5},$$

$$Pr(X\geq 5)=1-1.69\times 10^{-5}=0.99983.$$

Based on the Poisson pdf, the probabilities of interest can be found as

The Poisson distribution can be used as an approximation to the binomial distribution when the parameter p of the binomial distribution is small (e.g., when $p\leq 0.1$) and the parameter n is large. Here, the parameter of the Poisson distribution, μ, is substituted by np in the Poisson pdf. In addition, as μ increases, the Poisson distribution approaches the normal distribution with a mean and variance of μ. This asymptotical property is used as the normal approximation for the Poisson distribution.

Example 2.15

A radar system uses 650 similar electronic devices. Each device has a failure rate of 0.00015 failures per month. If all these devices operate independently, what is the probability that there are no failures in a year?

Solution:

The average number of failures of a device per year is

$$p=0.00015\cdot 12=0.0018.$$

The average number of failures of a radar system per year is

$$\mu = np = 0.0018 \cdot 650 = 1.17.$$

Finally, the probability of zero failures per year, according to the Poisson distribution, is

$$Pr(X=0) = \frac{1.17^0 e^{-1.17}}{0!} = 0.31.$$

2.7.1.3 Geometric Distribution

Consider a series of binomial trials with probability of failure per trial of p. Introduce an r.v., X, equal to the length of a series (number of trials) until the first failure is observed. The distribution of X is the geometric distribution which uses parameter p. The term $(1-p)^{x-1}$ is the probability that the failure will not occur in the first $(x-1)$ trials. When multiplied by p, it accounts for the probability of a failure in the xth trial. Table 2.5 shows important properties of a geometric distribution.

TABLE 2.5
Geometric Distribution Properties

Parameters & Description $X \sim geom(p)$		
p	$0 \le p \le 1$	Probability of outcome x in a single trial
Range	$x \in \{1,2,\ldots\}$	Number of trials at which the first failure occurs
Pdf, $f(x)$		
$f(x \mid p) = p(1-p)^{x-1}$		Excel: N/A Matlab: geopdf $(x-1, p)$ Python: geom.pmf $(x, p, 0)$

(*Continued*)

TABLE 2.5 (*Continued*)
Geometric Distribution Properties

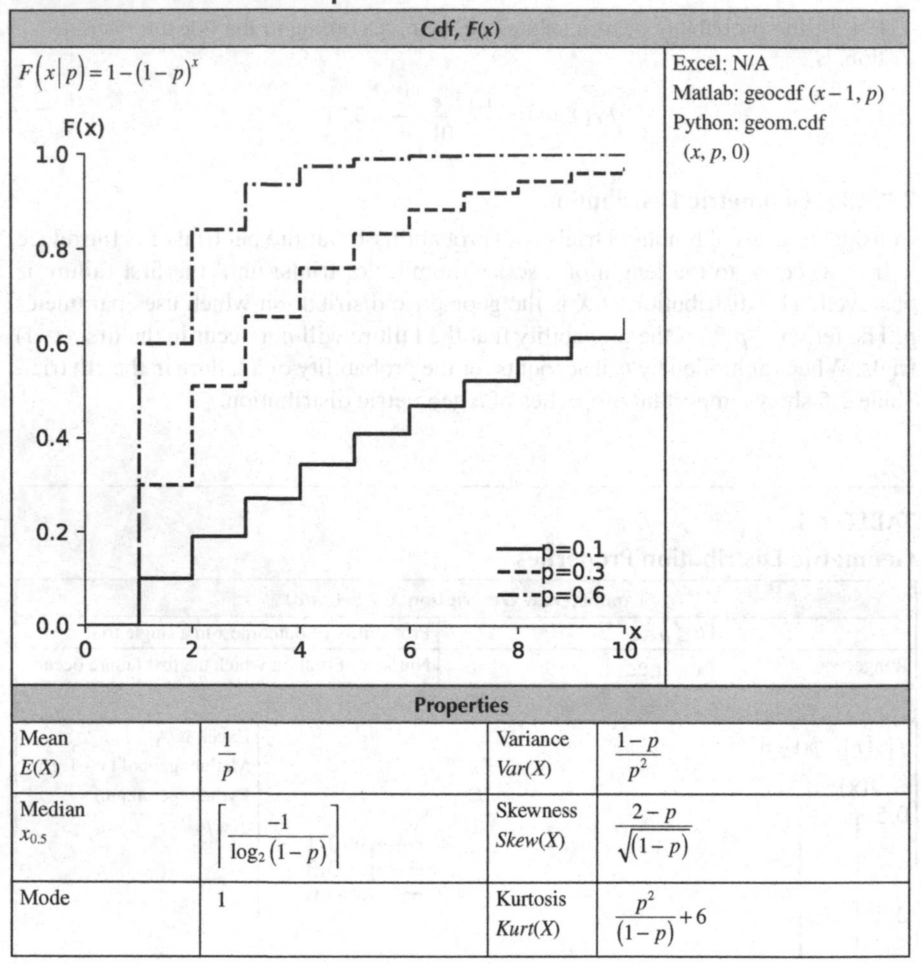

Cdf, $F(x)$	
$F(x\|p)=1-(1-p)^x$	Excel: N/A Matlab: geocdf $(x-1, p)$ Python: geom.cdf $(x, p, 0)$

Properties			
Mean $E(X)$	$\frac{1}{p}$	Variance $Var(X)$	$\frac{1-p}{p^2}$
Median $x_{0.5}$	$\left\lceil \frac{-1}{\log_2(1-p)} \right\rceil$	Skewness $Skew(X)$	$\frac{2-p}{\sqrt{(1-p)}}$
Mode	1	Kurtosis $Kurt(X)$	$\frac{p^2}{(1-p)}+6$

Example 2.16

In a nuclear power plant, diesel generators are used to provide emergency electric power to the safety systems. It is known that 1 out of 52 tests performed on a diesel generator results in a failure. What is the probability that the failure occurs at the tenth test?

Solution:

Using the geometric distribution with $x=10$ and $p=1/52=0.0192$ yields

$$Pr(x=10)=(0.0192)(1-0.0192)^9=0.0161.$$

2.7.2 Continuous Distributions

In this section we discuss examples of simple and commonly used continuous distributions. For a more exhaustive list of distributions see Johnson et al. (1995). In Chapter 2, we introduce these distributions, but focus mainly on applications of the exponential and normal distributions. More distributions are described in detail in Chapter 3.

2.7.2.1 Exponential Distribution

The exponential distribution was historically the first distribution used as a model of a time to failure distribution, and it is still widely used in reliability problems. The distribution has one parameter, a constant rate parameter λ, that is, $X \sim \exp(\lambda)$. In reliability engineering applications, the parameter λ represents the probability per unit of time that a device fails (called the failure rate of the device). A key characteristic of the exponential distribution is a constant failure rate over the period of interest.

The exponential distribution is closely associated with several distributions that are included in the exponential family of distributions. The most important is the Poisson distribution. Consider the following test. A unit is placed on a reliability test at $t=0$. When the unit fails, it is instantaneously replaced by an identical new one, which, in turn, is instantaneously replaced on its subsequent failures, and so on. The test is terminated at the known time t_{end}. It can be shown that if the number of failures during the test is distributed according to the Poisson distribution with a mean of $\mu = \lambda t_{\text{end}}$, then the time between successive failures (including the time to the first failure) has an exponential distribution with parameter λ. The test considered is an example of the homogeneous Poisson process (HPP) where the interarrival of failures is on average constant. We will elaborate more on the HPP in Chapter 7. Table 2.6 shows important properties of an exponential distribution.

TABLE 2.6
Exponential Distribution Properties

Parameters & Description $X \sim \exp(\lambda)$		
λ	$\lambda > 0$	Rate parameter λ representing the constant rate at which x occurs.
Range	$x \geq 0$	Typically, a time or distance

(Continued)

TABLE 2.6 (Continued)
Exponential Distribution Properties

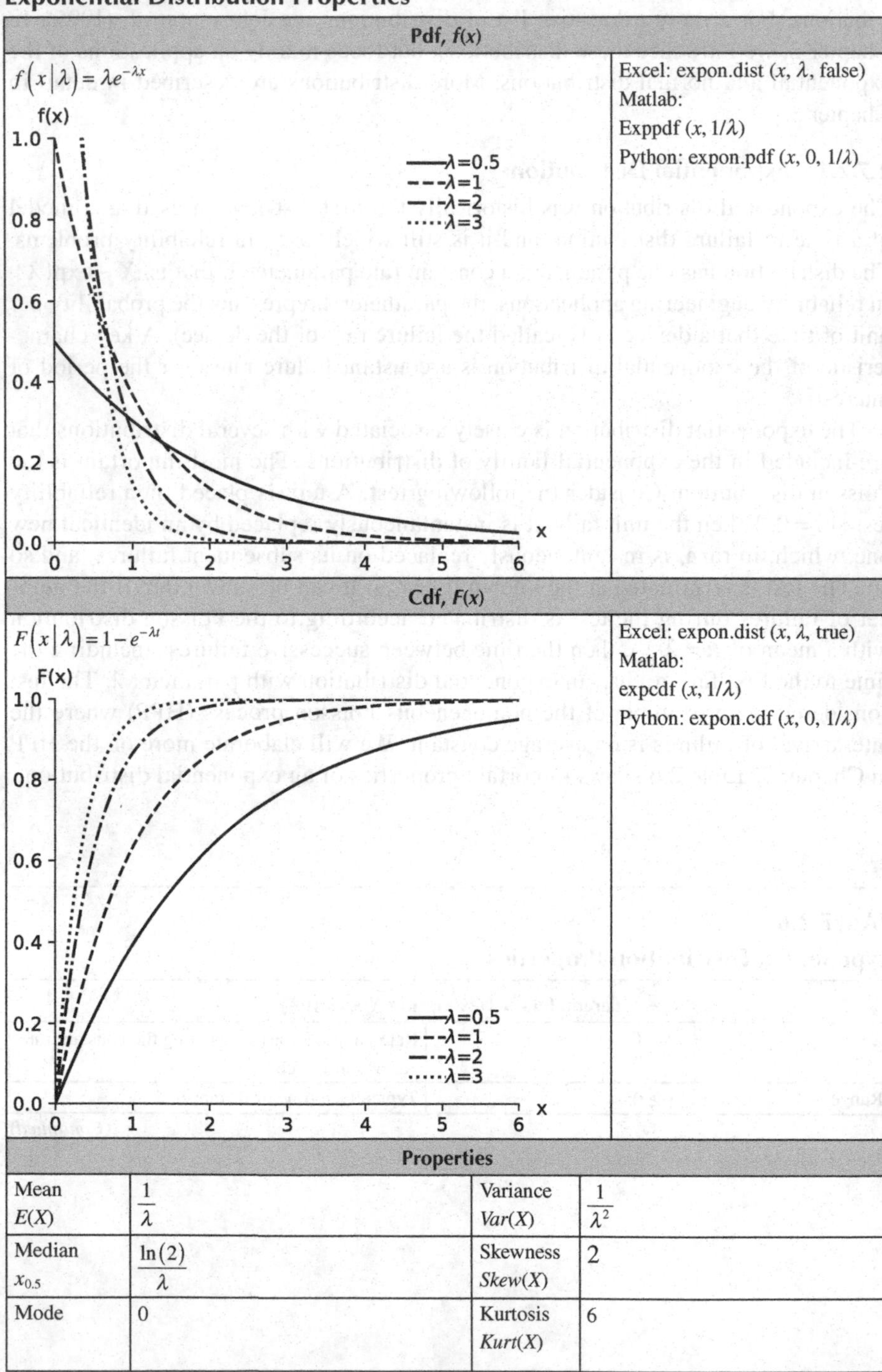

Pdf, $f(x)$	
$f(x\mid\lambda) = \lambda e^{-\lambda x}$	Excel: expon.dist (x, λ, false) Matlab: Exppdf (x, $1/\lambda$) Python: expon.pdf (x, 0, $1/\lambda$)

Cdf, $F(x)$	
$F(x\mid\lambda) = 1 - e^{-\lambda t}$	Excel: expon.dist (x, λ, true) Matlab: expcdf (x, $1/\lambda$) Python: expon.cdf (x, 0, $1/\lambda$)

Properties			
Mean $E(X)$	$\frac{1}{\lambda}$	Variance $Var(X)$	$\frac{1}{\lambda^2}$
Median $x_{0.5}$	$\frac{\ln(2)}{\lambda}$	Skewness $Skew(X)$	2
Mode	0	Kurtosis $Kurt(X)$	6

Example 2.17

A system has a constant failure rate of 0.001 failures/hour. What is the probability that the system will fail before t = 1,000 hours? Determine the probability that it works for at least 1,000 hours.

Solution:

Calculate the cdf for the exponential distribution at t = 1,000 hours:

$$Pr(t \le 1{,}000) = \int_0^{1{,}000} \lambda e^{-\lambda x} dt = -e^{-\lambda x}\Big|_0^{1{,}000},$$

$$Pr(t \le 1000) = 1 - e^{-1} = 0.632.$$

Therefore,

$$Pr(t > 1{,}000) = 1 - Pr(t \le 1{,}000) = 0.368.$$

2.7.2.2 Continuous Uniform Distribution

The continuous uniform distribution is also called the rectangular distribution due to the characteristic shape of its pdf. The uniform distribution is useful for understanding continuous distributions because of its straightforward equations and its mathematical simplicity. It is useful for random number generation, for representing bounds on variables, such as those coming from expert elicitation, and as a non-informative prior in Bayesian analysis. Table 2.7 shows important properties of a uniform distribution.

2.7.2.3 Normal Distribution

Perhaps the best known and most important continuous probability distribution is the normal distribution (also called a Gaussian distribution). The parameter μ is the mean of the distribution, and the parameter σ is the standard deviation. We denote a normally distributed random variable as $X \sim norm(\mu, \sigma)$. The normal distribution is of particular interest because of the *central limit theorem*, which states that if we take sufficiently large number (e.g., >30) of independent and identically distributed (i.i.d.) random samples with replacement from a parent population having the mean μ and standard deviation σ, then the distribution of the sample means will be approximately normally distributed. Table 2.8 shows important properties of a normal distribution.

Note that the integral in the normal cdf cannot be evaluated in a closed form, so the numerical integration and tabulation of normal cdf are required. However, it would be impractical to provide a separate table for every conceivable value of μ and σ. One way to get around this difficulty is to transform the normal distribution to the *standard normal distribution,* $\phi(z)$, using Equation 2.67, which has a mean of zero ($\mu = 0$) and a standard deviation of 1 ($\sigma = 1$).

The standard normal distribution has useful properties. Like all pdfs, the area under the curve is one. All three measures of central tendency are the same value: 0. At its mode, $f(z) = (2\pi)^{-.5} = 0.40$, and $F(z) = 0.5$. It is symmetric about the y-axis, and thus its skewness is zero. Like most continuous distribution, the normal distribution extends without limit.

TABLE 2.7
Uniform Distribution Properties

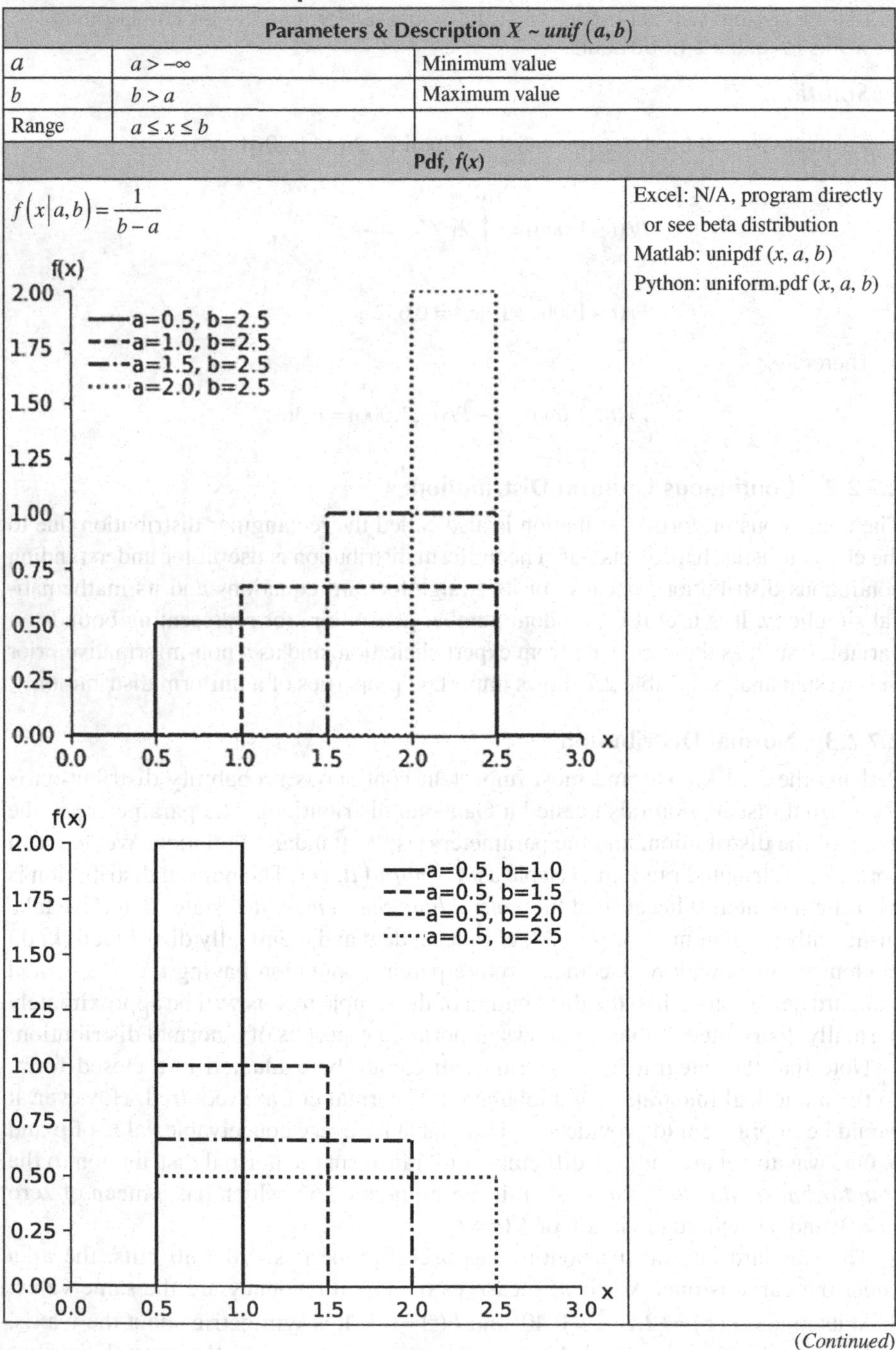

Parameters & Description $X \sim unif(a, b)$		
a	$a > -\infty$	Minimum value
b	$b > a$	Maximum value
Range	$a \le x \le b$	

Pdf, $f(x)$	
$f(x \mid a, b) = \frac{1}{b - a}$	Excel: N/A, program directly or see beta distribution Matlab: unipdf (x, a, b) Python: uniform.pdf (x, a, b)

(Continued)

TABLE 2.7 (*Continued*)
Uniform Distribution Properties

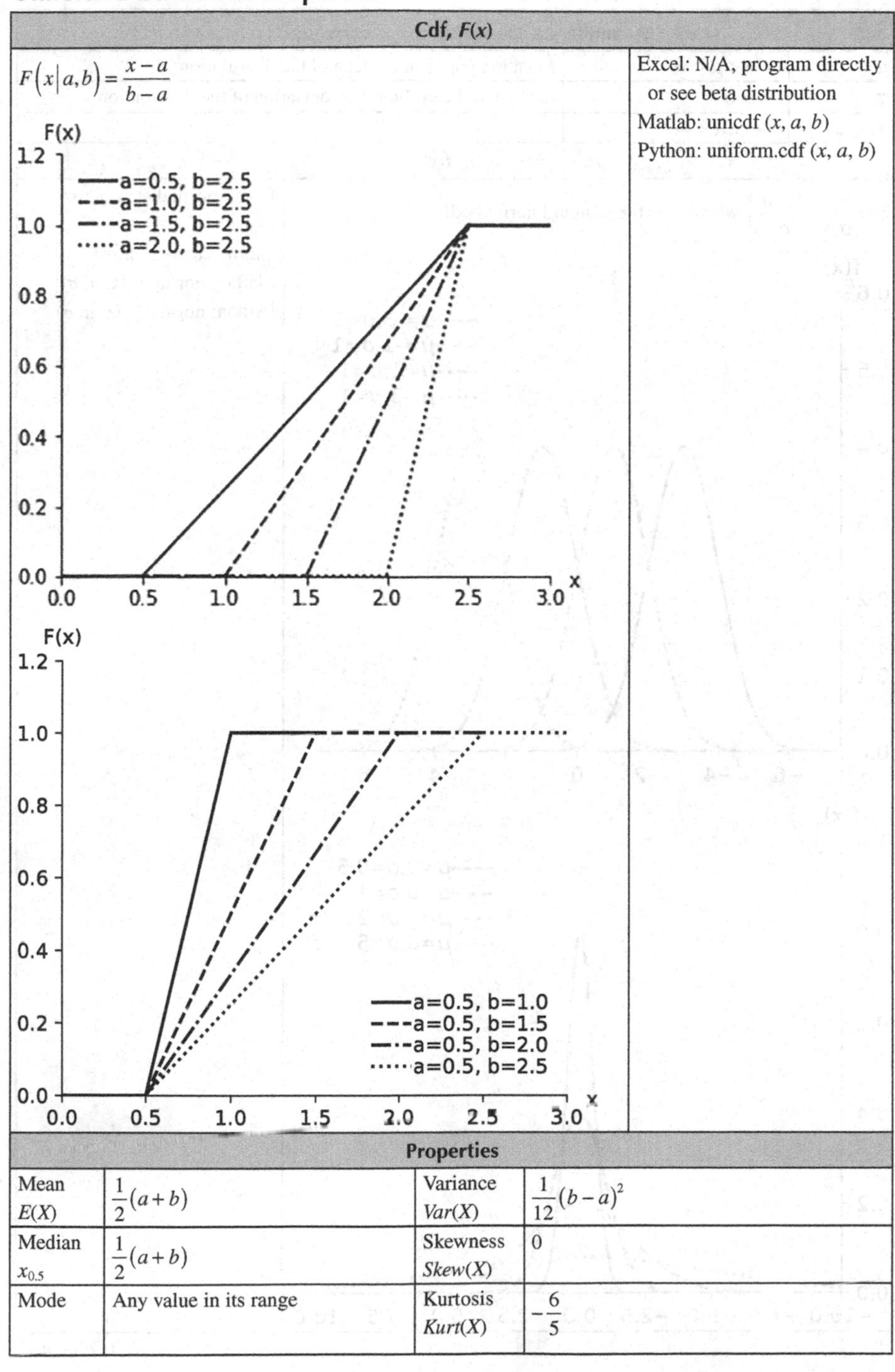

Cdf, $F(x)$	
$F(x \mid a,b) = \dfrac{x-a}{b-a}$	Excel: N/A, program directly or see beta distribution Matlab: unicdf (x, a, b) Python: uniform.cdf (x, a, b)

Properties			
Mean $E(X)$	$\frac{1}{2}(a+b)$	Variance $Var(X)$	$\frac{1}{12}(b-a)^2$
Median $x_{0.5}$	$\frac{1}{2}(a+b)$	Skewness $Skew(X)$	0
Mode	Any value in its range	Kurtosis $Kurt(X)$	$-\frac{6}{5}$

TABLE 2.8
Normal Distribution Properties

Parameters & Description $X \sim norm(\mu, \sigma)$		
μ	$-\infty < \mu < \infty$	Location parameter: Mean of the distribution
σ	$\sigma > 0$	Scale parameter: Standard deviation of the distribution
Range	$-\infty < x < \infty$	
Pdf, $f(x)$		
$S = \frac{1}{\sigma}\phi\left(\frac{x-\mu}{\sigma}\right)$ where: ϕ is the standard normal pdf. f(x) — μ=-3,σ=1; μ=-1,σ=1; μ=1,σ=1; μ=3,σ=1 f(x) — μ=0,σ=0.5; μ=0,σ=1; μ=0,σ=2; μ=0,σ=5		Excel: norm.dist (x, μ, σ, false) norm.s.dist (z, false) Matlab: normpdf (x, μ, σ) Python: norm.pdf (x, μ, σ)

(Continued)

TABLE 2.8 (*Continued*)
Normal Distribution Properties

<table>
<tr><th colspan="2">Cdf, $F(x)$</th></tr>
<tr><td>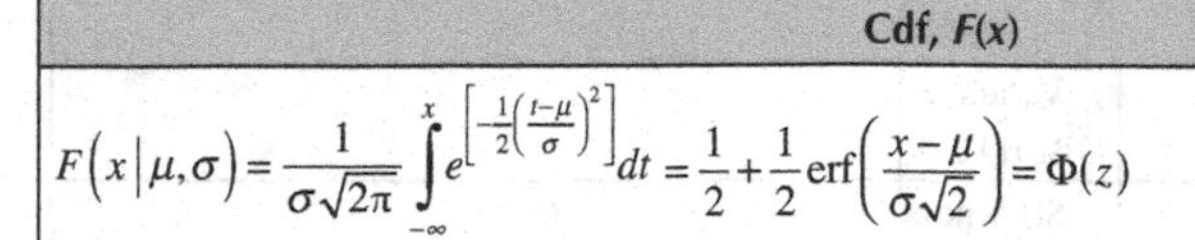

$$F(x\mid\mu,\sigma)=\frac{1}{\sigma\sqrt{2\pi}}\int_{-\infty}^{x}e^{\left[-\frac{1}{2}\left(\frac{t-\mu}{\sigma}\right)^2\right]}dt=\frac{1}{2}+\frac{1}{2}\operatorname{erf}\left(\frac{x-\mu}{\sigma\sqrt{2}}\right)=\Phi(z)$$

Where Φ is the standard normal cdf.

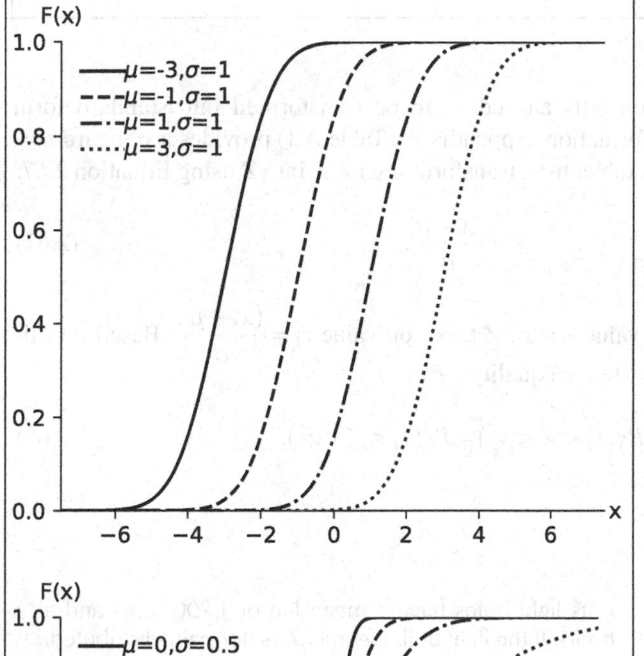

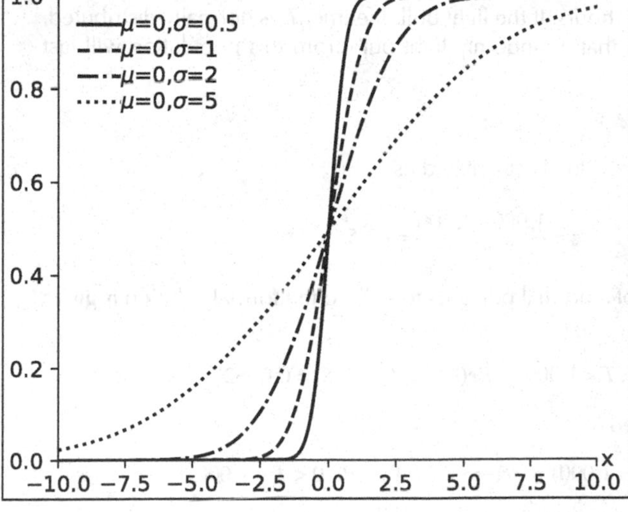
</td>
<td>Excel: norm.dist (x, μ, σ, true)
norm.s.dist (z, true)
Matlab: normcdf (x, μ, σ)
Python: norm.cdf (x, μ, σ)</td></tr>
</table>

(*Continued*)

TABLE 2.8 (*Continued*)
Normal Distribution Properties

Properties			
Mean $E(X)$	μ	Variance $Var(X)$	σ^2
Median 0.5	μ	Skewness $Skew(X)$	0
Mode	μ	Kurtosis $Kurt(X)$	0

All normal distribution pdfs and cdfs can be transformed into standard form through a simple Z transformation. Appendix A (Table A.1) provides a standard normal cdf table. To use this table, first, transform the r.v. X into Z using Equation 2.67:

$$Z = \frac{X - \mu}{\sigma}. \tag{2.67}$$

Therefore, if X takes on value $x = x_1$, Z takes on value $z_1 = \frac{(x_1 - \mu)}{\sigma}$. Based on this transformation, we can write the equality

$$Pr(x_1 < X < x_2) = Pr(z_1 < Z < z_2). \tag{2.68}$$

Example 2.18

A manufacturer states that its light bulbs have a mean life of 1,700 hours and a standard deviation of 280 hours. If the light bulb lifetime, T, is normally distributed, calculate the probability that a randomly light bulb from this population will last less than 1,000 hours.

Solution:

First, the corresponding z value is calculated as

$$z = \frac{1{,}000 - 1{,}700}{280} = -2.5.$$

Note that the lower tail of a normal pdf goes to $-\infty$, so the formal solution is given by

$$Pr(-\infty < T < 1{,}000) = Pr(-\infty < Z < -2.5) = 0.0062,$$

which can be represented as

$$Pr(-\infty < T < 1{,}000) = Pr(-\infty < T \le 0) + Pr(0 < T < 1{,}000).$$

The first term on the right side does not have a physical meaning since time cannot be negative. A more proper distribution, therefore, would be a truncated normal distribution; such a distribution will only exist in $t \geq 0$. Truncation will be covered later in this chapter. This is unnecessary for this problem because

$$Pr(-\infty < T < 0) = Pr(-\infty < Z) = Pr\left(-\infty < Z < -\frac{1,700}{280}\right) = 6.34 \times 10^{-10},$$

which can be considered as negligible. So, finally, one can write

$$Pr(-\infty < T < 1,000) \approx Pr(0 < T < 1,000) \approx Pr(-\infty < Z < -2.5) = 0.0062.$$

2.7.2.4 Lognormal Distribution

The lognormal distribution is closely related to the normal distribution. If r.v. X is lognormally distributed with parameters μ and σ, then the r.v. ln(X) is normally distributed with a mean of μ and a standard deviation of σ. That is, if $X \sim lognorm(\mu, \sigma)$ then $\ln(X) \sim norm(\mu, \sigma)$. Cautions: the parameters of the lognormal distribution are not the mean and variance of the lognormal distribution (or r.v. X); rather, they are the mean and variance of the normally distributed ln(X).

This distribution is often used to model an r.v. that can vary by several orders of magnitude. One major application of the lognormal distribution is to model an r.v. that results from multiplication of many independent variables. Table 2.9 shows important properties of the lognormal distribution.

Some applications of the lognormal distribution parameterize it in terms of its median and an error factor, EF,

$$EF = \sqrt{\frac{\text{95th percentile}}{\text{5th percentile}}} = \frac{\text{95th percentile}}{\text{50th percentile}} = \frac{\text{50th percentile}}{\text{5th percentile}}. \tag{2.69}$$

Like the normal distribution, the lognormal cdf does not have a closed form solution. It can be solved by transforming into Z space and using the standard normal distribution. Transforming the lognormal distribution into Z space using a slightly different Z:

$$Z = \frac{\ln x - \mu}{\sigma}. \tag{2.70}$$

Given the mean and variance of X, $E(X)$ and $var(X)$, respectively, you can transform those into the parameters of the lognormal distribution as such:

$$\mu = \ln\left(\frac{E(X)}{\sqrt{1 + \frac{Var(X)^2}{E(X)^2}}}\right), \tag{2.71}$$

and

$$\sigma = \sqrt{\ln\left(1 + \frac{Var(X)^2}{E(X)^2}\right)}. \tag{2.72}$$

TABLE 2.9
Lognormal Distribution Properties

Parameters & Description $X \sim lognorm(\mu, \sigma)$		
μ	$-\infty < \mu < \infty$	Scale parameter
σ	$\sigma > 0$	Shape parameter
Range	$x > 0$	

Pdf, $f(x)$	
$f(x\|\mu,\sigma) = \frac{1}{x\sigma\sqrt{2\pi}} e^{-\frac{1}{2}\left(\frac{\ln x-\mu}{\sigma}\right)^2} = \frac{1}{x\sigma}\phi\left(\frac{\ln x-\mu}{\sigma}\right)$ where $\phi(z)$ is the standard normal pdf f(x) 0.6 0.5 0.4 0.3 0.2 0.1 0.0; 0 2 4 6 8 10 μ=0.5,σ=1; μ=1,σ=1; μ=2,σ=1; μ=3,σ=1 f(x) 0.6 0.5 0.4 0.3 0.2 0.1 0.0; 0 2 4 6 8 10 μ=1,σ=0.5; μ=1,σ=1; μ=1,σ=2; μ=1,σ=5	Excel: lognorm.dist (x, μ, σ, false) Matlab: lognpdf (x, μ, σ) Python[a]: lognorm.pdf (x, σ, 0, np.exp(μ))

(*Continued*)

TABLE 2.9 (*Continued*)
Lognormal Distribution Properties

Cdf, *F*(*x*)	
$F(x\mid\mu,\sigma)=\frac{1}{\sigma\sqrt{2\pi}}\int_{-\infty}^{x}\frac{1}{x}e^{\left[-\frac{1}{2}\left(\frac{\ln t-\mu}{\sigma}\right)^{2}\right]}dt$ $=\frac{1}{2}+\frac{1}{2}\mathrm{erf}\left(\frac{\ln x-\mu}{\sigma\sqrt{2}}\right)=\phi\left(\frac{\ln x-\mu}{\sigma}\right),$ where (z) is the standard normal cdf.	Excel: lognorm.dist $(x, \mu, \sigma, \text{true})$ Matlab: logncdf (x, μ, σ) Python: lognorm.cdf $(x, \sigma, 0, \text{np.exp}(\mu))$

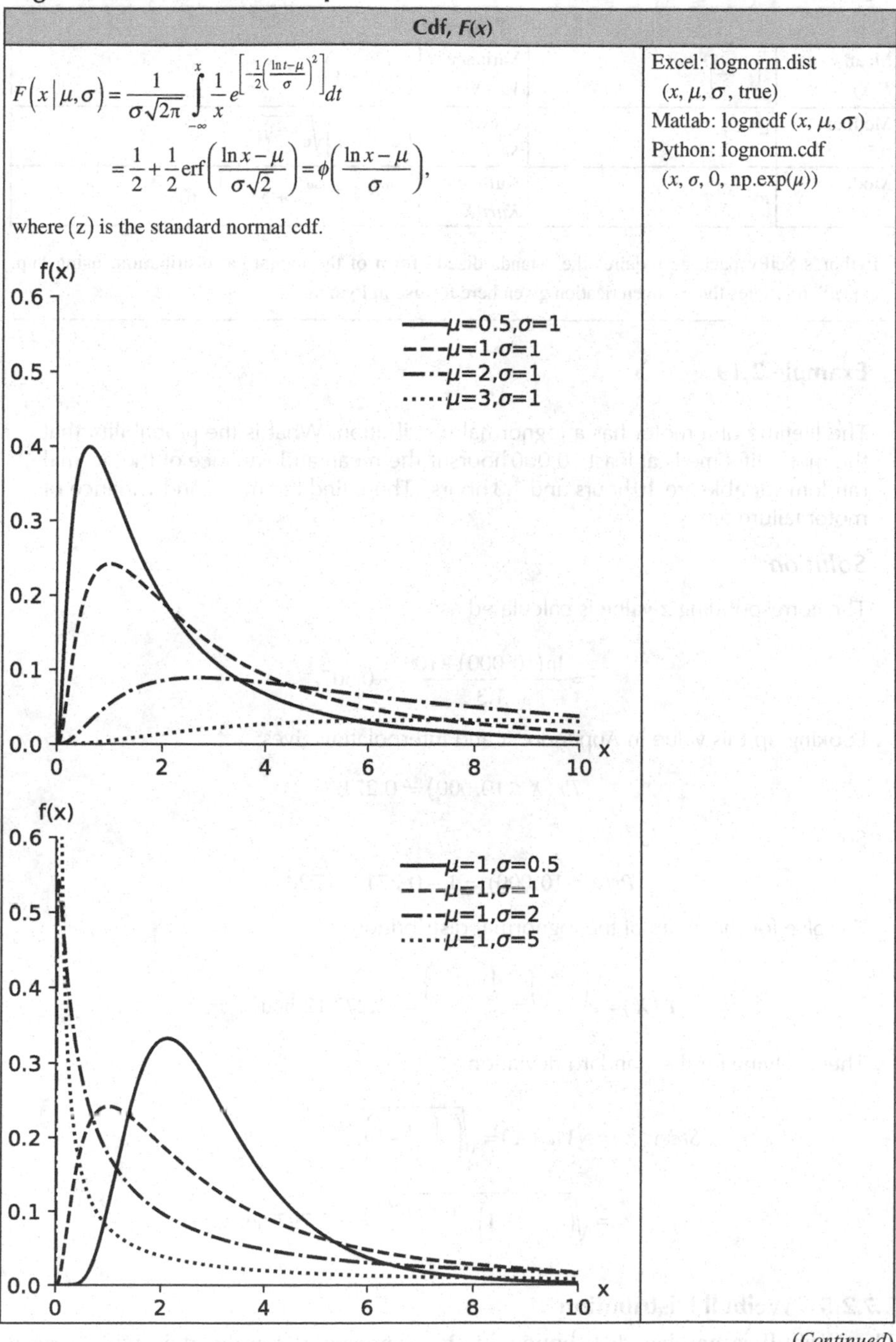

(*Continued*)

TABLE 2.9 (*Continued*)
Lognormal Distribution Properties

Properties			
Mean $E(X)$	$e^{\left(\mu-\frac{\sigma^2}{2}\right)}$	Variance $Var(X)$	$\left(e^{(\sigma^2)}-1\right)\cdot e^{(2\mu+\sigma^2)}$
Median 0.5	e^{μ}	Skewness $Skew(X)$	$\left(e^{(\sigma^2)}+2\right)\sqrt{e^{(\sigma^2-1)}}$
Mode	$e^{(\mu-\sigma^2)}$	Kurtosis $Kurt(X)$	$e^{(4\sigma^2)}+2e^{(3\sigma^2)}+3e^{(2\sigma^2)}-6$

[a] Python's SciPy package requires the "standardized" form of the lognormal distribution; using "np.exp(μ)" translates the parameterization given here for use in Python.

Example 2.19

The lifetime of a motor has a lognormal distribution. What is the probability that the motor lifetime is at least 10,000 hours if the mean and variance of the normal random variable are 10 hours and 1.3 hours? Then, find the mean and variance of motor failure times.

Solution:

The corresponding z value is calculated as:

$$z=\frac{\ln(10,000)-10}{1.3}=-0.607.$$

Looking up this value in Appendix A and interpolating gives:

$$Pr(X\le 10,000)=0.271.$$

So

$$Pr(X>10,000)=1-0.271=0.728.$$

To solve for the mean of the lognormal distribution:

$$E(X)=e^{\left(\mu+\frac{\sigma^2}{2}\right)}=e^{\left(10+\frac{1.3^2}{2}\right)}=51,277.12 \text{ hours.}$$

Then, solving for the standard deviation:

$$Stdev(X)=\sqrt{Var(X)}=\sqrt{\left(e^{(\sigma^2)}-1\right)e^{(2\mu+\sigma^2)}}$$

$$=\sqrt{\left(e^{(1.3^2)}-1\right)\cdot e^{(2\cdot 10+1.3^2)}}=107,797 \text{ hours.}$$

2.7.2.5 Weibull Distribution

The Weibull is another distribution in the exponential family. It is widely used to represent the time to failure or life duration of components as well as systems.

The parameter β is a shape parameter and α is a scale parameter, that is, $X \sim weibull(\alpha, \beta)$. Note that the two-parameter Weibull distribution has multiple parameterizations that are used in reliability. Additionally, there is a three-parameter version of the Weibull distribution which is not covered in this book.

Several properties of the Weibull distribution rely on a function known as the gamma function $\Gamma(x)$. See Section 2.7.2.6 for explanation of this function.

When $\beta = 1$, the Weibull distribution is reduced to the exponential distribution with $\lambda = 1/\alpha$, so the exponential distribution is a particular case of the Weibull distribution. For the values of $\beta > 1$, the distribution becomes bell-shaped with some skew. We will elaborate further on this distribution and its use in reliability analysis in Chapter 3. Table 2.10 shows important properties of the Weibull distribution.

2.7.2.6 Gamma Distribution (and Chi-Squared)

The gamma distribution can be thought of as a generalization of the exponential distribution. For example, if T_i, the time between successive failures of a system has an exponential distribution with (parameter λ), then an r.v. T such that $T = \sum_{i=1}^{\alpha} T_i$ follows a gamma distribution with parameters α and $\beta = 1/\lambda$. In the given context, T represents the cumulative time to the α th failure.

A different way to interpret this distribution is to consider a situation in which a system is subjected to shocks occurring according to the Poisson process (with parameter $\lambda = 1/\beta$). If the system fails after receiving α shocks, then the time to failure of such a system follows a gamma distribution. A simple example of this system is a redundant system with α standby parallel components. Failure of each standby component can be viewed as a shock. The system fails after the αth component fails (i.e., the last shock occurs). Standby systems will be discussed later in Chapter 6.

If $\alpha = 1$, the gamma distribution is reduced to an exponential distribution. Another important special case of the gamma distribution is the case when $\beta = 2$ and $\alpha = df/2$, where df is a positive integer, called the number of *degrees of freedom*. This one-parameter distribution is known as the chi-squared (χ^2) distribution. This distribution is widely used in reliability data analysis with some applications discussed in Chapter 4. Table 2.11 shows important properties of the gamma distribution.

The gamma distribution uses a function $\Gamma(n)$ called the *gamma function*.[2] The gamma function is:

$$\Gamma(n) = \int_0^{\infty} x^{n-1} e^{-x} dx. \tag{2.73}$$

For integer values of n, Γ(n)=(n-1)! where ! denotes factorial

2.7.2.7 Beta Distribution

The beta distribution is a useful model for r.v.s that are distributed in a finite interval. The pdf of the standard beta distribution is defined over the interval [0,1], making it particularly useful for Bayesian analysis. Like the gamma distribution, the cdf of

[2] Caution: do not get the gamma functions confused with the gamma distribution; we recommend being precise in using the term function vs. distribution to avoid the confusion.

TABLE 2.10
Weibull Distribution Properties

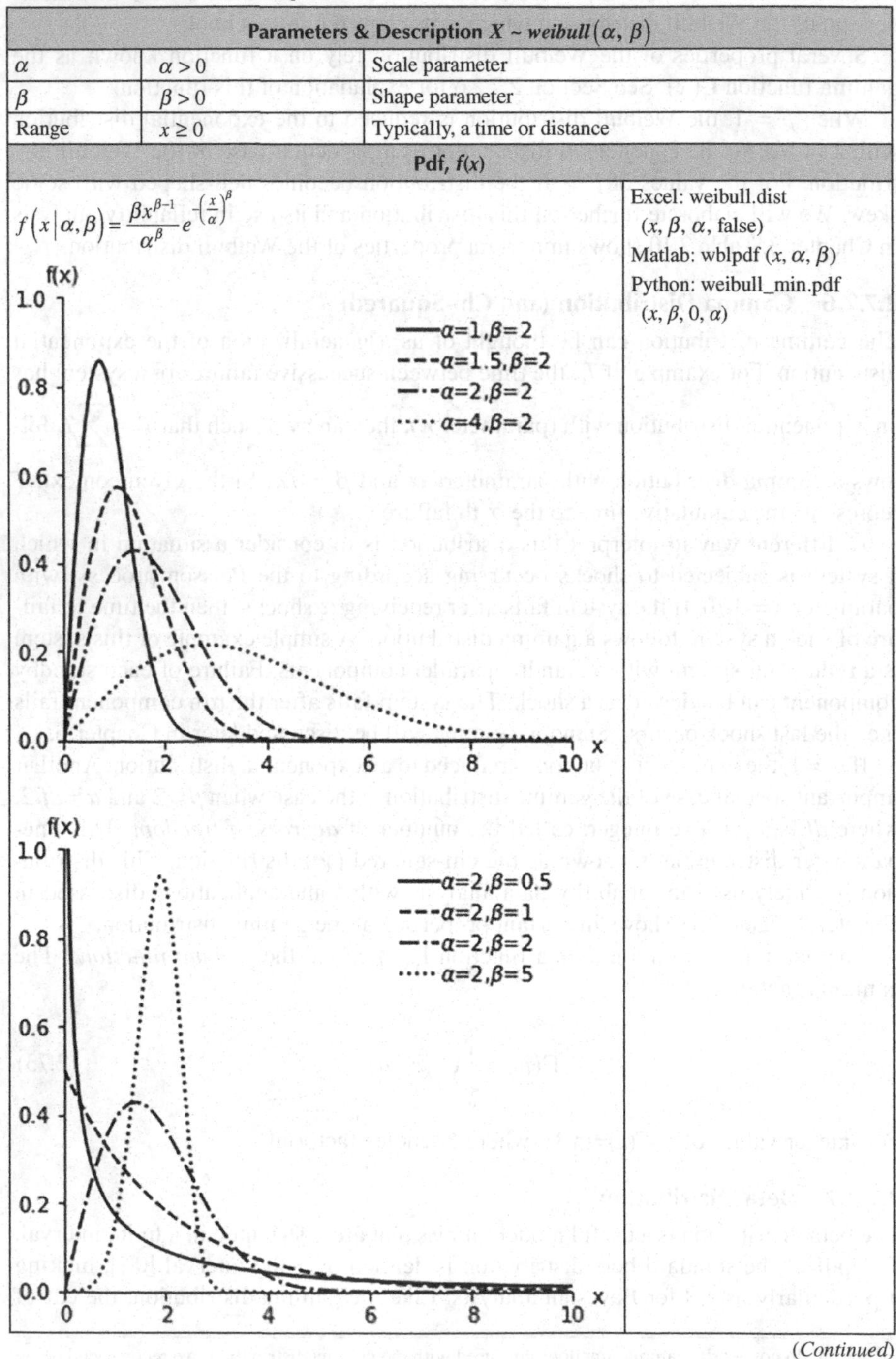

Parameters & Description $X \sim weibull(\alpha, \beta)$		
α	$\alpha > 0$	Scale parameter
β	$\beta > 0$	Shape parameter
Range	$x \geq 0$	Typically, a time or distance

Pdf, $f(x)$	
$f(x \mid \alpha, \beta) = \frac{\beta x^{\beta-1}}{\alpha^{\beta}} e^{-\left(\frac{x}{\alpha}\right)^{\beta}}$	Excel: weibull.dist $(x, \beta, \alpha, \text{false})$ Matlab: wblpdf (x, α, β) Python: weibull_min.pdf $(x, \beta, 0, \alpha)$

(*Continued*)

TABLE 2.10 (*Continued*)
Weibull Distribution Properties

Cdf, $F(x)$	
$F\left(x\,\middle|\,\alpha,\beta\right)=1-e^{-\left(\frac{x}{\alpha}\right)^{\beta}}$	Excel: weibull.dist (x, β, α, true) Matlab: wblcdf (x, α, β) Python: weibull_min.cdf (x, β, 0, α)

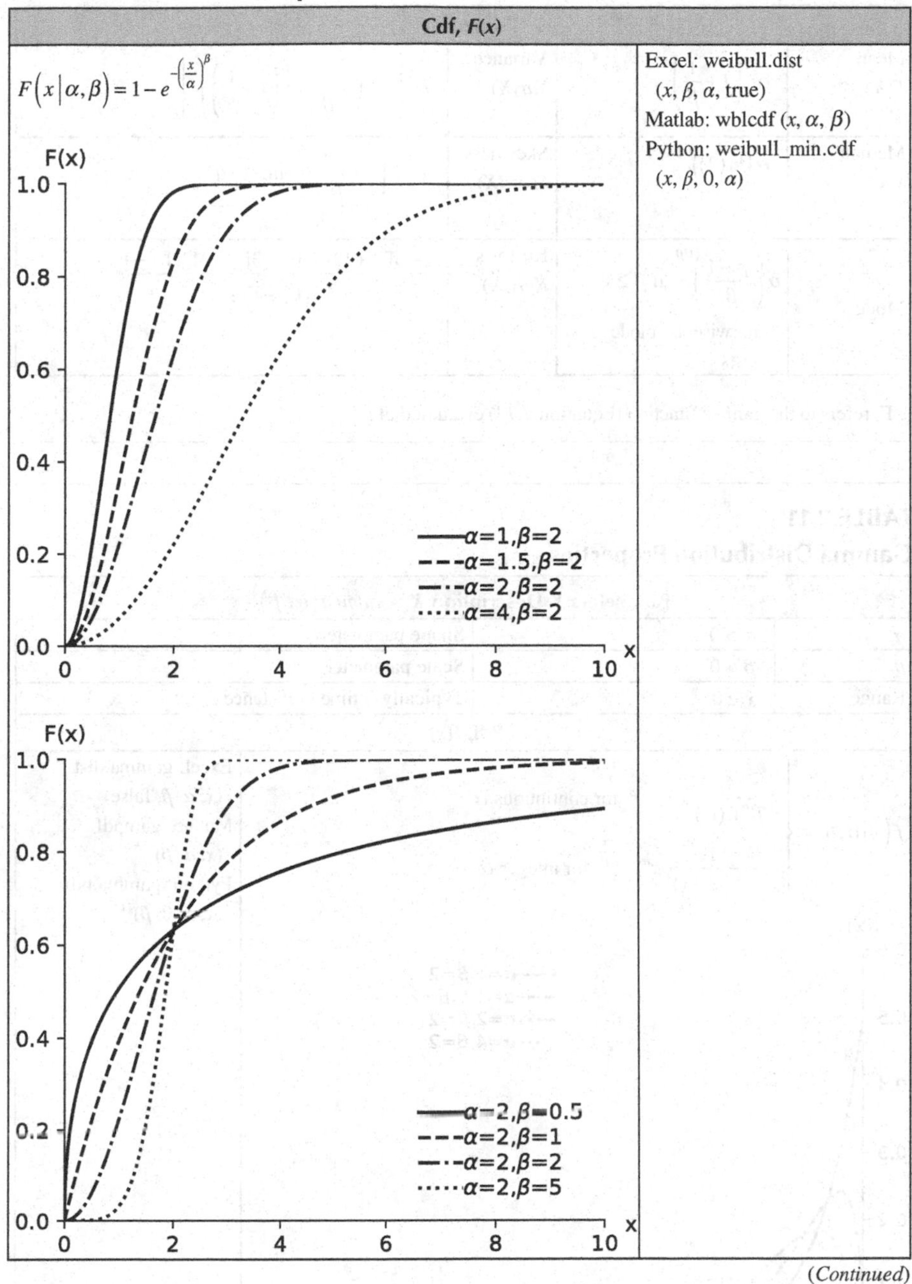

(*Continued*)

TABLE 2.10 (*Continued*)
Weibull Distribution Properties

Properties[a]			
Mean $E(X)$	$\alpha\Gamma\left(1+\frac{1}{\beta}\right)$	Variance $Var(X)$	$\alpha^2\left[\Gamma\left(1+\frac{2}{\beta}\right)-\left(\left(1+\frac{1}{\beta}\right)\right)^2\right]$
Median $x_{0.5}$	$\alpha\left(\ln(2)\right)^{1/\beta}$	Skewness $Skew(X)$	$\frac{\alpha^3\Gamma\left(1+\frac{3}{\beta}\right)-3\mu\sigma^2-\mu^3}{\sigma^3}$
Mode	$\alpha\left(\frac{\beta-1}{\beta}\right)^{1/\beta}$ if $\beta \geq 1$ Otherwise, no mode exists	Kurtosis $Kurt(X)$	$\frac{-6\Gamma_1^4+12\Gamma_1^2\Gamma_2-3\Gamma_2^2-4\Gamma_1\Gamma_3+\Gamma_4}{\left(\Gamma_2-\Gamma_1^2\right)^2}$

[a] Γ_i refers to the gamma function (Equation 2.73) evaluated at i.

TABLE 2.11
Gamma Distribution Properties

Parameters & Description $X \sim gamma(\alpha, \beta)$		
α	$\alpha > 0$	Shape parameter
β	$\beta > 0$	Scale parameter
Range	$x \geq 0$	Typically, a time or distance

Pdf, $f(x)$	
$f(x\mid\alpha,\beta)=\begin{cases}\frac{1}{\beta^\alpha\Gamma(\alpha)}x^{\alpha-1}e^{-\frac{x}{\beta}} & \text{for continuous } \alpha\\ \frac{1}{\beta^\alpha(\alpha-1)!}x^{\alpha-1}e^{-\frac{x}{\beta}} & \text{for integer } \alpha\end{cases}$	Excel: gamma.dist (x, α, β, false) Matlab: gampdf (x, α, β) Python: gamma.pdf (x, α, 0, β)[a]

(*Continued*)

TABLE 2.11 (*Continued*)
Gamma Distribution Properties

Pdf, *f*(*x*)	
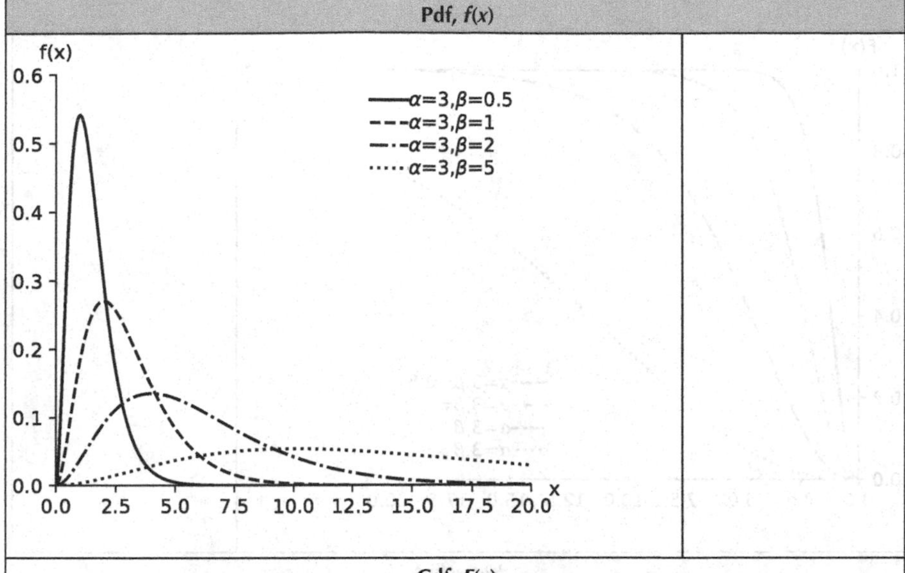	
Cdf, *F*(*x*)	
$F(x\|\alpha,\beta)=\begin{cases}\dfrac{1}{\beta^{\alpha}\Gamma(\alpha)}\displaystyle\int_0^x x^{\alpha-1}e^{-\frac{x}{\beta}}\,dx & \text{for continuous } \alpha\\ 1-e^{-\frac{x}{\beta}}\displaystyle\sum_{n=0}^{\alpha-1}\frac{x^n}{\beta^n n!} & \text{for integer } \alpha\end{cases}$ Cannot be written in closed form; often written as $F(x\|\alpha,\beta)=\frac{1}{\Gamma(\alpha)}\gamma\left(\alpha,\frac{x}{\beta}\right)$ where $\gamma\left(\alpha,\frac{x}{\beta}\right)$ is known as the lower incomplete gamma function. 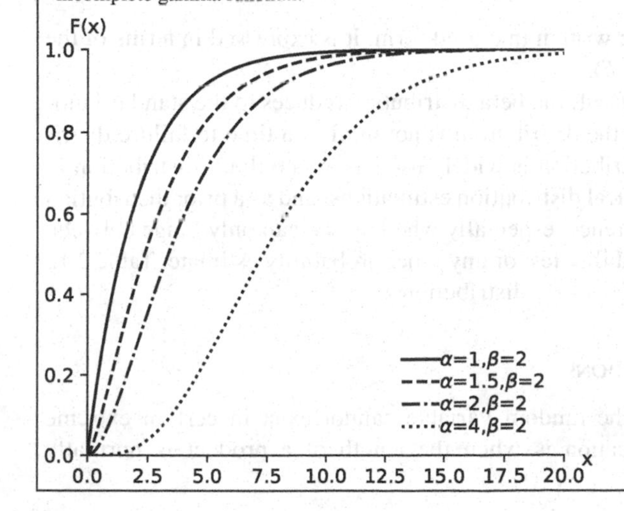	Excel: gamma.dist (x, α, β, true) Matlab: gamcdf (x, α, β) Python: gamma.cdf (x, α, 0, β)

(*Continued*)

TABLE 2.11 (*Continued*)
Gamma Distribution Properties

Cdf, *F*(*x*)

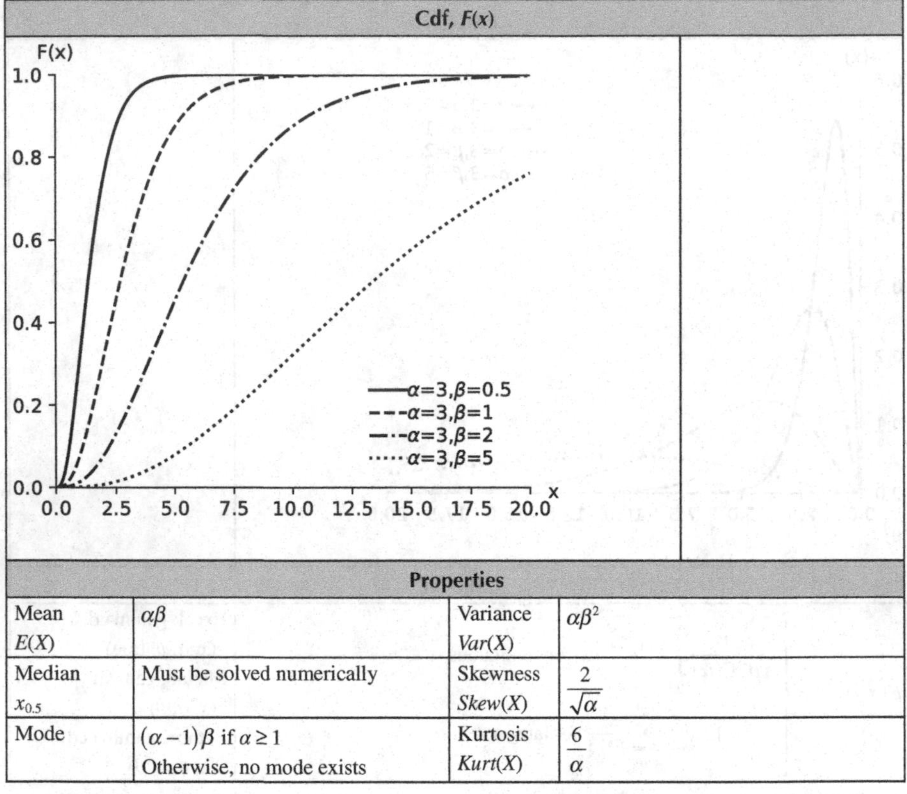

Properties			
Mean *E*(*X*)	$\alpha\beta$	Variance *Var*(*X*)	$\alpha\beta^2$
Median $x_{0.5}$	Must be solved numerically	Skewness *Skew*(*X*)	$\frac{2}{\sqrt{\alpha}}$
Mode	$(\alpha-1)\beta$ if $\alpha \geq 1$ Otherwise, no mode exists	Kurtosis *Kurt*(*X*)	$\frac{6}{\alpha}$

[a] Python's SciPy package uses the single-variable parameterization of the Gamma distribution. Using "scale=$1/\beta$" translates the parameterization given here for use in Python.

the beta distribution cannot be written in closed form. It is expressed in terms of the incomplete beta function, $I_t(\alpha, \beta)$.

For the special case of $\alpha=\beta=1$, the beta distribution reduces to the standard uniform distribution. Practically, the distribution is not used as a time to failure distribution. However, the beta distribution is widely used as an auxiliary distribution in nonparametric classical statistical distribution estimations, and as a prior distribution in the Bayesian statistical inference, especially when the r.v. can only range between 0 and 1: for example, in a reliability test or any other probability estimate. Table 2.12 shows important properties of the beta distribution.

2.7.3 Truncated Distributions

Truncation is needed when the random variable cannot exist in certain extreme ranges. An example of truncation is when the length of a product is normally

TABLE 2.12
Beta Distribution Properties

Parameters & Description $X \sim beta\ (\alpha, \beta)$			
α	$\alpha > 0$	Shape parameter	
β	$\beta > 0$	Shape parameter	
Range	$0 \le x \le 1$		
Pdf, $f(x)$			
$f(x \mid \alpha, \beta) = \frac{\Gamma(\alpha+\beta)}{\Gamma(\alpha)\Gamma(\beta)} x^{\alpha-1}(1-x)^{\beta-1}$ Where the beta function is $B(\alpha, \beta) = \frac{\Gamma(\alpha)\Gamma(\beta)}{\Gamma(\alpha+\beta)}$ f(x) 4.0 3.5 3.0 2.5 2.0 1.5 1.0 0.5 0.0 α=0.5,β=1 α=1,β=1 α=1.5,β=1 α=5,β=1 0.0 0.2 0.4 0.6 0.8 1.0 x f(x) 4.0 3.5 3.0 2.5 2.0 1.5 1.0 0.5 0.0 α=1,β=0.5 α=1,β=1 α=1,β=1.5 α=1,β=5 0.0 0.2 0.4 0.6 0.8 1.0 x			Excel: beta.dist (x, α, β, false) Matlab: betapdf (x, α, β) Python: beta.pdf (x, α, β, 0, 1)

(*Continued*)

TABLE 2.12 (*Continued*)
Beta Distribution Properties

Cdf, $F(x)$	
$F(x;\alpha,\beta)=\frac{\Gamma(\alpha+\beta)}{\Gamma(\alpha)\Gamma(\beta)}\int_0^x x^{a-1}(1-x)^{\beta-1}dx=\frac{\Gamma(\alpha,\beta)}{\Gamma(\alpha)\Gamma(\beta)}B(x\mid\alpha,\beta)$ where the integral above is $B(x\mid\alpha,\beta)$, the incomplete beta function. The cdf can also be written as $I_x(\alpha,\beta)$, the regularized incomplete beta function: $I_x(x\mid\alpha,\beta)=\frac{B(x\mid\alpha,\beta)}{B(\alpha,\beta)}$. $B(\alpha,\beta)$ is the marginalized version of the incomplete beta function.	Excel: beta.dist (x, α, β, true) Matlab: betacdf (x, α, β) Python: beta.cdf (x, α, β, 0, 1)

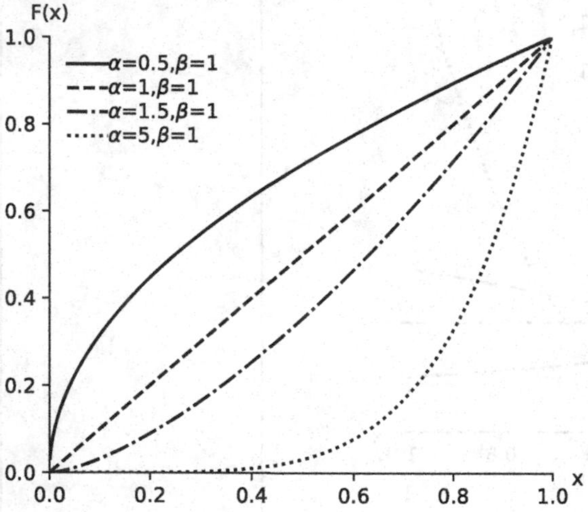

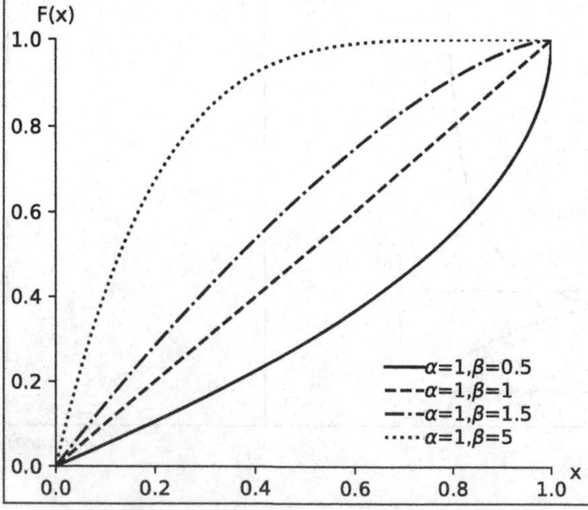

(*Continued*)

TABLE 2.12 (*Continued*)
Beta Distribution Properties

Properties			
Mean $E(X)$	$\frac{\alpha}{\alpha+\beta}$	Variance $Var(X)$	$\frac{\alpha\beta}{(\alpha+\beta)^2(\alpha+\beta+1)}$
Median $x_{0.5}$	$\approx \frac{\alpha-\frac{1}{3}}{\alpha+\beta-\frac{2}{3}}$ for $\alpha, \beta > 1$	Skewness $Skew(X)$	$\frac{2(\beta-\alpha)\sqrt{\alpha+\beta+1}}{(\alpha+\beta+2)\sqrt{\alpha\beta}}$
Mode	$\frac{\alpha-1}{\alpha+\beta-2}$ for $\alpha, \beta > 1$ See references for other combinations.	Kurtosis $Kurt(X)$	$\frac{6\left[(\alpha-\beta)^2(\alpha+\beta+1)-\alpha\beta(\alpha+\beta+2)\right]}{\alpha\beta(\alpha+\beta+2)(\alpha+\beta+3)}$

distributed about its mean. The normal distribution has a range from $-\infty$ to $+\infty$, however it is obviously not realistic for a physical dimension like length or time to be negative. Therefore, a truncated form of the normal distribution needs to be used if the normal distribution is used to model the length of a product. Another example of truncation is when the existence of a defect is unknown due to the defect's size being less than the inspection threshold.

A *truncated distribution* is therefore a conditional distribution that restricts the domain of another probability distribution. The following general equation forms apply to the truncated distribution, where $f_0(x)$ and $F_0(x)$ are the pdf and cdf of the non-truncated distribution.

The pdf of a truncated distribution can be expressed by

$$f(x) = \begin{cases} \dfrac{f_0(x)}{F_0(b) - F_0(a)} & a < x \le b, \\ 0 & \text{otherwise.} \end{cases} \tag{2.74}$$

As such the cdf of a truncated distribution would be

$$F(x) = \begin{cases} 0 & x \le a, \\ \dfrac{\int_a^x f_0(t)dt}{F_0(b) - F_0(a)} & a < x \le b, \\ 1 & x > b. \end{cases} \tag{2.75}$$

2.7.4 Multivariate Distributions

Thus far, we have discussed probability distributions that describe a single random variable and one-dimensional sample spaces. There exist, however, situations in which more than one possibly dependent r.v. is simultaneously measured and recorded. For example, in a study of human reliability in a control room situation, one can simultaneously estimate the r.v. T representing time that various operators spend to fulfill an emergency action, and the r.v. E representing the level of training the operators have had for performing these emergency actions. Since one expects E and T to have some relationship (e.g., better-trained operators act faster than less-trained ones), a joint distribution of both r.v.s T and E can be used to express their mutual dispersion.

Let X and Y be two r.v.s (not necessarily independent). The *joint pdf* (or bivariate pdf) of X and Y is denoted by $f(x, y)$. If X and Y are discrete, the joint pdf can be denoted by $Pr(X=x, Y=y)$, or simply $Pr(x, y)$, just as we described when we introduced the concept of joint probability. Thus, $Pr(x, y)$ gives the probability that the outcomes x and y occur simultaneously. For example, if r.v. X represents the number of electric circuits of a given type in a process plant and Y represents the number of failures of the circuit type in the most recent year, then $Pr(7, 1)$ is the probability that a randomly selected process plant has seven circuits and that one of them has failed once in the most recent year.

The function $f(x, y)$ is a joint pdf of continuous r.v.s X and Y if

$$1. \quad f(x, y) \geq 0, -\infty < x, y < \infty. \tag{2.76}$$

$$2. \quad \int_{-\infty}^{\infty}\int_{-\infty}^{\infty} f(x, y)\,dxdy = 1. \tag{2.77}$$

Similarly, the function $Pr(x, y)$ is a joint probability mass function of discrete r.v.s X and Y if

$$1. \quad Pr(x, y) \geq 0 \text{ for all values of } x \text{ and } y. \tag{2.78}$$

$$2. \quad \sum_{x}\sum_{y} Pr(x, y) = 1. \tag{2.79}$$

The probability that two joint r.v.s fall in a specified subset of the sample space is given by

$$Pr(x_1 \leq X \leq x_2, y_1 \leq Y \leq y_2) = \int_{x_1}^{x_2}\int_{y_1}^{y_2} f(x, y)\,dydx, \tag{2.80}$$

$$Pr(x_1 < X \leq x_2, y_1 < Y \leq y_2) = \sum_{x_1 < x \leq x_2,\ y_1 < y \leq y_2} Pr(x, y), \tag{2.81}$$

where Equation 2.80 is for continuous r.v.s. and Equation 2.81 is for discrete r.v.s.

The *marginal distribution* is the distribution of one r.v. independent of the other r.v.s. The marginal pdfs of X and Y are defined, respectively, as

$$f_X(x) = \int_{-\infty}^{\infty} f(x, y)dy, \tag{2.82}$$

and

$$f_Y(y) = \int_{-\infty}^{\infty} f(x, y)dx, \tag{2.83}$$

for continuous r.v.s, and as

$$Pr_X(x) = \sum_y Pr(x, y), \tag{2.84}$$

and

$$Pr_Y(y) = \sum_x Pr(x, y), \tag{2.85}$$

for discrete r.v.s. Note that the subscripts are often dropped from the notation.

Using Equation 2.12, the conditional probability of an event $Y=y$, given the event $X=x$, is

$$Pr(Y = y | X = x) = \frac{Pr(X = x \cap Y = y)}{Pr(X = x)} = \frac{Pr(x, y)}{Pr(x)}, \; Pr(x) > 0, \tag{2.86}$$

where X and Y are discrete r.v.s. Similarly, one can extend the same concept to continuous r.v.s X and Y and write

$$f(x | y) = \frac{f(x, y)}{f_Y(y)}, \; f_Y(y) > 0, \tag{2.87}$$

or

$$f(y | x) = \frac{f(x, y)}{f_X(x)}, \; f_X(x) > 0, \tag{2.88}$$

where Equations 2.86–2.88 are called the *conditional pdfs* of discrete and continuous r.v.s, respectively. The conditional pdfs have the same properties as any other pdf. Similar to the discussions earlier in this chapter, if X and Y are independent, then $f(x | y) = f(x)$. This would lead to the conclusion that for independent r.v.s X and Y,

$$f(x, y) = f_X(x) \cdot f_Y(y), \tag{2.89}$$

if X and Y are continuous, and

$$Pr(x, y) = Pr(x) \cdot Pr(y), \tag{2.90}$$

if X and Y are discrete.

Equation 2.89 and 2.90 can be expanded to a more general case as

$$f(x_1, \ldots, x_n) = f_1(x_1) \cdot \ldots \cdot f_n(x_n), \tag{2.91}$$

where $f(x_1, \ldots, x_n)$ is a joint pdf of $X_1, \ldots, X_n$ and $f_1(x_1), f_2(x_2), \ldots f_n(x_n)$ are marginal pdfs of $X_1, X_2, \ldots, X_n$ respectively.

For more general cases where $X_1, \ldots, X_n$ are not independent

$$\begin{aligned} f(x_1, x_2, \ldots, x_n) &= f(x_1 | x_2, \ldots, x_n) f(x_2, \ldots, x_n), \\ &= f(x_1) \cdot f(x_2 | x_1) \cdot \ldots \cdot f(x_{n-1} | x_1, \ldots, x_{n-2}) \cdot f(x_n | x_1, \ldots, x_n). \end{aligned} \tag{2.92}$$

The marginal pdf of the general joint continuous and discrete joint pdf can be, respectively, expressed as,

$$f(x_1) = \int_{x_n} \ldots \int_{x_3} \int_{x_2} f(x_1, x_2, \ldots, x_n) dx_2 dx_3 \ldots dx_n, \tag{2.93}$$

$$Pr(x_1) = \sum_{x_2} \ldots \sum_{x_3} \sum_{x_n} Pr(x_1, x_2, \ldots, x_n). \tag{2.94}$$

And the conditional pdfs are:

$$f(x_1 | x_2, \ldots, x_n) = \frac{f(x_1, x_2, \ldots, x_n)}{f(x_2, \ldots, x_n)} = \frac{f(x_1, x_2, \ldots, x_n)}{\int_{x_1} f(x_1, x_2, \ldots, x_n) dx_1}, \tag{2.95}$$

$$Pr(x_1 | x_2, \ldots, x_n) = \frac{Pr(x_1, x_2, \ldots, x_n)}{Pr(x_2, \ldots, x_n)} = \frac{Pr(x_1, x_2, \ldots, x_n)}{\sum_{x_1} Pr(x_1, x_2, \ldots, x_n)}. \tag{2.96}$$

Example 2.20

Let T_1 represent the time (in minutes) that a machinery operator spends to locate and correct a routine problem and let T_2 represent the length of time (in minutes) that the operator needs to spend reading procedures for correcting the problem. If T_1 and T_2 are represented by the joint probability distribution,

$$f(t_1, t_2) = \begin{cases} c\left(t_1^{1/3} + t_2^{1/5}\right), & 60 > t_1 > 0, 10 > t_2 > 0; \\ 0 & \text{otherwise.} \end{cases}$$

Find

a. The value of constant c.
b. The probability that an operator can correct the problem in less than 10 minutes. Assume that the operator in this accident should spend less than 2 minutes to read the necessary procedures.
c. Whether r.v.s T_1 and T_2 are independent.

Solution:

a. $Pr(t_1 < 60, t_2 < 10) = \int_0^{t_1=60} \int_0^{t_2=10} c\left(t_1^{\frac{1}{3}} + t_2^{\frac{1}{5}}\right) dt_1 dt_2 = 1,$

$c = 3.92 \times 10^{-4}.$

b. $Pr(t_1 < 10, t_2 < 2) = \int_0^{t_1=10} \int_0^{t_2=2} 3.92 \times 10^{-4}\left(t_1^{1/3} + t_2^{1/5}\right) dt_1 dt_2 = 0.02.$

c. $f(t_2) = \int_0^{t_1=60} f(t_1, t_2) dt_1 = 3.92 \times 10^{-4}\left(176.17 + 60t_2^{1/5}\right).$

Similarly, $f(t_1) = \int_0^{t_2=10} f(t_1, t_2) dt_2 = 3.92 \times 10^{-4}\left(10t_1^{1/3} + 13.21\right).$

Since $f(t_1, t_2) \neq f(t_1) \cdot f(t_2)$, T_1 and T_2 are not independent.

2.8 EXERCISES

2.1 Simplify the following Boolean functions

a. $\overline{((A \cap B) \cup C)} \cap \bar{B}.$

b. $(A \cup B) \cap \left(\bar{A} \cup (\bar{B} \cap \bar{A})\right).$

c. $\overline{A \cap B \cap B \cap C} \cap \bar{B}.$

2.2 Reduce the Boolean function $A \cap B \cap \left(\overline{C \cup \left(\bar{C} \cup A\right) \cup \bar{B}}\right).$

2.3 Simplify the following Boolean expression $\left[(A \cup B) \cap \bar{A}\right] \cup \left(\bar{B} \cap \bar{A}\right).$

2.4 Reduce the Boolean function $G = (A \cup B \cup C) \cap \left(\overline{A \cap \bar{B} \cap \bar{C}}\right) \cap \bar{C}$. If $Pr(A) = Pr(B) = Pr(C) = 0.9$, what is Pr(G)?

2.5 Simplify the following Boolean equations

a. $(A \cup B \cup C) \cap \left(\overline{A \cap \bar{B} \cap \bar{C}}\right) \cap \bar{C}.$

b. $(A \cup B) \cap \bar{B}.$

2.6 Reduce the Boolean equation $\left(A \cup (B \cap C)\right) \cap \left(B \cup (D \cap A)\right).$

2.7 Use both Equations 2.26 and 2.29 to find the $Pr(s)$ for the union of three events, $Pr(s) = Pr(E_1 \cup E_2 \cup E_3)$, with probabilities: $Pr(E_1)=0.8$; $Pr(E_2)=0.9$; $Pr(E_3)=0.95$. Which equation is preferred for numerical solution?

2.8 How many different license plates can be made if each consists of three numbers and three letters, and no number or letter can appear more than once on a single plate?

2.9 A guard works between 5 p.m. and 12 midnight; he sleeps an average of 1 hour before 9 p.m., and 1.5 hours between 9 and 12. An inspector finds him asleep, what is the probability that this happens before 9 p.m.?

2.10 An electronic assembly consists of two subsystems, *A* and *B*. Each assembly is given one preliminary checkout test. Records on 100 preliminary checkout tests show that subsystem *A* failed 17 times. Subsystem *B* alone failed 13 times. Both subsystems *A* and *B* failed together seven times.

   a. What is the probability of *A* failing, given that *B* has failed?
   b. What is the probability that *A* alone fails?

2.11 According to statistics from the U.S. National Eye Institute, 8% of Caucasian men and 0.5% of Caucasian women have some form of color vision deficiency (aka "color blindness"). Assume that women and men exist in equal number in the U.S.

   a. What is the probability that a random person, *X*, from this population is color blind?
   b. An eye doctor is seeing a Caucasian patient who is known to be color blind. Calculate the probability that this patient is male.

2.12 Suppose the lengths of the individual links of a chain are distributed with a uniform distribution, shown below. (Note: *m* is an unknown in this problem, not meters).

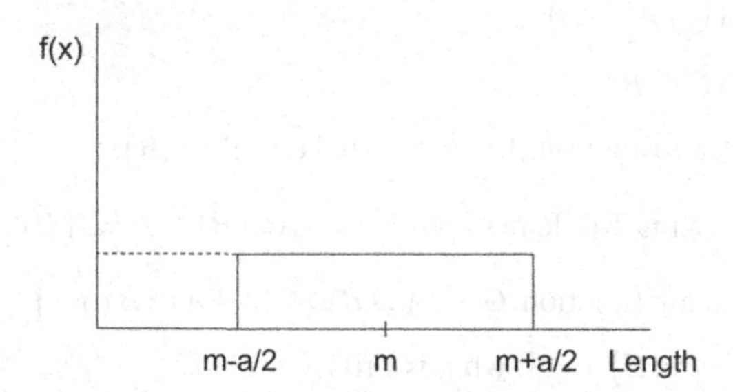

   a. What is the height of the rectangle?
   b. Find the cdf for the above distribution. Make a sketch of the distribution and label the axes.
   c. If numerous chains are made from two such links hooked together, what is the pdf of two link chains?
   d. Consider a 100-link chain. What is the probability that the length of the chain will be less than 100.5m if $a=0.1$m?

2.13 If $f(x, y) = 0.5xy^2 + 0.5yx^2$, $0 < x < 1$, $0 < y < 2$,

a. Show that $f(x, y)$ is a joint pdf.

b. Find $\Pr(x > y)$, $\Pr(y > x)$, and $\Pr(x = y)$.

2.14 Use Equations 2.54 and 2.59 to calculate the mean and variance of a Weibull distribution.

2.15 A random sample of ten resistors is to be tested. From experience, it is known that the probability of a given resistor being defective is 0.08. Let X be the r.v. for number of defective resistors.

a. What kind of distribution function would be recommended for modeling X?

b. Using the distribution function in (a), find the probability that the sample of ten resistors contains more than one defective resistor.

2.16 Given that $\Pr = 0.006$ is the probability of an engine failure on a flight between two cities, find the probability of:

a. No engine failure in 1,000 flights.

b. At least one failure in 1,000 flights.

c. At least two failures in 1,000 flights.

2.17 Suppose a process produces electronic components, 20% of which are defective. Find the distribution of X, the number of defective components in a sample size of five. Given that the sample contains at least three defective components, find the probability that four components are defective.

2.18 Between the hours of 2 and 4 p.m. the average number of phone calls per minute coming into an office is 2.5. Find the probability that during a particular minute, there will be more than five phone calls.

2.19 Assume that for a certain type of resistor, 1% are defective when purchased. What is the probability that a circuit with ten resistors has exactly one defective resistor?

2.20 The number of system breakdowns occurring at a constant rate in a given length of time has a mean value of two breakdowns. What is the probability that in the same length of time, two breakdowns will occur?

2.21 The number of cycles to failure, T, for an electronic device is known to follow an exponential distribution with $\lambda = 0.004$ failures/cycle,

a. Determine the mean cycles to failure for this device.

b. What is the probability that one of these devices will fail sometime after 1,000 cycles, if it has already been used for 250 cycles?

2.22 The consumption of maneuvering jet fuel in a satellite is known to be normally distributed with a mean of 10,000 hours and a standard deviation of 1,000 hours. What is the probability of being able to maneuver the satellite for the duration of a 1-year mission?

2.23 If the heights of 300 students are normally distributed with a mean of 68 inches and standard deviation of 3 inches, how many students have:

a. Heights of more than 70 inches?

b. Heights between 67 and 68 inches?

2.24 A presidential election poll shows one candidate leading with 60% of the vote. If the poll is taken from 200 random voters throughout the U.S., what is the probability that the candidate will get less than 50% of the votes in the election? (Assume the 200 voters sampled are true representatives of the voting profile.)

2.25 The life of a transformer is lognormally distributed with a mean of 7,500 hours, and standard deviation of 1,000 hours. Compute the probability that:

a. The transformer will survive at least 7,500 hours in operation.

b. A transformer that already has 5,000 hours in service will operate for at least 2,500 additional hours.

2.26 For a gamma distribution with the scale parameter of 400 and the shape parameter of 3.8, determine $Pr(x<200)$.

2.27 A company is studying the feasibility of buying an elevator for a building under construction. One proposal is a 10-passenger elevator that, on average, would arrive in the lobby once per minute. The company rejects this proposal because it expects an average of five passengers per minute to use the elevator.

a. Support the proposal by calculating the probability that in any given minute, the elevator does not show up and 10 or more passengers arrive.

b. Determine the probability that the elevator arrives only once in a 5-minute period.

2.28 A pipe manufacturer produces a model with, on average, three random flaws in each 100 m of pipe. If you purchase 200 m of pipe, what is the probability that it has at least two flaws?

2.29 The triangular distribution is a three-parameter distribution that is often used to model subjective information between bounds a and b and possessing a mode c; $a \le x \le b$.

$$f\left(x \middle| a, b, c\right) = \begin{cases} \dfrac{2(x-a)}{(b-a)(c-a)} & a \le x \le c, \\ \dfrac{2}{b-a} & x = c, \\ \dfrac{2(b-x)}{(b-a)(b-c)} & c \le x \le b, \\ 0 & \text{otherwise.} \end{cases}$$

where,

- Parameters a $-\infty \le a < b$ Minimum Value. a is the lower bound,
- Parameter b, $a < b < \infty$ Maximum Value. b is the upper bound,
- Parameter c $a \le c \le b$ *Mode Value*. c is the mode of the distribution (top of the triangle).

A large electronics store is returning a set of failed televisions and you need to know how many shipping containers to prepare. Assume however that all you know is that the minimum number of failed TV sets is 40, the maximum number that failed is 95, and the mode value of TV failures is 60. Draw what the triangular distribution of this failure model would look like. What is the probability that between 40 and 65 TVs are failed?

2.30 Derive the cdf for the triangular distribution given in Problem 2.29. Plot the cdf and pdf if $a = 15$, $b = 25$, $c = 50$.

REFERENCES

Cox, R. T, "Probability, Frequency and Reasonable Expectation." *American Journal of Physics*, 14(1), 1–13, 1946.

Hahn, G. J., and S. S. Shapiro, *Statistical Models in Engineering*. Wiley, New York, 1994

Johnson, N. L., S. T. Kotz, and N. Balakrishnan, *Continuous Univariate Distributions, Volume 2*. John Wiley & Sons, New York, 1995.

Nelson, W. B., *Applied Life Data Analysis*. John Wiley & Sons, New York, 2003.

O'Connor, A. N., M. Modarres, and A. Mosleh, *Probability Distributions Used in Reliability Engineering*, DML International, Washington, DC 2019.

3 Elements of Component Reliability

In this chapter, we discuss the basic elements of component reliability estimation. We start with a mathematical definition of reliability and define commonly used terms and metrics. These definitions also apply to physical items of various complexity (e.g., components, subsystems, systems). We then focus on the use of probability distributions in component reliability analysis in the rest of the chapter.

3.1 DEFINITIONS FOR RELIABILITY

Reliability has many connotations. In general, it refers to an item's ability to successfully perform an intended function during a mission. The longer the item performs its intended function without failure, the more reliable it is. There are both robustness in engineering design and probabilistic views of reliability. The robustness view deals with those engineering design and analysis activities that extend an item's life by controlling or eliminating its potential failure modes. Examples include designing stronger and more durable parts, reducing harmful environmental conditions, minimizing loads and stresses applied to an item during its use, and providing a condition monitoring, predictive or preventive maintenance program to minimize the occurrence of failures.

The probabilistic view measures the reliability of an item in terms of the probability of the successful achievement of the item's intended function. The probabilistic definition of reliability, given in Section 1.4, is the mathematical representation of this viewpoint. The right-hand side of Equation 1.1 denotes the probability that the r.v. T_{fail}, representing time to failure, exceeds a specified mission time or time of interest t_{interest} under specified operating stress conditions $c_1, \ldots, c_n$.

Other representations of the r.v. T_{fail} include the number of cycles to failure, distance to failure such as kilometers to failure that applies to vehicles, and so on. In the remainder of this book, we consider primarily the time to failure representation, although the same mathematical formulations and treatment equally apply to other representations. Expected design and operating conditions $c_1, \ldots, c_n$ are often implicitly considered and not explicitly shown; therefore, Equation 1.1 is written in a simplified form of Equation 1.2. We use Equation 1.2 in the remainder of this chapter. However, the conditions should be explicitly considered when stresses and operating conditions are relevant, such as in accelerated testing and causal modeling.

3.1.1 Reliability Function

Let us start with the formal definition of reliability given by Equation 1.1. Let $f(t)$ denote a pdf representing the item failure time r.v. T_{fail}. According to Equation 2.41, the probability of failure of the item as a function of time is defined by

DOI: 10.1201/9781003307495-3

$$Pr\left(T_{fail} \le t\right) = \int_0^t f(\theta)\mathrm{d}\theta = F(t) \text{ for } t \ge 0, \tag{3.1}$$

where $F(t)$ denotes the probability that the item will fail sometime up to the time of interest t. Equation 3.1 is the *unreliability* of the item. Formally, we can call $F(t)$ (which is the time to failure cdf) the *unreliability function*. Conversely, we can define the *reliability function* (or *the survivor* or *survivorship function*) as

$$R(t) = 1 - F(t) = \int_t^\infty f(x)\mathrm{d}x. \tag{3.2}$$

Let $R(t)$ be the reliability function of an item at time t. The probability that the item will survive for (additional) time x, given that it has survived for time t, is called the *conditional reliability function*, and is given by

$$R(t + x|t) = \frac{R(t+x)}{R(t)}. \tag{3.3}$$

Some literature defines x as a new random variable defined as the time after reference point t. Typically this framing is used with $t = t_0$ being a warranty or burn-in time. In that framing, $x = 0$ at t_0. In this case, Equation 3.3 becomes: $R(x \mid t_0) = \dfrac{R(x)}{R(t_0)}$.

For systems where their failure depends only on the number of times the item is used (rather than length of time) such as some demand items, the cycle-based reliability model may be used. Let N be the integer r.v. representing the cycle of use, then the pmf, cdf, and the reliability function at the point of n uses would be expressed, respectively, by

$$f(n) = Pr(N = n), \tag{3.4}$$

$$F(n) = Pr(N \le n) = \sum_{i=1}^{n} f(i), \tag{3.5}$$

$$R(n) = 1 - F(n) = 1 - \sum_{i=1}^{n} f(i). \tag{3.6}$$

Similarly, the conditional reliability if the item has survived n uses is

$$R(n + n'|n) = \frac{R(n+n')}{R(n)}. \tag{3.7}$$

Recall that integration is the continuous analog of summation, it is clear to see that Equations 3.1–3.3 apply to continuous variables whereas Equations 3.4–3.7 apply only to discrete variables.

3.1.2 MTTF, MRL, MTBF, and Quantiles

The basic characteristics of time to failure distribution and basic reliability measures can be expressed in terms of the pdf, $f(t)$, cdf, $F(t)$, or reliability function, $R(t)$. We can also describe reliability in terms of several types of characteristics, such as the central tendency. The *mean time to failure* (MTTF), for example, illustrates the expected time during which a nonrepairable item will perform its function successfully (sometimes called *expected life*). According to Equation 2.54,

$$\text{MTTF} = E(t) = \int_0^\infty tf(t)\text{d}t. \tag{3.8}$$

If $\lim_{t\to\infty} tf(t) = 0$, integrating by parts reduces Equation 3.8 to

$$E(t) = \int_0^\infty R(t)\text{d}t. \tag{3.9}$$

It is important to make a distinction between MTTF and the *mean time between failures* (MTBF). The MTTF is associated with nonrepairable components (i.e., replaceable components), whereas the MTBF is related to repairable components only. For MTBF, the pdf in Equation 3.8 can be the pdf of time between the first failure and the second failure, the second failure and the third failure, and so on for each repairable item. If we have surveillance, and the item is completely renewed through replacement, maintenance, or repair without delay, the MTTF coincides with MTBF. Theoretically, this means that the renewal process is assumed to be perfect. That is, the item that goes through repair or maintenance is assumed to exhibit the characteristics of a new item. In practice this may not be true. In this case, one needs to determine the MTBF for the item for each renewal cycle (each ith interval of time between two successive failures). However, the approach based on the *as good as new* assumption can be adequate for some reliability considerations. The mathematical aspects of MTBF will be discussed in Chapter 7.

The *mean residual life* (MRL) at time t is defined as the expected remaining life given the component has survived to time, t.

$$MRL\ (t) = \int_0^\infty R(x|t)dx = \frac{1}{R(t)}\int_t^\infty R(t')dt'. \tag{3.10}$$

where $t' = x + t$.

For a continuous r.v. T with cdf $F(t)$, the *p-level quantile* denoted as t_p, is expressed as $F(t_p) = p$, $0 < p < 1$. The *median* is defined as the quantile of the level of $p = 0.5$. A quantile is often called the "100p% point," or the "100pth *percentile*." In reliability studies, the 100pth percentile of time to failure is the point at which the probability of an item's failure is p. For example, the B_{10} life is the time at which 10% of the components are expected to fail. The most popular percentiles used in reliability are the first, fifth, tenth, and 50th percentiles.

Example 3.1

A device time to failure follows the exponential distribution. If the device has survived up to time t, determine its MRL.

Solution:

According to Equation 3.10,

$$\mathrm{MRL}(t) = \frac{\int_0^\infty \tau f(t+\tau)d\tau}{\int_t^\infty f(\tau)d\tau} = \frac{\int_0^\infty \tau\lambda e^{-\lambda(t+\tau)}d\tau}{\int_t^\infty \lambda e^{-\lambda\tau}d\tau} = \frac{e^{\lambda t}\int_0^\infty \tau\lambda e^{-\lambda\tau}d\tau}{e^{-\lambda t}} = \frac{1}{\lambda}.$$

This underlines a pitfall of the exponential distribution as the conditional MRL remains the same and independent of time, *t*, as the unconditional MTTF. This notion will be further discussed in Section 3.2.1.

3.1.3 Hazard Rate and Failure Rate

It is often useful to know the rate of failure at a particular time in an item's lifetime. The *hazard rate, h(t)*, is the *instantaneous failure rate* of an item of age *t*. The hazard rate is the conditional probability that an item which has survived up to time *t* will fail during the following small interval of time Δt, as Δt approaches zero.

$$h(t) = \lim_{\Delta t \to 0} \frac{1}{\Delta t} \frac{F(t+\Delta t) - F(t)}{R(t)} = \frac{f(t)}{R(t)}, \tag{3.11}$$

From Equation 3.11 it is evident that $h(t)$ is also a time to failure conditional pdf similar to Equation 3.3. The hazard rate is sometimes called the failure rate. In this book, we use the term failure rate only in the context of the constant failure rate.

The hazard rate can also be expressed in terms of the reliability function as

$$h(t) = -\frac{\mathrm{d}}{\mathrm{d}t}\left[\ln R(t)\right], \tag{3.12}$$

so that

$$R(t) = e^{-\int_0^t h(x)\mathrm{d}x}. \tag{3.13}$$

The integral of the hazard rate in the exponent is called the *cumulative hazard function, H(t)*:

$$H(t) = \int_0^t h(x)dx. \tag{3.14}$$

The expected value associated with the conditional pdf represented by the hazard rate is called the *residual* MTTF or simply MRL.

For a time interval, t, the mean hazard rate is given by

$$E[h(t)] = \frac{1}{t}\int_0^t h(x)dx, \tag{3.15}$$

therefore,

$$E[h(t)] = -\frac{\ln R(t)}{t}, \tag{3.16}$$

and

$$E[h(t)] = \frac{H(t)}{t}. \tag{3.17}$$

If we want to find the mean hazard rate over a life percentile, t_p, then $F(t_p) = 1 - R(t_p) = p$, and

$$E[h(t_p)] = -\frac{\ln(1-p)}{t_p}. \tag{3.18}$$

The hazard rate is an important function in reliability analysis since it shows changes in the probability of failure over the lifetime of a component. In practice, $h(t)$ often exhibits a bathtub shape and is called a *bathtub curve*. An example of a bathtub curve is shown in Figure 3.1.

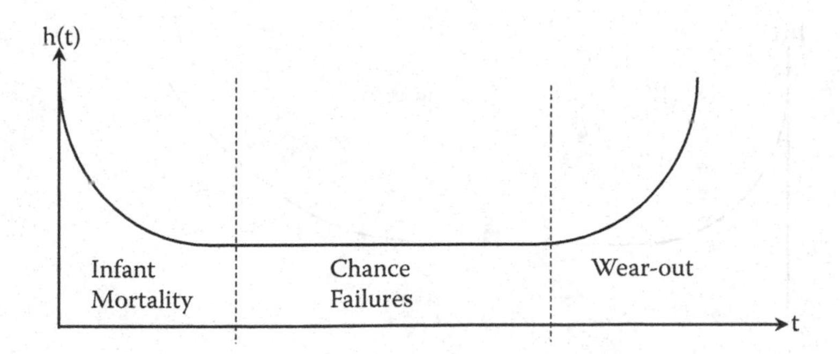

FIGURE 3.1 Example bathtub curve.

Generally, a bathtub curve can be divided into three regions. The *burn-in* or early failure region exhibits a *decreasing hazard rate (sometimes called decreasing (instantaneous) failure rate, DFR)*, characterized by early failures attributable to defects or poor quality control in materials, manufacturing, or construction of the components. Most components do not experience this early failure characteristic, so this part of the curve represents the population and not individual units. Often, warranties are designed to address concerns about early failures.

The *chance failure region* of the bathtub curve exhibits a reasonably *constant hazard rate (or constant failure rate, CFR)*, characterized by random failures of the component. In this period, many mechanisms of failure due to chance events, environmental conditions, or due to complex underlying physical, chemical, or nuclear phenomena give rise to this approximately constant failure rate. The third region, called the *wear-out region*, which exhibits an *increasing hazard rate (or increasing instantaneous failure rate, IFR)*, is characterized mainly by complex aging phenomena of an item approaching its end of useful life. Here, the component deteriorates (e.g., due to accumulated fatigue, corrosion, or other types of damage) and is more vulnerable to outside shocks. It is helpful to note that these three regions can be different for different types of components. There are different time to failure distributions that can properly model increasing, constant, or decreasing hazard rate, which we will come back to in Section 3.2.

There are drastic differences between the bathtub curves of various components. Figures 3.2 and 3.3 show typical bathtub curves for mechanical and electrical devices, respectively. These figures demonstrate that electrical items generally exhibit a relatively long chance failure period, followed by more abrupt wear-out, whereas mechanical items tend to wear-out over a longer period of time. Figure 3.4 shows the effect of various levels of stress on a component. That is, when under higher stress operation, the component will have shorter burn-in and chance failure regions and higher overall hazard rate values. As stress level increases, the chance failure region decreases and premature wear-out occurs. Therefore, it is important to minimize stress factors, such as a harsh operating environment, to maximize reliability.

For the cycle-based (discrete r.v.) items the corresponding hazard rate is defined by

$$h(n) = \frac{Pr(N = n)}{Pr(N \geq n)} = \frac{f(n)}{R(n-1)} = \frac{R(n-1) - R(n)}{R(n-1)}, \tag{3.19}$$

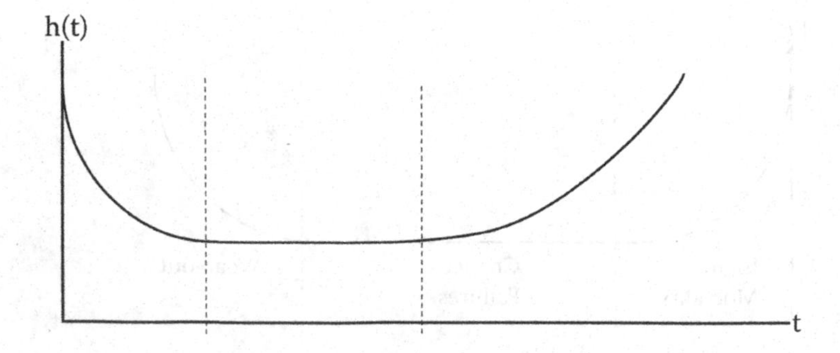

FIGURE 3.2 Typical bathtub curve for mechanical devices.

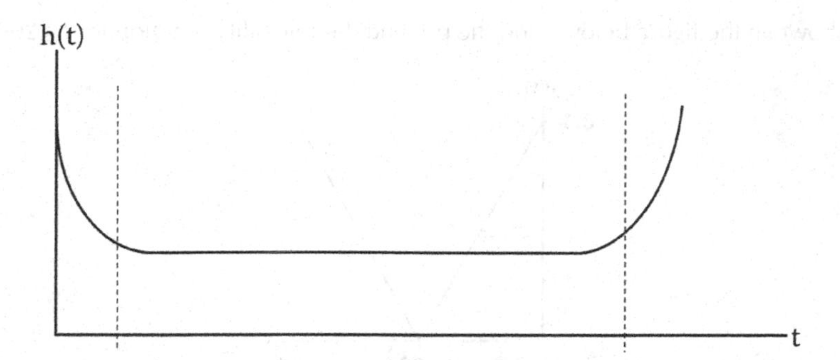

FIGURE 3.3 Typical bathtub curve for electrical devices.

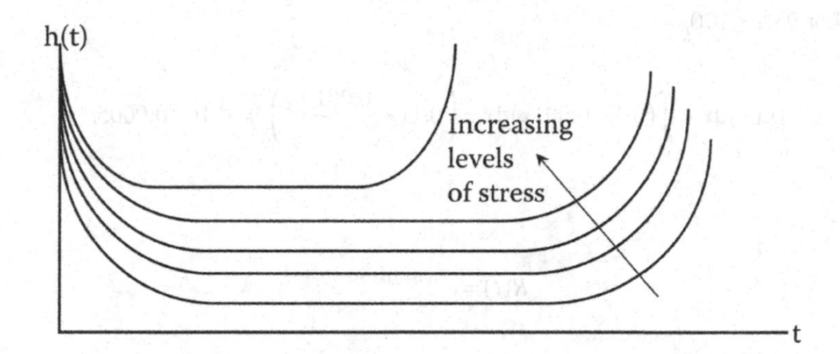

FIGURE 3.4 Effect of various stress levels on components.

where, *f(n)* is the pmf in Equation 3.4 and $R(n-1)$ is the reliability function in Equation 3.6 up to but excluding cycle *n*. Note that in the cycle-based reliability the cumulative hazard rate $H(n) = -\ln[R(n)] \neq \sum_{i=0}^{n} h(i)$. In a discrete counterpart of the continuous distribution the values of the hazard rate do not converge. To overcome this limitation, a function that maintains the monotonicity property called the second failure rate was proposed by Gupta et al. (1997) as

$$r(n) = \ln\frac{R(n-1)}{R(n)} = -\ln[1-h(n)]. \tag{3.20}$$

Example 3.2

The hazard rate *h*(*t*) of a device is approximated by

$$h(t) = \begin{cases} 0.1 - 0.001t, & 0 \le t \le 100, \\ -0.1 + 0.001t, & t > 100, \end{cases}$$

as shown in the figure below. Find the pdf and the reliability function for $t \leq 200$.

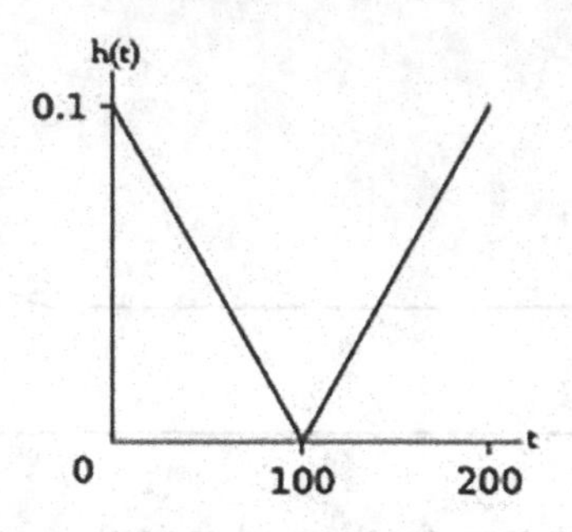

Solution:

For $0 \leq t \leq 100$,

$$\int_0^t h(x)\mathrm{d}x = \int_0^t (0.1 - 0.001x)\mathrm{d}x = \left(0.1x - \frac{0.001}{2}x^2\right)_0^t = 0.1t - 0.0005t^2,$$

thus

$$R(t) = e^{-0.1t + 0.0005t^2}.$$

Using Equation 3.4, one obtains

$$f(t) = (0.1 - 0.001t)e^{\left(-0.1t + 0.0005t^2\right)}.$$

Note that $R(100) = e^{-5} = 0.0067,$

For $t > 100$,

$$h(t) = -0.1 + 0.001t.$$

Accordingly,

$$R(t) = R(100) \cdot e^{\int_{100}^{t} (0.1 - 0.001x)dx}$$

$$= 0.0067e^{\left(0.1t - 0.0005t^2 - 5\right)},$$

$$f(t) = 0.0067(-0.1 + 0.001t)e^{\int_{100}^{t} (0.1 - 0.001x)dx}$$

$$= 0.0067(-0.1 + 0.001t)e^{\left(0.1t - 0.005t^2 - 5\right)}.$$

3.2 COMMON DISTRIBUTIONS IN COMPONENT RELIABILITY

This section discusses exponential, normal, lognormal, Weibull, gamma, and beta distributions commonly used as time to failure distribution models for components. Characteristics of these distributions were discussed in Chapter 2. Their hazard rates are discussed in this section, along with usage in component reliability analysis. Table 3.1 lists the hazard rate functions for continuous reliability models and summarizes applications for some important distributions used in reliability. Discrete reliability models are more involved, and the readers are referred to Bracquemond and Gaudoin (2003).

3.2.1 Exponential Distribution and Poisson Distribution

The exponential distribution is one of the most used distributions in reliability analysis. This can be attributed primarily to its simplicity and to the fact that it gives a constant hazard rate model. That is, for the exponential distribution, $h(t) = \lambda$.

In the bathtub curve, this distribution only represents the chance failure region. It is evident that this distribution might be adequate for components whose chance failure region is long in comparison with the other two regions. A constant failure rate is applicable for some electronic components and mechanical components, especially in certain applications when new components are screened out (e.g., during quality control) and only those that are determined to have passed the burn-in period are used. Furthermore, the exponential distribution is a possible model for representing more complex items, systems, and nonredundant components consisting of many interacting parts.

In Section 2.7.2.1, we noted that the exponential distribution can be introduced using the HPP. Now let us assume that each failure in this process is caused by a random shock, and the number of shocks occurs at a constant rate λ. The number of shocks in a time interval of length t is described by a Poisson distribution with the expected number of shocks equal to λt. Then, the random number of shocks, n, occurring in the interval $[0, t]$ is given by the Poisson distribution:

$$f(n) = Pr(X = n) = \frac{e^{-\lambda t}(\lambda t)^n}{n!}, \quad n = 0,1,2,\ldots > 0, \tag{3.21}$$

Since based on this model, the first shock causes component failure, then the component is functioning only when no shocks occur, that is, $n = 0$. Thus, one can write

$$R(t|\lambda) = Pr(X = 0) = e^{-\lambda t}. \tag{3.22}$$

Using Equation 3.2, the corresponding pdf can be obtained as

$$f(t) = \lambda e^{-\lambda t}. \tag{3.23}$$

Let us now revisit an interesting property of the exponential distribution in which *a failure process represented by the exponential distribution is memoryless.* The

TABLE 3.1
Hazard Rate and Applications in Reliability Engineering for Common Probability Distributions

Distribution	Hazard Rate $h(t)$	Major Applications in Component Reliability
Exponential	$h(t)=\lambda$	• Components past the burn-in period • Random shocks • Electronic components • Used in modeling complex systems due to mathematical simplicity
Weibull	$h(t)=\frac{\beta}{\alpha}\left(\frac{t}{\alpha}\right)^{\beta-1}$	• Used extensively in modeling components • Weakest link model • Corrosion modeling
Gamma	When α is continuous: $h(t)=\frac{t^{\alpha-1}e^{-\frac{t}{\beta}}}{\beta^{\alpha}\left[\Gamma(\alpha)-\int_0^t x^{\alpha-1}e^{-\frac{x}{\beta}}dx\right]}$ When α is an integer: $h(t)=\frac{t^{\alpha-1}}{\beta^{\alpha}\Gamma(\alpha)\sum_{n=0}^{\alpha-1}((t/\beta)^n/n!)}$	• Time between maintenance activities • Time to failure of systems with standby components • Prior distributions in Bayesian analysis
Normal	Solve from $h(t)=\frac{f(t)}{R(t)}$	• Life distributions of high stress components • Stress-strength analysis • Tolerance analysis
Lognormal	Solve from $h(t)=\frac{f(t)}{R(t)}$	• Size distributions of breaks (e.g., in pipes) • Maintenance activities • Prior distributions in Bayesian analysis
Beta	Solve from $h(t)=\frac{f(t)}{R(t)}$	• Prior distributions in Bayesian analysis
Uniform	$h(t)=\begin{cases}\frac{1}{b-t} & \text{for } a\le t\le b\\ 0 & \text{otherwise}\end{cases}$	• Prior distributions in Bayesian analysis • Random number generation • Expert elicitation
Binomial	$h(k)=\left[1+\frac{(1+\theta)^n-\sum_{i=0}^{k}\binom{n}{k}\theta^i}{\binom{n}{k}\theta^k}\right]^{-1}$ where $\theta=\frac{p}{1-p}$	• Demand-based failures • Number of failed items in a population
Poisson	$h(k)=\left[1+\frac{k!}{\mu^k}\left(e^{\mu}-1-\sum_{i=1}^{k}\frac{\mu^i}{i!}\right)\right]^{-1}$	• Homogeneous Poisson Processes • Repairable systems modeling • Rare event modeling
Geometric	$h(k)=\frac{p}{1-p}$	• Reliability test design • Maintenance planning

hazard rate is independent of operating time. Consider the law of conditional probability and assume that an item has survived after operating for a time t. The probability that the item will fail sometime between t and $t+\Delta t$ is

$$Pr\left(t \leq T \leq t+\Delta t \middle| T > t\right) = \frac{e^{-\lambda t} - e^{-\lambda(t+\Delta t)}}{e^{-\lambda t}} = 1 - e^{-\lambda \Delta t}, \tag{3.24}$$

which is independent of t. In other words, the component that has worked without failure up to time t has no memory of its past and remains as good as new. This property can also be easily described by the shock model. That is, at any point along time t, the rate at which fatal shocks occur is the same regardless of whether any shock has occurred up to time t.

For cycle-based (discrete failure) components the reliability of the exponential distribution representing the cycle to failure may be expressed as

$$R(n) = e^{-\alpha n} = (1-\left(1-e^{-\alpha}\right))^{n}. \tag{3.25}$$

where n is the cycle of use and α is the failure rate per cycle. Equation (3.25) is the reliability function of the geometric distribution with parameter $\left(1-e^{-\alpha}\right)$. Note that the failure rate of a geometric distribution is not equal to the failure rate of the corresponding exponential distribution.

3.2.2 Weibull Distribution

The Weibull distribution has a wide range of applications in reliability analysis. This distribution covers a variety of shapes. Due to its flexibility for describing hazard rates, all three regions of the bathtub curve can be represented by the Weibull distribution; by using three different Weibull distributions, each can represent one of the bathtub curve regions. It is possible to show that the Weibull distribution is appropriate for a system or complex component composed of a few components or parts whose failure is governed by the most severe defect or vulnerability of its components or parts (i.e., the *weakest link model*).

The functional form of the Weibull hazard rate is a power function: $h(t) = at^{b}$, although in a slightly modified form. Using Equation 3.11, the hazard rate, $h(t)$, for the Weibull distribution is:

$$h(t) = \frac{\beta}{\alpha}\left(\frac{t}{\alpha}\right)^{\beta-1}, \alpha, \beta > 0, t > 0. \tag{3.26}$$

Sometimes the transformation $\lambda = 1/\alpha^{\beta}$ is used. In this case Equation 3.26 transforms to $h(t) = \lambda\beta t^{\beta-1}$. This form will be used in Chapter 7.

Parameters α and β of the Weibull distribution are the *scale* and *shape* parameters, respectively. The shape parameter, β, also relates to the shape of the hazard rate function.

- If $0 < \beta < 1$, the Weibull distribution has a decreasing hazard rate that represents the burn-in (early) failure behavior.

- For $\beta = 1$, the Weibull distribution reduces to the exponential distribution with the constant failure rate $\lambda = 1/\alpha$.
- If $\beta > 1$, the Weibull distribution represents the wear-out (degradation) region of the bathtub curve having an increasing hazard rate. Different shapes of increasing hazard rate include:
- $1 < \beta < 2$: The hazard rate increases less as time increases
- $\beta = 2$: The hazard rate increases with a linear relationship to time
- $\beta > 2$: The hazard rate increases more as time increases
- $\beta < 3.447798$: The distribution is positively skewed (tail to right)
- $\beta \approx 3.447798$: The distribution is approximately symmetrical
- $3 < \beta < 4$: The distribution approximates a normal distribution
- $\beta > 10$: The distribution approximates a smallest extreme value distribution.

Applications of the Weibull distribution in reliability engineering include:

- Corrosion resistance studies.
- Time to failure of many types of hardware, including capacitors, relays, electron tubes, germanium transistors, photoconductive cells, ball bearings, and certain motors.
- Time to failure of basic elements of a system (components, parts, etc.).

Sometimes, a three-parameter Weibull distribution may also be used. This parameterization uses a third parameter, a location parameter, γ. The positive value of this parameter provides a measure of the earliest time at which a failure may be observed. That is, before the time γ the component is failure free. A negative γ shows that one or more failures could occur prior to the beginning of the reliability data collection. For example, this parameter could count failures occurring during production, in storage, in transit, or any period before the actual use.

The hazard rate for this three-parameter Weibull distribution would be represented by

$$h(t) = \frac{\beta}{\alpha}\left(\frac{t-\gamma}{\alpha}\right)^{\beta-1}, \ \beta, \alpha > 0, 0 < \gamma \leq t < \infty. \tag{3.27}$$

Accordingly, the pdf and reliability function become, respectively,

$$f(t) = \frac{\beta}{\alpha}\left(\frac{t-\gamma}{\alpha}\right)^{\beta-1} e^{-\left(\frac{t-\gamma}{\alpha}\right)^{\beta}}, \ t \geq \gamma, \tag{3.28}$$

and

$$R(t) = e^{\left[-\left(\frac{t-\gamma}{\alpha}\right)^{\beta}\right]}, \ t \geq \gamma. \tag{3.29}$$

There are different forms of discrete (cycle-based) Weibull models discussed in the literature. The form proposed by Padgett and Spurrier (1985) closely corresponds to the continuous Weibull distribution and is expressed as

$$f(n) = \left(1 - e^{-\alpha n^{\beta}}\right) e^{-\alpha \sum_{i=1}^{n-1} i^{\beta}}, \tag{3.30}$$

where α is the scale parameter and β is the shape parameter. Accordingly, the hazard rate and reliability functions are represented by, respectively,

$$h(n) = 1 - e^{-\alpha n^{\beta}}, \tag{3.31}$$

$$R(n) = e^{-\alpha \sum_{i=1}^{n-1} i^{\beta}}, \text{ with } R(0) = 1. \tag{3.32}$$

Note that when $\beta > 0$ we have an increasing hazard rate. For $\beta < 0$ Equation 3.31 is a decreasing hazard rate, and for $\beta = 0$ the distribution reduces to a geometric distribution. Typical values of the cycle-based Weibull parameters in components and devices are $\alpha = 0.001 - 0.1$, $\beta = 0.5 - 2$. Also note that α is related to the probability of failure at the first cycle. That is, $f(1) = 1 - e^{-\alpha}$. See Bracquemond and Gaudoin (2003) for other forms of discrete Weibull cycle-based life models.

Example 3.3

A printer has a time to failure distribution that can be represented by a Weibull distribution with $\alpha = 175$ days and $\beta = 4$. The warranty for the printer is 50 days.

a. Find the probability that the printer survives until the end of the warranty.
b. Now assume the printer has survived until the end of the warranty period. Find the conditional reliability of the printer at 70 days after the end of the warranty.
c. What region of the bathtub curve is this printer in?

Solution:

a. $R(t=50) = 1 - F(t=50) = 1 - Pr(t \leq 50)$.

For the Weibull distribution, $F(t|\alpha, \beta) = 1 - e^{-\left(\frac{t}{\alpha}\right)^{\beta}}$.
Using $\alpha = 175$ days and $\beta = 4$, we find

$$R(t = 50) = e^{\left(-\frac{50}{175}\right)^{\beta}} = 0.993.$$

b. Recall the expression for conditional reliability: $R(x+t|t)=\frac{R(t+x)}{R(t)}$.

Here, $t=50$ days, and thus $x=20$ days after warranty (so, 70 days total)

$$R(t>(20+50)|t>50)=\frac{Pr(t>70)}{Pr(t>50)}=\frac{0.975}{0.993}=0.981.$$

c. Since $\beta=4$, the hazard rate is increasing, and thus the printer is experiencing wear-out.

Similar to the three-parameter Weibull pdf (Equation 3.28), the *smallest extreme value distribution* and the *largest extreme value distribution* are sometimes used as two-parameter distributions. These pdfs take on the following forms.

The pdf of the smallest extreme value distribution is given by

$$f(t)=\frac{1}{\delta}e^{\left[\frac{1}{\delta}(t-\lambda)-e^{\frac{t-\lambda}{\delta}}\right]}, -\infty<\lambda<\infty,\ \delta>0, -\infty<t<\infty. \tag{3.33}$$

The parameter λ is the *location* parameter and can take on any value. The parameter δ is the *scale* parameter and is always positive. The hazard rate for the smallest extreme value distribution is

$$h(t)=\frac{1}{\delta}e^{\frac{t-\lambda}{\delta}}, \tag{3.34}$$

which is an increasing function of time, so the smallest extreme value distribution is an increasing hazard rate distribution that can be used as a model for component failures due to aging. In this model, the component's wear-out period is characterized by an exponentially increasing hazard rate. Clearly, negative values of t are not meaningful when representing time to failure.

The Weibull distribution and the smallest extreme value distribution are closely related to each other. If an r.v. X follows the Weibull distribution as given in Table 2.10, the r.v. T using the transformation $t=\ln(x)$ follows the smallest extreme value distribution with parameters

$$\lambda=\ln(\alpha),\ \delta=\frac{1}{\beta}. \tag{3.35}$$

The two-parameter largest extreme value pdf is given by

$$f(t)=\frac{1}{\delta}e^{\left[-\frac{1}{\delta}(t-\lambda)-e^{-\frac{t-\lambda}{\delta}}\right]}, -\infty<\lambda<\infty,\ \delta>0, -\infty<t<\infty. \tag{3.36}$$

The largest extreme value distribution, although not very useful for component failure behavior modeling, is useful for estimating natural extreme phenomena. For further discussions regarding extreme value distributions, see Johnson et al. (1995), Castillo (1988), and Gumble (1958).

Example 3.4

The maximum demand for electric power during a year is directly related to extreme weather conditions. An electric utility has determined that the distribution of maximum power demands, t, can be modeled by the largest extreme value distribution with $\lambda = 1{,}200$ MW and $\delta = 480$ MW. Determine the unreliability of the installed power, represented by the probability (per year) that the demand will exceed the utility's maximum installed power of 3,000 (MW).

Solution:

Since this is the largest extreme value distribution, we should integrate Equation 3.36 from 3,000 to $+\infty$:

$$Pr(t > 3{,}000) = \int_{3{,}000}^{\infty} f(t)\mathrm{d}t = 1 - e^{\left[-e^{\left(-\frac{t-\lambda}{\delta}\right)}\right]}.$$

Since

$$\frac{t-\lambda}{\delta} = \frac{3{,}000 - 1{,}200}{480} = 3.75,$$

then

$$Pr(t > 3{,}000) = 0.023.$$

3.2.3 Gamma Distribution

The gamma distribution was introduced in Chapter 2 as a generalization of the exponential distribution representing the sum of α independent exponential variables. Recalling the simple shock model considered in Section 3.2.1, one can expand this model for the case when a component fails after being subjected to α successive random shocks arriving at a constant rate (assuming integer α). In this case, the time to failure distribution of the component follows the gamma distribution.

Examples of its application include the distribution of times between recalibration of an instrument that needs recalibration after α uses, time between maintenance of items that require maintenance after α uses, and time to failure of a system with standby components, having the same exponential time to failure distribution with $\beta = 1/\lambda$ where λ is the exponential parameter. It can be seen that β is the mean time to occurrence of a single event.

The gamma distribution has two parameters, α (shape parameter) and β (scale parameter). The gamma cdf and reliability function, in the general case, do not have closed forms.

When the shape parameter α is an integer, the gamma distribution is known as the *Erlangian distribution*. Here, the reliability and hazard rate functions can be expressed in terms of the Poisson distribution as

$$R(t)=\sum_{n=0}^{\alpha-1}\frac{\left(\frac{t}{\beta}\right)^{n}e^{-\frac{t}{\beta}}}{n!}, \tag{3.37}$$

$$h(t)=\frac{t^{\alpha-1}}{\beta^{\alpha}\Gamma(\alpha)\sum_{n=0}^{\alpha-1}\frac{\left(\frac{t}{\beta}\right)^{n}}{n!}}. \tag{3.38}$$

In this case, α shows the number of shocks required before a failure occurs and β represents the mean time to occurrence of a shock.

The gamma distribution represents a decreasing hazard rate for $\alpha < 1$, a constant hazard rate for $\alpha = 1$, and an increasing failure rate for $\alpha > 1$. Thus, the gamma distribution can represent each of the three regions of the bathtub curve.

Example 3.5

The mean time to adjustment of an engine in an aircraft is $E(T)=100$ flight hours (assume time to adjustment follows the exponential distribution). Suppose there is a maintenance requirement to replace certain parts of the engine after three consecutive adjustments.

a. What is the distribution of the time to replace?
b. What is the probability that a randomly selected engine does not require part replacement for at least 200 flight hours?
c. What is the mean time to replace?

Solution:

a. Use gamma distribution for T with $\alpha = 3$, $\beta = 100$.

b. For the gamma distribution, $R(t)=\sum_{n=0}^{2}\frac{\left(\frac{t}{100}\right)^{n}e^{-\left(\frac{t}{100}\right)}}{n!}$,

$$=\frac{\left(\frac{200}{100}\right)^{0}e^{-2}}{0!}+\frac{(2)^{1}e^{-2}}{1!}+\frac{(2)^{2}e^{-2}}{2!},$$

$$=0.135+0.271+0.271=0.677.$$

c. Mean time to replace is $E(T)=\alpha\beta=3(100)=300$ flight hours.

3.2.4 Normal Distribution

The normal distribution is a basic distribution of statistics. The popularity of this distribution in reliability engineering can be explained by the *central limit theorem*. According to this theorem, the sum of a large number, n, of independent r.v.'s approaches the normal distribution as n approaches infinity. For example, consider a sample of many randomly generated observations such that the observations are independent of each other. If we generate multiple samples of this type, the central limit theorem says that as the number of these sample increase, the pdf representing the arithmetic mean of the samples will approach a normal distribution.

The normal distribution is an appropriate model for many practical engineering situations. Since a normally distributed r.v. can take on a value from the $(-\infty, +\infty)$ range, it has limited applications in reliability problems that involve time to failure estimations because time cannot take on negative values. However, for cases where the mean μ is positive and is larger than σ by several folds, the probability that the r.v. T takes negative values can be negligible. For cases where the probability that T takes negative values is not negligible, the truncated normal distribution can be used (see O'Connor et al., 2019).

The normal distribution hazard rate is always a monotonically increasing function of time, t, so the normal distribution is an increasing hazard rate distribution. Thus, the normal distribution can be used to represent the high stress wear-out region of the bathtub curve. The normal distribution is also a widely used model representing stress and/or strength in the framework of the *stress-strength* reliability models, which are time-independent reliability models. The normal distribution can also model simple repair or inspection tasks if they have a typical duration and variance symmetrical about the mean.

3.2.5 Lognormal Distribution

The lognormal distribution is widely used in reliability engineering. The lognormal distribution represents the distribution of an r.v. whose logarithm follows the normal distribution. The lognormal distribution is commonly used to represent the occurrence of certain events in time or space whose values span by several folds or more than one order of magnitude. For example, an r.v. representing the length of time required for a repair often follows a lognormal distribution, because depending on the skills of the repair technician the time to finish the job might be significantly different.

This model is suitable for failure processes that result from many small multiplicative errors. Specific applications of this distribution include time to failure of components due to fatigue cracks. Other applications of the lognormal distribution are associated with failures attributed to maintenance activities and distribution of cracks initiated and grown by mechanical fatigue. The distribution is also used as a model representing the distribution of particle sizes observed in breakage processes and the life distribution of some electronic components. Data that follows the lognormal distribution can typically be identified quickly as some data points may be orders of magnitudes apart. In Bayesian reliability analysis, the lognormal distribution is a popular model to represent the prior distributions. We discuss this topic further in Chapter 5.

The hazard rate for the lognormal distribution initially increases over time and then decreases. The rate of increase and decrease depends on the values of the parameters μ and σ. In general, this distribution is appropriate for representing time to failure for a component whose early failures (or processes resulting in failures) dominate its overall failure behavior.

Example 3.6

The time to failure (in hours) of an experimental laser device is given by the lognormal distribution, with parameters $\mu = 3.5$ hours and $\sigma = 0.9$. Find the MTTF for the laser device. Then, find the reliability of the laser device at 25 hours.

Solution:

The MTTF is found as the mean of the lognormal distribution with the given parameters:

$$\text{MTTF} = E(t) = e^{\left(\mu - \frac{\sigma^2}{2}\right)} = e^{\left(3.5 - \frac{0.9^2}{2}\right)} = 49.65 \text{ hours.}$$

Solving for reliability,

$$R(t = 25) = 1 - F(t = 25) = 1 - \Phi\left(\frac{\ln x - \mu}{\sigma}\right),$$

$$\Phi\left(\frac{\ln 25 - 3.5}{0.9}\right) = \Phi(-0.312),$$

From the lookup table in Appendix A, $\Phi^{-1}(-0.312) = 0.377$

Thus, $R(25) = 1 - 0.377 = 0.623$.

3.2.6 Beta Distribution

The beta distribution is often used to model parameters that are constrained in an interval. The distribution of a probability parameter $0 \le p \le 1$ is popular with the beta distribution. This distribution is frequently used to model proportions of a subclass of events in a set. An example of this is the likelihood ratios for estimating uncertainty. The beta distribution is often used as a conjugate prior in Bayesian analysis for the Bernoulli, binomial and geometric distributions to produce closed form posteriors.

The hazard rate for the Beta distribution may be expressed by

$$h(t) = \frac{t^{\alpha-1}(1-t)}{B(\alpha, \beta) - B(t|\alpha, \beta)}, \tag{3.39}$$

where $B(\alpha, \beta)$ is the beta function, $B(t|\alpha, \beta)$ is the incomplete beta function. The reliability function of the beta distribution is often used in reliability testing which shows the probability that a reliability target is met. It is also used in Bayesian analysis as the prior pdf of reliability during pass-fail reliability demonstration tests.

Example 3.7

The probability of failure on demand, p_{FOD}, for a backup diesel generator is modeled by the beta distribution with $\alpha = 3$ and $\beta = 150$.

a. Find the mean and the variance of the probability of failure on demand.
b. Use the mean value and find the probability that the generator fails exactly two times in the next 50 times it is demanded.

Solution:

a. Using the properties of the beta distribution from Chapter 2,

$$E(p_{FOD}) = \frac{\alpha}{\alpha+\beta} = \frac{3}{3+150} = 0.0196,$$

$$Var(p_{FOD}) = \frac{\alpha\beta}{(\alpha+\beta)^2(\alpha+\beta+1)} = \frac{3\cdot 150}{(153)^2(154)} = 1.25\times 10^{-4}.$$

b. The mean value from part (a) gives a probability of failure on demand. This becomes the parameter, p, of a binomial distribution. We are told we have $n=50$ demands, and want to know the probability that $x=2$ failures. Using the binomial pdf:

$$f(x|n, p) = \binom{n}{x} p^x (1-p)^{n-x} = f(x|n, p),$$

$$Pr(x=2) = \binom{50}{2} 0.0196^2(1-0.0196)^{50-2} = 1225\cdot 0.0196^2(0.9804)^{48} = 0.182.$$

3.3 EXERCISES

3.1 Assume that T, the random variable that denotes life in hours of a specified component, has a cumulative density function (cdf) of

$$F(t) = \begin{cases} 1-\dfrac{100}{t}, & t \geq 100, \\ 0, & t < 100. \end{cases}$$

Determine the following:

a. Pdf $f(t)$
b. Reliability function $R(t)$
c. MTTF

3.2 The hazard rate of a device is $h(t) = \frac{1}{\sqrt{t}}$. Find the following:
  a. Reliability function
  b. Probability density function
  c. MTTF
  d. Variance

3.3 Assume that 100 components are placed on test for 1,000 hours. From previous testing, we believe that the hazard rate is constant, and the MTTF = 500 hours. Estimate the number of components that will fail in the time interval of 100–200 hours. How many components will fail if it is known that 15 components failed in $T < 100$ hours?

3.4 For the following Rayleigh distribution,

$$f(t) = \frac{1}{\alpha^2} e\left(-\frac{t^2}{2\alpha^2}\right), t \geq 0, \alpha > 0.$$

  a. Find the hazard rate *h(t)* corresponding to this distribution.
  b. Find the reliability function *R(t)*.
  c. Find the MTTF. Note that $\int_0^\infty e\left(-\alpha x^2\right) = \frac{1}{2}\sqrt{\frac{\pi}{\alpha}}$
  d. For which part of the bathtub curve is this distribution adequate?

3.5 Show whether a uniform distribution represents an IFR, DFR or CFR.

3.6 Owing to the aging process, the failure rate of a nonrepairable item is increasing according to $\lambda(t) = \lambda\beta t^{\beta-1}$. Assume that the values of λ and β are estimate as $\hat{\beta} = 1.62$ and $\hat{\lambda} = 1.2 \times 10^{-5}$ hours. Determine the probability that the item will fail sometime between 100 and 200 hours. Assume an operation beginning immediately after the onset of aging.

3.7 If the time to failure of a component follows the symmetric pdf:

$$f(t) = \begin{cases} \alpha(t-\beta)^2, 0 < t < 10{,}000, \\ 0, \text{otherwise.} \end{cases}$$

  a. Determine α and β.
  b. Find the reliability function.
  c. Determine the hazard rate function and plot it vs. time.

3.8 The time to failure of a solid-state power unit has a hazard function in the form of $h(t) = ct^{1/2}$ for $t \geq 0$.
  a. What is the proper value of *c*?
  b. Compute the reliable life for 0.995.
  c. Compute the MTTF and the median life.
  d. If the unit has operated for 100 hours, what is the probability that the unit will operate for another 100 hours?
  e. What is the mean residual time to failure at $t = 100$ hours?

3.9 Consider a discrete r.v. with cdf $F(k) = \frac{1}{2^k}$, $k = 1, \ldots, n$.
  a. Derive expressions for the corresponding pdf, hazard function, mean *k* to failure, and reliability function.
  b. Compute $Pr(k > 3 \mid k > 2)$.

3.10 The reliability of a propellor can be represented by the following expression:

$$R(t) = \left(1 - \frac{t}{t_0}\right)^2, \; 0 \le t \le t_0$$

where t_0 is the maximum life of the propellor

a. What region of the bathtub curve is this propellor in?
b. Compute MTTF if $t_0 = 3{,}000$ hours.

3.11 The following is the time to failure pdf of a new appliance,

$$f(t) = 0.1(1 + 0.05t)^3, \; t \ge 0.$$

Determine the fraction of appliances that will fail up to the warranty period of 18 months.

3.12 Suppose X has the exponential pdf $f(x) = \lambda e^{-\lambda x}$ for $x > 0$, and $f(x) = 0$ for $x \le 0$. Find $Pr(x > (a+b)|x > a)$, a, $b > 0$.

3.13 A manufacturer uses the exponential distribution to model the number of cycles to for a product. The product has $\lambda = 0.003$ failures/cycle,

a. What is the mean cycle to failure for this product?
b. If the product survives for 300 cycles, what is the probability that it will fail sometimes after 500 cycles? If operational data show that 1,000 components have survived 300 cycles, how many of these would be expected to fail after 500 cycles?

3.14 Life of an aging device can be described by a Weibull distribution with the shape parameter of 2.15 and a MTTF of 25,500 hours. Assuming that a failed device is replaced with a new one that does not fail, determine how many devices among the 30 operating ones are expected to fail in 1,500 hours.

3.15 Time to failure of a relay follows a Weibull distribution with $\alpha = 10$ years, $\beta = 0.5$. Find the following:

a. *Pr*(failure after 1 year)
b. *Pr*(failure after 10 years)
c. MTTF

3.16 A medical device has a time to failure distribution given by a Weibull distribution with a shape parameter of 1.4 and $E(t) = 50{,}000$ hours.

a. Determine the reliability at 1 year.
b. Calculate the conditional reliability of the device at 15,000 hours if the device has already survived 5,000 hours.

3.17 The MTTF of a certain type of small motor is 10 years, with a standard deviation of 2 years. The manufacturer replaces free of charge all motors that fail while under warranty. If the manufacturer is willing to replace only 3% of the motors that fail, what warranty period should they offer? Assume that the time to failure of the motors follows a normal distribution.

3.18 A product's time to failure is lognormally distributed with 50% of the failures occurring between 200 and 300 hours of use. (Assume 30% of failures occur below 200 hours).

   a. Determine the mean and standard deviation of time to failure.
   b. Compute the probability that the product functions for an additional 100 hours, if the product has been operating for 300 hours without failure.

3.19 An electronic device has a time to failure modeled by the lognormal distribution with parameters $\mu = 5.8$ and $\sigma = 1.2$.

   a. Find the MTTF.
   b. If this device is used in an application which requires it to be replaced when its reliability falls below 0.9, when should the device be replaced?
   c. Find the hazard function for the device at the time calculated in (b).

3.20 The probability of failure of a component follows a log-logistic function (see O'Connor et al. or other sources for a description of this distribution).

   a. What is the pdf of the component?
   b. When the mean life of the component?
   c. What is the hazard function?
   d. What is the MRL at time t_s?

REFERENCES

Bracquemond, C., and O. Gaudoin, "A Survey on Discrete Lifetime Distributions." *International Journal of Reliability, Quality and Safety Engineering*, 10(1), 69–98, 2003.

Castillo, E., *Extreme Value Theory in Engineering*. Academy Press, San Diego, CA, 1988.

Gumble, E. J., *Statistics of Extremes*. Columbia University Press, New York, 1958.

Gupta, P. L., R. C. Gupta, and R. C. Tripathi, "On the Monotonic Properties of Discrete Failure Rates." *Journal of Statistical Planning and Inference*, 65(2), 255–268, 1997.

Johnson, N. L., S. T. Kotz, and N. Balakrishnan, *Continuous Univariate Distributions*. John Wiley & Sons, Vol. 2, New York, 1995.

O'Connor, A.N., M. Modarres, and A. Mosleh, Probability Distributions Used in Reliability Engineering, DML International, Washington, DC, 2019.

Padgett, W. J., and J. D. Spurrier, "On Discrete Failure Models." *IEEE Transactions on Reliability*, 34(3), 253–256, 1985.

4 Basic Reliability Mathematics
Statistics

4.1 INTRODUCTION

Statistics is the process of collecting, analyzing, organizing, and interpreting data. Data are used to support decisions about a *population* of interest. Data can be collected about a full population, but more often, the collected data are a *sample* from the population of interest. A set of observations from a distribution of an r.v. is called a sample. The number of observations in a sample is called the *sample size*. In the framework of classical statistics, a sample is usually composed of exchangeable, *independently, and identically distributed* (i.i.d.) observations. From a practical point of view, this assumption means that elements of a sample are obtained individually and without regards to other samples and under the same conditions.

Now, let's discuss two aspects of statistics of primary relevance. The first, descriptive statistics, is concerned with descriptive properties of observed data. The second, inferential statistics (or statistical inference) is concerned with using data to make conclusions about a population. In reliability, for example, we use descriptive statistics such as the mean and variance of observed data to characterize failure data, and inferential statistics to establish a probability distribution representing time to failure of a population of identical components. Caution: statistical terminology differs from similar terminology used in machine learning. In machine learning, the term inference is used to refer to making a prediction using a trained model; in this context, determining properties of the model is called learning (rather than inference), and using the model for prediction is called inference (rather than prediction).

In statistical inference, we can ask *deductive* questions, such as "given a population, what will a sample look like?" Alternatively, we can ask *inductive* questions, such as "given a sample, what can be inferred about a population?"

Since a sample is a subset of the population, it is necessary to make corrections based on sample characteristics, randomness, assumptions, and data collecting characteristics such as censoring. We will describe several corrections later in this chapter. A sample of data provides the basis for statistical inference about the underlying distribution of the parent population. Typically, the data come from special tests, experiments, operational data, or practical uses of a limited set of items. Generally, it is assumed that each sample point is i.i.d., but it is prudent to verify this before applying methods blindly. Each observed value is considered a *realization* (or *observation*) of some hypothetical r.v.: that is, a value that the r.v. can take on.

DOI: 10.1201/9781003307495-4

4.2 DESCRIPTIVE STATISTICS

Two important measurements related to observed samples are mean and variance. These are closely related to the concepts described for probability distributions in Section 2.6.2. One can define the *sample arithmetic mean* and *sample variance* as the respective expected values of a sample of size n from the distribution of X, namely, $x_1, \ldots, x_n$ as follows:

$$E(X) = \bar{x} = \frac{1}{n}\sum_{i=1}^{n} x_i, \tag{4.1}$$

and

$$Var(X) = s^2 = \frac{1}{n}\sum_{i=1}^{n}(x_i - \bar{x})^2. \tag{4.2}$$

When sample data vary by orders of magnitude, it is customary to use *geometric mean*. In this case, the logarithm of the sample data is used to represent x_i in Equations 4.1 and 4.2.

Equation 4.2 can be used to estimate the variance if the data represent the complete population. However, if the data come from a sample, the estimator of variance (Equation 4.2) is biased, since $\bar{x}$ is estimated from the same sample. It can be shown that this bias can be removed by multiplying Equation 4.2 by $n/(n-1)$:

$$Var(X) = s^2 = \frac{1}{n-1}\sum_{i=1}^{n}(x_i - \bar{x})^2. \tag{4.3}$$

The concept of median also applies to a sample, where the sample median is simply the midpoint of a set of data. Additional topics fall within the domain of descriptive statistics, including moments, sample covariance, and more. The reader is referred to statistics textbooks for full coverage.

Example 4.1

A sample of eight manufactured shafts is taken from a plant lot. The diameters of the shafts are 1.01, 1.08, 1.05, 1.01, 1.00, 1.02, 0.99, and 1.02 inches. Find the sample mean and variance.

Solution:

$$E(X) = \bar{x} = \frac{1}{n}\sum_{i=1}^{n} x_i = 1.0225.$$

Since these data are samples, we use the unbiased form the variance.

$$Var(X) = \frac{1}{n-1}\sum_{i=1}^{n}(x_i - \bar{x})^2 = 0.0085.$$

4.3 EMPIRICAL DISTRIBUTIONS AND HISTOGRAMS

When studying distributions, it is convenient to start with some preliminary procedures useful for data visualization, editing, and detecting outliers by constructing *empirical distributions* and *histograms*. Such preliminary data analysis procedures are useful to support understanding data, to illustrate the data (sometimes the data speak for themselves), to identify suitable distributions to consider, and to support other types of analysis (goodness of fit testing, for instance).

To illustrate some of these procedures, consider the data set composed of observed times to failure of 100 identical electronic devices, given in the first two columns of Table 4.1. Notice that the data are already aggregated into bins. The measure of interest here is the relative frequency associated with each interval of time to failure. This can be obtained using Equation 2.10, that is, by dividing each interval frequency by the total number of devices tested. Applying Equation 2.10 produces the empirical probability distribution in the third column of Table 4.1.

TABLE 4.1
Example of Binned Data with Corresponding Empirical Probability Distribution

Time Interval, *T* (hour)	Observed Frequency	Relative Frequency, $\frac{n_e}{n}$
0–100	35	0.35
100–200	26	0.26
200–300	11	0.11
300–400	12	0.12
400–500	6	0.06
500–600	3	0.03
600–700	4	0.04
700–800	2	0.02
800–900	0	0.00
900–1,000	1	0.01
Total:	100	1.0

Example 4.2

Consider the observed time to failure data for an electronic device below. It is believed that the data come from a time to failure process, *T*, that can be represented by an exponential distribution with parameter $\lambda = 0.005$ hour^{-1}. Determine the expected frequency of failures in each time interval.

Solution:

The probability (relative frequency) that T takes values between 0 and 100 hours is

$$Pr(0 < T < 100) = \int_0^{100} 0.005e^{-0.005t}dt,$$

$$Pr(0 < T < 100) = [1 - e^{-0.005t}]_0^{100} = 0.393.$$

Summing the observed frequency column shows that 100 data points were collected. By multiplying the probability by the total number of devices observed (100), we can determine the expected frequency for the interval 0–100.

The expected frequency for the 0–100 interval is $0.393 \cdot 100 = 39.3$.

The results for the rest of the intervals are shown below.

Interval, *T* (hour)	Observed Frequency	Exponential Expected Frequency (Count)	Exponential Expected Relative Frequency
0–100	35	39.3	0.393
100–200	26	23.8	0.238
200–300	11	14.5	0.145
300–400	12	8.8	0.088
400–500	6	5.3	0.053
500–600	3	3.2	0.032
600–700	4	2.0	0.020
700–800	2	1.2	0.012
800–900	0	0.7	0.007
900–1,000	1	0.4	0.004

A comparison of the observed and expected frequencies of each interval reveals differences as large as 4.3 failures.

Figure 4.1 illustrates the respective histogram of the data from Example 4.2 and its comparison to the exponential distribution with $\lambda = 0.005$ hour^{-1}.

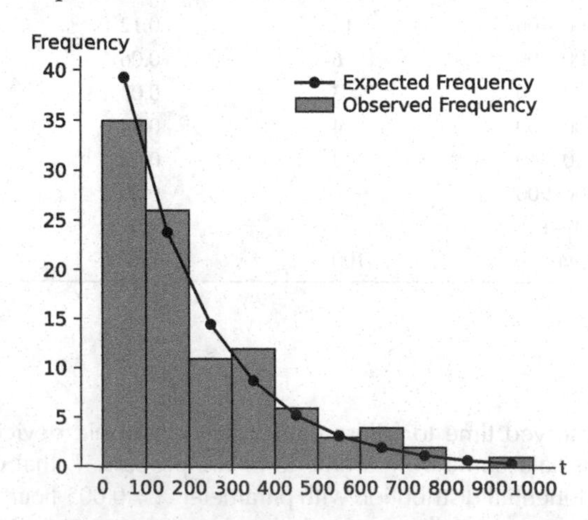

FIGURE 4.1 Observed frequencies (histogram) and expected frequencies from Example 4.2.

4.4 PARAMETER ESTIMATION: POINT ESTIMATION

Statistical inference involves using data to identify the distribution of an r.v. We can use a sample of failure times of an item, $t_1, \ldots, t_n$, to estimate, for example, λ, the parameter of an exponential distribution representing time to failure of the item. In this case, we are inferring a general distribution from a specific sample of data. Since a sample is one possible realization of data from the population, estimating the model parameter inherently involves uncertainty.

There are many methods for conducting this *parameter estimation*, some in the domain of frequentist statistics and others in the domain of Bayesian statistics. Let $f(x|\theta)$ denote the pdf of r.v. X where θ represents an unknown parameter or a vector of parameters. Let $x_1, \ldots, x_n$ denote a sample from $f(x|\theta)$. In frequentist statistics, the parameter θ is considered to be fixed but unknown, and a *random sample of data* is used to estimate it. In Bayesian statistics, the parameter θ is considered to be an r.v., and a *fixed set of data* (or evidence) is used to update our prior knowledge of the parameter. The contrasting viewpoints on data and parameters between the frequentist and Bayesian are critical in interpreting and formulating the likelihood function in the maximum likelihood estimation and Bayesian estimation procedures that will be discussed later in Sections 4.4.3 and 4.4.4.

Point estimation and *interval estimation* are the two basic kinds of estimation procedures. Point estimation uses a data set to obtain a single number that represents the most likely value of a parameter of the distribution function or other characteristic of the underlying distribution of interest. Point estimation does not provide information about the uncertainty in that number. Uncertainty is expressed using interval estimation, with either *confidence intervals (frequentist) or credible intervals (Bayesian)* which will be discussed later in this chapter.

Now, let's discuss terminology. Suppose we are interested in estimating a single-parameter distribution $f(x|\theta)$ based on a random sample $x_1, \ldots, x_n$. Let $g(x_1, \ldots, x_n)$ be a single-valued (simple) function of $x_1, \ldots, x_n$. It follows that $g(x_1, \ldots, x_n)$ is also an r.v., which is called a *statistic*. A point estimate of that function is obtained by using an appropriate statistic and calculating its value based on the sample data. The statistic (as a function) is called the *estimator* and its numerical value is called the *estimate*.

Consider the basic properties of point estimators. An estimator $g(x_1, \ldots, x_n)$ is said to be an *unbiased* estimator for θ if its expectation coincides with the value of the parameter of interest θ. That is, $E\left[g(x_1, \ldots, x_n)\right] = \theta$ for any value of θ. Thus, the *bias* in the estimator is the difference between the expected value of an estimator and the true parameter value itself. It is obvious that the smaller the bias, the better the estimator is.

Another desirable property of an estimator $g(x_1, \ldots, x_n)$ is the property of *consistency*. An estimator $g(x_1, \ldots, x_n)$ is said to be consistent if, for every $\varepsilon > 0$,

$$\lim_{n\to\infty} Pr\left(\left|g(x_1, \ldots, x_n) - \theta\right| < \varepsilon\right) = 1. \tag{4.4}$$

This property implies that as the sample size n increases, the estimator $g(x_1, \ldots, x_n)$ gets closer to the true value of θ. In some situations, several unbiased estimators can be found. A possible procedure for selecting the best one among the unbiased estimators can be based on choosing the one having the least variance. An unbiased estimator t of θ, having minimum variance among all unbiased estimators of θ, is called an *efficient* estimator.

Another estimation property is sufficiency. An estimator $g(x_1, \ldots, x_n)$ is said to be a *sufficient* statistic for the parameter if it contains all the information in the sample $x_1, \ldots, x_n$. In other words, the sample $x_1, \ldots, x_n$ can be replaced by $g(x_1, \ldots, x_n)$ without loss of any information about the parameter of interest.

Several methods of estimation are considered in statistics. In the following section, some of the most common methods are briefly discussed. Estimated parameters will be denoted with a hat, i.e., $\hat{\theta}$ denotes the estimate for parameter θ.

4.4.1 Method of Moments

The method of moments is an estimation procedure based on empirically estimated sample moments of the r.v. According to this procedure, the sample moments are equated to the corresponding distribution moments. That is, $\bar{x}$ and s^2 obtained from the equations in Section 4.2 can be used as the point estimates of the distribution mean, μ, and variance, σ^2. The solutions of these equalities provide the estimates of the distribution parameters.

4.4.2 Linear Regression

Another widely used method for parameter estimation is a nonstatistical optimization approach by linear regression, often used in conjunction with probability plotting methods. Because regression has many uses in addition to parameter estimation, we cover it separately in Section 4.7. We will return to the use of linear regression with the probability distribution plotting methods in Chapter 5.

4.4.3 Maximum Likelihood Estimation

This method is one of the most widely used methods of estimation. Consider a continuous r.v. X with pdf $f(X|\theta)$. Also consider a random set of data (evidence) E_i which provides information about a sample of values of X. The probability of occurrence of the sample evidence E_i depends on the parameter θ. The likelihood of this parameter is given by a *likelihood function*, $L(\theta)$ that is proportional to the probability of the observed evidence,

$$L(\theta) \propto \Pr\left(\bigcap_i E_i | \theta\right). \tag{4.5}$$

Let our evidence E_i be a random sample $x_1, \ldots, x_n$ of size n taken from the distribution of X. Assuming these observations are statistically independent, the likelihood of obtaining this particular set of sample values is proportional to the joint occurrence of n independent random events $x_1, \ldots, x_n$. That is, the likelihood function is proportional to the probability (for a discrete r.v.) or the relative frequency of the pdf (for a continuous r.v.) of the joint occurrence of $x_1, \ldots, x_n$:

$$L(\theta|x_1, \ldots, x_n) \propto \prod_{i=1}^{n} f(x_i|\theta) = f(x_1|\theta)\ldots f(x_n|\theta). \tag{4.6}$$

Note that the pdf may be viewed as a function of r.v. X with the fixed but unknown parameter θ, i.e., $f(x|\theta)$. In this case the likelihood function is written as $L(\theta|x_1, \ldots, x_n)$. However, the pdf can alternatively be viewed as a function of variable θ with fixed x, i.e., $f(\theta|x)$, in which case the likelihood function is written as $L(x_1, \ldots, x_n|\theta)$. The former view is how the MLE approach expresses the likelihood function, whereas the latter represents the Bayesian parameter estimation method's view of the likelihood function.

The *maximum likelihood estimate (MLE)*, is the parameter value that maximizes the likelihood function, $L(\theta|x_1, \ldots, x_n)$.

$$\hat{\theta} = \arg\max\left[L(\theta)\right], \\ \text{subject to } \theta \text{ contraints.} \tag{4.7}$$

The standard way to find a maximum of a parameter in MLE is to optimize the likelihood function. The simplest way to do this is to calculate the first partial derivative of $L(\theta)$ with respect to each parameter and equate it to zero. This yields the equation(s)

$$\frac{\partial L(\theta|x_1, \ldots, x_n)}{\partial \theta} = 0, \tag{4.8}$$

from which the MLE $\hat{\theta}$ can be obtained.

Due to the multiplicative form of Equation 4.6, in many cases it is more convenient to maximize the logarithm of the likelihood function, the *log-likelihood* $\Lambda(\theta|x)$:

$$\frac{\partial \log L(\theta|x_1, \ldots, x_n)}{\partial \theta} = \frac{\partial \Lambda(\theta)}{\partial \theta} = 0. \tag{4.9}$$

Since the logarithm is a monotonic transformation, the estimate of θ obtained from this equation is the same as that obtained from Equation 4.8. Sometimes, Equations 4.8 or 4.9 can be solved analytically, but often must be solved numerically. The optimum values of the unknown (but fixed) parameters, $\hat{\theta}$, are called *point estimates* or *best estimates* of the parameters in the classical (frequentist) treatments of probability.

For example, it can be shown that for complete data of n times to failure, t_i, believed to come from a normal distribution, the MLE parameters of the normal distribution are:

$$\hat{\mu} = \frac{1}{n}\sum_{i=1}^{n} t_i, \tag{4.10}$$

$$\hat{\sigma}^2 = \frac{1}{n}\sum_{i=1}^{n}(t_i - \mu)^2. \tag{4.11}$$

Usually, the unbiased estimator form of Equation 4.11 is used, as explained in Section 4.2:

$$\hat{\sigma}^2 = \frac{1}{n-1}\sum_{i=1}^{n}(t_i - \mu)^2. \tag{4.12}$$

To express the confidence over the estimation of the unknown fixed parameters, we should also express a *confidence interval* associated with a *confidence level* (likelihood) within which the true value of the unknown parameter resides. This concept will be discussed in Section 4.5.

Example 4.3

Consider a sample $t_1, \ldots, t_n$ of n times to failure of a component whose time to failure is known to be exponentially distributed. Find the MLE of the failure rate, λ.

Solution:

Using Equations 4.6 and 4.9, one can obtain

$$\mathcal{L} = L(\lambda \mid t_1, \ldots, t_n) = \prod_{i=1}^{n}\lambda\, e^{-\lambda t_i} = \lambda^n e^{-\lambda\sum_{i=1}^{n} t_i},$$

$$\Lambda = \ln \mathcal{L} = n(\ln \lambda) - \lambda\sum_{i=1}^{n} t_i,$$

$$\frac{\partial \Lambda}{\partial \lambda} = \frac{n}{\lambda} - \sum_{i=1}^{n} t_i = 0,$$

$$\hat{\lambda} = \frac{n}{\sum_{i=1}^{n} t_i}.$$

Recalling the second-order condition $\left(\frac{\partial^2 \Lambda}{\partial \lambda^2}\right)_{\hat{\lambda}} < 0$, since $-\frac{n}{\lambda^2} < 0$, the estimate $\hat{\lambda}$ is indeed the MLE for the problem considered.

4.4.4 Bayesian Parameter Estimation

Bayesian parameter estimation uses the subjective interpretation of probability and Bayes' Theorem to update the prior state of knowledge about the unknown parameter of interest, which is treated as an r.v. Recall Bayes' Theorem, which was introduced in Chapter 2 in the context of discrete variables as Equation 2.35. The continuous form of Bayes Theorem is:

$$f_1(\theta|E)=\frac{f_0(\theta)L(E|\theta)}{\int f_0(\theta)L(E|\theta)d\theta}, \tag{4.13}$$

where $f_0(\theta)$ is the prior pdf of r.v. θ, $f_1(\theta|E)$ is the posterior pdf of θ, and $L(E|\theta)$ is the likelihood function from Equation 4.6. The denominator of Equation 4.13, once integrated, is a constant.

Now, consider new evidence of the form of a random sample $x_1, \ldots, x_n$ of size n taken from the distribution of X. We want to use this evidence for estimating parameter(s) of a distribution. In this instance, Bayes' Theorem can be expressed by

$$f_1(\theta|x_1, \ldots, x_n)=\frac{f_0(\theta)L(x_1, \ldots, x_n|\theta)}{\int f_0(\theta)L(x_1, \ldots, x_n|\theta)d\theta}, \tag{4.14}$$

$$Pr_1(\theta|x_1, \ldots, x_n)=\frac{Pr_0(\theta)Pr(x_1, \ldots, x_n|\theta)}{\sum_{i=1}^{n} Pr_0(\theta_i)Pr(x_1, \ldots, x_n|\theta_i)}, \tag{4.15}$$

respectively for continuous or discrete distributions, where $f_0(\theta)$ and $Pr_0(\theta)$ are the prior pdfs, and $L(x_1, \ldots, x_n|\theta)$ and $Pr(x_1, \ldots, x_n|\theta)$ are the likelihood functions representing the observed data and information (evidence). The prior distribution is the probability distribution of the r.v. θ, which captures our state of knowledge of θ prior to the data $x_1, \ldots, x_n$ being observed. It is common for this distribution to represent soft evidence or probability intervals about the possible range of prior values of θ. If the distribution is dispersed it represents a case where little is known about the parameter θ. If the distribution is concentrated (has higher density) in an area, then it reflects a good knowledge about the likely values of θ. This Bayesian estimation process is illustrated in Figure 4.2. The relationship with MLE also becomes clear by observing Figure 4.2.

In Bayesian analysis we combine the prior distribution of the unknown random parameter(s) with the likelihood function representing the observed data (evidence) by considering the data as fixed evidence. We use it to obtain $f_1(\theta|x_1, \ldots, x_n)$ or $Pr_1(\theta|x_1, \ldots, x_n)$, the posterior pdf or probability distribution of the parameter(s). To obtain a single statistic akin to the MLE, we use either the mean or median of the distribution as the Bayesian estimate.

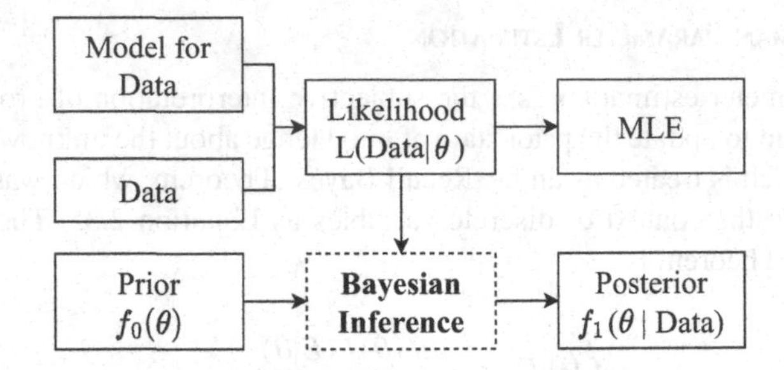

FIGURE 4.2 Bayesian inference method.

Note that we used $L(\theta | x_1, \ldots, x_n)$ in MLE but use $L(x_1, \ldots, x_n | \theta)$ in Bayesian estimation. As noted before, this is because the MLE approach is intended to obtain an estimation of the unknown but fixed parameters (regardless of any prior knowledge), using only observed random samples of data (evidence). In essence, because the frequentist approach does not consider the prior information, the MLE approach treats $L(\theta | E)$ and $L(E | \theta)$ as equal.

To express our uncertainties about the estimated parameters, Bayesian estimation develops the credible interval which shows the probability that the r.v. θ falls within an interval of values. This topic is further discussed next. Bayesian estimation will be examined in far more detail in Chapter 5 in the context of reliability parameter estimation.

4.5 PARAMETER ESTIMATION: INTERVAL ESTIMATION

Interval estimation is used to quantify the uncertainty about the parameter θ due to sampling error (e.g., a limited number of samples). Interval estimates are expressed as either *confidence intervals* (frequentist) or *credible intervals* (Bayesian). Readers should be careful in understanding the differences between the meaning and use of these intervals in decision making. Typical intervals of interest are 99%, 95%, and 90%. Note that interval estimation does not address uncertainty due to inappropriate model selection or invalid assumptions. Care must be taken to avoid these other issues by verifying assumptions and model correctness.

4.5.1 Confidence Intervals

The confidence interval is the frequentist approach to express uncertainties about the estimated parameters. The main purpose is for finding an interval with a high probability of containing the true but unknown value of the parameter θ. For the parameter θ describing the distribution of an r.v. X, a confidence interval is estimated based on a sample $x_1, \ldots, x_n$ of size n from the distribution of X. Consider two r.v.s $\theta_l(x_1, \ldots, x_n)$ and $\theta_u(x_1, \ldots, x_n)$ chosen in such a way that the probability that interval $[\theta_l, \theta_u]$ contains the true parameter θ is

$$Pr[\theta_l(x_1, \ldots, x_n) < \theta < \theta_u(x_1, \ldots, x_n)] = 1 - \gamma. \tag{4.16}$$

The interval $[\theta_l, \theta_u]$ is called a *k% confidence interval* (specifically, *a two-sided confidence interval*) for the parameter θ. The term $k\% = 100(1-\gamma)\%$ is the *confidence coefficient* or *confidence level*, and the endpoints θ_l, and θ_u are called the $k\%$ lower and upper confidence limits of θ, respectively. In the frequentist approach, these interval endpoints are estimated from sample data.

The confidence interval is interpreted as follows. If many confidence intervals are generated this way (i.e., from many sample sets from the same population), then $k\%$ of those intervals will contain the true value of the fixed parameter θ. Reversely, this also means that $100\gamma\%$ of the generated intervals will not contain the true value of θ. Therefore, we have a $k\%$ confidence that any generated interval contains the true value of θ. It is critical to note that the expressed confidence level $k\%$ is about the interval not the unknown parameter. See Figure 4.3 for an illustration of this concept.

Sometimes, frequentist intervals are sometimes misinterpreted to mean that there is a $k\%$ probability that the parameter will be in that interval. This probability is *not* given by the confidence interval (rather, it is given by the Bayesian credible interval). In the frequentist approach, it is not appropriate to make statements about the probability of a parameter; only the Bayesian approach does this.

Each row in Figure 4.3 is a sample $x_1, \ldots, x_n$ of n points from the distribution with unknown parameter θ (in this case, mean μ). The square indicates the sample mean and the whiskers represent the 70% confidence interval. The white squares indicate intervals that contain the true population mean, the black squares indicate intervals that do not. Because these are frequentist confidence intervals, 70% of the generated intervals contain the true population mean.

It is also possible to construct one-sided confidence intervals. The case when $\theta > \theta_l$ with the probability of $1-\gamma$, θ_l is called the *one-sided lower confidence limit*

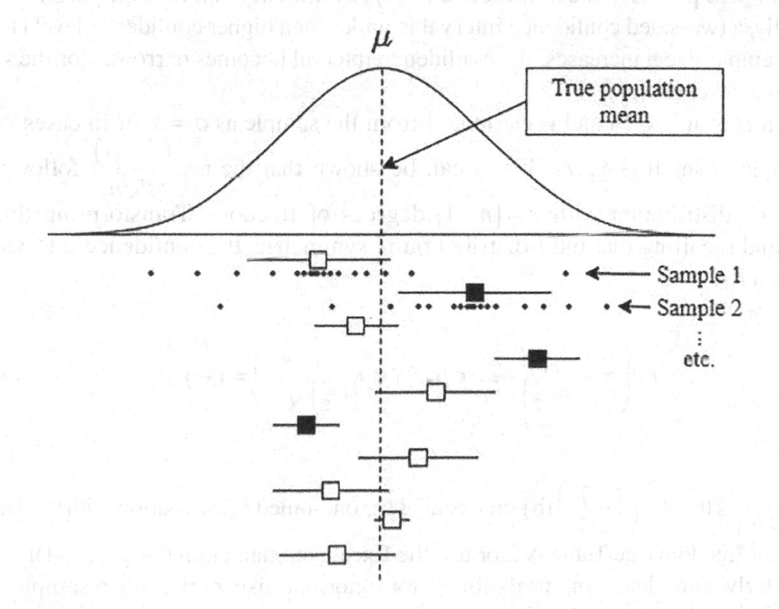

FIGURE 4.3 Illustration of how confidence intervals are generated from a set of samples.

for θ. The case when $\theta < \theta_u$ with probability of $1-\gamma$, θ_u is called the *one-sided upper confidence limit* for θ.

Consider a typical example illustrating the basic idea of confidence limits for the parameters of a normal distribution. First, consider a procedure for constructing confidence intervals for μ when σ is known. Let $x_1, \ldots, x_n$ be a random sample from the normal distribution, $X \sim N(\mu, \sigma)$. It can be shown that the sample mean $\bar{x}$ has the normal distribution $\bar{X} \sim N\left(\mu, \frac{\sigma}{\sqrt{n}}\right)$. By using the Z transformation, $\frac{(\bar{x}-\mu)}{\sigma/\sqrt{n}}$ has the standard normal distribution.

Using this distribution, one can write

$$Pr\left(z_{\left(\frac{\gamma}{2}\right)} \le \frac{\bar{x}-\mu}{\frac{\sigma}{\sqrt{n}}} \le z_{\left(1-\frac{\gamma}{2}\right)}\right) = 1-\gamma, \tag{4.17}$$

where $z_{(\gamma/2)}$ is the $100(\gamma/2)$ % value of the standard normal distribution, which can be obtained from Table A.1. After simple algebraic transformations, and recalling that the normal distribution is symmetric and thus $z_{(\gamma/2)} = -z_{(1-\gamma/2)}$, the inequalities inside the parentheses of Equation 4.17 can be rewritten as

$$Pr\left(\bar{x} - z_{\left(1-\frac{\gamma}{2}\right)}\frac{\sigma}{\sqrt{n}} \le \mu \le \bar{x} + z_{\left(1-\frac{\gamma}{2}\right)}\frac{\sigma}{\sqrt{n}}\right) = 1-\gamma. \tag{4.18}$$

Equation 4.18 provides the symmetric $(1-\gamma)$ confidence interval on parameter μ. Generally, a two-sided confidence interval is wider for a higher confidence level $(1-\gamma)$. As the sample size n increases, the confidence interval becomes narrower for the same confidence level $(1-\gamma)$.

When σ is unknown and is estimated from the sample as $\hat{\sigma} = s$, or in cases when the sample is small (e.g., $n < 30$), it can be shown that the r.v. $\frac{(\bar{x}-\mu)}{s/\sqrt{n}}$ follows the Student's t-distribution with $\nu = [n-1]$ degrees of freedom. Transforming this as above, and recalling that the t-distribution is symmetric, the confidence interval on μ is given by:

$$Pr\left(\bar{x} - t_{\left(1-\frac{\gamma}{2}\right)}\frac{s}{\sqrt{n}} < \mu < \bar{x} + t_{\left(1-\frac{\gamma}{2}\right)}\frac{s}{\sqrt{n}}\right) = 1-\gamma, \tag{4.19}$$

where $t_{\left(1-\frac{\gamma}{2}\right)}$ is the $100\left(1-\frac{\gamma}{2}\right)$th percentile of the one-tailed t-distribution with $\nu = [n-1]$ degrees of freedom (see Table A.2 or use the Excel command t.inv($1-\gamma/2$, n–1)).

Similarly, confidence intervals for σ^2 for a normal distribution for a sample with variance s^2 can be obtained as

$$\frac{(n-1)s^2}{\chi^2_{\left(1-\frac{\gamma}{2}\right)}[n-1]} < \sigma^2 < \frac{(n-1)s^2}{\chi^2_{\left(\frac{\gamma}{2}\right)}[n-1]}, \tag{4.20}$$

where $\chi^2_{\left(1-\frac{\gamma}{2}\right)}[n-1]$ is the $100\left(1-\frac{\gamma}{2}\right)$th percentile of a χ^2 distribution with $df=[n-1]$ degrees of freedom. They can be found using Table A.3, the Excel command chisq.inv($1-\gamma$, *df*), or in MATLAB using chi2inv($1-\gamma$, *df*). Confidence intervals for other distributions will be further discussed in Chapter 5.

4.5.2 Credible Intervals

The Bayesian analog of the classical confidence interval is known as the *credible interval* or *Bayesian probability interval*. To construct a Bayesian probability interval, the following relationship based on the posterior distribution is used:

$$Pr(\theta_l < \theta \le \theta_u) = 1-\gamma. \tag{4.21}$$

The interval $[\theta_l, \theta_u]$ is the $k\% = 100(1-\gamma)\%$ *credible interval* (specifically *a two-sided credible interval*) for the parameter θ, and θ_l and θ_u are the $k\%$ lower and upper credible limits of θ.

Interpretation of the credible interval is simple and intuitive. The Bayesian credible interval directly gives the probability that an unknown parameter θ falls into the interval $[\theta_l, \theta_u]$. At $k\%$ probability level, the credible interval gives a range which has a $k\%$ chance of containing θ, as shown in Figure 4.4. Like the confidence interval, the Bayesian credible interval can be calculated as a two-sided or one-sided interval. Note that the for the confidence interval, the $k\%$ confidence level is a property of the confidence interval estimation (not a property of θ the unknown parameter of interest), whereas the $k\%$ credible interval has direct relevance to the parameter θ.

The calculation of the credible interval follows directly from the posterior distribution. We will illustrate this in Example 4.4

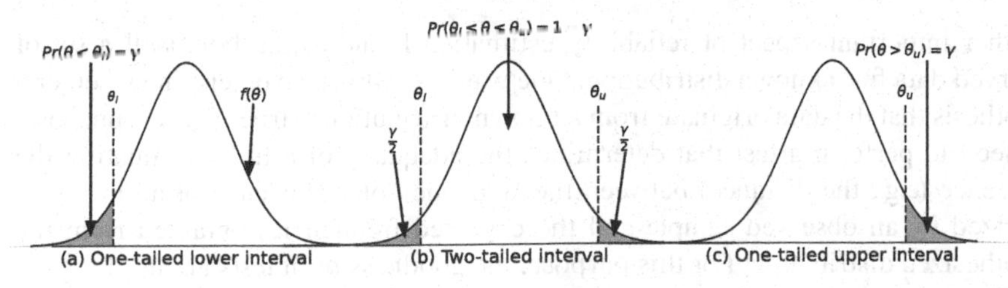

FIGURE 4.4 Illustration of Bayesian credible intervals for a parameter θ.

Example 4.4

A capacitor has a time to failure that can be represented by the exponential distribution with a failure rate λ. The mean value and uncertainty about λ are modeled through a Bayesian inference with the posterior pdf for λ expressed as a normal distribution with a mean of 1×10^{-3} and standard deviation of 2.4×10^{-4}. Find the 95% Bayesian credible interval for the failure rate.

Solution:

To obtain the 95% credible interval for the normal distribution, we find the 2.5th and 97.5th percents of the normal distribution representing uncertainty on the parameter.

$$\lambda_l = \Phi^{-1}\left(0.025, 1\times10^{-3}, 2.4\times10^{-4}\right) = 5.30\times10^{-4},$$

$$\lambda_u = \Phi^{-1}\left(0.975, 1\times10^{-3}, 2.4\times10^{-4}\right) = 1.47\times10^{-3}.$$

Therefore, $5.30\times10^{-4} < \hat{\lambda} < 1.47\times10^{-3}$. This is illustrated in the figure below.

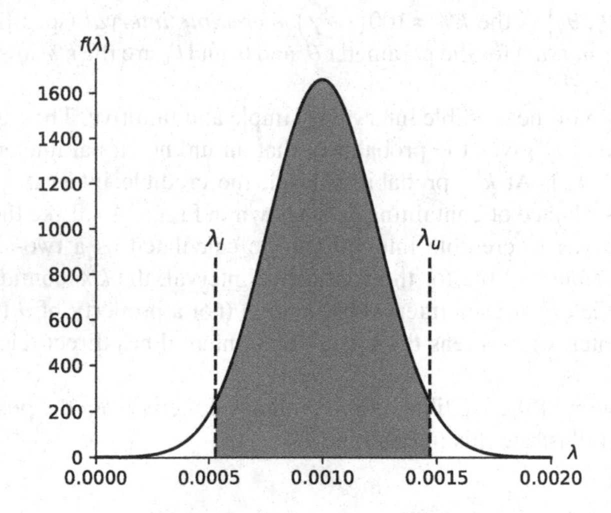

4.6 HYPOTHESIS TESTING AND GOODNESS OF FIT

Another important aspect of reliability estimation is indicating how well a set of observed data fits a known distribution. One way to do this is to determine whether a hypothesis that the data originate from a known distribution is true. For this purpose, we need to perform a test that determines the adequacy of a fit by evaluating the difference (e.g., the distance) between the frequency of occurrence of an r.v. characterized by an observed sample and the expected frequencies obtained from the hypothesized distribution. For this purpose, the goodness of fit tests are used.

It is necessary to calculate the expected frequencies of failures from the hypothesized distribution and measure the difference from the observed frequencies. Several methods exist to perform such a fit test. We discuss two of these methods in this book. For a more comprehensive discussion on goodness of fit tests the reader is referred to Rayner et al. (2009).

4.6.1 Hypothesis Testing Basics

Interval estimation and hypothesis testing may be viewed as mutually inverse procedures. Let us consider an r.v. X with a known pdf $f(x|\theta)$. Using a random sample from this distribution, one can obtain a point estimate $\hat{\theta}$ of parameter θ. Let θ have a hypothesized value of $\theta = \theta_0$. Under these conditions, the following question can be raised: is the $\hat{\theta}$ estimate compatible with the hypothesized value θ_0? In terms of *statistical hypothesis testing*, the statement $\theta = \theta_0$ is called the *null hypothesis*, which is denoted by H_0. For the case considered, it is written as

$$H_0\!: \theta = \theta_0. \tag{4.22}$$

The null hypothesis is always tested against an *alternative hypothesis*, denoted by H_1, which for the case considered might be the statement $\theta \neq \theta_0$, which is written as

$$H_1\!: \theta \neq \theta_0. \tag{4.23}$$

The null and alternative hypotheses are also classified as *simple* (or *exact*, when they specify exact parameter values) or *composite* (or *inexact*, when they specify an interval of parameter values). In the above equations, H_0 is simple and H_1 is composite. An example of a simple alternative hypothesis is

$$H_1\!: \theta = \theta^*. \tag{4.24}$$

For testing statistical hypotheses, *test statistics* are used. Often the test statistic is the point estimator of the unknown distribution. Here, as in the case of the interval estimation, one must obtain the distribution of the test statistic used.

Recall the example considered above. Let $x_1, \ldots, x_n$ be a random sample from the normal distribution, $X \sim N(\mu, \sigma)$, in which μ is an unknown parameter and σ^2 is assumed to be known. One must test the simple null hypothesis

$$H_0\!: \mu = \mu^*, \tag{4.25}$$

against the composite alternative

$$H_1\!: \mu \neq \mu^*. \tag{4.26}$$

As the test statistic in Equation 4.25, use the same sample mean, $\bar{x}$, which has the normal distribution, $\bar{X} \sim N\left(\mu, \frac{\sigma}{\sqrt{n}}\right)$. Having the value of the test statistic $\bar{x}$, one can construct the confidence interval using Equation 4.18 and see whether the value of

μ^* falls inside the interval. This is the test of the null hypothesis. If the confidence interval includes μ^*, then we can conclude the null hypothesis is *not rejected* at the *significance level* γ.

In terms of hypothesis testing, the confidence interval considered is called the *acceptance region*, and the upper and the lower limits of the acceptance region are called the *critical values*, and the significance level γ is called a probability of a *Type I error*. In deciding about whether to reject the null hypothesis, it is possible to commit these errors:

- Reject H_0 when it is true (Type I error).
- Do not reject H_0 when it is false (Type II error).

The probability of the Type II error is designated by β. These situations are traditionally represented by Table 4.2.

TABLE 4.2
Goodness of Fit Error Types

	State of Nature (True Situation)	
Decision	**H_0 is True**	**H_0 is False**
Reject H_0	Type I error	No error
Do not reject H_0	No error	Type II error

Increasing the acceptance region decreases γ and simultaneously results in increasing β. The traditional approach to this problem is to keep the probability of Type I errors at a low level (0.01, 0.05, or 0.10) and to minimize the probability of Type II errors. The probability of not making a Type II error is called the *power of the test*.

4.6.2 Chi-Squared Test

As the name implies, this test is based on a statistic that has an approximate χ^2 distribution. To perform this test, an observed sample taken from the population representing an r.v. X must be split into k nonoverlapping (mutually exclusive) intervals. The lower limit for the first interval can be $-\infty$, and the upper limit for the last interval can be $+\infty$. The assumed (hypothesized) distribution model is then used to determine the probabilities p_i that the r.v. X would fall into each interval i ($i=1,\ldots, k$). By multiplying p_i by the sample size n, we obtain the expected frequency for each interval i, denoted as e_i (see Example 4.2). It is obvious that $e_i = np_i$. If the observed frequency for each interval i of the sample is denoted by o_i, then the differences between e_i and o_i can characterize the goodness of the fit.

The χ^2 test uses the test statistic χ^2 or W, which is defined as

$$W = \chi^2 = \sum_{i=1}^{k} \frac{(o_i - e_i)^2}{e_i}. \tag{4.27}$$

The χ^2 statistic approximately follows the χ^2 distribution. If the observed frequencies o_i differ considerably from the expected frequencies e_i, then W will be large, and the fit is considered poor. A good fit would lead to not rejecting the hypothesized distribution, whereas a poor fit leads to the rejection. It is important to note that the hypothesis can be rejected, but cannot be positively affirmed. Therefore, the hypothesis is either *rejected* or *not rejected* as opposed to accepted or not accepted.

The chi-squared test has the following steps:

Step 1: Choose a hypothesized distribution for the sample.

Step 2: Select a significance level of the test, γ.

Step 3: Calculate the value of the χ^2 test statistic, W, using Equation 4.27.

Step 4: Define the critical value (edge of rejection region) $R \geq \chi^2_{(1-\gamma)}[df]$, where $\chi^2_{(1-\gamma)}[df]$ is $100(1-\gamma)\%$ of the χ^2 distribution with degrees of freedom $df = k - m - 1$, k is the number of intervals, and m is the number of parameters estimated from the sample. If the parameters of the distribution were estimated without using the sample, then $m = 0$.

Step 5: If $W > R$, reject the hypothesized distribution; otherwise, do not reject the distribution.

A few conditions must be met for the chi-squared test to be applicable. First, the data must be frequencies or counts, and they must be binned into mutually exclusive intervals. Generally, the value of the "expected frequency" cell should be greater than or equal to five in at least 80% of the cells. No cell should have an expected value of less than one. If these conditions are not met, either bin the data further, combine bins, or use another test where the assumptions are met.

It is important at this point to specify the role of γ in the χ^2 test. Suppose that the calculated value of W in Equation 4.27 exceeds the 95th percentile value, $\chi^2_{0.95}[\text{df}]$, given in Table A.3. This indicates that chances are lower than 5% that the observed data are from the hypothesized distribution. Here, the model should be rejected; by not rejecting the model, one would make the Type II error. But if the calculated value of W is smaller than $\chi^2_{0.95}(\cdot)$, chances are greater than 5% that the observed data match the hypothesized distribution model. In this case, the model should not be rejected; by rejecting the model, one would make a Type I error. It is critical to recognize that not rejecting a hypothesis is not equivalent accepting it.

One instructive step in χ^2 testing is to compare the observed data with the expected frequencies to note which classes (intervals) contributed most to the value of W. This can sometimes help to understand the nature of the deviations.

Example 4.5

The number of parts ordered per week by a maintenance department in a manufacturing plant is believed to follow a Poisson distribution. Use a χ^2 goodness of fit test to determine the adequacy of the Poisson distribution. Use the data found in the following table and a 0.05 significance level.

No. of Parts per Week, X	Observed Frequency, o_i (No. of Weeks)	Expected Frequency, e_i (No. of Weeks)		χ^2 Statistic $\frac{(o_i - e_i)^2}{e_i}$
0	18	15.783		0.311
1	18	18.818		0.036
2	8	11.219		0.923
3	5	4.459	6.1	0.588
4	2	1.329		
5	1	0.317		
Total:	52 weeks	52 weeks		$W = 1.86$

Solution:

Since under the Poisson distribution model, events occur at a constant rate, then a natural estimate (and MLE) of μ is

$$\hat{\mu} = \frac{\text{Number of parts used}}{\text{Number of weeks}} = \frac{62}{52} = 1.192 \text{ parts/week.}$$

From the Poisson distribution,

$$Pr(X = x_i) = \frac{\mu e^{-\mu}}{x_i!}.$$

Using $\hat{\rho} = 1.192$, one obtains $\Pr(X=0)=0.304$. Therefore, $e_i = 0.304 \cdot 52 = 15.783$. The expected frequency for other rows is calculated in the same way. Since we estimated one parameter (Γ) from the sample data, $m=1$. Notice that in the table above, we have binned the expected frequencies for 3, 4, and 5 parts per week. You can see that the original values did not meet the criterion that most bins have an expected value greater than 5. Thus, these three adjacent bins were combined into one that met the criterion. Therefore, $k=4$ bins.

Using Table A.3, we determine that $R = \chi^2_{0.95}[6-4-1] = 3.84$. Since $(W=1.86) < R$, there is no reason to reject the hypothesis that the data are from a Poisson distribution with $\mu = 1.192$. This is further supported by visual inspection of the figure below, which illustrates the similarity between observed and expected frequencies.

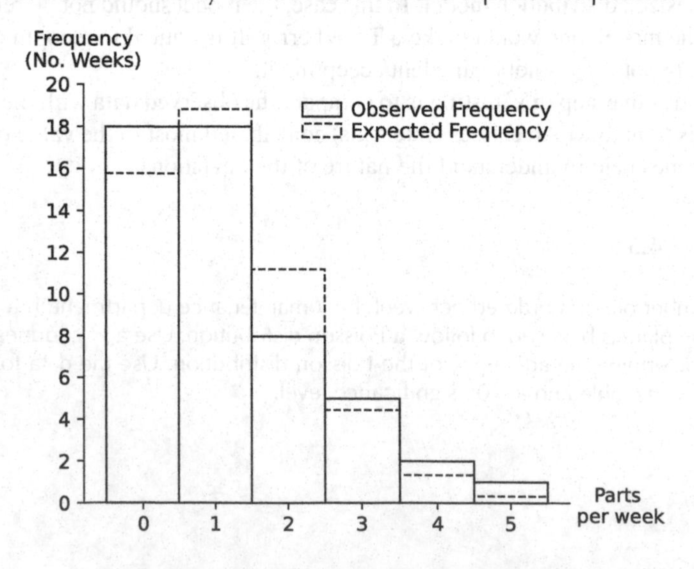

Example 4.6

The table below shows the accumulated mileage after 2 years in service for a sample of 100 vehicles. The mileage accumulation pattern is believed to follow a normal distribution. Use the χ^2 test to check this hypothesis at the 0.05 significance level.

32,797	38,071	16,768	26,713	25,754	37,603	39,485	15,261	45,283	41,064
47,119	35,589	43,154	35,390	32,677	26,830	25,056	20,269	16,651	27,812
33,532	44,264	22,418	40,902	29,180	25,210	28,127	14,318	27,300	28,433
55,627	20,588	14,525	22,456	28,185	16,946	29,015	19,938	36,837	36,531
11,538	25,746	52,448	35,138	22,374	30,368	10,539	32,231	21,075	45,554
34,107	28,109	28,968	27,837	41,267	24,571	41,821	44,404	27,836	8,734
26,704	29,807	32,628	28,219	33,703	43,665	49,436	32,176	47,590	32,914
9,979	16,735	31,388	21,293	36,258	55,269	37,752	42,911	21,248	28,172
10,014	28,688	26,252	31,084	30,935	29,760	43,939	18,318	21,757	26,208
22,159	22,532	31,565	27,037	49,432	17,438	27,322	37,623	17,861	24,993

Solution:

Using the MLE approach, first estimate the mean and standard deviation for the hypothesized normal distribution as $\hat{\mu} = 30{,}011$ and $\hat{\sigma} = 10{,}472$ miles.
Next, group the data into bins to calculate the observed frequencies. Using nine bins will give each bin an expected frequency of at least five in every interval. Use the normal distribution to find the probability in each interval, and multiply by 100 vehicles to get the expected frequency.

Mileage Interval				
Start	End	Observed Frequency, O_i	Expected Frequency, e_i	χ^2 Statistic $\frac{(o_i - e_i)^2}{e_i}$
0	13,417	5	5.446	0.0365
13,417	18,104	9	7.124	0.4942
18,104	22,791	13	11.750	0.1329
22,791	27,478	14	15.916	0.2306
27,478	32,165	20	17.704	0.2978
32,165	36,852	15	16.172	0.0849
36,852	41,539	8	12.131	1.4066
41,539	46,226	9	7.473	0.3122
46,226	>46,226	7	6.077	0.1401
	Total:	100	99.792[a]	$W=3.136$

[a] Note that the expected frequency column adds up to slightly less than 100. This is due to a small amount of density below 0 in the normal distribution; in this case it is negligible. In other cases, it may be necessary to use a truncated normal distribution.

Since both distribution parameters were estimated from the given sample, $m=2$. The critical χ^2 value of the statistic is, therefore, $\chi^2_{0.95}[9-2-1]=12.59$. This is higher than the test statistic $W=3.0758$; therefore, there is no reason to reject the hypothesis about the normal distribution at the 0.05 significance level. Visually assessing the plot of the observed frequencies vs. expected frequency further supports this conclusion.

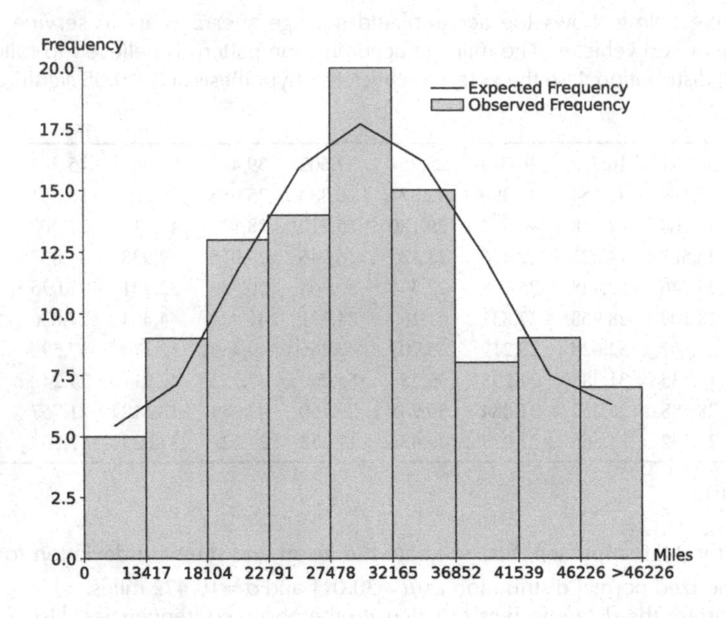

4.6.3 Kolmogorov-Smirnov (K-S) Test

In the framework of the K-S test, the individual sample items are treated without clustering them into intervals. Similar to the χ^2 test, a hypothesized cdf is compared with its estimate known as the *empirical cdf* or *sample cdf.*

A sample cdf is defined for an ordered sample $x_1 < x_2 < \cdots < x_n$ as

$$S_n(x) = \begin{cases} 0, & -\infty < x < x_1, \\ \dfrac{i}{n}, & x_i \le x < x_{(i+1)},\ i = 1,2,\ldots,n-1, \\ 1, & x_n \le x < \infty. \end{cases} \tag{4.28}$$

The K-S test statistic, D, measures the maximum difference between the sample cdf $S_n(x)$and a hypothesized cdf, $F(x)$. It is calculated as:

$$D = \max_i \left[\left| F(x_i) - S_n(x_i) \right|, \left| F(x_i) - S_n(x_{i-1}) \right| \right]. \tag{4.29}$$

Similar to the χ^2 test, the following steps compose the K-S test:

Step 1: Choose a hypothesized cumulative distribution $F(x)$ for the sample.

Step 2: Select a specified significance level of the test, γ.

Step 3: Define the rejection region $R > D_n(\gamma)$; critical values of the test statistic $D_n(\gamma)$ can be obtained from Table A.4.

Step 4: If $D > D_n(\gamma)$, reject the hypothesized distribution and conclude that $F(x)$ does not fit the data; otherwise, do not reject the hypothesis.

Example 4.7

Time to failure of a group of electronic devices is measured in a life test. The observed failure times are 254, 586, 809, 862, 1381, 1923, 2,542, and 4,211 hours. Use a K-S test with $\gamma = 0.05$ to assess whether the exponential distribution with $\lambda = 5\times10^{-4}$ is an adequate representation of this sample.

Solution:

For $\gamma = 0.05$, $D_8(0.05) = 0.454$. Thus, the rejection region is $D > 0.454$. We use Equation 4.29 to calculate the sample cdfs for Equation 4.29 as shown in the table below. For an exponential distribution with $\lambda = 5\times10^{-4}$, $F(t) = 1 - e^{-5\times10^{-4}t}$. Using this, we calculate the fitted cdfs in the table below. The elements of Equation 4.29 are shown in the last two columns. As can be seen the maximum value is obtained for $i = 7$ as $D = 0.156$, shown in bold in the table below. Since $(D = 0.156) < 0.454$, we should not reject the hypothesized exponential distribution model with the given λ. Visually assessing the plot of the sample cdf vs. hypothesized cdf further supports this conclusion.

		Empirical cdf		Hypothesized cdf		
Time to failure, t_i	i	$S_n(t_i)$	$S_n(t_{i-1})$	$F_n(t_i)$	$\lvert F_n(t_i) - S_n(t_i)\rvert$	$\lvert F_n(t_i) - S_n(t_{i-1})\rvert$
254	1	0.125	0.000	0.119	0.006	0.119
586	2	0.250	0.125	0.254	0.004	0.129
809	3	0.375	0.250	0.333	0.042	0.083
862	4	0.500	0.375	0.350	0.150	0.025
1,381	5	0.625	0.500	0.499	0.126	0.001
1,923	6	0.750	0.625	0.618	0.132	0.007
2,542	7	0.875	0.750	0.719	**0.156**	0.031
4,211	8	1.000	0.875	0.878	0.122	0.003

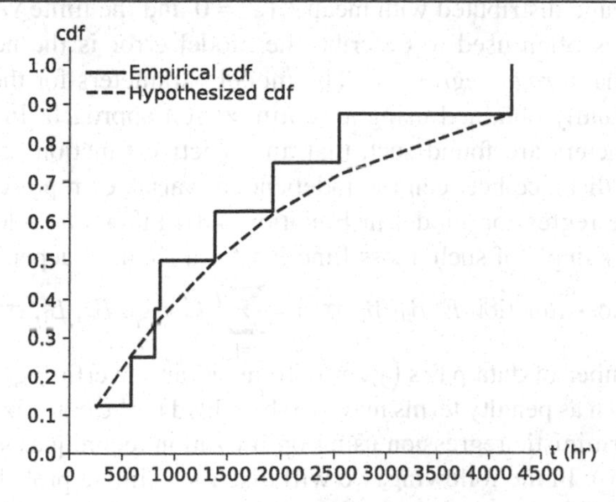

4.7 LINEAR REGRESSION

Reliability engineering and risk assessment problems often require relationships among several r.v.s or between random and nonrandom variables. For example, time to failure of an electrical generator can depend on its age, environmental temperature, applied mechanical stresses and power capacity. Here, we can consider the time to failure as an r.v. Y, which is a function of the r.v.s X_1 (age), X_2 (thermal and mechanical stresses), and X_3 (power capacity).

In regression analysis we refer to Y as the *dependent variable* and to $X_1, \ldots, X_j$ as the j *independent variables, explanatory variables*, or *features the latter is used* (in machine learning). Note that the explanatory variables X_i do not have to be independent of each other. Generally speaking, explanatory factors might be random or nonrandom variables whose values are known or chosen by the experimenter (in the case of the *design of experiments* (DoE)).

The conditional expectation of Y for any $X_1, \ldots, X_j$, $E\left(Y|X_1, \ldots, X_j\right)$ is known as the *regression* of Y on $X_1, \ldots, X_j$. In other words, regression analysis estimates the average value for the dependent variable corresponding to each value of the independent variable.

When the regression of Y is a linear function with respect to the explanatory variables $X_1, \ldots, X_j$ with values $x_1, \ldots, x_j$ it can be written in the form

$$E\left(Y|X_1, X_2, \ldots, X_n\right) = \beta_0 + \beta_1 x_1 + \cdots + \beta_j x_j. \tag{4.30}$$

The coefficients $\boldsymbol{\beta} = \left\{\beta_0, \ldots, \beta_j\right\}$ are called *regression coefficients* or *model parameters*. When the expectation of Y is nonrandom, the relationship Equation 4.30 is deterministic. The corresponding regression model for the r.v. $Y \sim f(y|\boldsymbol{x}, \boldsymbol{\beta}, \sigma^2)$, can be written in probabilistic form:

$$Y = \beta_0 + \beta_1 x_1 + \cdots + \beta_j x_j + \epsilon, \tag{4.31}$$

where ϵ is the *random model error*, assumed to be independent (for all combinations of X considered) and distributed with mean $E(\epsilon) = 0$ and the finite variance σ^2. One distribution that is often used to describe the model error is the normal distribution, known as the *normal regression*. The model parameters for the deterministic regression are usually obtained using an optimization approach. In this approach, the model parameters are found such that an objective function represented by a measure of total distance between the independent variables represented by values $x_1, \ldots, x_j$ and the regression model in Equation 4.31, known as a loss function, is minimized. An example of such a loss function for a single independent variable is the square error loss function $E\left(\beta_0, \beta_1, \sigma^2\right) = \sum_{i=1}^{n}\left(f\left(y|x_i; \beta_0, \beta_1, \sigma^2\right) - y_i\right)^2$, where n is the total number of data pairs $\left(y_i, x_i\right)$. To avoid any overfitting, there are additional terms known as penalty terms may also be added to the optimization. For more readings on deterministic regression using optimization techniques see Mendenhall and Sincich (2020). In the following, we will discuss a simple probabilistic solution to the estimation of the two-parameter linear regression models.

For the simple linear regression model consider the simple deterministic relationship between Y and one explanatory factor X:

$$y = \beta_0 + \beta_1 x. \tag{4.32}$$

Assume we have observed n pairs of data $(x_1, y_1), \ldots, (x_n, y_n)$. Now consider the probabilistic regression version shown by Equation 4.33,

$$Y \sim f(y|x, \beta_0, \beta_1, \sigma^2) = \beta_0 + \beta_1 x + \epsilon, \tag{4.33}$$

where the random error ϵ is normally distributed with mean zero and variance σ^2. Therefore, the r.v. Y will be represented by a normal distribution with mean $\beta_0 + \beta_1 x$ and variance σ^2. Also, suppose that for any values $x_1, \ldots, x_n$, values $y_1, \ldots, y_n$ are independent. For the above n pairs of observations, the joint pdf of $y_1, \ldots, y_n$ is given by

$$f_n(y|x, \beta_0, \beta_1, \sigma^2) = \frac{1}{\sigma^n \sqrt{(2\pi)^n}} e^{\left[-\frac{1}{2\sigma^2}\sum_{i=1}^{n}(y_i - \beta_0 - \beta_1 x_i)^2\right]}. \tag{4.34}$$

Equation 4.34 is the likelihood function (discussed in Section 4.4.3) for the parameters β_0 and β_1. Maximizing this function with respect to parameters β_0 and β_1 reduces the problem to minimizing the sum of squares:

$$S(\beta_0, \beta_1) = \sum_{i=1}^{n}(y_i - \beta_0 - \beta_1 x_i)^2, \tag{4.35}$$

with respect to β_0 and β_1.

Thus, the MLE of the parameters β_0 and β_1 is obtained by minimizing $S(\beta_0, \beta_1)$ by using the derivatives

$$\frac{\partial S(\beta_0, \beta_1)}{\partial \beta_0} = 0, \quad \frac{\partial S(\beta_0, \beta_1)}{\partial \beta_1} = 0. \tag{4.36}$$

The solution of the above equations yields the *least square point estimates* of the parameters β_0 and β_1 (denoted $\hat{\beta}_0$ and $\hat{\beta}_1$) as

$$\hat{\beta}_0 = \bar{y} - \hat{\beta}_1 \bar{x},$$
$$\hat{\beta}_1 = \frac{\sum_{i=1}^{n}(x_i - \bar{x})(y_i - \bar{x})}{\sum_{i=1}^{n}(x_i - \bar{x})^2}. \tag{4.37}$$

Note that the estimates are linear functions of the observations y_i, they are also unbiased and have the minimum variance among all unbiased estimates.

The estimate of the dependent variable variance, $Var(y) = \sigma^2$, can be found as

$$Var(y) = \sigma^2 = \frac{\sum_{i=1}^{n} (y_i - \hat{y}_i)^2}{n-2}, \tag{4.38}$$

where $\hat{y}_i = \hat{\beta}_0 + \hat{\beta}_1 x_i$, is predicted by the regression model values for the dependent variable and (n–2) is the number of degrees of freedom (and 2 is the number of the estimated parameters of the model). The estimate of the variance of Y from Equation 4.38 is called the *residual variance* and it is used as a measure of accuracy of model fit. The square root of *Var(y)* in Equation 4.38 is called the *standard error of the estimate* of y. As noted, the numerator in Equation 4.38 is called the *square error loss function*. For a more detailed discussion on the reliability applications of regression analysis, see Lawless (2002).

It is worth mention that alternatively one may minimize x residuals instead of y using

$$S(\beta_0, \beta_1) = \sum_{i=1}^{n} \left(x_i - \frac{y_i}{\beta_1} - \frac{\beta_0}{\beta_1} \right)^2, \tag{4.39}$$

in which case the corresponding estimates of the regression parameters are

$$\hat{\beta}_1 = \frac{n \sum y_i^2 - \left(\sum y_i^2 \right)^2}{n \sum x_i y_i - \left(\sum x_i \right) \left(\sum y_i \right)}, \tag{4.40}$$

and

$$\hat{\beta}_0 = \frac{\sum y_i}{n} - \hat{\beta}_1 \frac{\sum x_i}{n}. \tag{4.41}$$

Regression analysis is a major topic in machine learning. For more detailed methods and discussions refer to Bishop (2006).

Example 4.8

An electronic device was tested under the elevated temperatures of 50°C, 60°C, and 70°C. The test results as times to failure (in hours) for a sample of 30 tested items are given below.

Time to Failure, t (hr)			$\ln(t)$		
50°C	60°C	70°C	50°C	60°C	70°C
1,950	607	44	7.5756	6.4085	3.7842
3,418	644	53	8.1368	6.4677	3.9703
4,750	675	82	8.4659	6.5147	4.4067
5,090	758	88	8.5350	6.6307	4.4773
7,588	1,047	123	8.9343	6.9537	4.8122
10,890	1,330	189	9.2956	7.1929	5.2417
11,601	1,369	204	9.3588	7.2218	5.3181
15,288	1,884	243	9.6348	7.5412	5.4931
19,024	2,068	317	9.8535	7.6343	5.7589
22,700	2,931	322	10.0301	7.9831	5.7746
		$E[\ln(t)]$	8.9820	7.0549	4.9037

This is an example of accelerated life testing with the logarithm of time to failure, t, follows the normal distribution with the mean given by the Arrhenius model,

$$E(\ln t) = A + \frac{B}{T},$$

where T is the absolute temperature in Kelvin. Find estimates of parameters A and B.

Solution:

The equation above can be transformed to the simple linear regression (Equation 4.32),

$$y = \beta_0 + \beta_1 x,$$

using $y = \ln(t)$, $x = 1/T$, $\beta_0 = A$, and $\beta_1 = B$. Accordingly, from the data in the table.

$y = E[\ln(t)]$	T(°C)	T(K)	$x = \frac{1}{T}$
8.9820	50	323.15	0.0031
7.0549	60	333.15	0.0030
4.9037	70	343.15	0.0029

Using the transformed data above and the least squares estimates from Equation 4.37, we find the estimates of parameters A and B as

$$\hat{A} = -54.19,$$

$$\hat{B} = 20{,}392\ \text{K}.$$

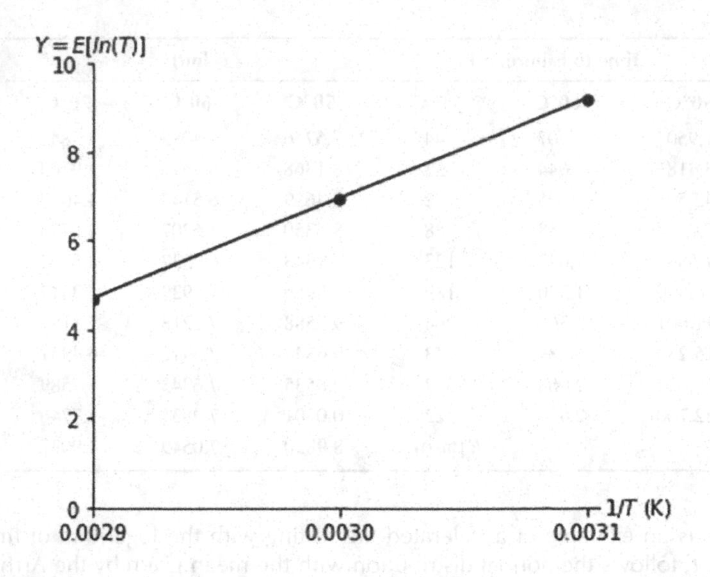

4.8 EXERCISES

4.1 The following data represent times to failure (in hours) for 20 mechanical devices.

a. Find the MTTF and variance of the failure times.

b. Develop an empirical probability distribution and histogram for the data.

4	7	8	12	19	27	50	65	66	69
71	73	75	91	107	115	142	166	184	192

4.2 The following data represent times to failure (in hours) for 50 mechanical devices.

a. Find the MTTF and variance of the failure times.

b. Develop an empirical probability distribution and histogram for the data.

493	510	849	901	955	967	1,035	1,068	1,078	1,135
1,199	1,210	1,214	1,325	1,361	1,457	1,464	1,473	1,539	1,541
1,570	1,584	1,632	1,639	1,689	1,859	1,860	1,872	1,876	1,893
1,931	1,961	1,982	2,020	2,113	2,147	2,153	2,168	2,259	2,267
2,367	2,408	2,449	2,515	2,524	2,529	2,567	2,789	2,832	2,882

4.3 The following data represent times to failure (in hours) for 50 mechanical devices.

a. Find the MTTF and variance of the failure times.

b. Develop an empirical probability distribution and histogram for the data.

Time Interval (hour)	Observed Frequency
0–100	3
100–200	5
200–300	9
300–400	11
400–500	6
500–600	4
600–700	2
Total:	40

4.4 A newspaper article reports that a New York medical team has introduced a new male contraceptive method. The effectiveness of this method was tested using a number of couples over a period of 5 years. The following statistics are obtained:

Year	Times Used	Unwanted Pregnancies (X)
1	8,200	19
2	10,100	18
3	2,120	1
4	6,120	9
5	18,130	30

a. Estimate the mean probability of an unwanted pregnancy per use. What is the standard deviation of the estimate?
b. What are the 95% upper and lower confidence limits of the mean and standard deviation?

4.5 The shaft diameters in a sample of 25 shafts are measured. The sample mean of diameter is 0.102 cm, with a standard deviation of 0.005 cm. What is the upper 95% confidence limit on the mean diameter of all shafts produced by this process, assuming shaft diameters are normally distributed?

4.6 The sample mean life of ten car batteries is of 102.5 months, with the standard deviation of 9.45 months. What are the 80% confidence limits for the mean and standard deviation of a pdf that represents these batteries?

4.7 The breaking strength X of five specimens of a rope of 1/4 inch diameter are 660, 460, 540, 580, and 550 lbs. Assume strength is normally distrib uted. Estimate the following.

a. The 95% confidence level on mean breaking strength.
b. The point estimate of strength at which only 5% of such specimens would be expected to break if μ is assumed to be an unbiased estimate of the true mean, and σ is assumed to be the true standard deviation.
c. The 90% confidence interval of the estimated standard deviation.

4.8 A manufacturer claims that certain machine parts will have a mean diameter of 4 cm, with a standard deviation of 0.01 mm. The diameters of five parts are measured and found to be (in mm): 39.98, 40.01, 39.96, 40.03, and 40.02. Would you accept this claim with a 90% confidence level? (Assume diameter is normally distributed).

4.9 The frequency distribution of time to establish the root causes of a failure by a group of experts is observed and given below. Test whether a normal distribution with known $\sigma = 10$ is an appropriate model for these data.

Time Interval (hour)	Obs. Freq.
45–55	7
55–65	18
65–75	35
75–85	28
85–95	12

4.10 A random number generator yields the following sample of 50 digits. Is there any reason to doubt the digits are uniformly distributed? (Use the chi-squared goodness of fit test.)

Digit	0	1	2	3	4	5	6	7	8	9
Frequency	4	8	8	4	10	3	2	2	4	5

4.11 A set of 40 high-efficiency pumps is tested, all of the pumps fail after 400 pumphours ($T=400$). It is believed that the time to failure of the pumps follows an exponential distribution. Using the following data and a goodness of fit method to determine if the exponential distribution is an appropriate model.

Time Interval (hour)	Obs. Freq.
0–2	6
2–6	12
6–10	7
10–15	6
15–25	7
25–100	2

4.12 Consider the following repair times. Use the chi-squared goodness of fit test to determine the adequacy of a lognormal distribution with known parameters $\mu=2.986$ and $\sigma=1.837$

a. For 5% level of significance.
b. For 1% level of significance.

Time Range (hour)		Frequency
0	4	17
4	24	41
24	72	12
72	300	7
300	5,400	9

4.13 Forty diesel generators were placed on an accelerated life test for 200 days and the following grouped failure times were reported. Perform a χ^2 goodness of fit test at the 5% significance level to determine whether an exponential distribution is a good fit for these data.

Interval (days)	Number of Failures
$0<T<50$	6
$50<T<75$	10
$75<T<100$	8
$100<T<150$	9
$150<T<200$	7

4.14 The following data represent time to failure data for a population of widgets. Perform a χ^2 goodness of fit test at the 10% significance level to assess the adequacy of the uniform distribution with $a=0$ and $b=500$ hours to represent the widget failure times.

Interval (hour)	Observed Failures
0–100	16
100–200	17
200–300	22
300–400	22
400–500	23

4.15 A manufacturing line is observed for a 60 days period. Each day, the number of line stoppages, X, is recorded. At the end of 60 days, the following data are presented to you. Use a chi-squared goodness of fit test to determine whether a Poisson distribution is a good fit to these data. Perform the test at a 5% significance level.

# of Stoppages (x)	Frequency (days)
0	16
1	24
2	15
3	5

4.16 Consider the following time to failure data with the ranked value of t_i. Use the K-S test to test the hypothesis that the data fit a normal distribution.

Event	1	2	3	4	5	6	7	8	9	10
Time to Failure (hour)	10.3	12.4	13.7	13.9	14.1	14.2	14.4	15.0	15.9	16.1

4.17 A manufacturer of light bulbs claims that their bulbs will last 50 weeks on average, with a standard deviation of 5 weeks. A random sample of seven light bulbs lasted 62, 41, 56, 51, 39, 58, and 48 weeks. Use a hypothesis test to evaluate the adequacy of the manufacturer's claim.

4.18 Calculate the regression coefficient and obtain the lines of regression for the following data.

X	1	2	3	4	5	6	7
Y	9	8	10	12	11	13	14

REFERENCES

Bishop, C. M., *Pattern Recognition and Machine Learning*. Vol. 4, New York: Springer, 2006.

Lawless, J. F., *Statistical Models and Methods for Lifetime Data*. 2nd Edition, Wiley, New York, 2002.

Mendenhall, W. and T. T. Sincich, *A Second Course in Statistics: Regression Analysis*. 8th edition, Pearson, Hoboken, NJ, 2020.

Rayner, J. C. W., O. Thas, and D. J. Best, *Smooth Tests of Goodness of Fit*. 2nd edition, John Wiley & Sons (Asia), Singapore, 2009.

5 Reliability Data Analysis and Model Selection

In Chapter 3, we discussed several common distribution models useful for reliability analysis of items. It is necessary at this point to discuss how field and reliability test data can support the selection and estimation of the parameters of a probability distribution model for reliability analysis. In this chapter, we first describe the types of data used in reliability model development, then we discuss the common procedures for selecting and estimating the model parameters using data and information. These procedures can be divided into two groups: nonparametric methods (that do not need a particular distribution function) and parametric methods (that are based on a selected distribution function). We discuss each in more detail. We will also discuss the methods of accounting for some uncertainties associated with model selection and estimated model parameters.

5.1 CONTEXT AND TYPES OF DATA

5.1.1 Types of Field and Test Data

Data and information are critical components of reliability analysis. Typically, reliability data in the form of failure reports, warranty returns, complaints, surveys and tests are gathered and used in reliability model developments. Reliability data are often limited, incomplete, and may be biased and subjective. Regardless, it is essential to gather and use this data despite the limitations, and appropriate mathematical techniques have been designed to enable this. In many cases, it is straightforward to adapt current field data collection programs to include information that is directly applicable to reliability reporting

Most of the reliability data and information come from one or more of the following sources:

- Reliability tests (prototype or production)
- Environmental tests
- Reliability growth tests
- Production returns including failure analysis
- Customer returns
- Surveillance, maintenances, and field service repairs
- Generic databases
- Customer/user surveys.

In life testing for reliability estimation, a sample of components from a sample population of such components is tested replicating the environment in which the

DOI: 10.1201/9781003307495-5

components are expected to work, and their times to failure (or censored times) are recorded. In general, two major types of tests are performed. The first is *testing with replacement* of the failed items, and the second is *testing without replacement* of the failed items. The test with replacement is sometimes called monitored testing.

Reliability data obtained from these sources can be divided into two types: complete data and censored data. In the following each type is discussed further.

5.1.2 Complete Data

When t_i, the exact time or cycle of failure for a specific failure mode is available for all items $(i = 1,\ldots,n)$, the data are complete. For example, consider an accelerated reliability test where the items under test are continuously monitored and when a failure occurs it is immediately recorded. If the test is run until all the items fail, and those times are recorded, this type of data is a complete failure data set.

5.1.3 Censored Data

In some field data gathering and testing, we reach the end of a test or our observation and some items may have not failed, or we detect failure events that are of no interest to us (e.g., the failure is not due to the failure mode of interest) or the exact time of failure is not known. This type of data is called censored data. There are three types of censored data: right censored (also called suspended data), interval censored, and left censored. In the following, we will further elaborate on each type of censoring.

5.1.3.1 Left, Right, and Interval Censoring

Censoring is a form of missing data characterization. It describes cases when complete failures do not occur, or the precise time of failure is not observed because, for example, a reliability test was terminated before some items failed. Censoring also includes cases when the failure event was of no interest (e.g., the failure was due to observing an unrelated failure mode), or the item was removed from the observation before failing. Censoring is very common in reliability data analysis.

Let n be the number of items in a sample and assume that all items of the sample are tested simultaneously. If in the test duration, t_{test}, only r of the n items have failed, the failure times are known, and when the failed items are not replaced, the sample is *singly censored on the right at* t_{test}. In this case, the only information we have about the $n - r$ non-failed (censored) items is that their failure times are greater than t_{test}, if they were allowed to continue working. Formally, an observation is *right censored at* t_{test}, if the exact time of failure is not known, but it is known that it is greater than or equal to t_{test}.

If the failure time for a failed item is not known but is known to be less than a given value, the failure time is called *left censored*. This type of censoring is not very common in reliability data collection, and so it is not discussed in detail, but will be introduced later in this chapter when the topic of likelihood functions of the reliability data including censored data is introduced.

If the only information available is that an item is failed in an interval (e.g., between successive inspections), the respective data are *grouped* or *interval censored*. Interval censoring commonly occurs when periodic data observation and collection (reliability test inspections) are used to assess if a failure event of interest has occurred.

A special kind of censoring, *right-truncated censoring*, refers to cases where failure times are truncated if they occurred after a set time t_U. Similarly, *left-truncated censoring* refers to cases where failure data are truncated for those failures that occurred before a set time t_L. Finally, *interval-truncated censoring* refers to cases where a failure occurred at time t_i should be truncated on both sides if the observations were between t_L and t_U. Again, these conditional censoring methods are not very common in reliability analysis.

It is important to understand the way—or *mechanism* by which—censored data are obtained. The basic discrimination is associated with *random* and *nonrandom* censoring, the simplest cases of which are discussed below.

5.1.3.2 Type I Censoring

Consider the situation of right censoring. If the test is terminated at a given nonrandom time, t_{test}, then r, the number of failures observed during the test period, is an r.v. These censored data are *Type I* or *time right-singly-censored* data, and the corresponding test is sometimes called *time terminated.* For the general case, Type I censoring is considered under the following scheme of observations. In the time-terminated life test, n items are placed on a test and the test is terminated after a predetermined time has elapsed. The number of components that failed during the test time and the corresponding time to failure of each component are recorded.

Let each item in a sample of n items be observed (during reliability testing or in the field) during different periods of time $L_1, \ldots, L_n$. The time to failure of an individual item, t_i, is a distinct value if it is less than the corresponding period, that is, if $t_i < L_i$. Otherwise, t_i is the *time to censoring*, which indicates that the time to failure of the ith item is greater than L_i. This is the case of *Type I multiply-censored* data; the singly-censored case considered above is its particular case, when $L_1 = \cdots = L_n = t_{test}$. Type I multiply-censored data are common in reliability testing. For example, a test may start with a sample size of n, but at some given times $L_1, \ldots, L_k$ (where $k < n$), a prescribed number of items are removed from (or placed on) the test.

Another source of censoring typically found in reliability data analysis when there are several failure modes whose times to failure must be estimated separately. The times to failure due to each failure mode are considered r.v.s having different distributions, while the whole system is called a *competing risks* (or *series*) system.

An example of this type of censoring is collecting failure data during the warranty service of a household product. The manufacturers are interested in reliability estimations and any defects in this product. There could be failures attributed to manufacturing, but failures could also be user-induced or caused by poor maintenance and repair. For estimating manufacturer-related failures, failures due improper use of the product or maintenance-related failures must be treated as time censored. Note that the situation might be opposite. For example, an organization (say, a regulatory agency) might be studying the safety performance of the product, so they would need to estimate the rates of any user-induced failures. In this case, the failures attributed to manufacturing causes should be censored. This differentiation in censored data is important in reliability database developments.

5.1.3.3 Type II Censoring

A test may also be terminated when a nonrandom number of failures (say, r), specified in advance, occurs. Here, the duration of the test is an r.v. This situation is known as *Type II right censoring* and the corresponding test is called a *failure-terminated test*. Under Type II censoring, only the r smallest times to failure $t_1 < \cdots < t_r$ out of the sample of n times to failure are known times to failure. The times to failure t_i $(i = 1, \ldots, r)$ are identically distributed r.v.s (as in the previous case of the Type I censoring).

In the failure-terminated life tests, n items are placed on test and the test is terminated when a predetermined number of component failures have occurred. The time to failure of each failed component, including the time of the last failure, is recorded. Type I and Type II life tests can be performed with replacement or without replacement.

5.2 RELIABILITY DATA SOURCES

Due to the lack of observed data, component reliability analysis may require the use of generic (base) failure data representing an average value over many organizations or industries adjusted for the factors that influence the failure rate for the component under analysis. Generally, these classes of influencing factors are considered.

1. *Environmental factors*: These factors affect the failure rate due to extreme mechanical, electrical, nuclear, and chemical environments. For example, a high-vibration environment would lead to high stresses that promote failure of components.
2. *Design and manufacturing factors*: These factors affect the failure rate due to the quality of material used or workmanship, material composition, functional requirements, geometry, and complexity.
3. *Operating factors*: These factors affect the failure rate due to the applied stresses resulting from operation, testing, repair, maintenance practices, and so on.

To a lesser extent, the *age factor* is used to correct for the infant and wear-out periods, and the *original factor* is used to correct for the accuracy and dependability of the data source (generic data). For example, obtaining data from observed failure records as opposed to expert judgment may affect the failure rate dependability.

The failure rate can be represented as

$$\lambda_a = \lambda_g K_E K_D K_O \ldots, \quad (5.1)$$

where λ_a is the actual (or adjusted) failure rate and λ_g is the generic (base) failure rate, and K_E, K_D, $K_O, \ldots$ are correction (adjusting) factors for the environment, design, and operation, respectively. Other multiplicative factors can be added to account for additional factors such as materials, installation, construction, and manufacturing. It is possible to subdivide each of the correction factors into their contributing subfunctions accordingly. For example, $K_E = f(k_a, k_b, \ldots)$, when k_a and k_b are factors such as vibration level, moisture, and pH level. These factors vary for different types of components.

This concept is used in the procedures specified in government contracts for determining the actual failure rate of electronic components. The procedure is summarized in MIL-HDBK-217. In this procedure, a base failure rate of the component is obtained from a table, and then they are multiplied by the applicable adjusting factors for each type of component. For example, the actual failure rate of a tantalum electrolytic capacitor is given by

$$\lambda_c = \lambda_b \left(\pi_E \cdot \pi_{SR} \cdot \pi_Q \cdot \pi_{CV}\right), \tag{5.2}$$

where λ_c is the adjusted (actual) component failure rate and λ_b is the base (or generic) failure rate, and the π factors are correction factors for the environment (E), series resistance (SR), quality (Q), and capacitance (CV). Values of λ_b and the correction factors are given in MIL-HDBK-217 for many types of electronic components. Generally, λ_b for electronic items is obtained from the Arrhenius empirical model:

$$\lambda_b = Ae^{\frac{E}{kT}}, \tag{5.3}$$

where E = activation energy for the process (treated as a model parameter), T = absolute temperature (K), $k = 1.38 \times 10^{-23}$ JK^{-1} is the Boltzmann constant, and A = a constant. The readers are referred to the software 217Plus (2015) for the most updated generic data on electronics components. Further, readers interested in the extension of the physics-based generic failure data for electronic components should refer to Salemi et al. (2008).

In MIL-HDBK-217 the Arrhenius model forms the basis for a large portion of electronic components subject to temperature degradation. However, care must be applied in using this database, especially because the data in this handbook are derived from repairable systems (and hence, apply to such systems) and assume constant failure rates. Also, application of the various correction factors can drastically affect the actual failure rates. Therefore, proper care must be applied to ensure correct use of the factors and to verify the adequacy of the factors. Also, the appropriateness of the Arrhenius model has been debated many times in the literature.

Many generic sources of data are available for a wide range of items. Among them are IEEE-500, CCPS Guidelines for Process Equipment Data, IEC/TR 62380 Reliability data handbook, and nuclear power plant Probabilistic Risk Assessment data sources such as NUREG/CR-4550 (Ericson et al., 1990) and NUREG/CR-6928 (Eide et al., 2007). For illustration, Appendix B shows a set of data from the 2020 update to NUREG/CR-6928 (Ma et al., 2021).

The following are some key reliability data sources:

- *MIL-HDBK-217F-Reliability Prediction of Electronic Equipment*: http://everyspec.com/MIL-HDBK/MIL-HDBK-0200-0299/MIL-HDBK-217F_14591/
- *EPRD - Electronic Parts Reliability Data (RIAC)*: https://www.quanterion.com/?s=riac
- *NPRD-95 Non-electronic Parts Reliability Data (RIAC)*: https://www.quanterion.com/product/publications/nonelectronic-parts-reliability-data-publication-nprd-2016/

- *FMD-97 Failure Mode/Mechanism Distributions (RIAC)*: https://www.quanterion.com/?s=riac
- *FARADIP (FAilure RAte Data In Perspective) https://www.m2k.com/FARADIP(FAilureRAteDataInPerspective)*
- *SR-332 Reliability Prediction for Electronic Equipment (Telcordia Technologies)*: http://www.ericsson.com/ourportfolio/telcordia_landingpage
- *FIDES (mainly electronic components)*: http://www.fides-reliability.org/
- *IEC/TR 62380 Reliability Data Handbook - Universal model for reliability prediction of electronics components, PCBs and equipment*: http://www.fides-reliability.org/
- EiReDA - European Industry Reliability Data Mainly components in nuclear power plants
- *OREDA - Offshore Reliability Data*: Topside and subsea equipment for offshore oil and gas production: http://www.oreda.com/
- *Handbook of Reliability Prediction Procedures for Mechanical Equipment* - Mechanical equipment - military applications
- T-Book (*Reliability Data of Components in Nordic Nuclear Power Plants*, ISBN 91-631-0426-1)
- *Reliability Data for Control and Safety Systems - PDS Data Handbook Sensors, detectors, valves & control logic*: http://www.sintef.no/Projectweb/PDS-Main-Page/PDS-Handbooks/
- *Safety Equipment Reliability Handbook (exida)*: Safety equipment (sensors, logic items, actuators): http://www.exida.com/
- *WellMaster (ExproSoft)*: Components in oil wells: http://www.exprosoft.com/
- *SubseaMaster (ExproSoft)*: Components in subsea oil/gas production systems: http://www.exprosoft.com/
- *PERD-ProcessEquipmentReliabilityData(AIChE)*:Processequipment:http://www.aiche.org/ccps/resources/process-equipment-reliability-database-perd
- *GIDEP (Government-Industry Data Exchange Program)*: http://www.gidep.org/
- CCPS Guidelines for Process Equipment Reliability Data, AIChE, 1989 Process equipment
- *IEEE Std. 500–1984*: IEEE Guide to the Collection and Presentation of Electrical, Electronic, Sensing Component, and Mechanical Equipment Reliability Data for Nuclear Power Generating Stations

5.3 NONPARAMETRIC AND PLOTTING METHODS FOR RELIABILITY FUNCTIONS

5.3.1 Nonparametric Procedures for Reliability Functions

The nonparametric approach, in principle, attempts to directly estimate the reliability characteristic of an item (e.g., the pdf, reliability, and hazard rates) from a sample. The shape of these functions, however, is often used as an indication of the most appropriate parametric distribution representation. Thus, such procedures can be considered tools for exploratory (preliminary) data analysis. It is important to mention a key assumption that each failure time must be considered an identical sample observation and independent of the other item failure times, thus meeting the identical and independently distributed (i.i.d.) criterion used in statistical analysis. With failure data from a repairable component, these methods can only be used if the item is assumed to be *as good as new* following repair or maintenance. Under the as-good-as-new assumption, each failure time can be considered an identical sample observation independent of the previously observed failure times, meeting the i.i.d. criterion used in statistical analysis. Therefore, n observed times to failure of such a repairable component is equivalent to putting n independent new components under life test

5.3.1.1 Nonparametric Component Reliability Estimation Using Small Samples

Suppose n times to failure makes a small sample (e.g., $n < 25$). Let the data be ordered such that $t_1 \leq \cdots \leq t_n$. Blom (1958) introduced the nonparametric estimators of the hazard rate function, reliability function and pdf as:

$$\hat{h}(t_i) = \frac{1}{(n - i + 0.625)(t_{i+1} - t_i)}, \quad i = 1, \ldots, n-1, \tag{5.4}$$

$$\hat{R}(t_i) = \frac{n - i + 0.625}{n + 0.25}, \quad i = 1, \ldots, n, \tag{5.5}$$

and

$$\hat{f}(t_i) = \frac{1}{(n + 0.25)(t_{i+1} - t_i)}, \quad i = 1, \ldots, n-1. \tag{5.6}$$

Although there are other estimators besides those above, Kimball (1960) concludes that estimators in Equations 5.4–5.6 have good properties and recommends their use. One should keep in mind that 0.625 and 0.25 in Equations 5.4–5.6, which are sometimes called the Kimball plotting positions, are correction terms of a minor importance, which result in a small bias and a small mean square error. Other popular estimators of $R(t_i)$ include the mean rank $(n - i + 1)/(n + 1)$ and the median rank $(n - i + 0.7)/(n + 0.4)$. See Kapur and Lamberson (1977) for more detail

Example 5.1

A high-pressure pump in a process plant has the failure times t_i (in thousands of hours, shown in the following table). Use a nonparametric estimator and the plotting positions in Equations 5.4–5.6 to calculate $\hat{h}(t_i), \hat{R}(t_i), \hat{f}(t_i)$. Plot $\hat{h}(t_i)$, and discuss the results.

Solution:

i	t_i	$t_{i+1}-t_i$	$\hat{h}(t_i)$	$\hat{R}(t_i)$	$\hat{f}(t_i)$
1	0.20	0.60	0.25	0.91	0.23
2	0.80	0.30	0.59	0.78	0.46
3	1.10	0.41	0.53	0.64	0.34
4	1.51	0.32	0.86	0.50	0.43
5	1.83	0.69	0.55	0.36	0.20
6	2.52	0.46	1.34	0.22	0.30
7	2.98	—	—	0.09	—

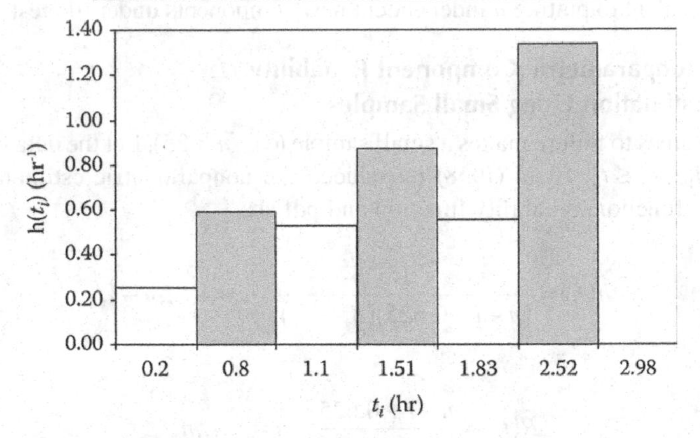

From the above histogram, one can conclude that the hazard rate is somewhat constant over the operating period of the component, with an increase toward the end. However, as a point of caution, although a constant hazard rate might be concluded, several other tests and additional observations may be needed to support this conclusion. Additionally, the histogram is only a representative of the case under study. An extension of the result to future times or other cases (e.g., other high-pressure pumps) may not be accurate.

5.3.1.2 Nonparametric Component Reliability Estimation Using Large Samples

Suppose n times to failure make a large sample. Suppose further that the times to failure in the sample are grouped into several equal increments, Δt. According to the definition of reliability, a nonparametric estimate of the reliability function is

$$\hat{R}(t_i)=\frac{N_s(t_i)}{N}, \tag{5.7}$$

where $N_s(t_i)$ represents the number of surviving components in the interval starting in t_i. This approach is sometimes called the Nelson-Aalen approach. Note that the estimator (Equation 5.7) is compatible with the empirical (sample) cdf (Equation 4.28) introduced in Chapter 4. Time t_i is usually taken to be the lower endpoint of each interval, although this may differ among practitioners. Similarly, the pdf is estimated by

$$\hat{f}(t_i)=\frac{N_f(t_i)}{N\Delta t}, \tag{5.8}$$

where $N_f(t_i)$ is the number of failures observed in the interval $(t_i, t_i + \Delta t)$. Finally, using Equations 5.7 and 5.8, one obtains

$$\hat{h}(t_i) = \frac{N_f(t_i)}{N_s(t_i)\Delta t}. \tag{5.9}$$

For $i = 1$, $N_s(t_i) = N$, and for $i > 1$, $N_s(t_i) = N_s(t_{i-1}) - N_f(t_i)$. Equation 5.9 gives an estimate of average hazard rate during the interval $(t_i, t_i + \Delta t)$. When $N_s(t_i) \to \infty$ and $\Delta t \to 0$, the estimate from Equation 5.9 approaches the true hazard rate $h(t)$.

In Equation 5.9, $\frac{N_f(t_i)}{N_s(t_i)}$ is the estimate of probability that the component will fail in the interval $(t_i, t_i + \Delta t)$, since $N_s(t_i)$ represents the number of components functioning at t_i. Dividing this quantity by Δt, the estimate of hazard rate (probability of failure per surviving item entering an interval Δt) is obtained. It should be noted that the accuracy of this estimate depends on Δt. Therefore, if smaller Δt values are used, we would, theoretically, expect to obtain a better estimation. However, the drawback of using smaller Δt values is the decrease in the amount of data for each interval to estimate $\hat{R}(t_i)$, $\hat{f}(t_i)$ and $\hat{h}(t_i)$. Therefore, selecting Δt requires consideration of both opposing factors.

Example 5.2

Times to failure for an electrical device are obtained during three stages of the component's life. The first stage represents the infant mortality of the component; the second stage represents chance failures; and the third stage represents the wear-out period. Plot the hazard rate for this component using the data provided below.

Solution:

Use the Nelson-Aalen approach to calculate the empirical hazard rate, reliability, and pdf.

Given Data				Calculated Data		
Interval (t_i)		Frequency				
Beginning	End	$N_f(t_i)$	$N_s(t_i)$	$\hat{h}(t_i)$	$\hat{R}(t_i)$	$\hat{f}(t_i)$
		Infant Mortality Stage				
0	20	79	150	0.02633	1.00000	0.02633
20	40	37	71	0.02606	0.47333	0.01233
40	60	15	34	0.02206	0.22667	0.00500
60	80	6	19	0.01579	0.12667	0.00200
80	100	2	13	0.00769	0.08667	0.00067
100	120	1	11	0.00455	0.07333	0.00033
120	>120	10	10	0.05000	0.06667	0.00333
		Total 150				

(Continued)

Given Data				Calculated Data		
Interval (t_i)		Frequency				
Beginning	End	$N_f(t_i)$	$N_s(t_i)$	$\hat{h}(t_i)$	$\hat{R}(t_i)$	$\hat{f}(t_i)$
Chance Failure Stage						
0	2,000	211	500	0.00021	1.00000	0.00021
2,000	4,000	142	289	0.00025	0.57800	0.00014
4,000	6,000	67	147	0.00023	0.29400	0.00007
6,000	8,000	28	80	0.00018	0.16000	0.00003
8,000	10,000	21	52	0.00020	0.10400	0.00002
10,000	>10,000	31	31	0.00050	0.06200	0.00003
		Total 500				
Wear-out Stage						
0	100	34	300	0.00113	1.00000	0.00113
100	200	74	266	0.00278	0.88667	0.00247
200	300	110	192	0.00573	0.64000	0.00367
300	>300	82	82	0.01000	0.27333	0.00273
		Total 300				

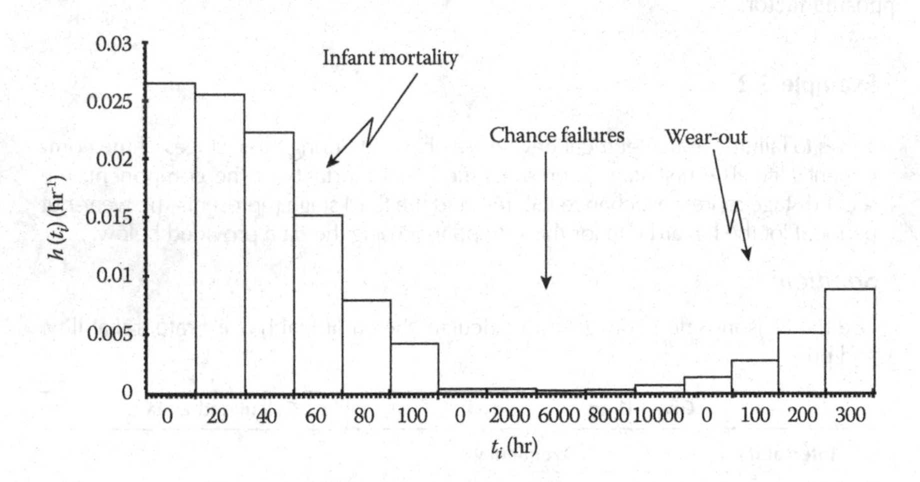

The graph above plots the estimated hazard rate functions for the three observation periods. Note that the three periods, each having a different scale, are combined on the same *x*-axis for demonstration purposes. The chance failure region represents most of the device's life.

5.3.2 Probability Distribution Plotting Using Life Data

Probability plotting is a simple nonparametric graphical method of displaying and analyzing observed data. The data are plotted on properly scaled probability papers or coordinates that transform a life cdf to plot as a straight line. Each type of distribution has its own probability plot scales. If a set of data is hypothesized to originate from a known distribution, the graph can be used to conclude whether the hypothesis might be rejected or not. From the plotted line, one can also roughly estimate the parameters of the hypothesized distribution. Probability plotting is often used in reliability analysis as an approximate first step to test the appropriateness of using known distributions to model a set of observed data. This method is used because it provides simple, visual representation of the data. Because of its approximate nature, the plotting method should be used with care and preferably as an exploratory data analysis procedure.

With modern computing techniques, the probability paper itself has become obsolete. However, the term "probability paper" is still used to refer to the graphical method of parameter estimation. It takes a simple linear regression method to find an algebraic expression for the model.

We will briefly discuss the scaled probability coordinates for the basic distributions considered in this book.

5.3.2.1 Exponential Distribution Probability Plotting

Taking the logarithm of both sides of the expression for the reliability function of the exponential distribution, one obtains

$$\ln R(t) = -\lambda t. \tag{5.10}$$

If $R(t)$ is plotted as a function of time, t, on semi-logarithmic scaled coordinates, according to Equation 5.10, the resulting plot will be a straight line with the slope of $(-\lambda)$ and a y-intercept of 0.

Consider the following n times to failure observed from a life test. Note that the data must be ordered such that: $t_1 \le \cdots \le t_n$. According to Equation 5.10, a nonparametric estimate of the reliability $R(t_i)$ can be made for each t_i. A crude, nonparametric estimate of $R(t_i)$ is $\left(1-\frac{i}{n}\right)$ (recall Equations 5.7 and 4.28). However, as noted in Section 5.3.1.1, the Kimball estimator in Equation 5.5 provides a better estimate for $R(t)$ for the Weibull distribution.

Graphically, on the semi-logarithmic scale the y-axis shows $R(t_i)$ and the x-axis shows t_i. The resulting points should reasonably fall on a straight line if these data can be described by the exponential distribution. Since the slope of $\ln R(t)$ vs. t is negative, it is often more convenient to plot $\ln(1/R(t))$ vs. t in which the slope is positive.

It is also possible to estimate the MTTF from the plotted line. For this purpose, at the level of $R = 0.368$ (or $1/R = e \approx 2.718$), a line parallel to the x-axis is drawn. At the intersection of this line and the fitted line, another line vertical to the x-axis is drawn. The value of t on the x-axis is an estimate of MTTF, and its inverse is $\hat{\lambda}$. Alternatively, a simple linear regression of $\ln(1/R(t))$ vs. t can be performed to obtain the parameter λ and data scatter around the fitted line (i.e., a measure of model error). Since the exponential distribution is a particular case of the Weibull distribution, the Weibull plotting paper may also be used for the exponential distribution.

Example 5.3

Nine times to failure of a diesel generator are recorded as 31.3, 45.9, 78.3, 22.1, 2.3, 4.8, 8.1, 11.3, and 17.3 days. If the diesel generator is restored to as good as new after each failure, determine whether the data represent the exponential distribution. That is, find $\hat{\lambda}$ and $\hat{R}$ (193 hours).

Solution:

First arrange the data in increasing order and then calculate the corresponding $\hat{R}(t_i)$.

i	T	$\hat{R}(t_i) = \frac{n-i+0.625}{n+0.25}$	$\ln\left(\frac{1}{\hat{R}(t_i)}\right)$
1	2.3	0.93	0.07
2	4.8	0.82	0.19
3	8.1	0.72	0.33
4	11.3	0.61	0.50
5	17.3	0.50	0.69
6	22.1	0.39	0.94
7	31.3	0.28	1.26
8	45.9	0.18	1.74
9	78.3	0.07	2.69

The figure below shows a plot of the above data on semi-logarithmic paper.

Using linear regression and setting the y-intercept to 0 produces an estimated parameter of $\hat{\lambda} = 0.0364$ failures/day. Similarly, implementing the least squares regression approach from Chapter 4 produces a parameter of $\hat{\lambda} = 0.0346$ failures/day and a small y-intercept of 0.0848.

Using the estimated parameter to estimate reliability at 52 days, we get

$$\hat{R}(52) = e^{-0.0364\times 52} = 0.15.$$

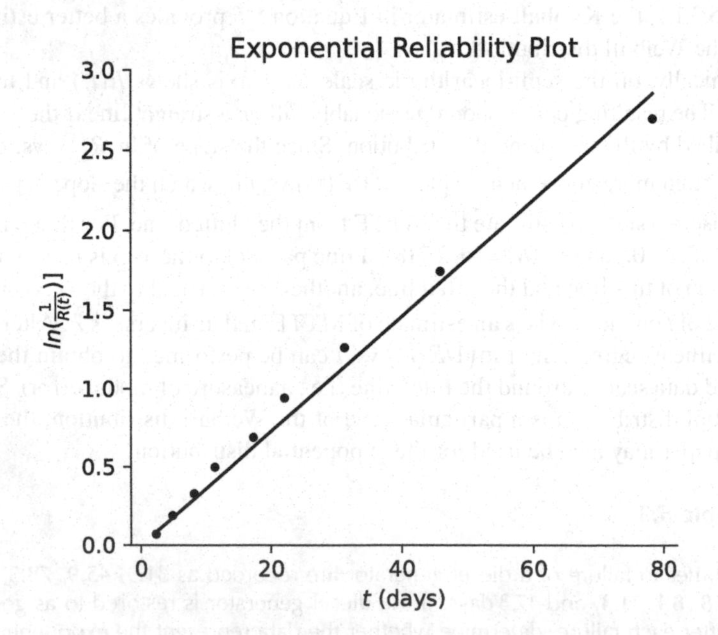

5.3.2.2 Weibull Distribution Probability Plotting

One can transform the Weibull cdf to a linear form by taking the logarithm of both sides twice:

$$\frac{1}{R(t)} = e^{\left(\frac{t}{\alpha}\right)^{\beta}},$$
$$\ln\left[\ln\left(\frac{1}{R(t)}\right)\right] = \beta \ln t - \beta \ln \alpha. \tag{5.11}$$

This linear relationship (in $\ln t$) provides the basis for the Weibull plots or regression analysis. The dependent term $\ln\left[\ln\left(\frac{1}{R(t)}\right)\right]$ plots as a straight line against the independent variable $\ln t$ with slope β and y-intercept $(-\beta \ln \alpha)$. Accordingly, the values of the Weibull parameters α and β can be obtained from the y-intercept and the slope of the graph or by using a regression approach to estimate these parameters.

As was mentioned earlier, several nonparametric estimators of $R(t)$ can be used to determine $\ln\left[\ln\left(\frac{1}{R(t)}\right)\right]$, including the recommended Kimball estimator (Equation 5.5). The corresponding plotting or linear regression procedure is straightforward.

The degree to which the plotted data or the regression analysis follows a straight line determines the conformance of the data to the Weibull distribution. If the data points approximately fall on a straight line, the Weibull distribution is a reasonable fit, and the shape parameter and the scale parameter β can be roughly estimated. Similar to the exponential distribution, it is straightforward to create a plot or perform least squares estimation to obtain α and β by using the linearized form of the Weibull cdf.

Example 5.4

Time to failure of a device is assumed to follow the Weibull distribution. Ten of these devices undergo a reliability test resulting in complete failures. The times to failure (in hours) are 89, 132, 202, 263, 321, 362, 421, 473, 575, and 663. If the Weibull distribution is the proposed model for these data, what are the parameters of this distribution? What is the reliability of the device at 1,000 hours?

Solution:

i	1	2	3	4	5	6	7	8	9	10
t_i	89	132	202	263	321	362	421	473	575	663
$\ln(t_i)$	4.49	4.88	5.31	5.57	5.77	5.89	6.04	6.16	6.35	6.50
$\hat{R}(t_i) = \frac{n-i+0.625}{n+0.25}$	0.939	0.841	0.744	0.646	0.549	0.451	0.354	0.256	0.159	0.061
$\ln\left[\ln\left(\frac{1}{\hat{R}(t_i)}\right)\right]$	−2.77	−1.76	−1.22	−0.83	−0.51	−0.23	0.04	0.31	0.61	1.03

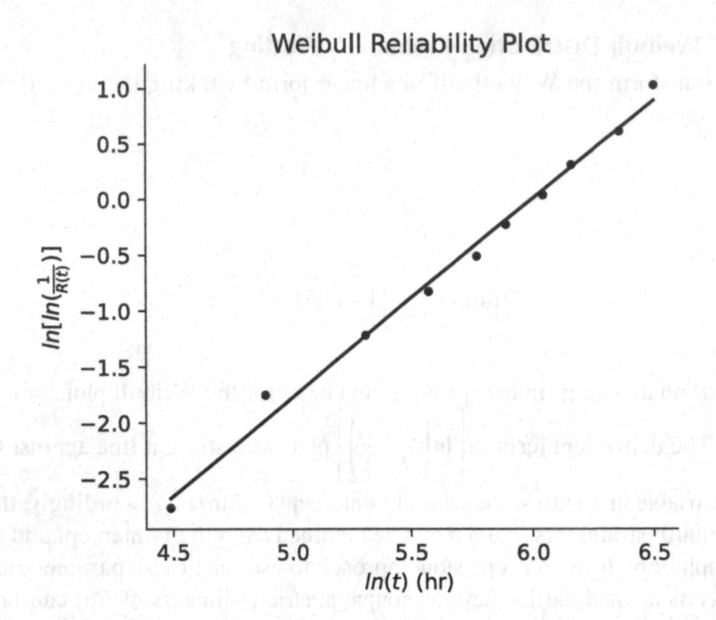

Based on the plot, the fit is reasonably good. The graphical estimate of β (the slope of the line) is approximately 1.78, and the estimate of α is approximately 402 hours. Therefore, at 1,000 hours, the reliability is 0.006 [i.e., $R(t=1.000)=0.006$].

In cases where the data do not fall on a straight line but are concave or convex in shape, it is possible to find a *location parameter* γ (i.e., to try using the three-parameter Weibull distribution (Equation 3.28)) that might straighten out the data fit. For this procedure, see Nelson (2003).

If the failure data are grouped, the class midpoints t_i' (rather than t_i) should be used for plotting, where $t_i'=(t_{i-1}+t_i)/2$. One can also use class endpoints instead of midpoints. Recent studies suggest that the Weibull parameters obtained by using class endpoints in the plot or regression match better with those of the MLE method.

5.3.2.3 Normal and Lognormal Distribution Probability Plotting

As before, properly scaled probability papers can be used for both normal and lognormal plots. In both cases, $F(t_i)$ is estimated by an appropriate nonparametric estimator. For the normal distribution, t_i is plotted on the x-axis, and for the lognormal distribution $\ln(t_i)$ is plotted. It is easy to show that the normal cdf can be linearized using the following transformation:

$$\Phi^{-1}\left[F(t_i)\right]=\frac{1}{\sigma}t_i-\frac{\mu}{\sigma}, \tag{5.12}$$

where $\Phi^{-1}(\cdot)$ is the inverse of the standard normal cdf. In the case of lognormal distribution, t_i in Equation 5.12 is replaced by $\ln(t_i)$. Some lognormal papers are logarithmic on the x-axis, in which case t_i can be directly expressed. If the plotted data fall on a straight line, a normal or lognormal distribution might be the appropriate life model.

Estimating the parameters from the plot follows a similar procedure to the other distributions we have discussed so far. Using linear regression, the parameter σ is the inverse of the line's slope, and μ is the y-intercept times σ. Another way to estimate μ is to use the x-intercept.

Example 5.5

The time it takes for a thermocouple to drift upward or downward to an unacceptable level is measured and recorded (see the following table, with times in months). Determine whether the time to failure (i.e., drifting to an unacceptable level) can be modeled by a normal distribution.

i	t_i	$\hat{F}(t_i)=\frac{i-0.375}{n+0.25}$	$\Phi^{-1}\left(\hat{F}(t_i)\right)$	i	t_i	$\hat{F}(t_i)=\frac{i-0.375}{n+0.25}$	$\Phi^{-1}\left(\hat{F}(t_i)\right)$
1	11.2	0.044	−1.71	8	17.2	0.535	0.09
2	12.8	0.114	−1.21	9	18.1	0.605	0.27
3	14.4	0.184	−0.90	10	18.9	0.675	0.45
4	15.1	0.254	−0.66	11	19.3	0.746	0.66
5	16.2	0.325	−0.45	12	20.0	0.816	0.90
6	16.3	0.395	−0.27	13	21.8	0.886	1.21
7	17.0	0.465	−0.09	14	22.7	0.956	1.71

Solution:

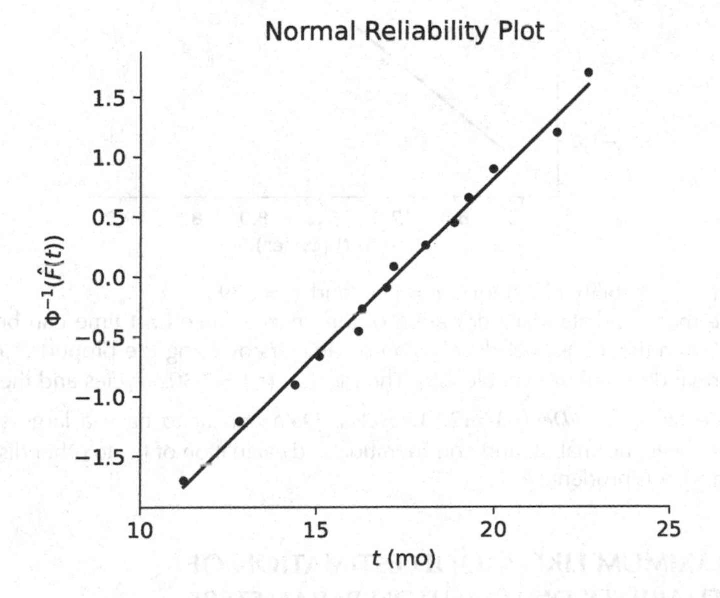

The resulting plot shows that the data reasonably follows the normal distribution, with $\mu = 17.21$ months and $\sigma = 3.42$ months.

Example 5.6

Five components undergo low-cycle fatigue crack tests. The cycles to failure for the five tests are: 363, 1,115, 1,982, 4,241, and 9,738. Determine if the lognormal distribution is a candidate model for this data. If it is, estimate the parameters of the lognormal distribution.

Solution:

i	t_i (cycles)	ln (t_i)	$\hat{F}(t_i)=\frac{i-0.375}{n+0.25}$	$\Phi^{-1}\left(\hat{F}(t_i)\right)$
1	363	5.89	0.119	−1.18
2	1,115	7.02	0.310	−0.50
3	1,982	7.59	0.500	0.00
4	4,241	8.35	0.690	0.50
5	9,738	9.18	0.881	1.18

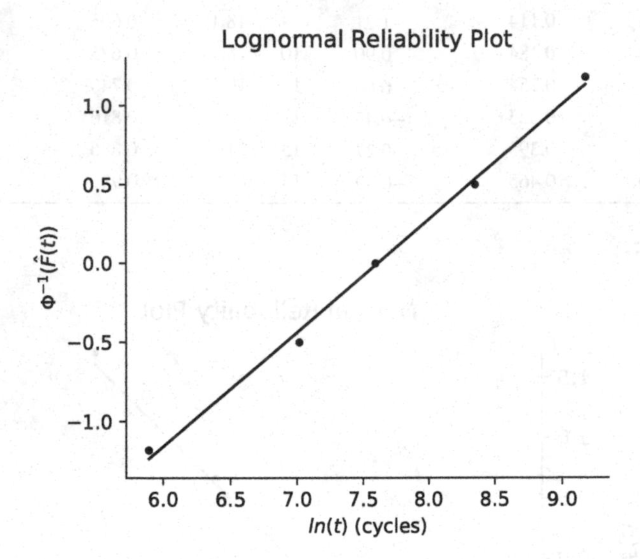

From the probability plot above $\mu = 7.61$ and $\sigma = 1.39$.

The mean and standard deviation of the unacceptable drift time can be estimated from the lognormal distribution parameters by using the properties of the lognormal distribution in Table 2.9). The mean, $E(t_i)$ =5,302 cycles and the standard deviation is $StDev(t_i) = 12{,}913$ cycles. Data appear to have a large scatter around the lognormal fit, and consideration and evaluation of fits to other distribution models is prudent.

5.4 MAXIMUM LIKELIHOOD ESTIMATION OF RELIABILITY DISTRIBUTION PARAMETERS

This section deals with the maximum likelihood estimation (MLE) method for estimating reliability distribution model parameters, such as parameter λ of the exponential distribution, μ and σ of the normal and lognormal distribution, p of the binomial distribution, and α and β of the Weibull distribution. The objective is to find a point estimate and a confidence interval for the parameters of interest. We briefly discussed the MLE concept which relies on the mathematical optimization discussed in Chapter 4 for estimating parameters of probability distribution function. In this approach, an objective function is expressed in terms of a likelihood function. This objective function is a metric expressing the likelihood that the data conform to a hypothesized distribution function. The optimization process finds the parameters of the hypothesized distribution function such that the objective function is maximized. In this section, we expand the application of the MLE to include estimation of parameters of distributions useful to reliability analysis.

It is important to appreciate why we need to consider confidence intervals or credible intervals (in case of Bayesian analysis) in this estimation process. In essence, when we have a limited amount of information (e.g., on times to failure), we cannot state our point estimation of the parameters of interest with certainty. Confidence intervals are the formal classical (frequentist) approach to express the uncertainty in our estimations attributed to the limited data set used. The confidence interval is highly influenced by the amount of data available. Of course, other factors, such as diversity and accuracy of the data sources and adequacy of the selected model, can also influence the state of our uncertainty regarding the estimated parameters. When discussing the goodness of fit tests in Chapter 4, we dealt with uncertainty due to the adequacy of the model by using the concept of levels of significance. However, uncertainty due to the diversity and accuracy of the data sources is a more difficult issue to deal with and motivates the use Bayesian approaches as well as appropriate design of data collection and post-processing practices.

The method of MLE-based parameter estimation discussed in this section is a more formal and accurate method for determining distribution parameters than the probability plotting and regression methods described previously.

5.4.1 Elements of MLE Using Reliability Data

Recall that the *likelihood functions* are objective functions indexes representing the likelihood of observing the reliability data, given the assumed (hypothesized) distribution with pdf $f(t;\theta)$ with parameter(s), θ. A useful feature of the likelihood function is that it can be used to determine if one distribution is a better fit for a particular set compared to another distribution by solving the likelihood functions for the distributions being compared. The distribution with the higher L value is the better fit. However, this approach may pose problems, because it will be biased toward pdf models with many parameters (due to overfit). The likelihood function is not dependent on the order of each event. Recall Equation 4.5 that defines a likelihood function for a sample of random data d_i as $L(\theta|\text{data}) \propto \prod_{\text{all } i} Pr(d_i|\theta)$. For complete data, the likelihood function using Equation 4.6 would be $L(\theta|x_1,\ldots,x_n) \propto \prod_{i=1}^{n} f(x_i|\theta) = f(x_1|\theta)f(x_2|\theta)\ldots f(x_n|\theta)$.

For censored data, consideration of Equation 4.5 makes it clear that for a left-censored data point, where failure occurred sometime before t_i, the probability would be $Pr(T \le t_i) = F(t_i)$. Similarly, for a right-censored data point for an item which

TABLE 5.1
Likelihood Functions for Different Types of Reliability Data

Type of Observation	Likelihood Function	Example Description
Exact lifetimes	$L_i(\theta\|t_i) = f(t_i\|\theta)$	Failure time is known.
Left censored	$L_i(\theta\|t_i) = F(t_i\|\theta)$	Component failed before time t_i.
Right censored	$L_i(\theta\|t_i) = 1 - F(t_i\|\theta) = R(t_i\|\theta)$	Component survived to time t_i.
Interval censored	$L_i(\theta\|t_i) = F(t_i^{RI}\|\theta) - F(t_i^{LI}\|\theta)$	Component failed between t_i^{LI} and t_i^{RI}.
Left truncated	$L_i(\theta\|t_i) = \dfrac{f(t_i\|\theta)}{R(t_L\|\theta)}$	Component failed at time t_i where observations are truncated before t_L.
Right truncated	$L_i(\theta\|t_i) = \dfrac{f(t_i\|\theta)}{F(t_U\|\theta)}$	Component failed at time t_i where observations are truncated after t_U.
Interval truncated	$L_i(\theta\|t_i) = \dfrac{f(t_i\|\theta)}{F(t_U\|\theta) - F(t_L\|\theta)}$	Component failed at time t_i where observations are truncated before t_L and after t_U.

has survived past time t_i, $Pr(t_i < T) = 1 - F(t_i|\theta)$. For other types of censoring, the contribution to the likelihood function also differs. Table 5.1 gives the constituent parts of the likelihood functions for different types of complete and censored data.

Using Table 5.1, we can write the total likelihood function representing multiple complete failure times and right-censored times as

$$L(\theta|D) = c\prod_i \left\{\left[f(t_i|\theta)\right]^{\delta_i} \times \left[1 - F(t_i|\theta)\right]^{1-\delta_i}\right\}, \tag{5.13}$$

where D stands for all data in the sample, c is a combinatorial constant which quantifies the number of combinations which the observed data could have occurred, and $\delta_i = 1$ for complete failure times and $\delta_i = 0$ for right-censored times. In MLE process to estimate parameters θ that maximizes Equation 5.13, the constant c will be omitted in the optimization. As we discussed in Chapter 4, a simple way to find the values of θ is to take the partial derivatives of $L(\theta|D)$ with respect to each parameter θ and equate it to zero. As such, we will find the same number of equations as unknowns that can produce the point estimates of parameters. Recall that for mathematical simplicity, a common practice is to maximize the log of the objective function under any constraints associated with the parameter(s) θ (e.g., $\theta > 0$). Finding parameters that maximize the logarithmic form of the likelihood function, shown in Equation 5.14 simplifies the computation and maintains the optimal results.

$$\hat{\theta} = \arg\max_\theta \ln[L(\theta|D)] = \arg\max_\theta \left\{\sum_i \left\{\delta_i \ln\left[f(t_i|\theta)\right] + (1-\delta_i)\ln\left[1 - F(t_i|\theta)\right]\right\}\right\}. \tag{5.14}$$

A convenient approach to find the confidence intervals of θ in Equation 5.13 is to use the *Fisher information matrix* approach. The approach has many uses besides finding the intervals of the parameters of reliability pdf, including application to the Jeffery's non-informative priors used as a non-informative prior distribution in the Bayesian parameters estimation.

The *Fisher information matrix* is obtained from using the log-likelihood function $\Lambda(\theta|D) = \ln\left[L(\theta|D)\right]$ and replacing the parameters by their point estimates $\hat{\theta}$.

For a single-parameter distribution, the Fisher information of the parameter θ is obtained from:

$$I(\hat{\theta}) = -\left[\frac{\partial^2 \Lambda(\theta|D)}{\partial \theta^2}\right]_{\theta=\hat{\theta}}. \tag{5.15}$$

The *observed Fisher information matrix* is the negative of the second derivative of the of the log-likelihood function (this second derivative is also called the Hessian matrix). So, for a pdf model with a vector of p parameters, $\boldsymbol{\theta}$, using the log-likelihood $\Lambda(\theta|D)$ for all data D discussed earlier, the observed Fisher information is expressed by a $p \times p$ symmetric matrix in the form of

$$I(\hat{\theta}) = \begin{bmatrix} -\frac{\partial^2 \Lambda(\theta|D)}{\partial \theta_1^2} & -\frac{\partial^2 \Lambda(\theta|D)}{\partial \theta_1 \partial \theta_2} & \cdots & -\frac{\partial^2 \Lambda(\theta|D)}{\partial \theta_1 \partial \theta_p} \\ -\frac{\partial^2 \Lambda(\theta|D)}{\partial \theta_2 \partial \theta_1} & -\frac{\partial^2 \Lambda(\theta|D)}{\partial \theta_2^2} & \cdots & -\frac{\partial^2 \Lambda(\theta|D)}{\partial \theta_2 \partial \theta_p} \\ \vdots & \vdots & \ddots & \vdots \\ -\frac{\partial^2 \Lambda(\theta|D)}{\partial \theta_p \partial \theta_1} & -\frac{\partial^2 \Lambda(\theta|D)}{\partial \theta_p \partial \theta_2} & \cdots & -\frac{\partial^2 \Lambda(\theta|D)}{\partial \theta_p^2} \end{bmatrix}_{\theta_i=\hat{\theta}_i} \tag{5.16}$$

When the inverse of the $I(\boldsymbol{\theta})$ matrix is evaluated at $\boldsymbol{\theta} = \hat{\boldsymbol{\theta}}$, we find the observed Fisher information matrix in Equation 5.16 produces the *variance-covariance matrix* in Equation 5.17:

$$Var(\hat{\theta}) = \left[I(\hat{\theta})\right]^{-1} = \begin{bmatrix} Var(\theta_1) & Cov(\theta_1,\theta_2) & \cdots & Cov(\theta_1,\theta_p) \\ Cov(\theta_2,\theta_1) & Var(\theta_2) & \cdots & Cov(\theta_2,\theta_p) \\ \vdots & \vdots & \ddots & \vdots \\ Cov(\theta_p,\theta_1) & Cov(\theta_p,\theta_2) & \cdots & Var(\theta_p) \end{bmatrix}. \tag{5.17}$$

The variance-covariance matrix provides information about the dependencies between parameters (in terms of their covariances), and its diagonal also shows the variance of each parameter. If we use this variance-covariance matrix with a large number of samples, the asymptotic normal property can be used to estimate confidence intervals. By connecting this to percentiles of the normal distribution, $100\,\gamma\%$ approximate confidence intervals are generated. If the range of a single parameter θ is unbounded $(-\infty, \infty)$ the approximate two-sided lower confidence limit θ_l and upper confidence limit θ_u are given by

$$\theta_l \approx \hat{\theta} - \Phi^{-1}\left(\frac{1+\gamma}{2}\right)\sqrt{Var\left(\hat{\theta}\right)}, \tag{5.18}$$

and

$$\theta_u \approx \hat{\theta} + \Phi^{-1}\left(\frac{1+\gamma}{2}\right)\sqrt{Var\left(\hat{\theta}\right)}. \tag{5.19}$$

If the range of θ is $(0, \infty)$ the corresponding approximate two-sided confidence limits are

$$\theta_l \approx \hat{\theta} \cdot e^{-\left[\frac{\Phi^{-1}\left(\frac{1+\gamma}{2}\right)\sqrt{Var\left(\hat{\theta}\right)}}{\hat{\theta}}\right]}, \tag{5.20}$$

and

$$\theta_u \approx \hat{\theta} \cdot e^{\left[\frac{\Phi^{-1}\left(\frac{1+\gamma}{2}\right)\sqrt{Var\left(\hat{\theta}\right)}}{\hat{\theta}}\right]}. \tag{5.21}$$

If the range of θ is $(0,1)$ the similar approximate two-sided confidence limits are

$$\theta_l \approx \hat{\theta} \cdot \left\{\hat{\theta} + \left(1-\hat{\theta}\right) e^{\left[\frac{\Phi^{-1}\left(\frac{1+\gamma}{2}\right)\sqrt{Var\left(\hat{\theta}\right)}}{\hat{\theta}\left(1-\hat{\theta}\right)}\right]}\right\}^{-1}, \tag{5.22}$$

and

$$\theta_u \approx \hat{\theta} \cdot \left\{\hat{\theta} + \left(1-\hat{\theta}\right) e^{-\left[\frac{\Phi^{-1}\left(\frac{1+\gamma}{2}\right)\sqrt{Var\left(\hat{\theta}\right)}}{\hat{\theta}\left(1-\hat{\theta}\right)}\right]}\right\}^{-1}. \tag{5.23}$$

The advantage of this approximate method is that it can be calculated for all distributions and is easy to determine. Thc disadvantage is that the assumption of a normal distribution is asymptotic and so sufficient data is required for the confidence interval estimate to be accurate. The number of samples needed for an accurate estimation changes from distribution to distribution. It also produces symmetrical confidence intervals which may be approximate.

As discussed in Chapter 4, understanding how to interpret the confidence intervals found from the classical statistical methods described in this section is critical. For a sample of data, it is possible to say that the probability that the associated confidence interval contains the true value of a parameter of interest is $(1-\gamma)$. Further, one can also interpret the interval such that if we obtained many similar samples to the one used to estimate the confidence interval, $100\,(1-\gamma)\%$ of the associated confidence intervals will contain the true value of the parameter. Therefore, the confidence level in the classical statistical estimation refers to the confidence over the interval not the parameter of interest.

5.4.2 Exponential Distribution MLE Point Estimation

The MLE point estimator for the exponential distribution parameter is

$$\hat{\lambda} = \frac{r}{TTT}, \tag{5.24}$$

where r is the number of failures observed and TTT is the *total time on test.* Correspondingly, the MLE for the MTTF is given by

$$\widehat{MTTF} = \frac{TTT}{r}. \tag{5.25}$$

For complete data, $TTT = \sum t_i$. For right-censored data, the total time on test differs depending on test design.

5.4.2.1 Type I Life Test with Replacement

Suppose n components are placed on reliability test with replacement (i.e., monitored and replaced when failed), and the test or field data collection is terminated after a specified time t_{end}. The total accumulated test time from both failed and right-censored components, TTT, (in hours or cycles), is given by

$$TTT = n\,t_{\text{end}}. \tag{5.26}$$

Equation 5.26 shows that at each time instant from the beginning of the test up to time t_{end}, exactly n components have been on test (i.e., under observation). Accordingly, if r failures have been observed up to time t_{end} for a component with times to failure following the exponential distribution, the maximum likelihood point estimate of the failure rate of the component can be found by using Equation 5.25.

The number of items tested during the test, n', is

$$n' = n + r. \tag{5.27}$$

5.4.2.2 Type I Life Test without Replacement

Suppose n components are placed on test without replacement, and the test (or field observation) is terminated after a specified time t_{end} during which r failures have occurred. The total time on test, for the failed and survived components is

$$TTT = \sum_{i=1}^{r} t_i + (n-r)t_{\text{end}}, \tag{5.28}$$

where $\sum_{i=1}^{r} t_i$ represents the accumulated time on test of the r failed components (note that r is a random variable here), and $(n-r)\, t_{\text{end}}$ is the accumulated time on test of the surviving (right-censored) components at the end of the test.

Since no replacement has taken place, the total number of components tested in the test is $n' = n$. Note that if items are right censored before the end of the test, their contributions to the total time on test are not reflected in Equation 5.28. To account for those events, the sum of the right-censored times for such items should be added to Equation 5.28.

5.4.2.3 Type II Life Test with Replacement

Consider a situation in which n components are being tested (or field observed) with replacement, and a component is replaced with an identical component as soon as it fails (except for the last failure). If the test is terminated after a time t_r when the rth failure has occurred (i.e., r is specified but t_r is random), then the total time on test associated with both the failed and right-censored components is given by

$$TTT = nt_r. \tag{5.29}$$

Note that t_r, unlike t_{end}, is an r.v.

The total number of items tested, n', is

$$n' = n + r - 1, \tag{5.30}$$

where $(r-1)$ is the total number of failed and replaced components. All failed components are replaced except the last one, because the test is terminated when the last component fails (i.e., the rth failure has been observed).

5.4.2.4 Type II Life Test without Replacement

Consider another situation when n components are being tested without replacement, that is, when a failure occurs, the failed component is not replaced by a new one. The test is terminated at time t_r when the rth failure has occurred (i.e., r is specified but t_r is random). The total time on test of both failed and right-censored components at the end of the test is obtained from

$$TTT = \sum_{i=1}^{r} t_i + (n-r)t_r, \tag{5.31}$$

where $\sum_{i=1}^{r} t_i$ is the accumulated time contribution from the failed components and $(n-r)t_r$ is the accumulated time contribution from the surviving components.

It should also be noted that the total number of items tested is

$$n' = n, \tag{5.32}$$

since no components are being replaced. Note that if items are right censored before the end of the test, their contributions to the total time on test are not reflected in Equation 5.31. To account for those events, the sum of the right-censored times for such items should be added to Equation 5.31.

Example 5.7

Ten light bulbs are placed under life test. The test is terminated at $t_{end} = 850$ hours. Eight components fail before 850 hours have elapsed. The failure times obtained are 183, 318, 412, 432, 553, 680, 689, and 748. Determine the total accumulated component hours and estimate of the failure rate and MTTF for these situations:

a. The components are replaced when they fail.
b. The components are not replaced when they fail.
c. Repeat a, assuming the test is terminated when the eighth component fails.
d. Repeat b, assuming the test is terminated when the eighth component fails.

Solution:

a. For a Type I test with replacement,
Using Equation 5.26, TTT = $10 \cdot 850$ = 8,500 component hours.

$$\hat{\lambda} = \frac{8}{8{,}500} = 9.4 \times 10^{-4}/\text{hour and } \widehat{\text{MTTF}} = \frac{1}{\hat{\lambda}} = 1{,}062.5 \text{ hours.}$$

b. For a Type I test without replacement

$$\sum_{i=1}^{r} t_i = 4{,}015 \text{ and } (n-r)t_{\text{end}} = (10-8)850 = 1{,}700.$$

Thus, $TTT = 4{,}015 + 1{,}700 = 5{,}715$ component hours.

$$\hat{\lambda} = \frac{8}{5715} = 1.4 \times 10^{-3} \text{ / hour and } \widehat{MTTF} = 714.4 \text{ hours.}$$

c. For a Type II test with replacement

Here, t_r is the time to the eighth failure, which is 748.

TTT = 10(748) = 7,480 component hours.

$$\hat{\lambda} = \frac{8}{7{,}480} = 1.1 \times 10^{-3} \text{ / hour and } \widehat{MTTF} = 935 \text{ hours.}$$

d. For a Type II test without replacement

$$\sum_{i=1}^{r} T_i = 4{,}015 \text{ and } (n-r)t_{end} = (10-8)748 = 1{,}496.$$

Thus, $TTT = 4{,}015 + 1{,}496 = 5{,}511$ component hours.

$$\hat{\lambda} = \frac{8}{5{,}511} = 1.5 \times 10^{-3} \text{ / hour and } \widehat{MTTF} = 688.8 \text{ hours.}$$

A simple comparison of the results shows that although the same set of data is used, the effect of the type of the test and of the replacement of the failed items could be significant.

5.4.3 Exponential Distribution Interval Estimation

In Section 5.4.2, we discussed the MLE approach to find the point estimate of parameter λ (i.e., failure rate) of the exponential distribution. This point estimator is $\hat{\lambda} = r/T$, where r is the number of failures observed and T is the total observed time (often, the total time on test, TTT). Epstein (1960) has shown that if the time to failure is exponentially distributed with parameter λ, the quantity $\frac{2r\lambda}{\hat{\lambda}} = 2\lambda \cdot TTT$ has the χ^2 distribution with $2r$ degrees of freedom for Type II censored data (failure-terminated test). Based on this information, one can construct the corresponding confidence intervals. Because uncensored data can be considered as a particular case of Type II right-censored data (when $r = n$), the same procedure applies to the complete (uncensored) sample.

Using the distribution of $2r\lambda/\hat{\lambda}$ at the $(1-\gamma)$ confidence level, one can write

$$Pr\left[\chi^2_{(\gamma/2)}[2r] \le \frac{2r\lambda}{\hat{\lambda}} \le \chi^2_{\left(1-\frac{\gamma}{2}\right)}[2r]\right] = 1-\gamma. \tag{5.33}$$

In Equation 5.33, the χ^2 is evaluated at the $2r$ degrees of freedom for a significance level of $\gamma/2$ for the lower limit, and $1-\gamma/2$ significance level for the upper limit. The values of χ^2 are available from Table A.3. By rearranging Equation 5.33 and using $\hat{\lambda} = r / TTT$, the two-sided confidence interval for the true value of λ can be obtained:

$$Pr\left[\frac{\chi^2_{\left(\frac{\gamma}{2}\right)}[2r]}{2\text{TTT}} \le \lambda \le \frac{\chi^2_{\left(1-\frac{\gamma}{2}\right)}[2r]}{2\text{TTT}}\right] = 1-\gamma. \tag{5.34}$$

The corresponding upper confidence limit (the one-sided confidence interval) is

$$Pr\left[0 \le \lambda \le \frac{\chi^2_{(1-\gamma)}[2r]}{2\text{TTT}}\right] = 1-\gamma. \tag{5.35}$$

Accordingly, confidence intervals for MTTF and $R(t)$ at a time $t = t_{\text{end}}$ can also be obtained as one-sided and two-sided confidence intervals from Equations 5.34 and 5.35. The results are summarized in Table 5.2.

TABLE 5.2
100 (1−γ)% Confidence Limits on λ, MTTF, and R(t)

	One-Sided Confidence Limits		Two-Sided Confidence Limits	
Parameter	**Lower Limit**	**Upper Limit**	**Lower Limit**	**Upper Limit**
	Type I (Time Terminated Test for Complete Data)			
λ	0	$\frac{\chi^2_{(1-\gamma)}[2r+2]}{2\text{TTT}}$	$\frac{\chi^2_{\left(\frac{\gamma}{2}\right)}[2r]}{2\text{TTT}}$	$\frac{\chi^2_{\left(1-\frac{\gamma}{2}\right)}[2r+2]}{2\text{TTT}}$
MTTF	$\frac{2\text{TTT}}{\chi^2_{(1-\gamma)}[2\text{r}+2]}$	∞	$\frac{2\text{TTT}}{\chi^2_{\left(1-\frac{\gamma}{2}\right)}[2\text{r}+2]}$	$\frac{2\text{TTT}}{\chi^2_{\left(\frac{\gamma}{2}\right)}[2\text{r}]}$
$R(t)$	$e^{-\left[\frac{\chi^2_{(1-\gamma)}[2r+2]}{2\text{TTT}}t_{\text{end}}\right]}$	1	$e^{-\left[\frac{\chi^2_{\left(1-\frac{\gamma}{2}\right)}[2r+2]}{2\text{TTT}}t_{\text{end}}\right]}$	$e^{-\left[\frac{\chi^2_{\left(\frac{\gamma}{2}\right)}[2r]}{2\text{TTT}}t_{\text{end}}\right]}$
	Type II (Failure Terminated Test for Complete Data)			
λ	0	$\frac{\chi^2_{(1-\gamma)}[2r]}{2\text{TTT}}$	$\frac{\chi^2_{\left(\frac{\gamma}{2}\right)}[2r]}{2\text{TTT}}$	$\frac{\chi^2_{\left(1-\frac{\gamma}{2}\right)}[2r]}{2\text{TTT}}$
MTTF	$\frac{2\text{TTT}}{\chi^2_{(1-\gamma)}[2r]}$	∞	$\frac{2\text{TTT}}{\chi^2_{\left(1-\frac{\gamma}{2}\right)}[2r]}$	$\frac{2\text{TTT}}{\chi^2_{\left(\frac{\gamma}{2}\right)}[2r]}$
$R(t)$	$e^{-\left[\frac{\chi^2_{(1-\gamma)}[2r]}{2\text{TTT}}t_{\text{end}}\right]}$	1	$e^{-\left[\frac{\chi^2_{\left(1-\frac{\alpha}{2}\right)}[2r]}{2\text{TTT}}t_{\text{end}}\right]}$	$e^{-\left[\frac{\chi^2_{\left(\frac{\gamma}{2}\right)}[2r]}{2\text{TTT}}t_{\text{end}}\right]}$

As opposed to Type II censored data, the corresponding exact confidence limits for Type I censored data are not available. However, the approximate two-sided confidence interval for failure rate λ for Type I (time-terminated test) data usually is constructed as

$$Pr\left[\frac{\chi^2_{(\gamma/2)}[2r]}{2\text{TTT}} \le \lambda \le \frac{\chi^2_{\left(1-\frac{\gamma}{2}\right)}[2r+2]}{2\text{TTT}}\right] = 1-\gamma. \tag{5.36}$$

The χ^2 in Equation 5.36 is evaluated at the $2r$ degrees of freedom for a significance level of $\gamma/2$ for the lower limit, and at the $(2r+2)$ degrees of freedom and $(1-\gamma/2)$ significance level for the upper limit.

The respective upper confidence limit (a one-sided confidence interval) is given by

$$Pr\left[0 \le \lambda \le \frac{\chi^2_{(1-\gamma)}[2r+2]}{2\text{TTT}}\right] = 1-\gamma. \tag{5.37}$$

If no failure is observed during a test, the formal estimation gives $\hat{\lambda}=0$, or $\text{MTTF}=\infty$. This cannot realistically be true, since we may have had a small or limited observations with no failures, or a failure may be about to occur before the test ended. Had the test been continued, eventually a failure would have been observed. Therefore, an upper confidence estimate for λ can be obtained for $r=0$. However, the lower confidence limit cannot be obtained with $r=0$. It is possible to relax this limitation by conservatively assuming that a single failure occurs exactly at the end of the observation period. Then $r=1$ can be used to evaluate the lower limit for the two-sided confidence interval. This conservative modification, although sometimes used to allow a complete statistical analysis, lacks a firm statistical basis. This limitation is relaxed in the Bayesian estimations of these parameters, as will be discussed later in this chapter.

Example 5.8

Twenty-five items are placed on a reliability test that lasts 500 hours. In this test, eight failures occur at 75, 115, 192, 258, 312, 389, 410, and 496 hours. The failed items are replaced. Assuming a constant failure rate, find $\hat{\lambda}$, the one-sided and two-sided confidence intervals for λ, and MTTF at the 90% confidence level; and one-sided and two-sided 90% confidence intervals for reliability at $t=1{,}000$ hours.

Solution:

This is a Type I test. The accumulated time TTT is obtained from Equation 5.26

$$\text{TTT} = 25 \cdot 500 = 12{,}500 \text{ hours}.$$

The point estimate of failure rate is

$$\hat{\lambda} = 8/12{,}500 = 6.4\times10^{-4}\,/\text{hour}.$$

One-sided confidence interval for λ is

$$0 \le \lambda \le \frac{\chi^2[2\cdot8+2]}{2\cdot12{,}500}.$$

Finding the inverse of the χ^2 distribution yields:

$$\chi^2_{0.9}(18) = 25.99,\; 0 \le \lambda \le 1.04\times10^{-3}\,/\text{ hour}.$$

Two-sided confidence interval for λ is

$$\frac{\chi^2_{0.05}[2\cdot8]}{2\cdot12{,}500} \le \lambda \le \frac{\chi^2_{0.95}[2\cdot8+2]}{2\cdot12{,}500}.$$

Finding the inverse of the χ^2 distribution yields:

$$\chi^2_{0.05}(16) = 7.96 \quad \text{and} \quad \chi^2_{0.95}(18) = 28.87.$$

Thus,

$$3.18\times10^{-4}\,/\text{hour} \le \lambda \le 1.15\times10^{-3}\,/\text{hour}.$$

One-sided 90% confidence interval for $R(1{,}000)$ is

$$e^{-(1.04\times10^{-3})(1{,}000)} \le R(1{,}000) \le 1,$$

or

$$0.35 \le R(1{,}000) \le 1.$$

Two-sided 90% confidence interval for $R(t)$ is

$$e^{-(1.15\times10^{-3})(1{,}000)} \le R(1{,}000) \le e^{-(3.18\times10^{-3})(1{,}000)},$$

or

$$0.32 \le R(1{,}000) \le 0.73.$$

5.4.4 Normal Distribution

The normal distribution was introduced in Chapter 4 in the context of complete data. The key equations are presented here again for completeness of the chapter.

For the normal distribution, the confidence interval for μ with σ known is:

$$Pr\left(-z_{\left(1-\frac{\gamma}{2}\right)} \le \frac{\bar{x}-\mu}{\frac{\sigma}{\sqrt{n}}} \le z_{\left(1-\frac{\gamma}{2}\right)}\right) = 1-\gamma, \tag{5.38}$$

where $z_{\left(1-\frac{\gamma}{2}\right)}$ is the $100(1-\gamma/2)\%$ of the standard normal distribution (which can be obtained from Table A.1) and $\bar{x}$ is the sample mean.

In the case when σ is unknown and is estimated as s, e.g., using Equation 4.3, the respective confidence interval on μ is given by:

$$Pr\left(\hat{\mu} - t_{\left(\frac{\gamma}{2}\right)}[n-1]\frac{s}{\sqrt{n}} < \mu < \hat{\mu} + t_{\left(\frac{\gamma}{2}\right)}[n-1]\frac{s}{\sqrt{n}}\right) = 1-\gamma, \tag{5.39}$$

where $t_{(\gamma/2)}$ is the percent of the one-sided student's t-distribution with $(n-1)$ degrees of freedom. Values of t_γ for different numbers of degrees of freedom are given in Table A.2.

Similarly, confidence intervals for σ^2 for a normal distribution can be obtained as

$$\frac{(n-1)s^2}{\chi^2_{\left(1-\frac{\gamma}{2}\right)}[n-1]} < \sigma^2 < \frac{(n-1)s^2}{\chi^2_{\left(\frac{\gamma}{2}\right)}[n-1]}, \tag{5.40}$$

where $\chi^2_{\left(1-\frac{\gamma}{2}\right)}[n-1]$ is the $100(1-\gamma/2)\%$ of the χ^2 distribution with $[n-1]$ degrees of freedom.

For a combination of complete data and right-censored data, the MLE parameters must be obtained numerically in a similar method to what is used for the lognormal distribution, which is explained in the next section.

5.4.5 Lognormal Distribution

By taking the natural logarithm of the data, lognormal distribution is reduced to the case of the normal distribution, so the MLE point estimates for the two parameters of the lognormal distribution for a complete sample of size n can be obtained from

$$\hat{\mu} = \frac{1}{n}\sum_{i=1}^{n} \ln t_i, \tag{5.41}$$

$$\hat{\sigma}^2 = \frac{1}{n-1}\sum_{i=1}^{n}\left(\ln t_i - \hat{\mu}\right)^2, \tag{5.42}$$

where $\hat{\mu}$ and $\hat{\sigma}^2$ are the point estimates of the mean and variance of the logarithm of times to failure t_i, respectively. The two-sided confidence interval for μ is given by,

$$Pr\left[\hat{\mu} - \frac{\hat{\sigma}t_{(\gamma/2)}[n-1]}{\sqrt{n}} \le \mu \le \hat{\mu} + \frac{\hat{\sigma}t_{(\gamma/2)}[n-1]}{\sqrt{n}}\right] = 1-\gamma, \tag{5.43}$$

where $t_{\gamma/2}[n-1]$ is the $100(\gamma/2)\%$ of the one-sided student's t-distribution with $[n-1]$ degrees of freedom.

The respective confidence interval for σ_t^2 is:

$$Pr\left[\frac{\hat{\sigma}^2(n-1)}{\chi^2_{\left(1-\frac{\gamma}{2}\right)}[n-1]} \le \sigma_t^2 \le \frac{\hat{\sigma}^2(n-1)}{\chi^2_{\left(\frac{\gamma}{2}\right)}[n-1]}\right] = 1-\gamma, \tag{5.44}$$

where $\chi^2_{\left(1-\frac{\gamma}{2}\right)}[n-1]$ is the $100(1-\gamma/2)\%$ of a χ^2 distribution with $[n-1]$ degrees of freedom.

With censored data, the corresponding statistical estimation turns out to be much more complicated, see Nelson (2003) and Lawless (2011) for a comprehensive treatment. Below, we cursorily summarize cases involving right-censored data.

If right-censored data exist, then the MLE of parameters for the distributions discussed before are obtained by maximizing the likelihood function

$$L = \prod_{i=1}^{r} f\left(t_i|\mu,\sigma\right)\prod_{j=1}^{n-r}\left[1-F\left(t_j|\mu,\sigma\right)\right], \tag{5.45}$$

where μ, σ are unknown parameters to be estimated, n is the total number of items, and r is the number of items that failed at times t_i, and $n-r$ items were right censored at times t_j. $f(\cdot)$ is the pdf of time to failure and $F(\cdot)$ is the cdf. For lognormal distribution, the log-likelihood function and its partial derivatives with respect to μ_t, σ_t take the following forms:

$$\Lambda = \ln L = \sum_{i=1}^{n}\ln f\left(t_i|\mu,\sigma\right) - \sum_{j=1}^{n-r}\ln F\left(t_j|\mu,\sigma\right) \tag{5.46}$$

$$\frac{1}{\hat{\sigma}^2}\sum_{i=1}^{r}\left[\ln(t_i) - \hat{\mu}\right] + \frac{1}{\hat{\sigma}}\sum_{j=1}^{n-r}\frac{\phi\left[\ln(t_j) - \hat{\mu}/\hat{\sigma}\right)}{1-\Phi\left[\ln(t_j) - \mu/\sigma\right]} = 0 \tag{5.47}$$

$$\sum_{i=1}^{r}\left\{\frac{\left[\ln(t_i)-\hat{\mu}\right]^2}{\hat{\sigma}^3}-\frac{1}{\hat{\sigma}}\right\}+\frac{1}{\hat{\sigma}}\sum_{j=1}^{n-r}\frac{\left[\ln(t_j)-\frac{\hat{\mu}}{\hat{\sigma}}\right]\phi\left[\frac{\ln(t_j-\hat{\mu})}{\hat{\sigma}}\right]}{1-\Phi\left[\frac{\ln(t_j-\hat{\mu})}{\hat{\sigma}}\right]}=0 \tag{5.48}$$

where, $\phi(x)$ and $\Phi(x)$ are the pdf and cdf of the standard normal distribution.

Solving the system of two equations with two unknowns $\hat{\mu}$ and $\hat{\sigma}$ in Equations 5.47 and 5.48 requires a numerical solution. For estimations on the confidence intervals involving right-censored data, the approximate method using the variance-covariance matrix described in Section 5.4.1 can be used.

Example 5.9

a. Estimate the two parameters of the lognormal distribution for the following time to failure data (with no censored items) using the MLE approach.

Item No.	**1**	**2**	**3**	**4**	**5**	**6**
Time of failure (hours)	144	385	747	1,144	1,576	2,612

b. Now assume that at $t=3{,}000$ hr, five items were still operating without failure (i.e., they are censored). Repeat the estimation of the two parameters and compare the difference.

Solution:

a. Using Equations 5.41 and 5.42

$$\hat{\mu}=6.635,\ \hat{\sigma}=1.044.$$

Using properties of the lognormal distribution, the actual mean and standard deviation for the time to failure is $E(t)=1{,}312.9$ and St Dev.(t) $=1{,}844.6$ hours, respectively.

b. Using Equations 5.47 and 5.48 with five censored items at 3,000 hours, the new estimates are:

$$\hat{\mu}=7.800,\ \hat{\sigma}=0.158.$$

5.4.6 Weibull Distribution

The Weibull distribution can be used for data assumed to be from increasing hazard rate, decreasing hazard rate, or constant failure rate distributions. Similar to the lognormal distribution, the Weibull distribution is a two-parameter distribution (although a three-parameter version also exists). The estimation of Weibull distribution parameters using MLE, even with complete data, is not a trivial problem.

For complete data, the parameters of the Weibull distribution using MLE are solved first by solving Equation 5.49 numerically with respect to $\hat{\beta}$:

$$\frac{\sum_{i=1}^{n}(t_i)^{\hat{\beta}}\ln t_i}{\sum_{i=1}^{n}(t_i)^{\hat{\beta}}}-\frac{1}{\hat{\beta}}=\frac{1}{n}\sum_{i=1}^{n}\ln t_i. \tag{5.49}$$

And then substituting into Equation 5.50 to find $\hat{\alpha}$:

$$\hat{\alpha}=\left(\frac{\sum_{i=1}^{n}(t_i)^{\hat{\beta}}}{n}\right)^{1/\hat{\beta}}. \tag{5.50}$$

It can be shown that, in the special situation when r items fail without replacement out of n items placed on test or under observation for time t_{end} (i.e., Type I test), the MLEs of α and β parameters of the Weibull distribution $f(t|\alpha,\beta)$ can be obtained as a solution for the following system of nonlinear equations. In this case $(n-r)$ items are right censored at the end of the test (i.e., no censoring occurred during the test):

$$\frac{\sum_{i=1}^{r}(t_i)^{\hat{\beta}}\ln t_i+(n-r)t_{\text{end}}^{\hat{\beta}}\ln t_{\text{end}}}{\sum_{i=1}^{r}(t_i)^{\hat{\beta}}+(n-r)t_{\text{c}}^{\hat{\beta}}}-\frac{1}{\hat{\beta}}=\frac{1}{r}\sum_{i=1}^{r}\ln t_i, \tag{5.51}$$

and

$$\hat{\alpha}=\left(\frac{\sum_{i=1}^{n}(t_i)^{\hat{\beta}}}{n}+(n-r)t_{\text{end}}^{\hat{\beta}}\right)^{1/\hat{\beta}}. \tag{5.52}$$

The system of Equations 5.51 and 5.52 can be solved using an appropriate numerical procedure. The corresponding confidence estimation is also complicated. See Nelson (2003), Bain (2017), and Mann et al. (1974) for further discussions.

If right-censored data exist before the end of the test due to failed item removals unrelated to the failure mode observed, then the MLEs of parameters α and β are obtained by finding the partial derivative of the log-likelihood function

$$\Lambda=\sum_{i=1}^{r}\ln\left[\frac{\beta}{\alpha}\left(\frac{t_i}{\alpha}\right)^{\beta-1}e^{-\left(\frac{t_i}{\alpha}\right)^{\beta}}\right]-\sum_{j=1}^{n-r}\left(\frac{t_j}{\alpha}\right)^{\beta}. \tag{5.53}$$

After simplification, the partial derivatives lead to the two nonlinear equations

$$\frac{r}{\hat{\beta}}+\sum_{i=1}^{r}\ln\left(\frac{t_i}{\hat{\alpha}}\right)-\sum_{i=1}^{r}\left(\frac{t_i}{\hat{\alpha}}\right)^{\hat{\beta}}\ln\left(\frac{t_i}{\hat{\alpha}}\right)-\sum_{j=1}^{n-r}\left(\frac{t_j}{\hat{\alpha}}\right)^{\hat{\beta}}\ln\left(\frac{t_j}{\hat{\alpha}}\right)=0, \tag{5.54}$$

$$\frac{r\hat{\beta}}{\hat{\alpha}}+\frac{\hat{\beta}}{2}\sum_{i=1}^{r}\left(\frac{t_i}{\hat{\alpha}}\right)^{\hat{\beta}}\ln\left(\frac{t_i}{\hat{\alpha}}\right)+\frac{\hat{\beta}}{\hat{\alpha}}\sum_{j=1}^{n-r}\left(\frac{t_j}{\hat{\alpha}}\right)^{\hat{\beta}}=0, \tag{5.55}$$

where n is the total number of items observed under test or from the field, and r is the number of failed items. Thus, $(n-r)$ would be the number of censored items, t_i the time that item i failed, and t_j the time that item j was right censored. Note that if items remaining at the end of the test, then for those items $t_j = t_{end}$. Equations 5.54 and 5.55 can be solved for the two unknowns $\hat{\alpha}$ and $\hat{\beta}$.

The approximate confidence interval of the $\hat{\alpha},\hat{\beta}$ parameters of the distributions, especially for censored data, may be obtained from using the Fisher information matrix:

$$\hat{\beta}e^{\left(-Z_{1-\frac{\gamma}{2}}\frac{\sqrt{Var(\hat{\beta})}}{\hat{\beta}}\right)}\le\beta\le\hat{\beta}e^{\left(Z_{1-\frac{\gamma}{2}}\frac{\sqrt{Var(\hat{\beta})}}{\hat{\beta}}\right)}, \tag{5.56}$$

$$\hat{\alpha}e^{\left(-Z_{1-\frac{\gamma}{2}}\frac{\sqrt{Var(\hat{\alpha})}}{\hat{\alpha}}\right)}\le\alpha\le\hat{\alpha}\,e^{\left(Z_{1-\frac{\gamma}{2}}\frac{\sqrt{Var(\hat{\alpha})}}{\hat{\alpha}}\right)}, \tag{5.57}$$

where $Var\left(\hat{\beta}\right)$ and $Var\left(\hat{\alpha}\right)$ can be found from the observed Fisher information matrix in Equation 5.58 with the z_γ values obtained from the standard normal distribution using $\left(1-\gamma\right)=1/\sqrt{2\pi}\int_{z_\gamma}^{\infty}e^{-t^2/2}\,dt$ for the $100\left(1-\gamma\right)\%$ confidence level.

$$\begin{bmatrix} Var\left(\hat{\beta}\right) & Cov\left(\hat{\alpha},\hat{\beta}\right) \\ Cov\left(\hat{\beta},\hat{\alpha}\right) & Var\left(\hat{\alpha}\right)\end{bmatrix}=\begin{bmatrix} -\dfrac{\partial^2\Lambda}{\partial\beta^2} & -\dfrac{\partial^2\Lambda}{\partial\alpha\,\partial\beta} \\ \dfrac{\partial^2\Lambda}{\partial\beta\,\partial\alpha} & -\dfrac{\partial^2\Lambda}{\partial\alpha^2}\end{bmatrix}^{-1}_{\alpha=\hat{\alpha},\beta=\hat{\beta}}. \tag{5.58}$$

Example 5.10

a. Use the MLE approach to estimate the parameters of Weibull distribution using the times to failure from Example 5.9.
b. Then assume that three items have been censored at 192, 323, and 685 hours. What are the new estimates for the parameters $\hat{\alpha}$ and $\hat{\beta}$ and their 90% confidence intervals?

The solution requires iterative numerical calculations such as a Monte Carlo simulation.

Solution:

c. Using the numerical procedure, systems of Equations 5.51 and 5.52 are solved, which results in estimates of $\hat{\beta} = 2.092, \hat{\alpha} = 395.92$ hours. A comparison of these results with the plot from Example 5.9 is reasonable, but it illustrates the approximate nature of these data analysis methods and demonstrates the importance of using more than one method discussed.

d. For the second part, the MLE Equations 5.54 and 5.55 need to be solved again considering the three censored data in addition to the failures. The iterative numerical solution results in estimates of $\hat{\alpha}$ and $\hat{\beta}$ as well as their 90% confidence intervals:

Parameter	5%	Mean	95%
$\hat{\alpha}$	355.20	463.30	604.28
$\hat{\beta}$	1.30	1.97	2.99

5.4.7 Binomial Distribution

When the data are in the form of failures occurring on demand, that is, r failures observed in n trials, there is a constant probability of failure (or success), and the binomial distribution can be used as an appropriate model. This is often the situation for standby components (or systems). For instance, a redundant pump is demanded for operation n times in a period of test or observation.

The MLE estimator for p is given by the formula

$$\hat{p} = \frac{r}{n}. \tag{5.59}$$

The lower and the approximate upper confidence limits for p can be found, using the Clopper–Pearson procedure.

$$p_l = \left\{1 + \frac{(n-r+1)}{r} F_{\left(1-\frac{\gamma}{2}\right)}[2n-2r+2;2r]\right\}^{-1}, \tag{5.60}$$

$$p_u = \left\{1 + \frac{n-r}{(r+1)F_{\left(1-\frac{\gamma}{2}\right)}[2r+2;2n-2r]}\right\}^{-1}, \tag{5.61}$$

where $F_{1-\frac{\gamma}{2}}[v_1;v_2]$ is the $100(1-\gamma/2)\%$ of the F-distribution with v_1 degrees of freedom for the numerator, and v_2 degrees of freedom for the denominator. Table A.5 contains some percentiles of the F-distribution. Note that the Poisson distribution can be used as an approximation for the binomial distribution when the parameter, p, of the binomial distribution is small and the parameter n is large, for example, $r < \frac{n}{10}$, which means that approximate confidence limits can be constructed using Equation 5.36 with $TTT = n$. The binomial distribution model will be re-examined in more detail later in this chapter under the nonparametric life testing and reliability estimation.

Example 5.11

An emergency pump in a nuclear power plant is in a standby mode. There have been 563 start tests for the pump, and only three failures have been observed. No degradation in the pump's physical characteristics or changes in operating environment are observed. Find the 90% confidence interval for the probability of failure per demand.

Solution:

With $n = 563$, $r = 3$. Using Equations 5.59–5.61, find $\hat{p} = \mathbf{3/563 = 0.0053}$.

$$p_l = \left\{1 + \frac{563 - 3 + 1}{3} F_{0.95}\left[2 \cdot 563 - 2 \cdot 3 + 2; 2 \cdot 3\right]\right\}^{-1} = 0.00145,$$

where $F_{0.95}[1122; 6] = 3.67$ from Table A.5.
Similarly,

$$p_u = \left\{1 + \frac{(563 - 3)}{(3 + 1) F_{0.95}\left[2 \cdot 3 + 2; 2 \cdot 563 - 2 \cdot 3\right]}\right\}^{-1} = 0.0137.$$

Therefore,

$$Pr(0.00145 \le p \le 0.0137) = 90\%.$$

Example 5.12

In a commercial nuclear plant, the performance of the emergency diesel generators has been observed for about 5 years. During this time, there have been 35 demands with four observed failures. Find the 90% confidence limits and point estimate for the probability of failure per demand. What would the error be if we used Equation 5.36 instead of Equations 5.60 and 5.61 to solve this problem?

Solution:

Here, $x = 4$ and $n = 35$. Using Equation 5.59,

$$\hat{p} = \frac{4}{35} = 0.114.$$

To find lower and upper limits, use Equations 5.60 and 5.61. Thus,

$$p_l = \left\{1 + \frac{35-4+1}{4} F_{0.95}[2\cdot 35 - 2\cdot 4 + 2; 2\cdot 4]\right\}^{-1} = 0.04.$$

$$p_u = \left\{1 + \frac{35-4}{(4+1)F_{0.95}[2\cdot 4+2; 2\cdot 35 - 2\cdot 4]}\right\}^{-1} = 0.243.$$

If we used Equation 5.36,

$$p_l = \frac{\chi^2_{0.05}(8)}{2\times 35} = \frac{2.733}{70} = 0.039.$$

$$p_u = \frac{\chi^2_{0.95}(10)}{2\times 35} = \frac{18.31}{70} = 0.262.$$

The error due to this approximation is

$$\text{Lower limit error} = \frac{|0.04 - 0.039|}{0.04} \times 100 = 2.5\%.$$

$$\text{Upper limit error} = \frac{|0.243 - 0.262|}{0.243} \times 100 = 7.8\%.$$

Note that this is not a negligible error, and Equation 5.36 should not be used. Since $x > n/10$, Equation 5.36 is not a good approximation.

5.5 CLASSICAL NONPARAMETRIC DISTRIBUTION ESTIMATION

We have already established that any reliability measure or index can be expressed in terms of the time to failure cdf or reliability function. Thus, the problem of estimating these functions is important. We will now turn to how to achieve this using nonparametric methods. The commonly used estimate of the cdf is the *empirical (or sample) distribution function* (edf) introduced for uncensored data in Chapter 4 (see Equation 4.28) in the context of the K-S test. In this section, we consider some other nonparametric point and confidence estimation procedures applicable for censored data.

5.5.1 Confidence Intervals for cdf and Reliability Function for Complete and Censored Data

Constructing an edf requires a complete sample, but an edf can also be constructed for the right-censored samples for the failure times which are less than the last time to failure observed ($t<t_r$). The edf is a random function, since it depends on the sample items. For any point, t, the edf, $S_n(t)$, is the fraction of sample items that failed before t.

The edf is, in a sense, the estimate of the probability, $p = F(t)$, in a binomial trial. It is possible to show that the MLE of the binomial parameter p coincides with the sample cdf $F_n(t) = i/n$ (Equation 4.28) and that $F_n(t)$ is a consistent estimator of the cdf, $F(t)$.

Using the relationship between reliability and the edf, one can obtain an estimate of the reliability function. This estimate, called the *empirical (or sample) reliability function,* is

$$R_n(t) = \begin{cases} 1, & 0 < t < t_1; \\ 1 - \dfrac{i}{n}, & t_i \le t < t_{i+1}, i = 1,2,\ldots,n-1; \\ 0, & t_n \le t < \infty; \end{cases} \tag{5.62}$$

where $t_1 < \cdots < t_n$ are the ordered failure data.

The mean number of failures observed during time t is $E(r) = np = nF(t)$, and so the mean value of the proportion of items failed before t is $E(r/n) = p = F(t)$. The variance of this fraction is given by

$$Var\left(\frac{r}{n}\right) = \frac{p(1-p)}{n} = \frac{F(t)[1-F(t)]}{n}. \tag{5.63}$$

For practical problems, Equation 5.63 is used by replacing $F(t)$ with $F_n(t)$. As the sample size, n, increases, the binomial distribution can be approximated by a normal distribution with the same mean and variance [i.e., $\mu = np$, $\sigma^2 = np(1-p)$]. This approximation provides reasonable results if both np and $n(1-p)$ are ≥ 5. Using this approximation, the $100(1-\gamma/2)\%$ confidence interval for the unknown cdf, $F(t)$, at any point t can be constructed as:

$$F_n(t) - z_{\left(1-\frac{\gamma}{2}\right)}\sqrt{\frac{F_n(t)[1-F_n(t)]}{n}} \le F(t) \le F_n(t) + z_{\left(1-\frac{\gamma}{2}\right)}\sqrt{\frac{F_n(t)[1-F_n(t)]}{n}}, \tag{5.64}$$

where $z_{\left(1-\frac{\gamma}{2}\right)}$ is the inverse of the probability of level $\left(1-\frac{\gamma}{2}\right)$ of the standard normal distribution.

The corresponding estimate for the reliability (survivor) function is $R_n = 1 - F_n(t)$.

Example 5.13

Using the data from Example 5.4, find the nonparametric point estimate and the 95% confidence interval for the cdf, $F(t)$ at $t = 350$ hours.

Solution:

Using Equation 5.59 the point estimate for $F(350) = F_n(350) = 5/10$.

The respective approximate 95% confidence interval based on Equation 5.64 is

$$0.5-1.96\sqrt{\frac{0.5(1-0.5)}{10}} \leq F(350) \leq 0.5+1.96\sqrt{\frac{0.5(1-0.5)}{10}}.$$

Therefore,

$$Pr(0.190 < F(350) < 0.8099) = 0.95.$$

Using complete or right-censored reliability data from an unknown cdf, one can also obtain the strict *confidence intervals for the unknown cdf, F(t)*. This can be done using the same Clopper–Pearson procedure for constructing the approximate confidence intervals for the binomial parameter p, using Equations 5.60 and 5.61. However, these limits can also be expressed in more compact form in terms of the incomplete beta function as follows.

The lower confidence limit, $F_l(t)$, at the point t where $S_n(t) = i/n$ ($r = 0,1,\ldots, n$), is the largest value of p that satisfies the following inequality

$$I_p(r, n-r+1) \leq \frac{\gamma}{2}, \tag{5.65}$$

and the upper confidence limit, $F_u(t)$, at the same point is the smallest p satisfying the inequality

$$I_{1-p}(n-r, r+1) \leq \frac{\gamma}{2}, \tag{5.66}$$

where

$$I_p(\alpha,\beta) = \frac{B(p;\, \alpha,\beta)}{B(1;\, \alpha,\beta)} \text{ with B}(\cdot) \text{ defined as } B(p;\, \alpha,\beta) = \int_0^p u^{\alpha-1}(1-u)^{\beta-1}\, du.$$

is known as $I_p(\alpha,\beta)$ the *regularized incomplete beta function* which represents the cdf of the beta distribution. The regularized incomplete beta function is difficult to tabulate; however, its numerical approximations are available online.

Example 5.14

For the data from Example 5.13, find the 95% confidence interval for the cdf, $F(t)$ at $t=350$ hours, using Equations 5.65 and 5.66.

Solution:

Using Equation 5.65, the lower confidence limit is found from

$$I_p(5, 10-5+1) \leq 0.025.$$

Solving for the largest p, the inequality above yields a value for $p_l = 0.1871$ and, using Equation 5.66, the inequality for the upper confidence limit below leads to $p_u = 0.8131$.

$$I_p(5,6) \le 0.975.$$

Therefore, the confidence interval is

$$Pr(0.1871 < F(350) < 0.8131) = 0.95,$$

which is reasonably close to the approximate interval obtained in the previous example.

Another typical reliability estimation problem that can be solved using this nonparametric approach is the estimation of the lower confidence limit for the reliability function using the same type of data. This can be accomplished using Equation 5.66, in which $1-p=1-F(t)$ is replaced by the reliability function, $R(t)$. Accordingly, one obtains

$$I_R(n-r, r+1) \le \gamma. \tag{5.67}$$

This procedure is illustrated by the following example.

Example 5.15

Twenty-two identical components were reliability tested for 1,000 hours, and no failure was observed. Find the lower confidence limit, $R_l(t)$, for the reliability function at $t=1{,}000$ hours and for confidence probability $1-\gamma=0.90$. Repeat the result for the case where one failure occurred.

Solution:

We need to find the largest value of R satisfying Equation 5.67. For the problem considered $\gamma=0.1$, $r=0$, and $n=22$; therefore,

$$I_R(22,1) \le 0.1.$$

Solving for R, one obtains $R_l \approx 0.90$. If we repeat this for the case of observing one failure $I_R(21,2) \le 0.1$, $R_l \approx 0.834$.

Another possible application of Equation 5.67 is for planning reliability demonstration tests when the lower $1-\gamma$ confidence limit for reliability, R_l, is given with acceptable number of failures, r, during the test duration. The problem is to optimize for the minimum sample size, n, to be tested with the number of failures not exceeding a preset value r. Accordingly, that quantities n, r, and R_l, would satisfy Equation 5.67. For given values of R_l and γ, the minimum sample size for which Equation 5.67 can be satisfied corresponds to $r=0$. This is illustrated by the following example.

Example 5.16

The reliability test on a component must demonstrate the lower limit of reliability of 86% with 90% confidence. If no failure is tolerated, what sample size must be tested to satisfy the above requirements? Repeat this question with the minimum sample size if up to 1 failure is allowed.

Solution:

When $r=0$, Equation 5.67 can be written as $I_{0.86}(\mathrm{n}-1,1)\leq 0.1$. The resulting value is $n=16$. For $r=1$ Equation 5.67 can be written as $I_{0.86}(n-1,2)\leq 0.1$ which yields the minimum number of items needed for test as $n=26$.

5.5.2 Confidence Intervals of Reliability Function for Censored Data

The point and confidence estimates considered so far do not apply when censored items occur during the test. For such samples, the *Kaplan-Meier* or *product-limit* estimate, which is the MLE of the cdf (unreliability function), can be used.

Suppose we have a sample of n times to failure, among which there are only k distinct failure times. Consider the ordered times of failure as $t_1 \leq \cdots \leq t_k$, and let $t_0=0$. Let n_j be the number of items under observation just before t_j. Assume that the time to failure pdf is continuous and there is only one failure at a time. Then, $n_{j+1}=(n_j+1)$. Under these conditions, the Kaplan-Meier estimate is given by

$$F_n(t)=1-R_n(t)=\begin{cases} 0, & 0\leq t<t_1, \\ 1-\prod_{j=1}^{i}\dfrac{n_j-1}{n_j}, & t_i\leq t<t_{i+1},\quad i=1,\ldots,m-1, \\ 1, & t\geq t_m, \end{cases} \tag{5.68}$$

where integer $m=k$, if $k<n$, and $m=n$, if $k=n$.

It is possible to show that for complete data samples, the Kaplan-Meier estimate coincides with the edf given by Equation 4.28.

In the general case when there are multiple failures, d_j at time t_j or during an interval, the Kaplan-Meier estimate is given by

$$F_n(t)=1-R_n(t)=\begin{cases} 0, & 0\leq t<t_1, \\ 1-\prod_{j=1}^{i}\dfrac{n_j-d_j}{n_j}, & t_i\leq t<t_{i+1}, i=1,\ldots,m-1, \\ 1, & t\geq t_m, \end{cases} \tag{5.69}$$

For estimation of variance of S_n (or R_n), Greenwood's formula (Lawless, 2011) is used:

$$Var\left[\hat{F}_n(t)\right]=Var\left[\hat{R}_n(t)\right]=\sum_{j:t_j<t}\frac{d_j}{n_j(n_j-d_j)} \tag{5.70}$$

Using Equation 5.70, one can construct the corresponding approximate confidence limits for the reliability or the cdf of interest similar to Equation 5.64.

If the data are shown in intervals or are grouped, then it is possible to use an adjusted number of items at risk, if censoring time occurs uniformly over the interval. Suppose F_i = number of failures in the ith interval, C_i = number of censored items in the ith interval, and H_i = number of items at risk at the beginning of the interval t_i (and thus $H_i = H_{i-1} - F_{i-1} - C_{i-1}$), then the effective number of items in interval i is $H'_i = H_i - (C_i/2)$. Therefore, the conditional probability of an item failure in the ith interval given survival to time t_{i-1} is (F_i/H'_i), and the conditional probability of survival is $p_i = 1-(F_i/H'_i)$. Therefore, the unconditional probability of survival is $\hat{R}_i = p_i \cdot \hat{R}_{i-1}$ with $R_0 = 1$.

Example 5.17

The table below shows censored test data of a mechanical component (censored data points are shown with +). Find the Kaplan-Meier estimate of the time to failure cdf. Plot the data on a Weibull paper and estimate the parameters of the distribution.

i	**Ordered Failure Time, t_i, or Censoring time, t_i+**	$F_n[t_i]$
0	0	0
1	32	1 − 15/16 = 0.06255
2	41	1 − (15/16)(14/15) = 0.125
3	58	1 − (15/16)(14/15)(13/14) = 0.187
4	64+	-
5	66	1 − (13/16)[(13 − 1 − 1)/12] = 1 − 0.745 = 0.255
6	72	1 − 0.745(10/11) = 0.323
7	74+	-
8	76+	-
9	83+	-
10	88+	-
11	92+	-
12	100+	-
13	104	0.492
14	108+	-
15	109	0.746
16	121+	-

Solution:

The third column contains the product-limit estimate of the cdf as a function of time, calculated using Equation 5.69. To clarify the calculations, the detailed calculations for the first seven lines of the table are also given. Further, the right-censored items are not failure events and thus do not change the preceding nonparametric estimate of $F_n[t_i]$. The parameters are then calculated using the probability plot of the data or with MLE.

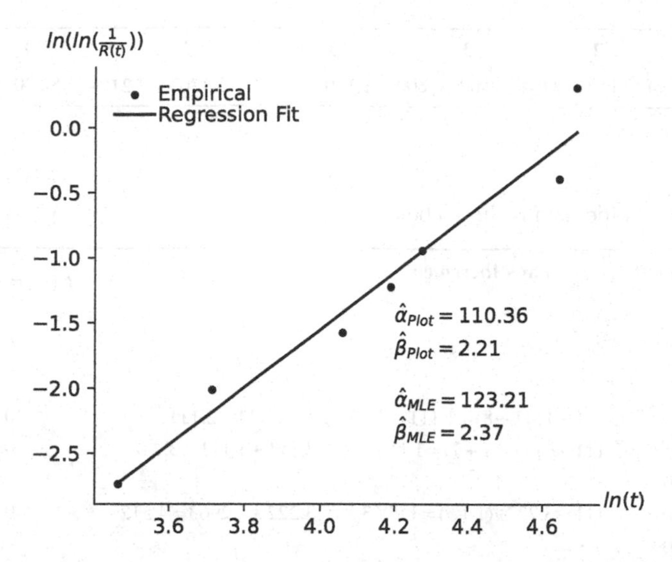

Another useful and more accurate method for estimating the cdf and reliability function for multiply-censored data is the method of *rank adjustment*. Here, if n items are observed, then the initial ordered ranks, i, are rank adjusted by using

$$j_{t_i} = j_{t_{i-1}} + \text{rank increment}, \tag{5.71}$$

where, j_{t_i} is the adjusted rank corresponding to failure time t_i.

Here, rank increment is computed from:

$$\text{rank increment} = \frac{\left[(n+1) - j_{t_{i-1}}\right]}{(1+\text{m})}, \tag{5.72}$$

where m is the number of items survived, including and beyond the ith item.

A plot of the rank order against any plotting position point provides the basis for estimating the nonparametric reliability function and cdf. The first failure takes on the rank increment of 1 unless if an item was censored before the occurrence of the first failure.

The adjusted rank j_{t_i} is calculated only for the non-censored items (with i_{t_i} still being the initial rank for all ordered times). A convenient form of the adjusted rank is:

$$j_{t_i} = j_{t_{i-1}} + \frac{n+1-j_{t_{i-1}}}{n-i+2}. \tag{5.73}$$

Example 5.18

Consider the failure cycles in a low-cycle fatigue test in the table below. Using the rank adjustment method, find the corresponding two-parameter Weibull cdf of cycles to failure and associated parameters.

i	1	2	3	4	5	6	7	8	9	10
c_i (cycles)	150	340+	560	800	1,130+	1,720	2,470+	4,210+	5,230	6,890

Solution:

Evaluation of the data is shown below.

i	t_i (Cycles)	Rank Increment	j_{t_i}	$\hat{F}(t_i) = \frac{j_{t_i} - 0.375}{n + 0.25}$
1	150	1	1	0.061
2	340+	—	—	—
3	560	$(11-1)/(1+8)=1.111$	$1+1.111=2.111$	0.169
4	800	$(11-2.111)/(1+7)=1.111$	$2.111+1.111=3.222$	0.278
5	1,130+	—	—	—
6	1,720	$(11-3.222)/(1+5)=1.2963$	$3.222+1.2963=4.512$	0.404
7	2,470+	—	—	—
8	4,210+	—	—	—
9	5,230	$(11-4.518)/(1+2)=2.16$	$4.518+2.160=6.679$	0.615
10	6,890	$(11-6.679)/(1+1)=2.16$	$6.679+2.160=8.839$	0.826

5.6 BAYESIAN ESTIMATION PROCEDURES

In Chapter 4, we introduced Bayesian parameter estimation in terms of both point estimates and credible intervals. In the framework of the Bayesian approach, the parameters of interest are treated as r.v.s which are unknown and thus estimated from data. Thus, a distribution can be assigned to represent both the parameter and the uncertainty around that parameter, including the mean (and other moments) of the distribution for the parameter. The pdf of a parameter in Bayesian terms is given directly by the posterior pdf.

Bayesian analysis has a clear advantage: the ability to incorporate related prior knowledge of the parameter or phenomenon being modeled. This is essential in reliability engineering because our data are often limited. Using prior information allows us to combine experience, physical theory, partially relevant data, reliability tests, and generic data from sources such as those discussed in Section 5.2. Bayesian estimation allows the use of small amounts of data that are often deemed insufficient for classical techniques. Another advantage of Bayesian estimation is that it models uncertainty inherently as part of the parameter estimation process rather than requiring separate methods.

To implement Bayesian estimation, we begin with the prior pdf. The prior pdf is used to represent the relevant prior knowledge, including subjective judgment and generic information regarding the characteristics of the parameter and its distribution. The prior knowledge is combined with other relevant information (such as statistics obtained from reliability tests and operational data), it is combined with the prior through a Bayesian updating process. The resulting posterior distribution for the parameter is one which better represents the parameter of interest. Since the selection of the prior often involve subjective judgments, the Bayesian estimation is sometimes called the subjectivist approach to parameter estimation. As noted in Chapter 2, the Bayesian interpretation of probability is the degree of belief placed in a proposition. In this interpretation, the prior belief is combined with any specific evidence to find a new belief, and this process can be done iteratively as new data emerge. Since the parameter of interest in the Bayesian interpretation is viewed as a random variable, the prior and posterior distributions show the probability mass or density of the distribution describing the parameter.

The basic concept of Bayes' theorem was discussed in Chapters 2 and 4. To estimate parameters of interest, such as parameters of a pdf model, this theorem can be written in one of three forms: discrete, continuous, or mixed (Equations 4.13–4.15). The continuous and mixed forms, which are the common forms used for parameter estimation in reliability and risk analysis, are briefly discussed below.

Let $\boldsymbol{\theta}$ be a vector of parameters of interest. Examples of a parameter are the reliability (e.g., as a binomial distribution parameter), MTTF, failure rate or hazard rate, shape parameter of a Weibull pdf and so on. Suppose parameters in vector $\boldsymbol{\theta}$ are continuous r.v.s, so the prior and posterior distributions of $\boldsymbol{\theta}$ can be represented by continuous pdfs. In this section, we use $\pi(t)$ refer to the pdf of an unknown probability distribution. Let $\pi_0(\boldsymbol{\theta})$ be the continuous prior joint pdf of parameters in vector $\boldsymbol{\theta}$, and let $L(t|\theta)$ be the likelihood function based on sample data, t. Recall that in Bayesian estimation, the data are viewed as constant and the parameters are treated as random. Thus, we express the likelihood function as the probability of data, t, conditioned on the pdf parameters $\boldsymbol{\theta}$. Then the posterior pdf of $\boldsymbol{\theta}$, $\pi_1(\boldsymbol{\theta})$ is given by

$$\pi_1(\boldsymbol{\theta}) = \frac{\pi_0(\boldsymbol{\theta})L(t|\boldsymbol{\theta})}{\int_{\theta} \pi_0(\boldsymbol{\theta})L(t|\boldsymbol{\theta})d\boldsymbol{\theta}}. \tag{5.74}$$

Equation 5.74 is Bayes' theorem for a continuous r.v.

The Bayesian inference process includes these steps:

1. Constructing the likelihood function based on the distribution model of interest and type of the evidence available (complete data, censored data, grouped data, etc.). This function depicts the likelihood of the evidence (data) given the distribution model and the vector of parameters being estimated.
2. Identifying and quantifying the prior information about the parameters of interest in the form of a prior pdf.
3. Calculating the posterior pdf using the Bayesian inference in Equation 5.74.
4. Using the posterior distribution to calculate the mean and probability (credible) intervals of the vector of parameters.

The posterior mean of $\boldsymbol{\theta}$ is,

$$E(\boldsymbol{\theta}) = \mu_{\theta} = \int \boldsymbol{\theta}\pi_1(\boldsymbol{\theta}|t)d\boldsymbol{\theta}. \tag{5.75}$$

Recall from Chapter 4 that, for constructing a Bayesian two-sided credible interval, the following relationship for each parameter based on the posterior distribution is used:

$$Pr(\theta_l < \theta \leq \theta_u) = 1 - \gamma. \tag{5.76}$$

The following are examples of the parametric Bayesian estimation in reliability analysis. Martz and Waller (1982), Gelman et al. (2013), and Kelly and Smith (2011) have significantly elaborated on the concept of the Bayesian technique and its application to reliability analysis.

5.6.1 Estimation of the Parameter of Exponential Distribution

Consider the case where the exponential distribution describes the time to failure. The problem is to estimate the parameter λ of the exponential distribution using Bayesian estimation and reliability test data.

Consider a test of n items which results in r distinct times to failure $t_1 < t_2 < \cdots < t_r$ and $n-r$ times to right censoring $t_{c1}, t_{c2}, \ldots, t_{c(n-r)}$, so that the total time on test is

$$\text{TTT} = \sum_{i=1}^{r} t_i + \sum_{j=1}^{n-r} t_{cj}. \tag{5.77}$$

Now, suppose a gamma distribution describes the prior distribution of parameter λ. Recall from Chapter 2 that the pdf of the gamma distribution is a function of two parameters, α, β. So far, we have used this as a function of time to failure, t. However, it can also be seen as a function of λ. If rewrite this prior pdf as a function of an r.v. λ, and transform its parameter slightly, we have a prior pdf for λ:

$$\pi_0(\lambda|\alpha,\beta) = \frac{1}{\beta^\alpha \Gamma(\alpha)} \lambda^{\alpha-1} exp\left(-\frac{\lambda}{\beta}\right), \quad \lambda > 0, \beta > 0, \alpha > 0. \tag{5.78}$$

In the Bayesian context, the parameters α and β, the parameters of the prior distribution, are called *hyperparameters*. Selection of the hyperparameters is discussed later; but for the time being, suppose these parameters are known based on prior data and information. Now, let us parameterize the gamma distribution slightly differently, using $\rho = \frac{1}{\beta}$

$$\pi_0(\lambda|\alpha,\rho) = \frac{1}{\Gamma(\alpha)} \rho^\alpha \lambda^{\alpha-1} e^{-\lambda\rho}, \quad \lambda > 0, \rho > 0, \alpha > 0. \tag{5.79}$$

For the available data and the exponential time to failure distribution, using Equation 5.45, the likelihood function can be written as

$$L(\lambda|t) = \prod_{i=1}^{r} \lambda e^{-\lambda t_i} \prod_{j=1}^{n-r} e^{-\lambda t_{cj}} = \lambda^r e^{-\lambda \text{TTT}}, \tag{5.80}$$

where TTT is the total time on test given by Equation 5.77.

Using the prior distribution Equation 5.79, the likelihood function 5.80 and Bayes' theorem in the form of Equation 5.74, one can find the posterior pdf of the parameter λ as

$$\pi_1(\lambda|\text{TTT},\alpha,\rho)=\frac{e^{-\lambda(\text{TTT}+\rho)}\lambda^{r+\alpha-1}}{\int_0^{\infty} e^{-\lambda(\text{TTT}+\rho)}\lambda^{r+\alpha-1}\,d\lambda}. \tag{5.81}$$

Recalling the definition of the gamma function, it is easy to show that the integral in the denominator of Equation 5.81 is

$$\int_0^{\infty}\lambda^{r+\alpha-1}e^{-\alpha(\text{TTT}+\rho)}\,d\lambda=\frac{\Gamma(r+\alpha)}{(\text{TTT}+\rho^{r+\alpha})}. \tag{5.82}$$

Finally, the posterior pdf of λ can be written as

$$\pi_1(\lambda|\text{TTT})=\frac{(\text{TTT}+\rho)^{r+\alpha}}{\Gamma(r+\alpha)}\lambda^{r+\alpha-1}e^{-\lambda(\text{TTT}+\rho)}. \tag{5.83}$$

A comparison with the prior pdf (Equation 5.79) shows that the posterior pdf (Equation 5.83) is also a gamma distribution with parameters $\alpha' = r+\alpha$ and $\rho' = \text{TTT}+\rho$. Prior distributions that result in posterior distributions of the same family are called *conjugate prior distributions* or simply *conjugate distributions.* Table 5.3 shows examples of conjugate distributions. For more examples and applications of conjugate distributions, see O'Connor et al. (2019).

The mean estimate of λ is the expected value of λ using the posterior gamma distribution with parameters α' and ρ', which we know from the properties of the gamma distribution to be:

$$E(\lambda)=\frac{\alpha'}{\rho'}=\frac{r+\alpha}{TTT+\rho}. \tag{5.84}$$

The corresponding credible intervals can be obtained using Equation 5.76. For example, the $(1-\gamma)$ probability level the upper one-sided Bayesian credible interval for

TABLE 5.3
Examples of Conjugate Distributions

Prior Distribution	Likelihood Function	Posterior Distribution
Beta	Binomial	Beta
Gamma	Poisson or Exponential	Gamma
Normal	Normal	Normal
Lognormal	Lognormal	Lognormal

λ can be obtained from the following equation based on the posterior distribution Equation 5.83:

$$Pr(\lambda < \lambda_u) = 1 - \gamma. \tag{5.85}$$

Similarly, the two-sided Bayesian credible interval can be obtained from the following equation using the posterior distribution:

$$Pr(\lambda_l < \lambda < \lambda_u) = 1 - \gamma. \tag{5.86}$$

Example 5.19

A sample of identical items was reliability tested. The time to failure distribution for these items is assumed to be exponential. Six failures were observed during the test with a total time on test of 1,440 hours. Assume that the prior distribution for the exponential failure rate can be represented by a gamma distribution $\lambda \sim \text{gamma}(\alpha = 11.1, \rho = 1{,}110 \text{ hours})$. Find the posterior distribution for gamma and the two-sided 90% credible interval for λ.

Solution:

The respective parameters of the distribution are $\alpha = 11.1$ and $\rho = 1{,}110$ hours. This corresponds to a prior mean of $E(\lambda_0) = \dfrac{11.1}{1{,}110} = 0.01$ failures/hour.
The posterior distribution is a conjugate of the prior, so the resulting posterior distribution will be a gamma distribution, with:

$$\alpha' = r + \alpha = 11.1 + 6 = 17.1 \text{ failures},$$

$$\rho' = \rho + \text{TTT} = 1{,}110 + 1{,}440 = 2{,}550 \text{ hours},$$

$$\lambda_1 \sim \text{gamma}(\alpha = 17.1, \rho = 2{,}550 \text{ hours}).$$

Using Equation 5.84, the point posterior estimate (the mean of the posterior distribution) of the hazard rate is evaluated as

$$E(\lambda_1) = \frac{17.1}{2{,}550} = 6.71 \times 10^{-3}/\text{hour}.$$

Calculating the inverse of the gamma distribution at the 5% and 95% levels, we obtain the Bayesian credible interval as

$$4.28 \times 10^{-3} \text{ hr}^{-1} < \lambda < 9.58 \times 10^{-3} \text{ hr}^{-1}.$$

The prior and the posterior distribution of λ are plotted below. The 90% credible interval on the posterior is shaded in gray.

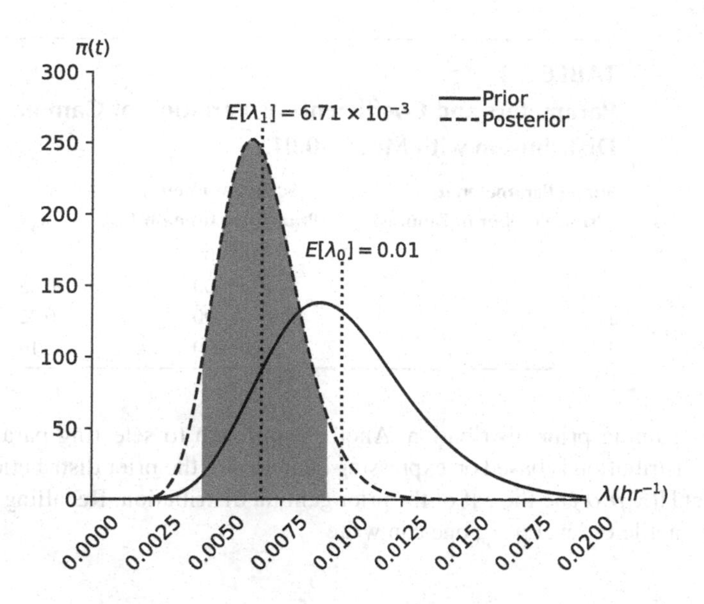

5.6.1.1 Selecting Hyperparameters for the Gamma Prior Distribution

In the previous section, the gamma distribution was chosen as the prior distribution for simplicity and performance. Now let us consider the reliability interpretation of the Bayesian estimate $E(\lambda)$, obtained in Equation 5.84.

The parameter α is a prior number of failures observed during a prior test that has ρ as the total time on test (or taken from a reliability database with similar interpretation). So, intuitively, one would choose the prior estimate of λ as the ratio α/ρ, which coincides with the *mean value* of the prior gamma distribution used (Equation 5.79).

The corresponding practical situation is sometimes the opposite—perhaps one has a prior belief of λ; meanwhile, the parameters α and ρ must be found. Given the prior point estimate λ_p, one can only estimate the ratio $\frac{\alpha}{\rho} = \lambda_p$. To estimate these parameters separately, additional information about the degree of belief or accuracy of this prior estimate is required. Since the variance of the gamma distribution is α/ρ^2, the coefficient of variation of the prior distribution is $\frac{1}{\sqrt{\alpha}}$ (recall from Chapter 2 that the coefficient of variation is the ratio of standard deviation to mean). The coefficient of variation can be used as a measure of relative accuracy of the prior point estimate of λ_p.

Thus, given the prior point estimate, λ_p, and the relative accuracy of this estimate, one can estimate the corresponding parameters of the prior gamma distribution. To get a sense of the scale of these errors, consider the following example. Let the prior point estimate λ_p be 0.01 (in some arbitrary units). The corresponding values of the coefficient of variation, expressed in percentages, for different values of the parameters α and ρ are given in Table 5.4.

This approach is a simple and convenient way of expressing prior information (e.g., knowledge or degree of belief expressed by the subject matter experts) in terms

TABLE 5.4
Parameters and Coefficients of Variation of Gamma Distribution with Mean = 0.01.

Shape Parameter, α (Prior Number of Failures)	Scale Parameter, ρ (Prior Total Time on Test)	c_v
1	100	1.00
5	500	0.45
10	1,000	0.32
100	10,000	0.10

of the gamma prior distribution. Another approach to selecting parameters of the prior distribution is based on expressing quantiles of the prior distribution. For example, let $\Pi(\lambda|\rho,\alpha)$ be the cdf of the prior gamma distribution. Recalling the definition of a quantile of level p, λ_p, one can write

$$\Pi_0\left(\lambda_p|\rho,\alpha\right)=\int_0^{\lambda_p}\Pi_0\left(\lambda\right)\mathrm{d}\lambda=p, \tag{5.87}$$

where $\pi_0\left(\lambda\right)$ is given by Equation 5.79.

Given a pair of subjective quantiles, say, λ_{p1} and λ_{p2} of levels p_1 and p_2, and using Equation 5.86, one obtains two equations with two unknowns. These equations uniquely determine the parameters of the prior distribution. Practically, the procedure is as follows.

An expert or prior data specifies the values p_1, λ_{p1}, p_2, and λ_{p2} such that

$$\begin{aligned}\Pi_0\left(\lambda_{p1},\rho,\alpha\right)&=Pr\left(\lambda<\lambda_{p1}\right)=p_1,\\ \Pi_0\left(\lambda_{p2},\rho,\alpha\right)&=Pr\left(\lambda<\lambda_{p2}\right)=p_2.\end{aligned} \tag{5.88}$$

For example, the expert specifies that there is 90% probability that the parameter is < 0.1 and 50% probability that λ is < 0.01. Then, a numerical procedure is used to solve the system of equations to find the values of the parameters α and ρ.

Usually, the value of p_1 is chosen as 0.90 or 0.95, and the value of p_2 as 0.5 (the median value) or 0.05 and 0.1 for the lower limit of the unknown variable. In Martz and Waller (1982), a special table and graphs are provided for solving the system of equations above.

Example 5.20

A sample of identical items was reliability tested. Two failures were observed during the test. The total time on test was 100 hours. The time to failure distribution is assumed to be exponential. The gamma distribution with a mean of 0.01 hour^{-1} and with a coefficient of variation of 30% was selected as a prior distribution to represent the parameter of interest, λ. Find the posterior distribution and point estimate for λ.

Solution:

For the prior gamma distribution representing λ, $c_v = \frac{1}{\sqrt{\alpha}}$, which is given in the problem statement as 0.30% Solving for α:

$$\alpha = \left(\frac{1}{0.30}\right)^2 = 11.1.$$

Then, using the mean of the gamma distribution:

$$E(\lambda_0) = 0.01 = \frac{\alpha}{\rho} = \frac{11.1}{\rho},$$

which yields $\rho = 1{,}110$ hours.

Since the prior gamma distribution is conjugate with the exponential likelihood, the posterior is also a gamma distribution. The posterior is:

$$\lambda_1 \sim \text{gamma}(\alpha' = 11.1 + 2, \rho' = 1{,}110 + 100 \text{ hours}).$$

And the posterior point estimate (the mean of the posterior distribution) is:

$$E(\lambda_1) = \frac{13.1}{1{,}210} = 1.08 \times 10^{-2}/\text{hour}.$$

5.6.1.2 Uniform Prior Distribution

The uniform distribution has a very simple form and is a non-informative prior distribution, so it is convenient to use as an expression of prior information. Consider the prior uniform pdf in the form

$$\pi_0(\lambda|a, b) = \begin{cases} \frac{1}{b-a}, & a < \lambda < b; \\ 0, & \text{Otherwise.} \end{cases} \tag{5.89}$$

Using the likelihood function, Equation 5.80, the posterior pdf can be written as

$$\pi_1(\lambda|t) = \frac{\lambda^r e^{-\lambda t}}{\int_a^b \lambda^r e^{-\lambda t} d\lambda}, \quad a < \lambda < b. \tag{5.90}$$

A key property of the uniform prior distribution (and other non-informative priors) is that they will cancel out in the posterior distribution because the prior distribution $\pi_0(\theta)$ becomes a constant in both the numerator and denominator, as seen in Equation 5.90.

Example 5.21

An electronic component has an exponential time to failure distribution. The uniform prior distribution of λ is given by $a = 1\times10^{-6}$ and $b = 5\times10^{-6}$ /hour. A life test of the component results in $r = 30$ failures in a total time on test of 10^7 hours. Find the mean and 90% two-sided Bayesian credible interval for the failure rate λ

Solution:

Using Equation 5.90, the point estimate $E(\lambda_1) = 3.1\times10^{-6}$ /hour and the 90% two-sided Bayesian credible interval is $\left(2.24\times10^{-6} < \lambda_1 < 4.04\times10^{-6}\right)\text{hour}^{-1}$. The prior and posterior distributions are plotted here.

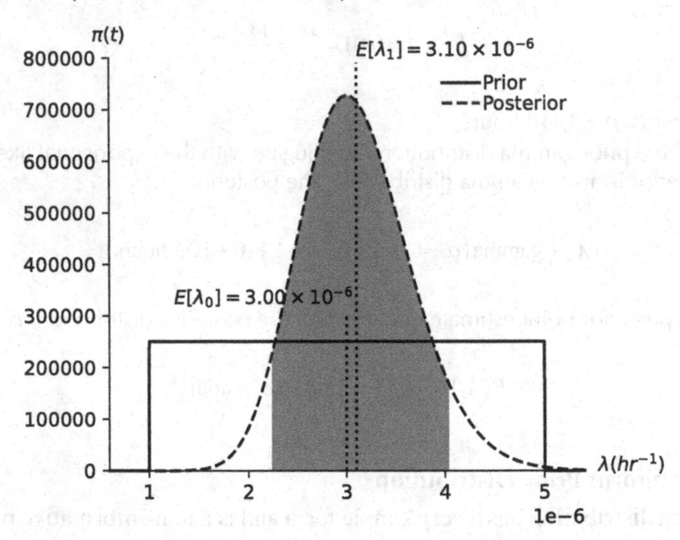

5.6.1.3 Jeffrey's Prior Distribution

This prior distribution is a non-informative prior distribution defined as $\pi_0(\theta) = \sqrt{\det(I_\theta)}$ where I_θ is the Fisher information matrix of Equation 5.16. This derivation is motivated by the fact that it is not dependent upon the set of parameter variables chosen to describe parameter space. Jeffrey suggested the need to make ad hoc modifications to the prior to avoid problems in multi-dimensional distributions. Jeffrey's prior is normally improper, meaning that it does not have a finite integral and is therefore not a pdf. For the exponential distribution, the Jeffrey's prior is expressed as

$$\pi_0(\lambda) = \sqrt{\left(\frac{1}{\lambda}\right)}. \tag{5.91}$$

As with the uniform distribution, a key property of non-informative prior distributions, such as Jeffery's prior, is that they cancel out in the posterior distribution. See O'Connor et al. (2019) for the corresponding Jeffery's prior for various distributions.

5.6.2 Bayesian Estimation of the Parameter of Binomial Distribution

Suppose that n identical items have been placed on test (without replacement of the failed items) for a specified time, t, and that the test yields r failures. The number of failures, r, can be considered as a discrete r.v. having the binomial distribution with parameters n and $p(t)$, where p is the constant probability of failure of a single item during time t. As discussed in Section 5.5, $p(t)$, as a function of time, is the time to failure cdf, and $1 - p(t)$ is the reliability or survivor function.

A straightforward application of the binomial distribution is the modeling of failures to start on demand for a redundant item. The probability of failure in this case might be time-independent. Thus, one should keep in mind two possible applications of the binomial distribution:

- The survivor (reliability) function or time to failure cdf.
- The binomial distribution itself.

The MLE of the parameter p is the ratio r/n, which is widely used as the classical estimate. To obtain a Bayesian estimation for the reliability (survivor) function, let us consider p as the survivor probability (i.e., reliability) in a single Bernoulli trial (so, now "success" means surviving). If the number of items placed on test, n, is fixed in advance, the probability distribution of x, the number of items surviving during the test, is given by the binomial distribution:

$$f(x|n,p) = \frac{n!}{(n-x)!x!} p^x (1-p)^{n-x}, \; 0 < p < 1. \tag{5.92}$$

The corresponding likelihood function, given that the test results in x items surviving out of n items tested, can be written as

$$L(p|x, n) = cp^x (1-p)^{n-x}, \tag{5.93}$$

where c is a constant that does not depend on the parameter of interest, p. For any continuous prior distribution with pdf $\pi_0(p)$, the corresponding posterior pdf can be written as

$$\pi_1(p|x, n) = \frac{p^x (1-p)^{n-x} \pi_0(p)}{\int_0^1 p^x (1-p)^{n-x} \pi_0(p)\, dp} \tag{5.94}$$

5.6.2.1 Standard Uniform Prior Distribution

Consider the particular case of the uniform distribution, $p \sim unif(a,b)$, which in the Bayesian context represents a state of total ignorance. While this seems to have little practical importance when $a=0$ and $b=1$, it is interesting, nevertheless, from a methodological point of view. For this case, one can write the prior and posterior distributions of p as, respectively,

$$\pi_0(p) = \begin{cases} 1, & 0 \le p \le 1; \\ 0, & \text{Otherwise.} \end{cases} \tag{5.95}$$

and

$$\pi_1(p|x) = \frac{p^{(x+1)-1}(1-p)^{(n-x+1)-1}}{\int_0^1 p^{(x+1)-1}(1-p)^{(n-x+1)-1}\,dp}. \tag{5.96}$$

The integral in the denominator can be expressed as

$$\int_0^1 p^{(x+1)-1}(1-p)^{(n-x+1)-1}\,dp = \frac{\Gamma(x+1)\Gamma(n-x+1)}{\Gamma(n+2)}. \tag{5.97}$$

By substituting $\alpha = x+1$ and $\beta = n-x+1$ into the posterior cdf, it can be easily recognized as the pdf of the beta distribution: $f(p|\alpha,\beta) = \frac{\Gamma(\alpha+\beta)}{\Gamma(\alpha)\Gamma(\beta)} p^{\alpha-1}(1-p)^{\beta-1}$.

Thus, the posterior is a beta distribution $\pi_1(p|x+1, n-x+1)$. Recalling the expression for the mean value of the beta distribution, the Bayesian point estimate of p can be written as:

$$E(p_1) = \frac{\alpha}{\alpha+\beta} = \frac{x+1}{n+2}. \tag{5.98}$$

Note that the estimate differs from the MLE (i.e., x/n), but for large sample sizes, the two estimates are close.

Since the cdf of the beta distribution is expressed in terms of the incomplete beta function (see Chapter 2), the $(1-\gamma)$ two-sided Bayesian credible interval for p can be obtained by solving the following equations

$$Pr(p < p_l) = I_{p_l}(x+1, n-x+1) = \frac{\gamma}{2},$$

$$Pr(p > p_u) = I_{p_u}(x+1, n-x+1) = 1 - \frac{\gamma}{2}. \tag{5.99}$$

Example 5.22

Calculate the point estimate and the 95% two-sided Bayesian probability (credible) interval for the reliability of a new component based on a life test of 300 components, out of which four have failed. Suppose that for this component no historical information is available. Accordingly, its prior reliability estimate may be assumed to be uniformly distributed between 0 and 1.

Solution:

Using Equation 5.98, we find

$$R = 1 - p_B = 1 - \frac{4+1}{300+2} = 0.9834.$$

Using Equation 5.99, the 95% lower and upper probability limits are evaluated as 0.9663 and 0.9946, respectively.

It is interesting to compare the above results with the classical results. The point estimate of the reliability is $R=1-p=1-4/300=0.9867$, and the 95% lower and upper limits, according to Equations 5.63 and 5.64, are 0.9662 and 0.9964, respectively.

5.6.2.2 Beta Prior Distribution

The most widely used prior distribution for the parameter p of the binomial distribution is the beta distribution. The beta prior distribution is a conjugate prior distribution for the estimation of the parameter p of the binomial distribution of interest. The standard uniform distribution is a particular case of the beta distribution, wherein $\alpha = 1$ and $\beta = 1$.

Since the beta distribution is conjugate with the binomial distribution, for data with r failures in n trials, the posterior distribution for p is a beta distribution with the parameters

$$\alpha' = \alpha_0 + r,$$

$$\beta' = \beta_0 + n - r. \tag{5.100}$$

That is, $p_1 \sim \text{beta}(\alpha', \beta')$

Recall the MLE estimate for the binomial distribution representing r failures out of n trials is $\hat{p} = \frac{r}{n}$, and the mean of the beta distribution is $E(p) = \frac{\alpha}{\alpha+\beta}$. It becomes evident that one interpretation of this beta distribution is for a system with $n=\alpha+\beta$ tests, where $\alpha=n-r$ items survive and $\beta=r$ items fail (or vice versa). If tests or field data are collected, the parameters α and β are directly obtained from these test or field data. Otherwise, if expert elicitation or generic data is used, α and β can be interpreted as a pseudo number of identical items that survive and fail a pseudo test of $\alpha+\beta$ items during pseudo time t.

The corresponding $(1-\gamma)$ two-sided Bayesian credible interval for p can be obtained as the solutions of the following equations

$$Pr(p < p_l) = I_{p_l}(\alpha_0 + r, \beta_0 + n - r) = \frac{\gamma}{2},$$

$$Pr(p > p_u) = I_{p_u}(\alpha_0 + r, \beta_0 + n - r) = 1 - \frac{\gamma}{2}. \tag{5.101}$$

On the other hand, experts could evaluate the prior mean, that is, the ratio x_0/n_0, and their degree of belief in terms of standard deviation or coefficient of variation of the prior distribution. For example, if the coefficient of variation is used, it can be treated as a measure of uncertainty (relative error) of the prior assessment.

Let p be the prior mean and c_v be the coefficient of variation of the prior beta distribution. It can be shown that $c_v = \sqrt{\frac{\beta}{\alpha(\alpha+\beta+1)}}$. The corresponding parameters can be found as a solution of the following equation system

$$p = \frac{\alpha}{\alpha+\beta},$$
$$\beta = \frac{1-p}{pc_v^2} - 1 - \alpha. \tag{5.102}$$

Example 5.23

Let the prior mean of the reliability function of a new item be chosen as $p_{pr} = x_0/n_0 = 0.9$. Select the parameters α_0 and β_0 of a prior beta distribution.

Solution:

The choice of the parameters α_0 and β_0 can be based on the values of the coefficient of variation used as a measure of dispersion (or accuracy) of the prior point estimate p_{pr}. Some values of the coefficient of variation and the corresponding values of the parameters x_0 and n_0 for $p_{pr} = x_0/n_0 = 0.9$ are given in the table below.

α_0	β_0	$n=\alpha+\beta$	Coefficient of Variation
0.9	0.1	1	0.2357
9	1	10	0.1005
90	10	100	0.0332

Note that as n approaches infinity, the Bayesian estimate approaches the MLE, x/n, Equation 5.59. In other words, the classical inference will coincide with the Bayesian inference as the amount of data increases.

Example 5.24

A design engineer assesses the reliability of a new component at the end of its useful life (T=10,000 hours) as $R = 0.75 \pm 0.19$. A sample of 100 new components were tested for 10,000 hours, and 29 failures were recorded. Given the test results, find the posterior mean and the 90% Bayesian credible interval for the component reliability if the prior distribution of the component reliability is assumed to be a beta distribution.

Solution:

The prior mean is $E(R)=0.75$, and the coefficient of variation is $c_v = 0.19/0.75 = 0.25$. Using Equation 5.100, the parameters of the prior distribution are found to be α_0=3.15 and β_0=1.02.

Thus, the posterior parameters are: $\alpha' = (3.15 + 71) = 74.15$ and $\beta' = (1.02 + 29) = 30.02$.

Thus, according to Equation 5.98, the posterior point estimate of the new component reliability is:

$$E(R) = \frac{74.15}{74.15 + 30.02} = 0.712.$$

According to Equation 5.101, the 90% lower and upper credible limits are 0.637 and 0.782, respectively. The figure below shows the prior and posterior distributions of $R = 1 - p$.

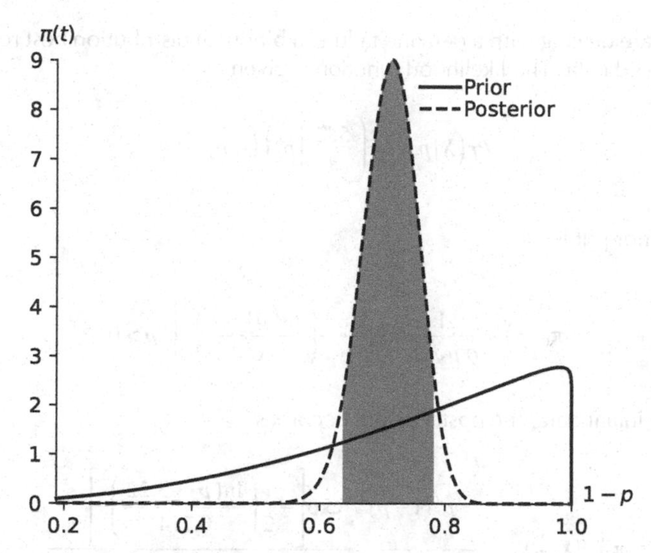

5.6.2.3 Jeffrey's Prior Distribution

This distribution provides a non-informative prior for the parameter p in the form of a beta distribution with specific parameter values,

$$\frac{1}{\sqrt{p(1-p)}} = \pi_0(p|0.5, 0.5), \tag{5.103}$$

which represent a pseudo-case with $r = 0.5$ failures out of $n = 0.5 + 0.5 = 1$ trials. The posterior will be a conjugate beta distribution

$$\pi_1(p|r,n) = \text{Beta}(p|\alpha_0 + 0.5, \beta_0 + 1 - 0.5). \tag{5.104}$$

5.6.2.4 Lognormal Prior Distribution

The following example illustrates the case when the prior distribution and the likelihood function do not result in a conjugate posterior distribution, and the posterior distribution obtained cannot be expressed in terms of standard function. This is the case when numerical integration is required.

Example 5.25

For a diesel generator, the number of failures to start on demand, X, has a binomial distribution with parameter p. The prior data on the performance of a similar diesel are obtained from field data, and p is assumed to follow the lognormal distribution with known parameters $\mu=-3.22$ and $\sigma=0.714$. A limited test of the diesel generators of interest shows that eight failures are observed in 582 demands. Calculate the Bayesian point estimate of p (mean and median) and the 90% credible interval of p. Compare these results with the corresponding values for the prior distribution.

Solution:

Since we are dealing with a demand failure, a binomial distribution best represents the observed data. The likelihood function is given by

$$Pr(X|p,n)=\binom{582}{8}p^{8}(1-p)^{574},$$

and the prior pdf is

$$\pi_0(p)=\frac{1}{\sigma p\sqrt{2\pi}}\exp\left[-\frac{1}{2}\left(\frac{\ln(p)-\mu}{\sigma}\right)^2\right],\ p>0.$$

Using the initial data, the posterior pdf becomes

$$\pi_1(p|X,n)=\frac{p^{7}(1-p)^{574}\exp\left[-\frac{1}{2}\left(\frac{\ln(p)+3.22}{0.714}\right)^2\right]}{\int_0^1\left(p^{7}(1-p)^{574}\exp\left[-\frac{1}{2}\left(\frac{\ln(p)+3.22}{0.714}\right)^2\right]dp\right)}.$$

It is evident that the denominator cannot be expressed in a closed form, so a numerical integration must be applied. The table below shows results of a numerical integration used to find the posterior distribution. In this table many values of p are arbitrarily selected between 1.23×10^{-8} and 1.78×10^{-1}.

i	Failure Probability p_i	Prior pdf	Likelihood	Prior × Likelihood	Posterior pdf	Posterior cdf
0	1.23×10^{-8}	0.00	1.63×10^{-46}	0.00	0.00	0.00
1	1.80×10^{-3}	2.49×10^{-2}	1.21×10^{-5}	3.01×10^{-7}	1.30×10^{-5}	1.77×10^{-9}
2	3.60×10^{-3}	5.27×10^{-1}	1.10×10^{-3}	5.79×10^{-4}	2.51×10^{-2}	8.21×10^{-6}
3	5.39×10^{-3}	2.03	1.00×10^{-2}	2.03×10^{-2}	8.79×10^{-1}	5.09×10^{-4}

(Continued)

i	Failure Probability p_i	Prior pdf	Likelihood	Prior × Likelihood	Posterior pdf	Posterior cdf
4	7.19×10^{-3}	4.34	3.53×10^{-2}	1.54×10^{-1}	6.65	6.07×10^{-3}
5	8.99×10^{-3}	7.01	7.44×10^{-2}	5.22×10^{-1}	$2.26\times10^{+1}$	3.07×10^{-2}
6	1.08×10^{-2}	9.64	1.13×10^{-1}	1.09	4.71×10^{1}	9.27×10^{-2}
7	1.26×10^{-2}	1.20×10^{1}	1.36×10^{-1}	1.63	7.08×10^{1}	2.00×10^{-1}
8	1.44×10^{-2}	1.40×10^{1}	1.39×10^{1}	1.94	8.43×10^{1}	3.41×10^{-1}
9	1.62×10^{-2}	1.55×10^{1}	1.25×10^{-1}	1.94	8.42×10^{1}	4.94×10^{-1}
10	1.80×10^{-2}	1.66×10^{1}	1.02×10^{-1}	1.69	7.34×10^{1}	6.37×10^{-1}
11	1.98×10^{-2}	1.74×10^{1}	7.63×10^{-2}	1.33	5.75×10^{1}	7.55×10^{-1}
12	2.16×10^{-2}	1.78×10^{1}	5.34×10^{-2}	9.52×10^{-1}	4.13×10^{1}	8.44×10^{-1}
13	2.34×10^{-2}	1.80×10^{1}	3.52×10^{-2}	6.35×10^{-1}	2.75×10^{1}	9.05×10^{-1}
14	2.52×10^{-2}	1.80×10^{1}	2.21×10^{-2}	3.98×10^{-1}	1.73×10^{1}	9.45×10^{-1}
15	2.70×10^{-2}	1.78×10^{1}	1.33×10^{-2}	2.37×10^{-1}	1.03×10^{1}	9.69×10^{-1}
16	2.88×10^{-2}	1.75×10^{1}	7.72×10^{-3}	1.35×10^{-1}	5.84	9.83×10^{-1}
17	3.06×10^{-2}	1.70×10^{1}	4.33×10^{-3}	7.37×10^{-2}	3.19	9.91×10^{-1}
18	3.24×10^{-2}	1.65×10^{1}	2.36×10^{-3}	3.89×10^{-2}	1.69	9.96×10^{-1}
19	3.42×10^{-2}	1.60×10^{1}	1.25×10^{-3}	1.99×10^{-2}	8.64×10^{-1}	9.98×10^{-1}
20	3.60×10^{-2}	1.54×10^{1}	6.46×10^{-4}	9.92×10^{-3}	4.30×10^{-1}	9.99×10^{-1}
21	3.78×10^{-2}	1.48×10^{1}	3.27×10^{-4}	4.82×10^{-3}	2.09×10^{-1}	1.00
22	3.96×10^{-2}	1.41×10^{1}	1.62×10^{-4}	2.29×10^{-3}	9.91×10^{-2}	1.00
23	4.14×10^{-2}	1.35×10^{1}	7.89×10^{-5}	1.06×10^{-3}	4.61×10^{-2}	1.00
24	4.32×10^{-2}	1.29×10^{1}	3.77×10^{-5}	4.86×10^{-4}	2.10×10^{-2}	1.00
25	4.49×10^{-2}	1.23×10^{1}	1.78×10^{-5}	2.18×10^{-4}	9.44×10^{-3}	1.00
…	…	…	…	…	…	…
91	1.62×10^{-1}	5.07×10^{-1}	1.45×10^{-33}	7.35×10^{-34}	3.18×10^{-32}	1.00
92	1.64×10^{-1}	4.86×10^{-1}	4.62×10^{-34}	2.24×10^{-34}	9.72×10^{-33}	1.00
93	1.65×10^{-1}	4.67×10^{-1}	1.46×10^{-34}	6.83×10^{-35}	2.96×10^{-33}	1.00
94	1.67×10^{-1}	4.48×10^{-1}	4.63×10^{-35}	2.07×10^{-35}	8.99×10^{-34}	1.00
95	1.69×10^{-1}	4.30×10^{-1}	1.46×10^{-35}	6.27×10^{-36}	2.72×10^{-34}	1.00
96	1.71×10^{-1}	4.13×10^{-1}	4.58×10^{-36}	1.89×10^{-36}	8.19×10^{-35}	1.00
97	1.73×10^{-1}	3.96×10^{-1}	1.43×10^{-36}	5.68×10^{-37}	2.46×10^{-35}	1.00
98	1.74×10^{-1}	3.81×10^{-1}	4.47×10^{-37}	1.70×10^{-37}	7.37×10^{-36}	1.00
99	1.76×10^{-1}	3.66×10^{-1}	1.39×10^{-37}	5.07×10^{-38}	2.20×10^{-36}	1.00
100	1.78×10^{-1}	3.52×10^{-1}	4.29×10^{-38}	1.51×10^{-38}	6.54×10^{-37}	1.00
			Sum 1.28×10^{1}			

Then, the numerator and denominator of the posterior pdf are calculated. A comparison of the prior and posterior is given below.

	Prior	Posterior
Mean	0.0516	0.0170
Median	0.0399	0.0158
5%	0.0123	0.0089
95%	0.1293	0.0267

The point estimate of the actual data using the classical inference is

$$\hat{p} = \frac{8}{582} = 0.0137.$$

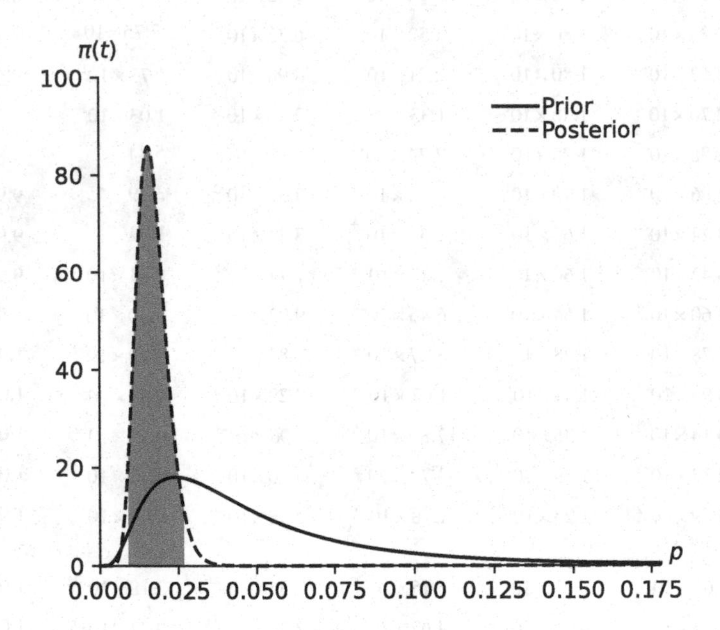

5.6.3 Bayesian Estimation of Other Distributions

The Bayesian procedures considered in the previous sections are related to the binomial and exponential distributions. In the framework of the Bayesian approach, both the exponential and binomial distributions have their respective conjugate prior distributions, which makes their practical use more convenient.

In the case of distributions with two or more parameters, the respective Bayesian estimation procedures become more complicated because they require multi-dimensional prior distributions and multi-dimensional integration in the denominator of the Bayesian inference in Equation 5.74. The methods required to solve problems with these procedures are beyond the scope of this introductory book and typically rely on Markov Chain Monte Carlo (MCMC) numerical simulations. The reader is referred to Kelly and Smith (2011) for further information.

5.7 EXERCISES

5.1 A test was run on ten electric motors under high temperature. The test was run for 60 hours, during which six motors failed. The failures occurred at the following times: 37.5, 46.0, 48.0, 51.5, 53.0, and 54.5 hours. We do not know whether an exponential distribution or a Weibull distribution model is better for representing these data. Use the plotting method as the main tool to discuss the appropriateness of these two models.

5.2 One-hundred and twenty-four devices are placed on an overstress test with failures occurring at the following times.

Time (hour)	0.4	1	2	5	8	12	25
Total number of failures	1	3	5	15	20	30	50

a. Plot the data on Weibull probability paper.
b. Estimate the shape parameter.
c. Estimate the scale parameter.
d. What other distributions may also represent these failure data?
e. Is the exponential distribution a good fit for this data?

5.3 A new device was tested for the wear failure mechanism over 200 cycles. When a given undesirable wear depth reached, we considered the units as failed. Further, assume that the life distribution can best match a Weibull distribution. We obtained the following data.

Cycles	1–50	50–100	100-150	150-200	200+
Number of failures	0	3	3	1	1

a. Plot the data on Weibull probability paper. Is it a good fit?
b. Estimate the parameters of the distribution using the plot.
c. Find the 10% life of this device.
d. Find the mean life of the device.

5.4 A life test of six identical home appliances produced the following data.

Appliance	1	2	3	4	5	6
Failure time (Cycles)	660	711	798	911	1,012	1,191

a. Create a Weibull plot of the test data and use it to estimate the pdf parameters.
b. Estimate the warranty life if we expect to see no more than 5% of the appliances fail during the warranty period.

5.5 Eight handheld devices were put through tumble reliability testing. Failures were found after the inspections at 150 (1 failure), 200 (1 failure), 300 (3 failures), 400 (1 failure), Two devices had not failed at the last inspection 500 hours. Use these data to estimate the life distribution under the test conditions:

a. Plot a nonparametric estimate of the cdf for time to failure at the test conditions.
b. Propose a hazard rate.
c. Estimate the reliability at 50 hours under this test condition.

5.6 The following time to failure data are found when 158 transformer units are put under test. Use a nonparametric method to estimate $f(t)$, $h(t)$, and $R(t)$ of the transformers. No failures are observed prior to 1,750 hours.

Time Interval		Number
1,750	2,250	17
2,250	2,750	54
2,750	3,250	27
3,250	3,750	17
3,750	4,250	19
4,250	4,750	24

5.7 A test of 25 integrated circuits over 500 hours yields the following data. Plot the pdf, hazard rate, and reliability function for each interval of these integrated circuits using a nonparametric method.

Time Interval	Number of Failures
0–100	10
100	7
200	3
300	3
400	2

5.8 By placing 10 units on a 700-day life test we observed these failure times: 311, 380, 417, 518, 611, with 4 units still operating at 700 days.

a. Perform a Weibull plot and show whether a two-parameter Weibull pdf is a good fit. If so, estimate the parameters.
b. Find an empirical hazard rate using an appropriate nonparametric method.
c. How does the Weibull hazard rate from (a) compare with the empirical hazard rate (b)?

5.9 The time to first failure of the lube oil system in 15 ships are:

Aircraft	1	2	3	4	5	6	7	8	9	10	11	12	13	14	15
Time (hour)	88	69	56	24	101	50	366	56	140	492	223	111	102	406	203

Based on these data, is the hazard rate increasing, decreasing, or constant? Use graphical nonparametric analysis to support your conclusion.

5.10 The discrete life distribution of an appliance is given by the U-quadratic pdf $f(N) = \alpha(N-\beta)^2$, where α and β are constants and $1 < N < 10{,}000$. Use the MLE method to determine the value of α and β if we have observed the following cycles to failure data: $N_1 = 1{,}202, N_2 = 2{,}345$, $N_3 = 8{,}660, N_4 = 9{,}230$, and $N_5 = 9{,}802$.

5.11 Total test time for a device is 50,000 hours. The test is terminated after the first failure. If the pdf of the device time to failure is known to be exponentially distributed, what is the probability that the estimated failure rate is not greater than 4.6×10^{-5} per hour?

5.12 Over a three and half years of operation in the field, the following failurc times of an item were observed: 918, 1,996, 10,052, 25,312, 29,990 hours.
 a. Assuming a lognormal pdf, find the point estimate of the log-mean and log-standard deviation of the pdf.
 b. Determine the 90% one-sided upper confidence interval of μ.

5.13 Seven pumps have failure times (in months) of 15.1, 10.7, 8.8, 11.3, 12.6, 14.4, and 8.7. Assume an exponential distribution.
 a. Find a point estimate of the MTTF.
 b. Estimate the reliability of a pump for $t = 12$ months.
 c. Calculate the 95% two-sided interval of λ.

5.14 In an accept-reject reliability test of 1,000 units, 5 units were failed. Determine:
 a. A point estimate of the reliability of the unit.
 b. The upper one-sided 95% confidence interval for expected failure probability per unit.
 c. The 95% two-sided confidence interval for expected failure probability per unit.

5.15 For an experiment, 25 relays are allowed to run until the first failure occurs at $t = 15$ hours. At this point, the experimenters decide to continue the test for another 5 hours. No failures occur during this extended period, and the test is terminated. Using the 90% confidence level, determine the following:
 a. Point estimate of MTTF.
 b. Two-sided confidence interval for MTTF.
 c. Two-sided confidence interval for reliability at $t = 25$ hours.

5.16 A locomotive control system fails 15 times out of the 96 times it is activated to function. Determine the following:
 a. A point estimate for failurc probability of the system.
 b. 95% two-sided confidence intervals for the probability of failure. (Assume that after each failure, the system is repaired and put back in an as good as new state.)

5.17 Over an accept-reject replaceable reliability test period of 500 hours, one failure is observed in 100 units tested.
 a. Assuming an exponential distribution, estimate of the failure rate of this item.
 b. Determine the 95% confidence interval (two-sided) of the failure rate.
 c. How do you interpret the results found in (b)?

5.18 Time to failure data from eight devices placed on an accelerated test is shown below.

i	1	2	3	4	5	6	7	8
Time to failure	65	85	90	95	340	405	555	575

 a. Use probability plotting to estimate the parameter of the exponential distribution.
 b. Discuss the suitability of the exponential distribution for this data.

5.19 During a factory acceptance test of 150 components, five are found to be defective.
  a. Estimate the probability that a randomly selected component is defective.
  b. Estimate the 90% confidence interval for the probability of being defective.

5.20 In flaw measurement experiment, a sample of ten measurements of a fatigue crack gives a mean of 0.483 mm, with a standard deviation of 0.01 mm. What is the 99% confidence interval of the true mean and standard deviation for:
  a. A normal pdf model.
  b. A lognormal pdf model.
  c. Explain how you would choose between the two distribution models.

5.21 Eighteen units were reliability tested until eight failed at the following times (in hours): 73, 290, 333, 501, 676, 869, 990, 1,215. Assuming a constant failure rate, estimate the failure rate and 90% two-sided confidence interval of the failure rate. What is the reliable life at 100 hours?

5.22 In an accelerated test, failures of 8 units are recorded after 8, 17, 21, 21, 22, 39, 42, and 47 days after starting in operation. The other two units were operating after 50 days.
  a. Perform a Weibull probability plot of these data.
  b. Is an exponential distribution appropriate for modeling this unit?
  c. Assuming an exponential distribution is appropriate, estimate the failure rate using MLE.
  d. Determine an approximate 95% confidence interval for λ.
  e. Determine the approximate 99% one-sided confidence interval of reliability at 50 days.

5.23 Time to failures (in days) of ten components in a reliability tests are: 122, 154, 192, 198, 199, 208, 210, 216. Two units did not fail after 220 days.
  a. Find the nonparametric estimates of reliability and hazard rate.
  b. Assume the Weibull distribution is an appropriate model for this data. What is the Weibull log-likelihood function for this data?
  c. Use MLE to estimate the point and confidence intervals for both parameters of the Weibull distribution.

5.24 Right-censored failure cycles of a device are shown below. Which distribution best represents these data? At what time will 5% of units fail? To solve this problem use both the Kaplan-Meier and Rank Increment methods with probability plotting to compare and discuss your results.

8+, 8.2, 12, 15.5+, 18, 25.7, 26+, 33+, 34.5, 39.7.

5.25 Use the cycle to failure data: 8 +, 8.2, 12, 15.5 +, 18, 25.7, 26 +, 33 +, 34.5, 39.7. Find a point estimate of a Weibull model cycle to failure using:
  a. Probability plotting and the rank adjustment method.
  b. Maximum likelihood estimation.

5.26 An accelerated reliability test of 8 components having an exponential pdf is performed and the following right-censored data were found.
   a. Determine the likelihood function for the failure rate, λ.
   b. Use MLE to determine the point estimate of the failure rate, λ.
   c. Use the Rank Increment method to show appropriateness of the exponential pdf. Propose an alternate more representative pdf, if needed.

Test number	1	2	3	4	5	6	7	8
Time to failure	811	907+	1,099	2,290	2,900+	3,000+	3,300	4,100+

5.27 Consider the following data.

Time Interval (Year)	Number of Items Failed	Number of Items Censored
$1 < t < 2$	2	4
$2 < t < 3$	2	3
$3 < t < 4$	5	8
$4 < t < 5$	8	3
$5 < t < 6$	8	4
$6 < t$		13

   a. Use the Kaplan-Meier method to assess appropriateness of a Weibull pdf for this data.
   b. Estimate the parameters of the Weibull distribution.
   c. Estimate the 90% reliable life.
   d. Find the probability the component will fail after 6 years.

5.28 An accelerated test of nine devices produced failure and censored times as follows: 20, 50, 50, 65+, 75, 125, 150+, 150+, 150+. Use the rank adjustment method and plotting to assess the suitability of the lognormal distribution. Estimate the parameters of the distribution. Is this an appropriate fit?

5.29 You are to design a life test experiment to estimate the failure rate of a new device. Your boss asks you to make sure that the 80% upper and lower limits of the estimate interval (two-sided) do not differ by more than a factor of 2. Due to cost constraints, the components will be tested until they fail. Determine how many components should be put on test.

5.30 The specification for space electronics requires devices to have a reliability of 0.98 at after 6 months of operation. Five hundred transistors are placed on replaceable test for 2,000 hours with 12 failures observed. Has this specification been met?

5.31 Historical data show that the failure rate of a component falls uniformly between 10^{-2} and 10^{-3} failures/hour. Ten new, slightly revised version of this component are tested over 200 hours and no failures are observed. Use Bayesian methods to determine the mean and variance of the posterior pdf of λ. Also find the 90% credible interval for λ.

5.32 In the reactor safety study, the failure rate of a diesel generator can be described as having a lognormal distribution with the upper and lower 90% credible bounds of 3×10^{-2} and 3×10^{-4}, respectively. If a given nuclear plant experiences two failures in 8,760 hours of operation, determine the upper and lower 90% credible bounds given this plant experience. (Consider the reactor safety study values as prior information.)

5.33 A consumer electronics company releases a slightly revised design of a device each fall. From its past experience, the company is confident that the reliability of the revised design will exceed 60% over its designated life. Only two prototypes of the new design were available for reliability testing, and they met the reliability requirements without a failure after the test.

a. Propose a prior distribution for the reliability. Then estimate the posterior pdf of the designated life reliability.
b. Later, the company conducted another test using five new prototypes and observed one failure. What is the new estimate of the device's reliability?
c. What is the one-sided 90% credible interval for the reliability from (b)?

5.34 For a component that follows exponential time to failure model with the failure rate λ, suppose we have observed failures of k such components at times: $t_1,\ldots,t_k$.

a. Write the likelihood function for the observed data.
b. Assume a Jeffrey's prior and find the posterior pdf of λ. What is the posterior mean and 100p% two-sided credible interval?

5.35 In a pass or fail acceptance test, 1 of the 25 systems tested failed. Estimate the reliability and 90% credible interval of the reliability assuming a uniform prior unreliability in the range of 0.002–0.01. Compare the results with the MLE estimate of the mean and confidence intervals of the reliability. Using the results, compare and contrast the two methods.

5.36 Engineering judgment for the reliability of a new device over its designed lifetime is 0.85. Propose an appropriate prior pdf that describes this judgment. If we performed accelerated tests replicating the device lifetime on ten new devices and observed no failures, assess the posterior mean and 90% two-side credible interval of the reliability.

5.37 Assume the exponential distribution has been shown to be a suitable model for X, the size of embedded cracks in pipes. The prior on the crack growth parameter λ is expressed by a gamma pdf with parameters $\alpha=52$, $\beta=2.68$ mm. After inspecting eight pipes, we measured the following crack sizes (in mm): 0.95, 1.00, 1.12, 1.98, 2.02, 2.10, 2.13, 2.82.

a. Find the pdf of the posterior distribution for λ.
b. Find the mean and 95% credible interval of crack sizes.

REFERENCES

217Plus, Quanterion Solutions, 2015. https://www.quanterion.com/products-services/tools/217plus/

Bain, L. J., *Statistical Analysis of Reliability and Life-Testing Models: Theory and Methods*, 2nd Edition, Routledge, New York, 2017.

Blom, G., *Statistical Estimates and Transformed Beta-Variables*. Wiley, New York, 1958.

Center for Chemical Process Safety of the American Institute of Chemical Engineer, *Guidelines for Process Equipment Data*, Center for Chemical Process Safety of the American Institute of Chemical Engineer, New York, 1989.

Eide, S. A. et al., *Industry-Average Performance for Components and Initiating Events at U.S. Commercial Nuclear Power Plants*, NUREG/CR-6928, U.S. Nuclear Regulatory Commission, Washington, DC, Feb. 2007.

Ericson, D. M. et al., *Analysis of Core Damage Frequency: Internal Events Methodology*. Vol. 1, NUREG/CR-4550. U.S. Nuclear Regulatory Commission, Washington, DC, 1990.

Epstein, B., "Estimation from Life Test Data." *Technometrics*, 2, 447, 1960.

Gelman, A., J. B. Carlin, H. S. Stern, D. B. Dunson, A. Vehtari, and D. B. Rubin, *Bayesian Data Analysis*, 3rd edition, Chapman & Hall/CRC, New York, 2013.

IEEE Std. 500, *Guide to the Collection and Presentation of Electrical, Electronic, Sensing Component and Mechanical Equipment Reliability Data for Nuclear Power Generating Stations*. IEEE Standards, New York, 1984.

IEC/TR 62380, *Reliability Data Handbook—Universal Model for Reliability Prediction of Electronics Components, PCBs and Equipment*. International Electrotechnical Commission, Geneva, 2004.

Kapur, K. C. and L. R. Lamberson, *Reliability in Engineering Design*. John Wiley and Sons, New York, 1977.

Kelly, D. and C. Smith, *Bayesian Inference for Probabilistic Risk Assessment: A Practitioner's Guidebook*, Springer-Verlag, London, 2011.

Kimball, B. F., "On the Choice of Plotting Positions on Probability Paper." *Journal of the American Statistical Association*, 55(291), 546–560, 1960.

Lawless, J. F., *Statistical Models and Methods for Lifetime Data*. Wiley, New York, 2011.

Ma, Z., T. E. Wierman, and K. J. Kvarfordt, *Industry-Average Performance for Components and Initiating Events at U.S. Commercial Nuclear Power Plants: 2020 Update*. INL/EXT-21-65055, Idaho National Laboratory, Nov. 2021.

Martz, H. F. and R. A. Waller, *Bayesian Reliability Analysis*. Wiley, New York, 1982.

Mann, N. R., R. E. Schafer, and N. D. Singpurwalla, *Methods for Statistical Analysis of Reliability and Life Data*. Wiley, New York, 1974.

MIL -217 (1991) *Military Handbook: Reliability Prediction of Electronic Equipment*. MIL-HDBK-217F. 2 December 1991. Department of Defense, Washington DC.

Nelson, W. B., *Applied Life Data Analysis*. John Wiley & Sons, New York, 2003.

O'Connor, A. N., M. Modarres, and A. Mosleh, *Probability Distributions Used in Reliability Engineering*, DML International, Washington, DC, 2019.

Salemi, S., L. Yang, J. Dai, J. Qin, and J. Bernstein, *Physics-of-Failure Based Handbook of Microelectronic Systems*. Reliability Information Analysis Center, Utica, NY, 2008.

6 System Reliability Analysis

Recall from Chapter 1 that a *system* is a collection of items (subsystems, structures, components, software, human operators, etc.) whose proper, coordinated operation leads to the proper functioning of the system.

Several aspects must be considered in system reliability analysis. One is the reliability of the components that comprise the system, using the techniques described in Chapters 2–5. Another relevant consideration is the physical configuration of the system and its components. A third consideration is the failure mode of an item discussed in Chapter 1 and how that item's failure mode affects the system.

For example, consider a pumping system that draws water from two independent storage tanks—one primary and one backup tank, each equipped with a supply valve. If the supply valve on one tank fails closed (i.e., it does not open when demanded), the pumping system can still operate because of the second tank. Thus, it is important to model the configuration of the system in addition to modeling the reliability of each component. Now, consider the effect of failure modes. If the second tank's supply valve fails in an open position, the pump can still draw water and successfully operate. However, if the supply valve fails in a closed position, the pump cannot operate, and the pumping system fails. So, the failure mode of the valves is highly relevant for system reliability analysis.

Thus, in this chapter, we will be describing functions such that the reliability of a system with n components is described as a function of the reliability of those components:

$$R_{\text{system}}(t) = f\left(R_{\text{comp } 1}(t),\ R_{\text{comp } 2}(t),\ \ldots,\ R_{\text{comp } n}(t)\right). \tag{6.1}$$

Chapter 1 discussed the major steps in conducting system reliability modeling, drawing on techniques discussed in every chapter. In this chapter we will explain several important methods:

- *Reliability block diagrams:* A success-oriented model illustrating how the functional arrangement of items in the system affects overall system performance.
- *Fault trees, event trees & event sequence diagrams (ESDs):* A set of Boolean logic-based methods for modeling complex systems.
- *FMEA and FMECA (Failure modes and effects analysis and failure modes, effects, and criticality analysis):* A qualitative technique used to identify potential failure modes and their effects on the system. The FMECA is a simple semi-quantitative system reliability assessment approach.

DOI: 10.1201/9781003307495-6

6.1 IMPORTANT NOTES AND ASSUMPTIONS

In some aspects of this chapter, we assume that items composing a system are statistically independent (according to the definition provided in Chapter 2). This is a critical assumption—and reliability engineers must verify that this assumption is appropriate if working with a real-world system. We also assume that all failure modes of a component are mutually exclusive. In Chapter 8, we will elaborate on system reliability considerations when there are statistically dependent components.

Regarding notation, in this chapter, sometimes we will abbreviate $R(t)$ as simply R. When this is the case, all reliabilities or probabilities should be evaluated for the same time t. We also remind the reader that notation is highly variable in the field. In this book, usually we use the notation A to denote failure of an item A and $\bar{A}$ to mean success (non-failure) of item A. However, we may sometimes swap this notation to permit easier communication, most often when describing models or concepts that work in failure space. In this chapter, we may also use the notation R_A or A_s denote reliability or success of item A, and F_A or A_f to denote failure of item A when the overbar notation becomes burdensome.

Sometimes, this notation is reversed or relaxed, especially in problems involving success space and in engineering drawings. We strongly encourage readers to be mindful of this. *Always document the notation and conventions used at the beginning of your work.*

6.2 RELIABILITY BLOCK DIAGRAM METHOD

Reliability block diagrams (RBDs) are a success-focused method frequently used to model simple systems. They provide an effective starting point for our discussions on system reliability. The RBD shows the functional configuration of the system by depicting which components must function to achieve the system mission or function. A separate RBD should be drawn for each function if a system has several functions.

In RBDs, blocks often correspond to the physical arrangement of some items in the system. However, in some instances, they may not correspond. For example, consider two valves structurally in a series configuration in a gas pump (Figure 6.1a), each of which has a failure mode "failure to open on demand." Suppose these gas are supply valves and we are modeling the reliability of this system as to whether the system reliability delivers fluid when available. In that case, the RBD is shown

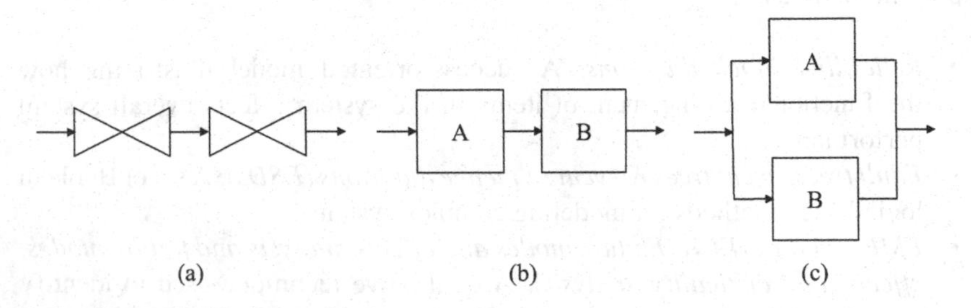

FIGURE 6.1 (a) Two valves in series may become either (b) a series RBD or (c) a parallel RBD, depending on the failure logic of the system.

in Figure 6.1b, which denotes that *both* Valve A and Valve B must operate reliably to deliver the fluid. Thus, this valve arrangement is a series system with respect to fluid delivery.

Instead, consider these valves as emergency relief valves designed to open and vent to atmosphere if the supply line becomes over-pressurized. If one valve successfully opens, then the fluid can be released from the system. Considering the valve failure mode of "failure to open on demand," the RBD looks different, as shown in Figure 6.1c. This would be a parallel RBD structure because only one valve needs to be reliable for the emergency relief function to be achieved. Thus, even though these valves are physically in series, they act as a parallel system regarding reliability.

For instance, when two resistors are in a parallel configuration, the system fails if one resistor fails short. Therefore, the RBD of this system for the "fail short" mode of failure would be composed of two series blocks, denoting that both resistors must succeed for the system to succeed. However, for other failure modes of one item, such as the "open" failure mode, the RBD comprises two parallel blocks. In the remainder of this section, we discuss the reliability of the system for several types of the system's functional configurations.

6.2.1 Series System

A reliability block diagram is in the series configuration when the failure of any one item (in the failure mode of each item considered when the RBD is developed) fails the system. Accordingly, for the functional success of a series system, all its blocks (items) must successfully function during the intended mission time of the system. Figure 6.2 shows the RBD of a series system consisting of n blocks.

The system, A, in Figure 6.2 is reliable when all n items successfully operate (are reliable) during mission time t. If we write this in event notation, using Boolean logic, the event "System A succeeds," A_S, has the following logic:

$$A_S = R_1 \cap R_2 \cap R_3 \cap \ldots \cap R_n, \tag{6.2}$$

meaning that system A succeeds if item 1 works, item 2 works, and so on; that is, all n items succeed during the intended mission time t.

Thus, the reliability of the system is the probability of the joint success of all n items:

$$R_A(t) = Pr(R_1 \cap R_2 \cap R_3 \cap \ldots \cap R_n). \tag{6.3}$$

Probabilistically, the system reliability $R_s(t)$, which represents the intersection of events wherein items 1 through n work, is calculated from the chain rule of probability:

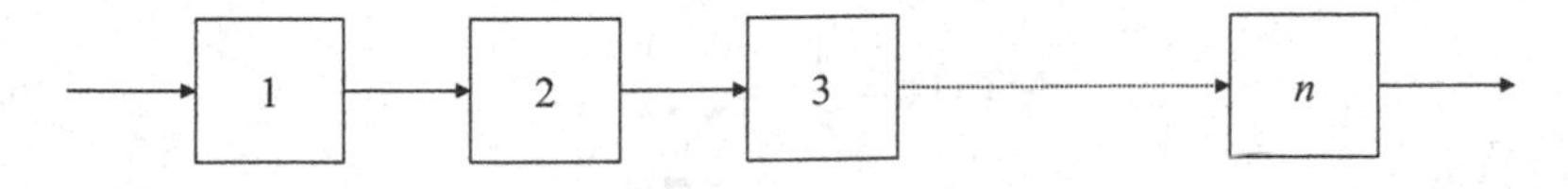

FIGURE 6.2 Series system reliability block diagram.

$$R_s(t) = Pr(R_1) \cdot Pr(R_2|R_1) \cdot Pr(R_3|R_2 \cap R_1) \cdot \ldots \cdot Pr(R_n|R_{n-1} \cap \ldots \cap R_1). \quad (6.4)$$

This expression is relatively complex; however, for independent items, it reduces readily.

The system reliability $R_s(t)$ for n *independent* items (indicated as $1 \perp 2 \perp \ldots \perp n$) is obtained from:

$$R_s(t) = R_1(t) \cdot R_2(t) \cdot \ldots \cdot R_n(t) = \prod_{i=1}^{n} R_i(t). \quad (6.5)$$

where $R_i(t)$ represents the reliability of the ith block or item.

If the items are not independent, Equation 6.4 can be calculated exactly for small numbers of dependent items. In Chapter 8 we will discuss some advanced methods for handling dependency.

Intuition: Upon examining Equation 6.5, it is seen that the upper bound on system reliability in a series system is that of the least reliable component. Thus, the system's reliability can be no greater than the reliability of the least reliable component.

The hazard rate for a series system of independent items is also a convenient expression. Since $h(t) = -d \ln R(t)/dt$, applying this to Equation 6.5, the hazard rate of the system, $h_s(t)$, is:

$$h_s(t) = -\frac{d \ln \prod_{i=1}^{n} R_i(t)}{dt} = \sum_{i=1}^{n} -\frac{d \ln R_i(t)}{dt} = \sum_{i=1}^{n} \lambda_i(t). \quad (6.6)$$

Let us further assume a constant failure rate model for each block (i.e., assume an exponential time to failure for each independent item i), that is, $h_i(t) = \lambda_i$. Then, according to Equation 6.6, the system failure rate for such a system is also constant and is given by:

$$h_s = \lambda_s = \sum_{i=1}^{n} \lambda_i. \quad (6.7)$$

The reliability expression for a system with independent items and CFR can also be easily obtained from Equation 6.5 using the constant failure rate reliability model for each item, $R_i(t) = e^{-\lambda_i t}$. Then the system reliability and MTTF are given by Equations 6.8 and 6.9, respectively.

$$R_s(t) = \prod_{i=1}^{n} e^{-\lambda_i t} = e^{\left(-t \sum_{i=1}^{n} \lambda_i\right)} = e^{-\lambda_s t}, \quad (6.8)$$

$$\text{MTTF}_s = \frac{1}{\lambda_s} = \frac{1}{\sum_{i=1}^{n} \lambda_i}. \quad (6.9)$$

Example 6.1

A system consists of three items whose reliability block diagram is in series. The failure rate for each item is constant: $\lambda_1 = 4.0 \times 10^{-6}$/hour, $\lambda_2 = 3.2 \times 10^{-6}$/hour, and $\lambda_3 = 9.8 \times 10^{-6}$/hour. Determine these parameters of the system:

a. λ_s
b. R_s (1,000 hours)
c. MTTF_s

Solution:

a. According to Equation 6.7, $\lambda_s = (4.0 \times 10^{-6}) + (3.2 \times 10^{-6}) + (9.8 \times 10^{-6})$
$= 1.7 \times 10^{-5}$/hour.

b. $R_s(t = 1{,}000) = e^{-\lambda_s t} = e^{-1.7 \times 10^{-5} \times 1000} = 0.983$, or unreliability of $F_s(1{,}000) = 0.017$.

According to Equation 6.9, $\text{MTTF}_s = \dfrac{1}{\lambda_s} = \dfrac{1}{1.7 \times 10^{-5}} = 58{,}824$ hours.

6.2.2 Parallel Systems

In an RBD with a parallel configuration, system A succeeds when any of its n items succeed during the mission time t. This parallel configuration is a form of active redundancy. Figure 6.3 shows the RBD of a parallel system consisting of n blocks.

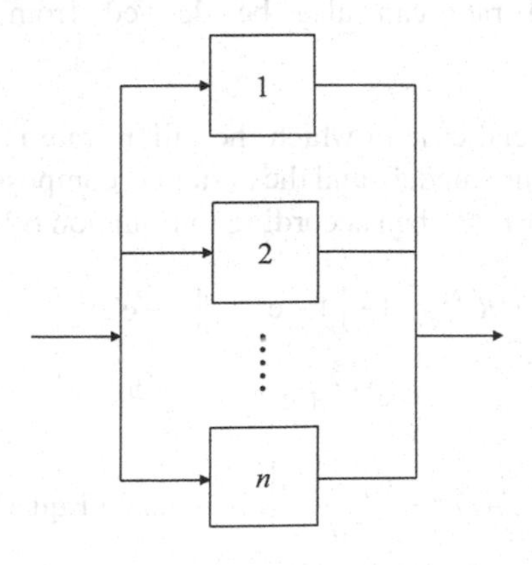

FIGURE 6.3 Parallel system reliability block diagram.

If we write this in event notation using Boolean logic, the event "System A succeeds," A_S, has the following logic:

$$A_S = R_1 \cup R_2 \cup R_3 \cup \ldots \cup R_n, \tag{6.10}$$

meaning system A succeeds if item 1 works or item 2 works or item 3 works, and so on. Notice the difference between the logic for the series and the parallel system. The success of only one item will ensure the system's success if the items are arranged in parallel.

Accordingly, the parallel (or active redundant) system S fails only if *all* of its n items fail. In Boolean logic, "system A fails" has the following logic:

$$A_f = \overline{A_s} = \overline{R_1} \cap \overline{R_2} \cap \overline{R_3} \cap \ldots \cap \overline{R_n}, \tag{6.11}$$

where the overbar denotes "not." Thus, the unreliability of a parallel system F_S, is:

$$F_s(t) = Pr(F_1 \cap F_2 \cap F_3 \cap \ldots \cap F_n). \tag{6.12}$$

For a system of n *independent items*, $1 \perp 2 \perp \ldots \perp n$, the expressions for reliability and unreliability can be directly established from Equation 6.12 using the probability laws discussed in Chapter 2.

$$F_s(t) = F_1(t) \cdot F_2(t) \cdot \ldots \cdot F_n(t) = \prod_{i=1}^{n} F_i(t). \tag{6.13}$$

Since $R_i(t) = 1 - F_i(t)$, then

$$R_s(t) = 1 - F_s(t) = 1 - \prod_{i=1}^{n} [1 - R_i(t)]. \tag{6.14}$$

If items are not independent, the corresponding expressions need to be derived using the chain rule of probability or some advanced methods discussed in Chapter 8.

The system hazard rate can also be derived from Equation 6.14 using $h(t) = -\dfrac{d \ln R(t)}{dt}$.

Let us analyze a special case in which the failure rate is constant for each item (exponential time to failure model), and the system is composed of only two independent items. Since $R_i(t) = e^{-\lambda_i t}$, then according to Equation 6.14,

$$\begin{aligned} R_s(t) &= 1 - \left[1 - \mathrm{e}^{-\lambda_1 t}\right]\left[1 - \mathrm{e}^{-\lambda_2 t}\right] \\ &= \mathrm{e}^{-\lambda_1 t} + \mathrm{e}^{-\lambda_2 t} - \mathrm{e}^{\left[-(\lambda_1 + \lambda_2)t\right]}. \end{aligned} \tag{6.15}$$

Since $\lambda_s(t) = \dfrac{f_s(t)}{R_s}$ and $f_s(t) = -\dfrac{d\left[R_s(t)\right]}{dt}$, then using Equation 6.15,

$$f_s(t) = \lambda_1 e^{-\lambda_1 t} + \lambda_2 e^{-\lambda_2 t} - (\lambda_1 + \lambda_2) e^{\left[-(\lambda_1 + \lambda_2)t\right]}. \tag{6.16}$$

Thus,

$$\lambda_s(t)=\frac{\lambda_1 e^{-\lambda_1 t}+\lambda_2 e^{-\lambda_2 t}-(\lambda_1+\lambda_2)e^{[-(\lambda_1+\lambda_2)t]}}{e^{-\lambda_1 t}+e^{-\lambda_2 t}-e^{[-(\lambda_1+\lambda_2)t]}}. \quad (6.17)$$

The MTTF of the system can also be obtained as

$$\text{MTTF}_s=\int_0^{\infty} R_s(t)dt=\int_0^{\infty} e^{-\lambda_1 t}+e^{-\lambda_2 t}-e^{[-(\lambda_1+\lambda_2)t]}dt$$
$$=\frac{1}{\lambda_1}+\frac{1}{\lambda_2}-\frac{1}{\lambda_1+\lambda_2}. \quad (6.18)$$

For the system of n parallel items, we can use the binomial expansion to derive the MTTF:

$$\text{MTTF}_s=\left(\frac{1}{\lambda_1}+\frac{1}{\lambda_2}+\ldots+\frac{1}{\lambda_n}\right)-\left(\frac{1}{\lambda_1+\lambda_2}+\frac{1}{\lambda_1+\lambda_3}+\ldots+\frac{1}{\lambda_{n-1}+\lambda_n}\right)$$
$$+\left(\frac{1}{\lambda_1+\lambda_2+\lambda_3}+\ldots+\frac{1}{\lambda_{n-2}+\lambda_{n-1}+\lambda_n}\right)-\ldots+(-1)^{n+1}\frac{1}{(\lambda_1+\lambda_2+\ldots+\lambda_n)}. \quad (6.19)$$

In the special case where all items are independent and identically distributed (i.i.d.) with a constant failure rate λ, Equation 6.14 simplifies to the following form:

$$R_s(t)=1-\left[1-e^{-\lambda t}\right]^n, \quad (6.20)$$

and from Equation 6.19,

$$\text{MTTF}_s=\text{MTTF}_i\left(1+\frac{1}{2}+\frac{1}{3}\ldots+\frac{1}{n}\right). \quad (6.21)$$

Example 6.2

A system is composed of the same items as in Example 6.1. However, these items are in parallel. Find the time to failure cdf (unreliability) for a mission time of 1,000 hours and MTTF_s.

Solution:

For a parallel system,

$$F_s(t)=\left[1-e^{-\lambda_1 t}\right]\left[1-e^{-\lambda_2 t}\right]\left[1-e^{-\lambda_3 t}\right],$$

$$F_s(1{,}000) = \left(1 - e^{\left[-4.0\times10^{-6}(1{,}000)\right]}\right)\left(1 - e^{\left[-3.2\times10^{-6}(1{,}000)\right]}\right)\left(1 - e^{\left[-9.8\times10^{-6}\,(1{,}000)\right]}\right) = 1.25\times10^{-7}.$$

$$\text{MTTF}_s = \left(\frac{1}{\lambda_1}+\frac{1}{\lambda_2}+\frac{1}{\lambda_3}\right)-\left(\frac{1}{\lambda_1+\lambda_2}+\frac{1}{\lambda_1+\lambda_3}+\frac{1}{\lambda_2+\lambda_3}\right)+\left(\frac{1}{\lambda_1+\lambda_2+\lambda_3}\right)$$

$$= 4.35\times10^5 \text{ hour.}$$

Intuition: It can be seen from Equation 6.14 that in the design of parallel redundant systems, the reliability of the system exceeds the reliability of an individual item. However, as seen in Equation 6.21, the contribution to the MTTF of the system from the second item, the third item, and so on would have a diminishing return as the number of items, n, increases. Therefore, there would be an optimum number of parallel items by which a designer can maximize the reliability and when trying to minimize the life cycle cost.

6.2.3 *k*-Out-of-*n* Redundant Systems

Let us consider a more general structure of series and parallel systems: the *k-out-of-n system*. In this type of system, if any combination of k out of n items works, it guarantees the success of the system. Clearly, the series system is an example of an n-out-of-n system, and the parallel system is a 1-out-of-n structure.

Suppose these n items are independent and have identical reliabilities at a given time, t, by viewing each item as an independent trial. Thus, we can see that reliability of these k-out-of-n systems of i.i.d. items can be modeled with the binomial distribution. Thus, the binomial distribution can easily represent the probability that the system functions:

$$R_s(t) = \sum_{x=k}^{n}\binom{n}{x}\left[R_i(t)\right]^x\left[1-R_i(t)\right]^{n-x}, \quad (6.22)$$

Which can also be rearranged to solve as such:

$$= 1-\sum_{x=0}^{k-1}\binom{n}{x}\left[R_i(t)\right]^x\left[1-R_i(t)\right]^{n-x}. \quad (6.23)$$

Example 6.3

How many components should be used in an active redundancy design to achieve a system reliability of 0.999 such that, for successful system operation, at least two components are required? Assume a mission of $t=720$ hours for a set of identical and independent components that have a constant failure rate of 0.00015 failures per hour.

Solution:

For each component, $R_i(t) = e^{-\lambda t} = e^{(-0.00015 \times 720)} = 0.8976$.

For $k = 2$ components, and using Equation 6.23,

$$R_s(t = 720) = 0.999 = 1 - \sum_{x=0}^{1} \binom{n}{x} (0.8976)^x (0.1024)^{n-x}$$

$$= 1 - (0.1024)^n - n(0.8976)(0.1024)^{n-1}.$$

From the above equation, $n = 5$, so at least five components should be used to achieve the desired reliability over the specified mission time.

6.2.4 Standby Systems

A system is called a *redundant standby system* when some of its independent items remain idle until they are called for service by a sensing and switching device. For simplicity, let us consider a situation where only one item operates actively while the others (blocks 2–n) are on standby, as shown in Figure 6.4.

This differs from active redundancy parallel systems, where parallel items both operate actively. In the redundant standby configuration, item 1 constantly operates until it fails. The sensing and switching device recognizes an item failure in the system and switches to another identical item. This process continues until all standby items have failed, at which point the system fails. Since items 2 to n don't operate constantly, we would expect them to fail at a much slower rate (i.e., to have higher reliability in standby than in active use; we will denote the reliability in standby as R'). This is because the failure rate for components is usually higher when operating than when they are idle or dormant.

Here, system reliability depends on the reliability of the sensing and switching device and the reliability of the items. Consider the reliability of a two-item redundant standby system. We recognize two possible scenarios. The first scenario is the

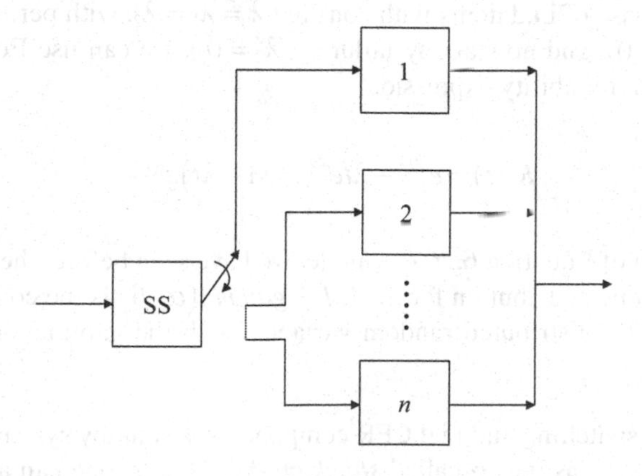

FIGURE 6.4 Redundant standby system reliability block diagram.

event of item 1 working over the entire mission time t (i.e., the probability that item 1 succeeds the whole mission time). The second scenario finds the probability of an event wherein: (1) item 1 fails at time t_1 before mission time t, (2) the probability that the sensing and switching item does not fail by t_1, (3) the probability that standby item 2 does not fail by t_1 (in the standby mode), and (4) the probability that standby item 2 successfully functions for the remainder of the mission. These scenarios are mutually exclusive, and the system reliability (probability of no failure) associated with each scenario can be added.

Mathematically, the reliability function for a two-item standby device with independent items can be obtained as:

$$R_s(t) = R_1(t) + \int_0^t f_1(t_1)dt_1 \cdot R_{ss}(t_1) \cdot R_2'(t_1) \cdot R_2(t - t_1). \tag{6.24}$$

Equation 6.24 is constructed as follows: $R_1(t)$ is the probability that item 1 does not fail by the mission time t; $f_1(t_1)dt_1$ is the probability of item 1 failing at an arbitrary time t_1 within the small time interval dt_1; $R_{ss}(t_1)$ is the probability that the switch has not failed by time t_1; $R_2'(t_1)$ is the probability that item 2 has not failed by time t_1 while in standby mode; and $R_2(t-t_1)$ is the probability that item 2 does not fail after time t_1 to some variable time, t, in the future. Since t_1 is an arbitrary time, then it is integrated over its possible range of 0–t (i.e., to the current or mission time, t).

Let us consider a simple case where the time to failure of both independent items and the switching system follows an exponential distribution.

$$\begin{aligned} R_s(t) &= e^{-\lambda_1 t} + \int_0^t \lambda_1 e^{-\lambda_1 t_1} e^{-\lambda_{ss} t_1} e^{-\lambda_2' t_1} e^{[-\lambda_2(t-t_1)]} dt_1, \\ &= e^{-\lambda_1 t} + \frac{\lambda_1 e^{-\lambda_2 t}}{\lambda_1 + \lambda_{ss} + \lambda_2' - \lambda_2}\left(1 - e^{-(\lambda_1 + \lambda_{ss} + \lambda_2' - \lambda_2)t}\right). \end{aligned} \tag{6.25}$$

For the special case of i.i.d items with constant $\lambda = \lambda_1 = \lambda_2$, with perfect sensing and switching ($\lambda_{ss} = 0$), and no standby failures ($\lambda_2' = 0$), we can use Equation 6.25 to derive the system reliability expression:

$$R_s(t) = e^{-\lambda t} + \lambda t e^{-\lambda t} = (1 + \lambda t)e^{-\lambda t}. \tag{6.26}$$

A generalization of Equation 6.26 is a model we have seen before: the gamma distribution. The gamma distribution for r.v. T, $T \sim gamma(\alpha, \beta)$ is the convolution (sum) of α exponentially distributed random variables with the gamma shape parameter $\beta = \frac{1}{\lambda}$.

With perfect switching and i.i.d CFR components, a standby system possesses the same characteristics as the so-called *shock model*. That is, one can assume that the

nth shock (i.e., the nth item failure) causes the system to fail. Thus, a gamma distribution can represent the time to failure of the system such that α = number of units and $\beta = \frac{1}{\lambda}$.

$$R_s(t) = 1 - \int_0^t \frac{\lambda^n}{\Gamma(n)} x^{n-1} e^{-\lambda x} dx$$

$$= e^{-\lambda t}\left[1 + \lambda t \frac{(\lambda t)^2}{2!} + \ldots + \frac{(\lambda t)^{n-1}}{(n-1)!}\right]. \tag{6.27}$$

Accordingly, the MTTF of the above system is given by the mean of the gamma distribution:

$$\text{MTTF}_S = \frac{n}{\lambda}, \tag{6.28}$$

which is n times the MTTF of a single item. This expression explains why high reliability can be achieved through a standby system when the switching is perfect, and no failures occur during the standby period.

When more than two items are on standby, the function describing the reliability of the standby redundant system becomes more difficult, but the concept remains the same. For example, for three items with perfect switching,

$$R_s(t) = R_1(t) + \int_0^t f_1(t_1)dt_1 \cdot R_2(t - t_1) + \int_0^t f_1(t_1)dt_1 \cdot \int_0^t f_2(t_2)dt_2 \cdot R_3(t - t_1 - t_2). \tag{6.29}$$

If the sensing and switching devices are not perfect, appropriate terms should be added to Equation 6.29 to account for their unreliability, similar to Equation 6.24.

Example 6.4

Consider two identical independent items with $\lambda = 0.01$ per hour, operating for a mission time $t = 24$ hours. Compare the reliability of a system made of these items if they are placed in

a. Parallel configuration.
b. Series configuration.
c. Standby configuration with perfect switching.
d. Standby configuration with imperfect switching and standby failure rates of $\lambda_{ss} = 1 \times 10^{-6}$ and $\lambda' = 1 \times 10^{-5}$ per hour, respectively.

Solution:

From the exponential time to failure model for each item, we find:

$$R_i(t=24)=e^{-\lambda t}=e^{[-0.01(24)]}=0.787.$$

Then

a. For the parallel system, $R_s(t)=1-\prod_{i=1}^{n}[1-R_i(t)]$,

$$R_s(t=24)=1-(1-0.787)^2=0.954.$$

b. For the series system, $R_s(t)=\prod_{i=1}^{n}R_i(t)$.

$$R_s(t=24)=0.787(0.787)=0.619.$$

c. For the standby system with perfect switches, $R_s(t)=(1+\lambda t)e^{-\lambda t}$,

$$R_s(t=24)=(1+0.24)e^{[-0.01(24)]}=0.976.$$

d. For the standby system with imperfect switching and standby failure rate using Equation 6.25,

$$R_s(t=24)=0.787+\frac{0.01(0.787)}{1.1\times10^{-5}}\left(1-e^{[-1.1\times10^{-5}(24)]}\right)=0.976.$$

6.2.5 Load-Sharing Systems

A *load-sharing system* refers to a system whose items equally share or contribute to the system function. For example, if a set of two identical parallel pumps together delivers x gpm of water to a reservoir, each pump delivers $x/2$ gpm. If at least x gpm is always required, and one pump fails at a given arbitrary time, t_0, then the other pump's speed should be increased to provide x gpm alone. Other examples of load-sharing are multiple load-bearing items (such as columns in a bridge) and load-sharing multiunit electric power plants. In these cases, when one item fails, the others should carry the full load. Since these other items would then work under a more stressful condition, they would experience a higher failure rate.

Load-sharing system reliability models can be divided into two groups: time-independent models (where stress, not time, is the agent of failure) and time-dependent models (where both stress and time are agents of failure). The time-independent reliability models are considered in the framework of stress-strength analysis. While most of the reliability models discussed in this book are time-dependent, for conceptual clarity in this section we start with the time-independent. Historically, the first time-independent load-sharing system model is known as the Daniels model

(Daniels, 1945). This model was initially applied to textile strength problems; now, it is also applied to composite materials.

To illustrate the basic ideas associated with this model, consider a simple parallel system composed of two identical components. Let $F(s)$ be the failure probability for the component subjected to load (stress) s. The failure probability for a parallel system of two identical components, denoted by $F_2(s)$, can be found. The reliability function of the system, $R_2(s)$, is $1-F_2(s)$. Initially, both components are subjected to an equal load s. When one component fails, the non-failed component takes on the full load, $2s$.

The probability of the system failure, $F_2(s)$, can be modeled. Let A be the event when the first component fails under load s and the second component fails under load $2s$; let B be the event in which the second component fails under load s, and the first component fails under load $2s$. Finally, let $A \cap B$ be the event that both components fail under load s:

$$Pr(A \cup B) = Pr(A) + Pr(B) - Pr(A \cap B). \tag{6.30}$$

It is evident that

$$Pr(A) = Pr(B) = F(s) \cdot F(2s),\ Pr(A \cap B) = F^2(s), \tag{6.31}$$

hence

$$F_2(s) = 2F(s) \cdot F(2s) - F^2(s), \tag{6.32}$$

and

$$R_2(s) = 1 - 2F(s) \cdot F(2s) + F^2(s). \tag{6.33}$$

A similar equation for the reliability of a three-component load-sharing system contains seven terms, and the problem becomes more difficult as the number of components increases. For such situations, different recursive procedures have been developed (Crowder et al., 1991).

Now, consider a simple example of the time-dependent load-sharing system model. Let us assume again that two components share a load (i.e., each component carries half the load) and that the time to failure distribution for both components is $f_h(s, t)$. When one component fails (i.e., one component carries the full load), the time to failure distribution is $f_f(2s, t)$. Let us also assume that the corresponding reliability functions during full-load and half-load operation are $R_f(2s, t)$ and $R_h(s, t)$, respectively. The system will succeed if both components carry half the load, or if component 1 fails at time t_1 and component 2 carries a full load after that, or if component 2 fails and component 1 takes the full load thereafter. Accordingly, the system reliability function $R_s(t)$ can be obtained from Kapur and Lamberson (1977):

$$R_s(t) = \left[R_h(s,t)\right]^2 + 2\int_0^t f_h(s,t_1) R_h(s,t_1) R_f(2s,t-t_1)\,dt_1. \tag{6.34}$$

In Equation 6.34, the first term shows the contribution from both components working successfully, with each carrying a half load; the second term represents the two equal probabilities that component 1 fails first and component 2 takes the full load at time t_1 or vice versa. The second term is constructed as follows: $f_h(s, t_1)dt_1$ is the probability that one component fails around the arbitrary time t_1 within the short time interval, dt_1; $R_h(s, t_1)$ is the probability that the other component survives at half load until time t_1; $R_f(2s, t-t_1)$ is the probability that the surviving component works at full load from time t_1 until a future (mission) time, t. So, if we integrate the second term over all possible values of the arbitrary time, t_1, the system reliability as expressed in Equation 6.34 can be found.

Suppose switching or control mechanisms are involved to shift the total load to the non-failed component when one component fails. Then, similar to Equation 6.24, the reliability of the switching mechanism can be incorporated into Equation 6.34.

In the special situation where exponential time to failure models with failure rates λ_f and λ_h can be used for the two components under full and half loads, respectively, then Equation 6.34 simplifies to:

$$R_s(t) = e^{-2\lambda_h t} + \frac{2\lambda_h e^{-\lambda_f t}}{(2\lambda_h - \lambda_f)}\left\{1 - e^{-(2\lambda_h - \lambda_f)t}\right\}. \tag{6.35}$$

The reader is referred to Crowder et al. (1991) for a review of more sophisticated time-dependent load-sharing models.

6.3 COMPLEX SYSTEM EVALUATION METHODS

Most practical systems are neither parallel nor series. Often, systems exhibit some hybrid combination of the two. These systems are called *parallel-series systems*. Figure 6.5 shows an example of such a system. Another type of complex system is

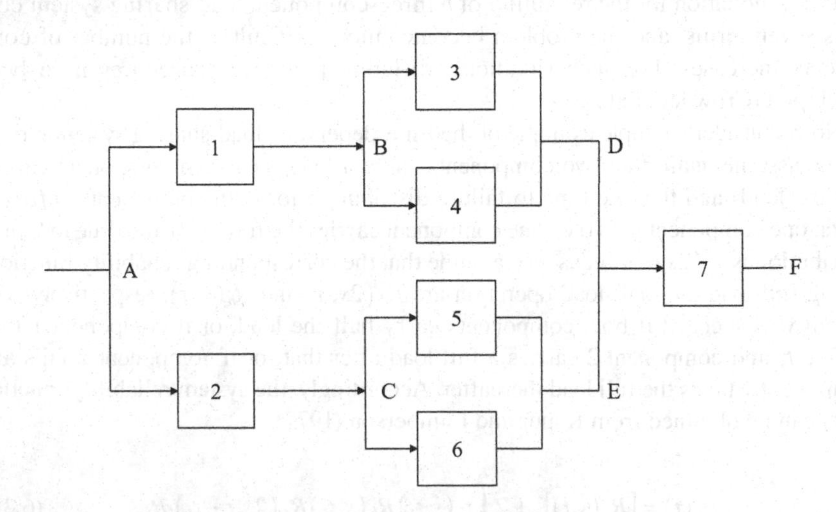

FIGURE 6.5 Complex parallel-series system reliability block diagram.

one that is neither series nor parallel alone, nor parallel-series. Figure 6.6 shows an example of such a system.

A parallel-series system can be analyzed by dividing it into its basic parallel and series modules and then determining the reliability function for each module separately. The process can be continued until a reliability function for the whole system is determined. Shooman (1990) describes several analytical methods for complex systems, including the inspection method, event space method, path tracing method, and decomposition method to analyze all types of complex systems. However, these methods are suitable only when there are not many items in the system. For analysis of complex systems with many items, fault trees are more appropriate.

In cases of very complex systems with multiple failure modes for each item and complex physical and operational interactions, using RBDs becomes difficult. The logic-based methods such as the fault tree and success tree analyses are more appropriate in this context. We will elaborate on this topic in Section 6.4. However, first we will describe two methods, the decomposition method and two path tracing methods, that are useful for qualitative and quantitative analysis of both RBDs and logic-based methods like fault trees.

6.3.1 Decomposition Method

The *decomposition method* uses the conditional probability concept to simplify the system and uses the Boolean expression $R = [R \cap X] \cup [R \cap \bar{X}]$. Accordingly, the reliability of a system is equal to the reliability of the system given that a chosen item works multiplied by the reliability of the item, plus the reliability of the system given the item's failure multiplied by the unreliability of the item. For example, using item 3 in Figure 6.6,

$$R_s(t) = R_s(t|\text{item 3 works}) \cdot R_3(t) + R_s(t|\text{item 3 fails})[1 - R_3(t)], \quad (6.36)$$

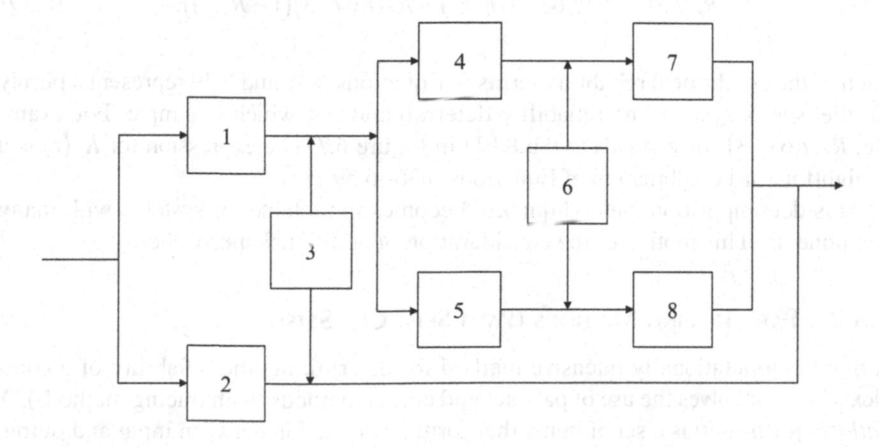

FIGURE 6.6 Complex nonparallel-series system reliability block diagram.

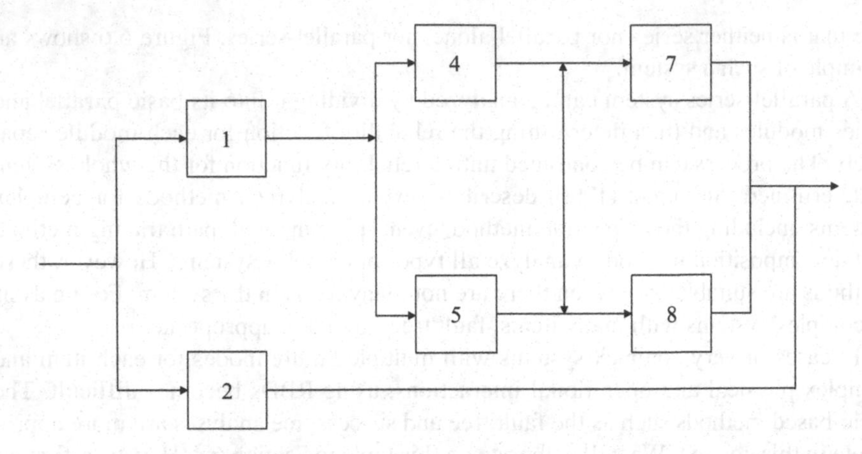

FIGURE 6.7 RBD Representation of $R_S\left(t|\overline{6}\cap 3\right)$.

which in the common notation used in this book is:

$$R_S(t) = R_s\left(t|\overline{3}\right)R_3(t) + R_s(t|3)\left[1 - R_3(t)\right]. \tag{6.37}$$

If Equation 6.36 is applied to all items that make the system a nonparallel-series (such as items 3 and 6 in Figure 6.6), the system will reduce to a simple parallel-series system. Thus, for Figure 6.6 and for the conditional reliability terms in Equation 6.37, it follows that

$$R_S\left(t|\overline{3}\right) = R_S\left(t|\overline{6}\cap\overline{3}\right)R_6(t) + R_S\left(t|6\cap\overline{3}\right)\left(1 - R_6(t)\right), \tag{6.38}$$

and

$$R_s(t|3) = R_S\left(t|\overline{6}\cap 3\right)R_6(t) + R_S(t|6\cap 3)\left(1 - R_6(t)\right). \tag{6.39}$$

Each of the conditional reliability terms in Equations 6.38 and 6.39 represent a purely parallel-series system, the reliability determination of which is simple. For example, $R_S\left(t|\overline{6}\cap 3\right)$ corresponds to the RBD in Figure 6.7. The expression for $R_S(t)$ is a straightforward combination of Equations 6.36–6.39.

This decomposition method quickly becomes intractable for systems with many components. This motivates the consideration of additional approaches.

6.3.2 Path Tracing Methods (Path Sets, Cut Sets)

A more computationally intensive method for determining the reliability of a complex system involves the use of path set and cut set methods (path tracing methods). A *path set* (or *tie set*) is a set of items that form a connection between input and output when traversed in the direction of the RBD arrows. Thus, a path set merely represents a "path" through the graph, and therefore successful system operation. A *minimal path set* (or minimal tie set) is a path set containing the minimum number of items needed to guarantee a connection between the input and output points. If any item were removed from the path set, it would no longer be a path set. For example, in Figure 6.6, path set $P_1 = (\bar{1}, \bar{3})$ is a minimal path set, and $P_2 = (\bar{1}, \bar{3}, \bar{6})$ is also a path set, but it is not a minimal path set, since the success of items 1 and 3 are sufficient to guarantee a path.

The *order* of a set describes how many elements are in that set. For example, in Figure 6.6, the minimal path sets are $P_1 = (\bar{2})$, $P_2 = (\bar{1}, \bar{3})$, $P_3 = (\bar{1}, \bar{4}, \bar{7})$, $P_4 = (\bar{1}, \bar{5}, \bar{8})$, $P_5 = (\bar{1}, \bar{4}, \bar{6}, \bar{8})$, *and* $P_6 = (\bar{1}, \bar{5}, \bar{6}, \bar{7})$; this represents a system with one path set of order one, one path set of order two, two path sets of order three, and two path sets of order four.

A *cut set* is a set of items that interrupts all possible connections between the input and output points, and therefore guarantees system failure. A *minimal cut set* is a cut set that contains only the set of items needed to guarantee an interruption of flow. Minimal cut sets show a combination of item failures that cause a system to fail. Thus, cut sets contain elements corresponding to item failure, whereas path sets contain elements corresponding to item success. The minimal cut sets for Figure 6.6 are $C_1 = (1, 2)$, $C_2 = (2, 3, 4, 5)$, $C_3 = (2, 3, 7, 8)$, $C_4 = (2, 3, 4, 6, 8)$, and $C_5 = (2, 3, 5, 6, 7)$.

Notice that each element of a path set represents the success of an item operation, whereas each element of a cut set represents the failure of an item. Thus, for probabilistic evaluations, the reliability function of each item should be used in connection with path set evaluations. In contrast, the unreliability function should be used in connection with cut set evaluation—and notational consistency is important.

Significant qualitative insight can be gained from the path sets and cut sets—even without quantifying reliability numerically. For example, identifying cut sets of order 1 can illustrate whether a system has any single point failures. One can also evaluate how many cut sets or path sets a component appears in, to understand the component's importance to the system reliability. In Chapter 8 we will discuss some importance measures that relate to the structure of the system alone.

Since many path sets may exist, the union of all these sets gives all possible events for the successful operation of the system. The probability of this union represents the reliability of the system. More simply, if a system has n minimal path sets denoted by $P_1, \ldots, P_n$, then the system reliability is given by:

$$R_S(t) = Pr(P_1 \cup P_2 \cup \ldots \cup P_n), \tag{6.40}$$

where each minimal path set, P_i, represents the event that items in the path set survive during the mission time t. This guarantees the success of the system.

It should be noted that the minimal path sets P_i are not necessarily mutually exclusive or independent. This poses a problem for determining the right side of Equation 6.40. In Section 6.4, we will explain formal methods to deal with this problem. However, an upper bound on the system reliability may be obtained by assuming that the minimal path sets P_i are mutually exclusive. Thus,

$$R_s \leq Pr(P_1) + Pr(P_2) + \cdots + Pr(P_n). \quad (6.41)$$

Equation 6.41 yields better answers when dealing with very small reliability values. Since this is not usually the case, Equation 6.41 is a poor bound to use in practical applications. This occurs because Equation 6.40 results in additional terms combinatorial terms, even if we assume independence among path sets (e.g., combinatorial terms such as $Pr(P_1 \cup P_2) = Pr(P_1) + Pr(P_2) - Pr(P_1)Pr(P_2)$. The final multiplicative term is clearly not negligible if reliability values are large.

Similarly, system reliability can be determined through the minimal cut sets. If the system has n minimal cut sets denoted by $C_1, \ldots, C_n$, then the system unreliability is the union of the cut sets; reliability is obtained from system unreliability:

$$F_s(t) = Pr(C_1 \cup C_2 \cup \ldots \cup C_n), \quad (6.42)$$

$$R_s(t) = 1 - Pr(C_1 \cup C_2 \cup \ldots \cup C_n), \quad (6.43)$$

where C_i represents the event that items in the cut set fail before the mission time t. This guarantees system failure. The Pr(·) term on the right-hand side of Equation 6.43 shows the probability that at least one of all possible minimal cut sets exists before time t.

Similar to the union of minimal path sets, the union of minimal cut sets is not a disjoint (mutually exclusive) function, and the cut sets are not always independent. So this complicates solving Equation 6.43. However, Equation 6.43 can be written in the form of a lower bound on system reliability:

$$R_s(t) \geq 1 - \left(Pr(C_1) + Pr(C_2) + \ldots + Pr(C_n)\right). \quad (6.44)$$

In practice, the bounding technique used in Equation 6.44 yields a much better representation of the reliability of the system than Equation 6.41 because most engineering items have reliability greater than 0.9 over their mission time, making the error from the use of Equation 6.44 considerably smaller than from the use of Equation 6.41. Similar to the path sets, assuming independence between cut sets, the following relation is applied: $Pr(C_1 \cup C_2) = Pr(C_1) + Pr(C_2) - Pr(C_1)Pr(C_2)$. The additional multiplicative terms will be small if the reliability is large. Therefore, they can be considered negligible. Also, this means that the cut sets method is a more conservative estimate of the reliability of the system due to these additional terms being ignored. This treatment is termed *rare event approximation*, which we introduced in Chapter 2 and will be further discussed in Section 6.4.2.1.

Both cut set and path set approaches are approximations; the exact system reliability will be somewhere between the two approaches, $R_S^{\text{cutset}} \leq R_S \leq R_S^{\text{pathset}}$.

Example 6.5

Consider the RBD in Figure 6.6. Determine the lower bound of the system reliability function using the cut set approach. The constant failure rates of each item are $\lambda_1, \ldots, \lambda_8$.

Solution:

Using the system cut sets discussed earlier, and the cut set method,

$$\left(R_s(t) \geq 1 - \left(Pr(C_1) + Pr(C_2) + \ldots + Pr(C_5)\right)\right).$$

Each cut set is quantified as such:

$Pr(C_1) = \left[1 - e^{-\lambda_1 t}\right]\left[1 - e^{-\lambda_2 t}\right]$, and so on. Therefore,

$$R_s(t) \geq 1 - \Big\{\left[1-e^{-\lambda_1 t}\right]\left[1-e^{-\lambda_2 t}\right] + \left[1-e^{-\lambda_2 t}\right]\left[1-e^{-\lambda_3 t}\right]\left[1-e^{-\lambda_4 t}\right]\left[1-e^{-\lambda_5 t}\right] + \left[1-e^{-\lambda_2 t}\right]$$
$$\left[1-e^{-\lambda_3 t}\right]\left[1-e^{-\lambda_7 t}\right]\left[1-e^{-\lambda_8 t}\right] + \left[1-e^{-\lambda_2 t}\right]\left[1-e^{-\lambda_3 t}\right]\left[1-e^{-\lambda_4 t}\right]\left[1-e^{-\lambda_6 t}\right]$$
$$\left[1-e^{-\lambda_8 t}\right] + \left[1-e^{-\lambda_2 t}\right]\left[1-e^{-\lambda_3 t}\right]\left[1-e^{-\lambda_5 t}\right]\left[1-e^{-\lambda_6 t}\right]\left[1-e^{-\lambda_7 t}\right]\Big\}$$

For some typical values of λ, the lower bound for $R_S(t)$ can be compared to the exact value of $R_S(t)$. Here, "exact" means that the cut sets are not assumed disjoint. For example, plotting both shows the exact and the lower probability bound of system reliability for $\lambda_1 = 1\times10^{-6}$/hour, $\lambda_2 = 1\times10^{-5}$/hour, $\lambda_3 = 2\times10^{-5}$/hour and $\lambda_4 = \lambda_5 = \lambda_6 = \lambda_7 = \lambda_8 = 1\times10^{-4}$/hour. The plot illustrate that as time increases, the reliability of the system decreases (item failure probability increases), causing Equation 6.44 to yield a poorer approximation. It is more appropriate to use Equation 6.41. Again, notice that Equations 6.41 and 6.44 assume the path sets and cut sets are disjoint.

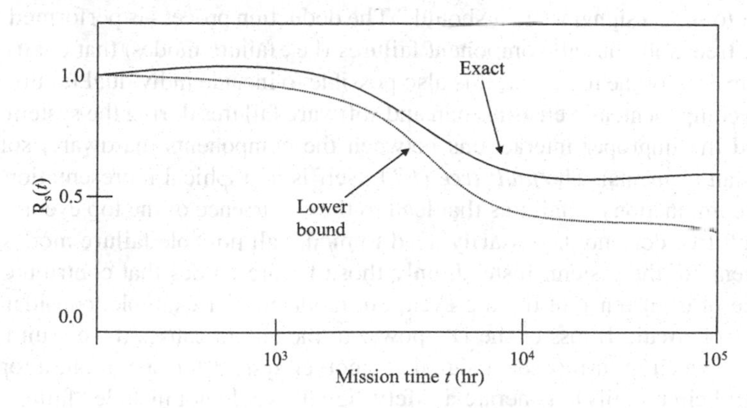

In very complex systems, especially those involving redundant items, components having multiple failure modes, and complex physical and operational interactions, RBDs become challenging to build and analyze. The logic-based methods such as fault tree and success tree analyses are more appropriate in this context. We will elaborate on this topic in the next section.

6.4 FAULT TREE AND SUCCESS TREE METHODS

Logic trees are useful approaches for modeling the reliability of complex systems and are widely used in applications where RBDs become too impractical to use. The operation of a system can be considered from two opposite viewpoints: the various ways that a system fails or the multiple ways that a system works. Most of the construction and analysis methods used are, in principle, the same for both fault trees and success trees. First, we will discuss the fault tree method and then describe the success tree method. Success tree analysis follows the same process and uses the same symbols and evaluation methods.

In general, conducting *fault tree analysis (FTA)* includes these steps, discussed in the following sections.

1. Define the system to be analyzed
2. Define the top event
3. Construct the fault tree
4. Perform qualitative evaluation (logic evaluation)
5. Perform quantitative evaluation (compute the probability of the top event).

6.4.1 Fault Tree Method

The fault tree method is a deductive process through which an undesirable event, called the *top event*, is postulated. The possible ways for this event to occur are systematically deduced. For example, a typical top event might be, "Failure of control circuit *A* to send a signal when it should." The deduction process is performed so that the fault tree embodies all component failures (i.e., failure modes) that contribute to the occurrence of the top event. It is also possible to include individual failure modes for each component, as well as human and software failures during the system operation (and the improper interactions between the components' hardware, software, and human elements). The *fault tree (FT)* itself is a graphical representation of the various combinations of failures that lead to the occurrence of the top event.

A fault tree does not necessarily need to model all possible failure modes of the components of the system. Instead, only those failure modes that contribute to the existence or occurrence of the top event are modeled. For example, consider a fail-safe control circuit. If loss of the DC power to the circuit causes the circuit to open a contact, which in turn sends a signal to another system for operation, a top event of "control circuit fails to generate a safety signal" would not include "failure of DC power source" as one of its events, even though the DC power source (e.g., batteries) is part of the control circuit. This is because the top event would not occur due to the loss of the DC power source.

The postulated fault or failure events that appear on the fault tree may not be exhaustive. Only those events considered important need be included. However, the decision to include specific failure events is not arbitrary; it is influenced by the fault tree construction procedure, system design and operation, operating history, available failure data, and the experience and best judgment of the analyst. At each intermediate point, the postulated events represent the immediate, necessary, and sufficient causes for the occurrence of the intermediate (or top) events. Multiple fault trees must be developed if multiple top events need to be considered.

The fault tree itself is a Boolean logic model and thus represents the qualitative characterization of the system logic. There are many quantitative algorithms to evaluate fault trees. For example, the concept of cut sets discussed earlier can also be applied to fault trees using the Boolean algebra method. Finding the probability of occurrence of the top event $Pr(C_1 \cup C_2 \cup \cdots \cup C_n)$, where $C_1, \ldots, C_n$ are the cut sets using Equation 6.42, is a simple quantitative approach.

To understand the symbology of logic trees, including fault trees, consider Figure 6.8. There are three types of symbols used in fault trees: events, gates, and transfers. Basic events, undeveloped events, conditioning events, and external events are sometimes called primary events. When postulating events in the fault tree, it is important to include the undesired component states (e.g., applicable failure modes) and when they occur.

Primary Event Symbols

BASIC EVENT—A basic event requiring no further development

CONDITIONING EVENT—Specific conditions or restrictions that apply to any logic gate (used primary with PRIORITY AND and INHIBIT gate)

UNDEVELOPED EVENT—An event which is not further developed either because it is of insufficient consequence or because information is unavailable

EXTERNAL EVENT—An event which is normally expected to occur

Intermediate Event Symbols

INTERMEDIATE EVENT—An event that occurs because of one or more antecedent causes acting through logic gates

FIGURE 6.8 Primary event, gate, and transfer symbols used in fault trees and success trees.

(Continued)

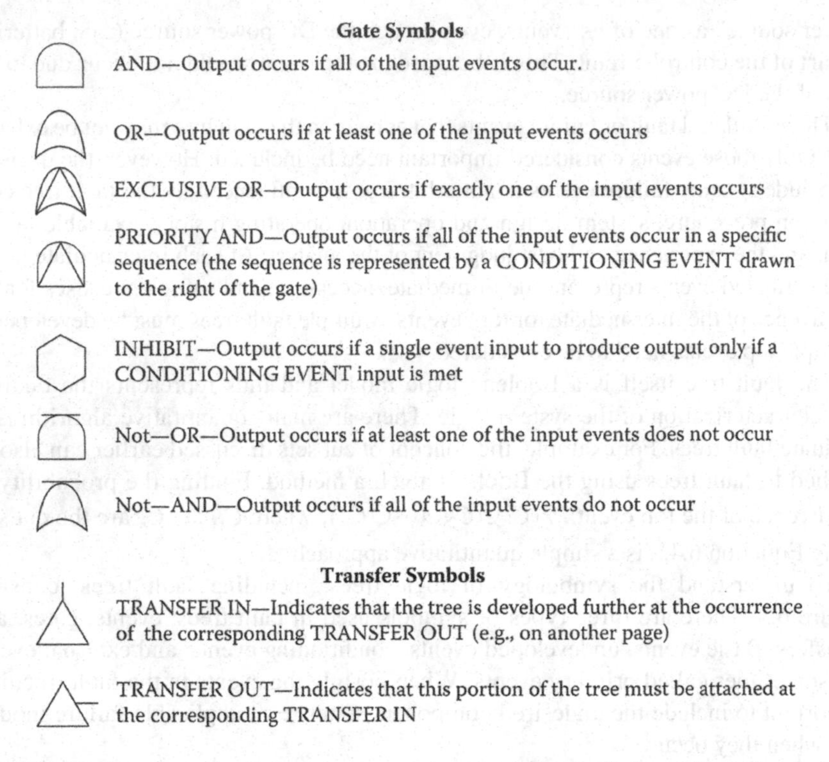

Gate Symbols
AND—Output occurs if all of the input events occur.
OR—Output occurs if at least one of the input events occurs
EXCLUSIVE OR—Output occurs if exactly one of the input events occurs
PRIORITY AND—Output occurs if all of the input events occur in a specific sequence (the sequence is represented by a CONDITIONING EVENT drawn to the right of the gate)
INHIBIT—Output occurs if a single event input to produce output only if a CONDITIONING EVENT input is met
Not—OR—Output occurs if at least one of the input events does not occur
Not—AND—Output occurs if all of the input events do not occur
Transfer Symbols
TRANSFER IN—Indicates that the tree is developed further at the occurrence of the corresponding TRANSFER OUT (e.g., on another page)
TRANSFER OUT—Indicates that this portion of the tree must be attached at the corresponding TRANSFER IN

FIGURE 6.8 (*Continued*) Primary event, gate, and transfer symbols used in fault trees and success trees.

To better understand the fault tree concept, consider the complex block diagram in Figure 6.9. Let us also assume that the block diagram models a circuit in which the arrows indicate the direction of the electric current. A top event of "no current at point *F*" is selected, and all events that cause this top event are deductively postulated. An easy way to model a fault tree deductively is through failure tracing. That is to start from the system's output (sink), whose failure represents the top event, and trace failure events and modes back to the input points (source). This is done by systematically identifying applicable failure modes and events of each component and failures due to its connections and interactions with other components and events (such as human actions). At each point, you ask what failure mode, human error, software fault, spatial and environmental interactions, or input from a directly connected component could lead to the failure of this component. The process should continue until all system components are modeled in the fault tree. Figure 6.9 shows the results of conducting this process on Figure 6.5.

As an example, consider the pumping system in Figure 6.10. Sufficient water is delivered from the water source *T-1* when at least one of the two pumps, *P1* or *P2*, works. If one of the two pumps fails to start or fails during operation, the mission is still considered successful if the other pump functions properly. All the valves, *V1* through *V5*, are normally open. The sensing and control system *S* senses the demand for the pumping system and automatically starts both *P1* and *P2*. If one of the two

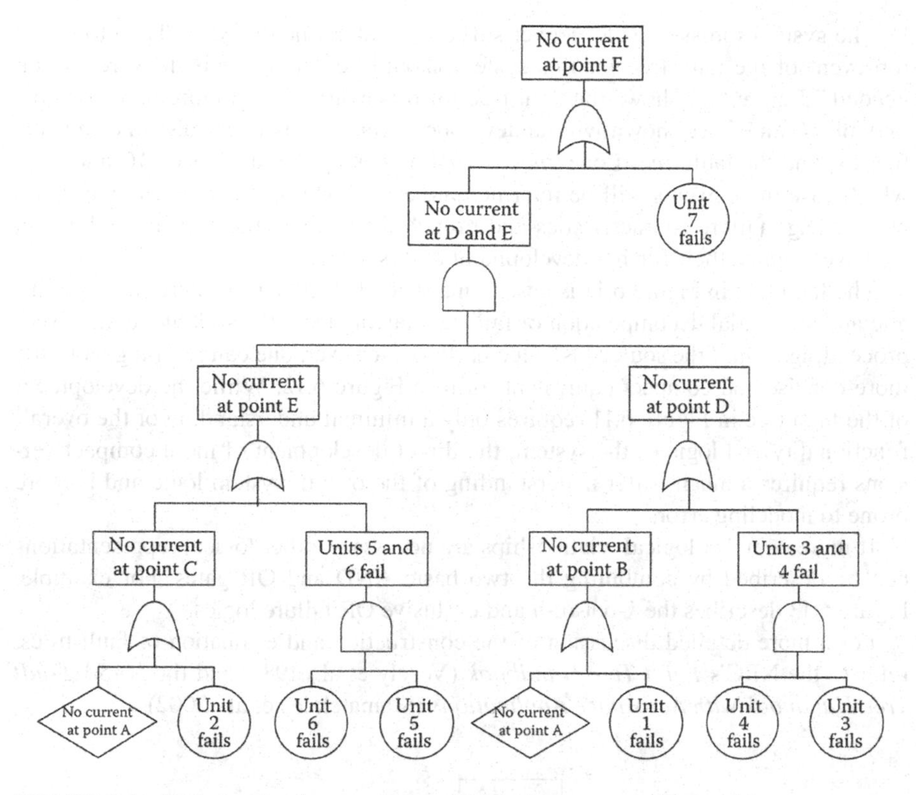

FIGURE 6.9 Fault tree for the complex parallel-series system in Figure 6.5.

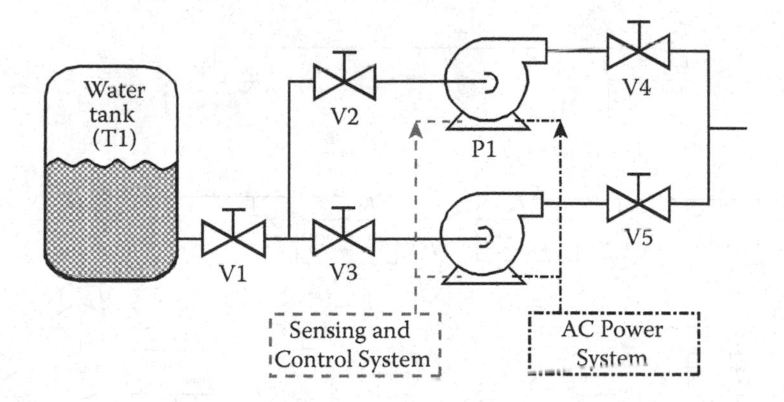

FIGURE 6.10 Example of a pumping system.

pumps fails to start or fails during operation, the mission is still considered successful if the other pump functions properly. The two pumps and the sensing and control system use the same AC power source, *AC*. Assume the water content in *T*1 is sufficient and available, there are no human errors, and failure in the pipe connections is considered out of scope.

The system's mission is to deliver sufficient water when needed. Therefore, the top event of the fault tree for this system should be "no water is delivered when needed." Figure 6.11 shows the fault tree for this example. In Figure 6.11, the failures of *AC* and *S* are shown with undeveloped events. This is because one can further expand the fault tree if one knows what makes up the failures of *AC* and *S*, in which case these events will be intermediate events. However, since enough information (e.g., failure characteristics and probabilities) about these events is known, we have stopped their further development at this stage.

The fault tree in Figure 6.11 is based on a strict deductive procedure (i.e., systematic and sequential decomposition of failures starting from the sink and deductively proceeding toward the source, as noted earlier). However, one can rearrange it to the more concise and compact equivalent form in Figure 6.12. While the development of the fault tree in Figure 6.11 requires only a minimal understanding of the overall functionality and logic of the system, the direct development of more compact versions requires a much better understanding of the overall system logic and is more prone to modeling error.

If more complex logical relationships are necessary, other logical representations can be described by combining the two basic AND and OR gates. For example, Figure 6.13 describes the *k*-out-of-*n* and exclusive OR failure logics.

For a more detailed discussion of the construction and evaluation of fault trees, refer to the NRC's *Fault Tree Handbook* (Vesely et al., 1981) and the *NASA Fault Tree Handbook with Aerospace Applications* (Stamatelatos et al., 2002).

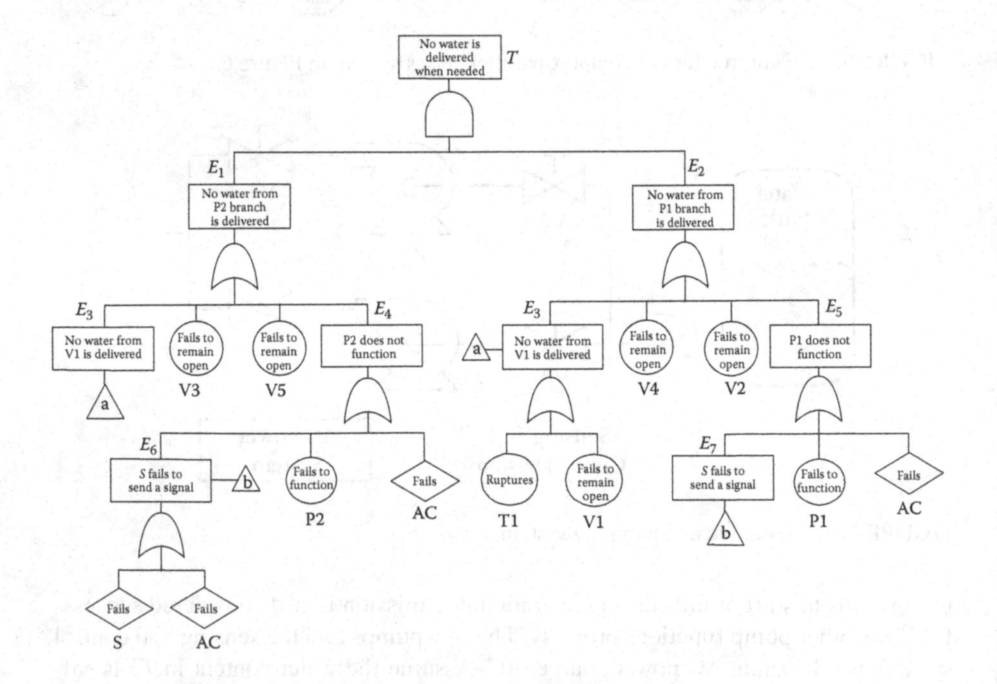

FIGURE 6.11 Fault tree for the pumping system in Figure 6.10.

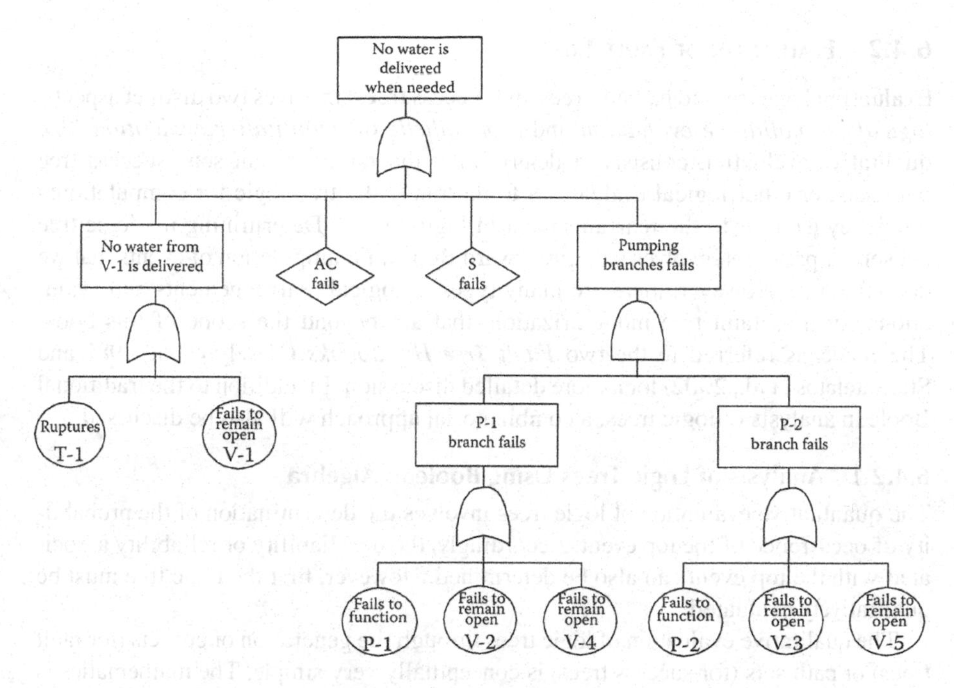

FIGURE 6.12 More compact form of the fault tree in Figure 6.11.

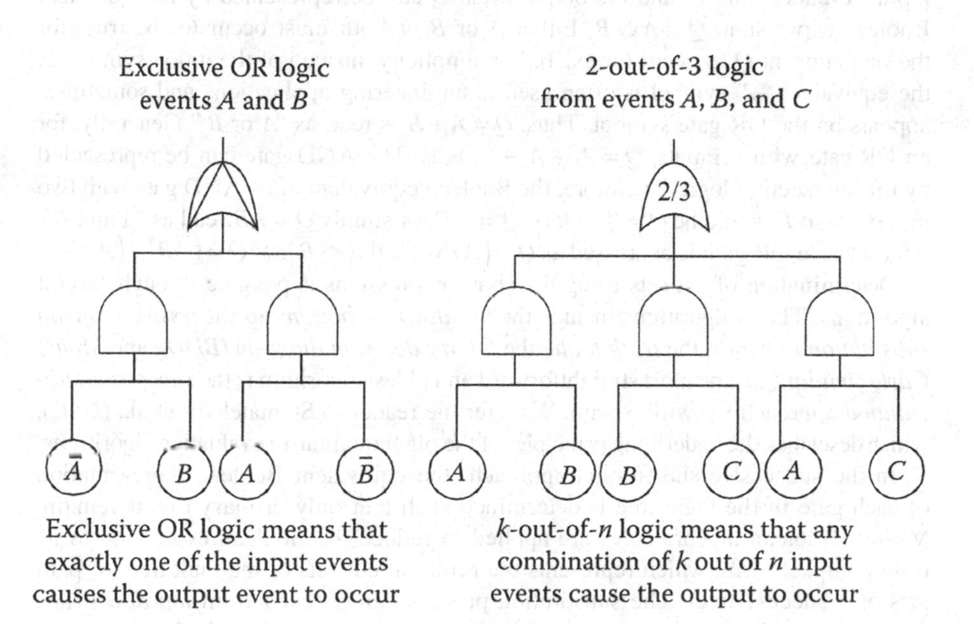

FIGURE 6.13 Exclusive OR and k-out-of-n logics.

6.4.2 Evaluation of Fault Trees

Evaluating logic trees (e.g., fault trees and success trees) involves two distinct aspects: *logical* or *qualitative evaluation* and *probabilistic* or *quantitative evaluation*. The qualitative evaluation consists of determining the fault tree cut sets, success tree path sets, or other logical evaluations to rearrange the tree logic for computational efficiency (similar to the rearrangement in Figure 6.12). Determining the logic tree cut sets or path sets involves straightforward Boolean manipulation of events that we describe here. However, there are many types of logical rearrangements and evaluations, such as fault tree modularization, that are beyond the scope of this book. The reader is referred to the two *Fault Tree Handbooks* (Vesely et al., 1981 and Stamatelatos et al., 2002) for a more detailed discussion. In addition to the traditional Boolean analysis of logic trees, a combinatorial approach will also be discussed.

6.4.2.1 Analysis of Logic Trees Using Boolean Algebra

The quantitative evaluation of logic trees involves the determination of the probability of occurrence of the top event. Accordingly, the unreliability or reliability associated with the top event can also be determined. However, first the logic tree must be qualitatively evaluated.

The qualitative evaluation of logic trees through the generation of cut sets (for fault trees) or path sets (for success trees) is conceptually very simple. The mathematics is the same for a fault tree, a success tree, or when connected to an event tree. The tree OR gate logic represents the union of the input events. That is, any of the input events, if they occur, will cause the output event to occur. For example, an OR gate with two inputs, events A and B, and the output event Q can be represented by its equivalent Boolean expression, $Q = A \cup B$. Either A or B or both must occur (or be true) for the output event Q to occur (or exist). For simplicity, instead of the union symbol $\cup$, the equivalent "+" symbol is often used in engineering applications and sometimes appears on the OR gate symbol. Thus, $Q = A + B$ is read as "A or B." Generally, for an OR gate with n inputs, $Q = A_1 + A_2 + \ldots + A_n$. The AND gate can be represented by the intersection logic. Therefore, the Boolean equivalent of an AND gate with two inputs A and B would be $Q = A \cap B$ or $Q = A{\cdot}B$, or simply $Q = AB$,read as "A and B." An exclusive OR gate is expressed as $Q = (\bar{A} \cap B) \cup (A \cap \bar{B})$, or $Q = (\bar{A}B) + (A\bar{B})$.

Determination of cut sets using the above expressions is possible through several algorithms. These algorithms include the *top-down* or *bottom-up successive Boolean substitution methods*, the *truth table*, the *binary decision diagram (BDD)*, and *Monte Carlo simulation*. The most straightforward and oldest algorithm is the *successive substitution* approach we will explain. We refer the reader to Stamatelatos et al., (2002), which describes the underlying principles of the other qualitative evaluation algorithms.

In the successive substitution approach, the equivalent Boolean representation of each gate in the logic tree is determined such that only primary events remain. Various Boolean algebra rules are applied to reduce the Boolean expression to its most compact form, which represents the minimal cut sets of the fault tree (or path sets of a success tree). The substitution process can proceed from *top-down* (i.e., from the top of the tree to the bottom) or *bottom-up* (starting at the bottom of the tree and working to the top). Depending on the logic tree and its complexity, either approach, or a combination of the two, can be used.

For example, let us consider the fault tree in Figure 6.11. Each node represents a failure. The step-by-step, top-down Boolean substitution of the top event is presented below.

Step 1: $T = E_1E_2$,

Step 2: $E_1 = E_3 + V_3 + V_5 + E_4, \quad E_2 = E_3 + V_4 + V_2 + E_5$,

$$T = E_3 + V_3V_4 + V_3V_2 + V_5V_4 + V_5V_2 + E_4V_4 + E_4V_2 + E_4E_5 + V_3E_5 + V_5E_5.$$

(Note: In this step, T has been reduced by using the Boolean identities

$$E_3E_3 = E_3,\ E_3 + E_3X = E_3, \text{ and } E_3 + E_3 = E_3.)$$

Step 3:

$$E_3 = T_1 + V_1,\ E_4 = E_6 + P_2 + AC, \quad E_5 = E_6 + P_1 + AC,$$

$$T = T_1 + V_1 + AC + V_3V_4 + V_3V_2 + V_5V_4 + V_5V_2 + V_4P_2 + P_2V_2 + E_6 + P_2P_1 + V_3P_1 + V_5P_1.$$

(Again, identities such as $AC + AC = AC$ and $E_6 + V_3E_6 = E_6$

have been used to reduce T.)

Step 4:

$$E_6 = AC + S,$$

$$T = AC + S + T_1 + V_1 + V_3V_4 + V_3V_2 + V_5V_4 + V_5V_2 + V_4P_2 + P_2V_2 + P_2P_1 + V_3P_1 + V_5P_1.$$

The Boolean expression obtained in Step 4 represents four cut sets of order 1 (meaning the cut set contains only one failure mode or event) and nine cut sets of order 2. Notice that the non-minimal cut sets have been reduced from the final expression. The order one cut sets are the occurrences of failure events S, AC, T_1, and V_1. The order two cut sets are events V_3 and V_4, V_3 and V_2, V_5 and V_4, V_5 and V_2, V_4 and P_2, P_2 and V_2, P_2 and P_1, V_3 and P_1, and V_5 and P_1. A simple examination of each cut set shows that its occurrence guarantees the occurrence of the top event (failure of the system). For example, the cut set V_5 and P_1, which represents the simultaneous failure of valve V_5 and pump P_1, causes the two flow branches of the system to be lost, which in turn disables the system.

The same substitution approach can be used to determine the path sets. Here, the events are success events representing the adequate realization of the described functions.

Evaluating a large logic tree by hand can be a formidable job. Several computer programs are available for the analysis of logic trees. For an overview of fault tree software tools see Shen et al. (2022).

Quantitative evaluation of the cut sets or path sets was introduced earlier in this chapter. Equation 6.42 forms the basis for quantitative evaluation of the cut sets. That is, the probability that the top event, T, occurs in a mission time t is

$$Pr(T) = Pr(C_1 \cup C_2 \cup \ldots \cup C_n). \tag{6.45}$$

The probability of the fault tree top event in a system reliability framework is the system's unreliability. To understand the complexities discussed earlier regarding the determination of $Pr(T)$, let us consider the case where these two minimal cut sets are obtained:

$$C_1 = AB,$$

$$C_2 = AC.$$

Then, $Pr(T) = Pr(AB + AC)$.

According to Equation 6.14,

$$\begin{aligned} Pr(T) &= Pr(AB) + Pr(AC) - Pr(ABAC) \\ &= Pr(AB) + Pr(AC) - Pr(ABC). \end{aligned} \tag{6.46}$$

If A, B, and C are independent, then

$$Pr(T) = Pr(A) \cdot Pr(B) + Pr(A) \cdot Pr(C) - Pr(A) \cdot Pr(B) \cdot Pr(C). \tag{6.47}$$

Determining the cross-product terms, such as $Pr(A)Pr(B)Pr(C)$ in Equation 6.47, poses a dilemma in the quantitative evaluation of cut sets, especially when the number of cut sets is large. There are 2^{n-1} such terms in cut sets. For example, in the 13 cut sets generated for the pumping example, there are 8,191 such terms. This can be a formidable job for large logic trees, even for powerful computers.

Fortunately, when dealing with cut sets, evaluation of these cross-product terms is often unnecessary due to their small values, and the boundary approach in Equation 6.44 is adequate. This is true whenever we are dealing with small probabilities, which is often the case for the probability of failure events. In these cases, for example, in Equation 6.47, $Pr(A)Pr(B)Pr(C)$ is substantially smaller than $Pr(A)Pr(B)$ and $Pr(A)Pr(C)$. Thus, the bounding result can also approximate the system's true reliability or unreliability value. This is often called the *rare event approximation*. Let us assume that

$$Pr(A) = Pr(B) = Pr(C) = 0.1. \tag{6.48}$$

Then,

$$Pr(A)Pr(B) = Pr(A)Pr(C) = 0.01, \tag{6.49}$$

and

$$Pr(A)Pr(B)Pr(C) = 0.001. \tag{6.50}$$

The latter is smaller than the former by an order of magnitude. Although $Pr(T) = 0.019$, the rare event approximation yields $Pr(T) \approx 0.02$. Obviously, smaller probabilities of the events lead to better approximations.

For another example, consider the simple RBD in Figure 6.14, representing a system with three paths from point X to point Y. The equivalent fault tree is shown in Figure 6.15.

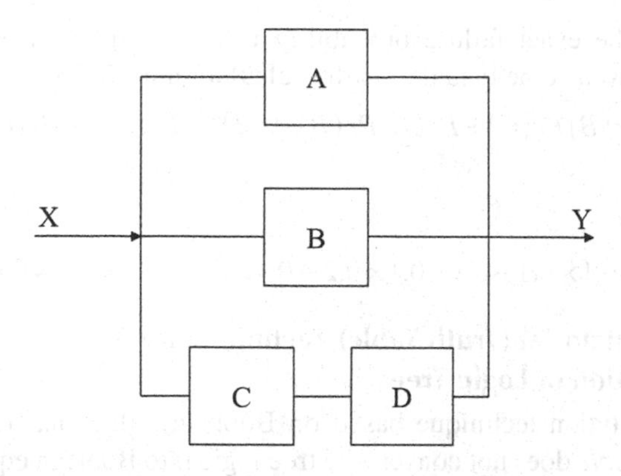

FIGURE 6.14 Simple RBD example.

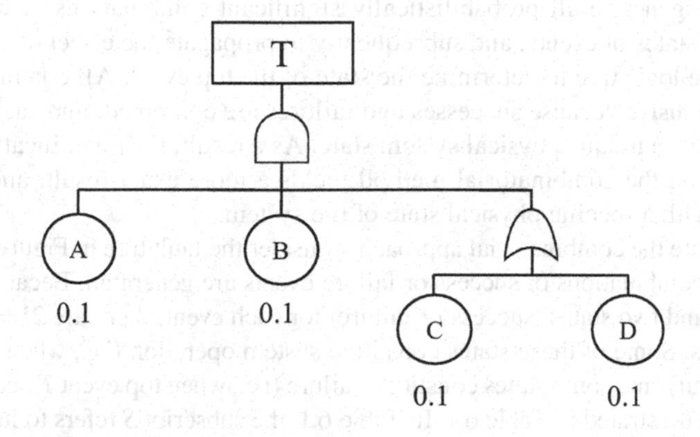

FIGURE 6.15 Fault tree representation of Figure 6.14.

The top-down Boolean substitution proceeds as follows:

$$T = ABG_1,$$
$$G_1 = C + D,$$
$$T = AB(C + D),$$
$$T = ABC + ABD.$$

If the probability of events A, B, and C is 0.1, and the probability of event D is 0.2, when all events are independent, the top event probability is evaluated as follows. Using the rare event approximation discussed earlier,

$$Pr(T) \approx Pr(A)Pr(B)Pr(C) + Pr(A)Pr(B)Pr(D). \tag{6.51}$$

Therefore,

$$Pr(T) \approx 0.1 \times 0.1 \times 0.1 + 0.1 \times 0.1 \times 0.2 = 0.003. \tag{6.52}$$

Note that the terms ABC and ABD are not mutually exclusive and, therefore, the value of $Pr(T)$ is approximate since the rare event approximation has been used.

To calculate the exact failure probability using minimal cut sets, their cross-product terms must also be included in the calculation of $Pr(T)$,

$$Pr(T) = Pr(A)Pr(B)Pr(C) + Pr(A)Pr(B)Pr(D) - Pr(A)Pr(B)Pr(C)Pr(D). \tag{6.53}$$

Accordingly,

$$Pr(T) = 0.1 \times 0.1 \times 0.1 + 0.1 \times 0.1 \times 0.2 - 0.1 \times 0.1 \times 0.1 \times 0.2 = 0.0028. \tag{6.54}$$

6.4.2.2 Combinatorial (Truth Table) Technique for Evaluation of Logic Trees

Unlike the substitution technique based on Boolean reduction, the *combinatorial method* or *truth table* does not convert the tree logic into Boolean equations to generate cut sets or path sets. Instead, this method relies on a combinatorial algorithm to exhaustively generate all probabilistically significant combinations of both failure and success states of events, and subsequently to propagate the effect of each combination on the logic tree to determine the state of the top event. All combinations are mutually exclusive because successes and failures are combined, and each combination represents a unique physical system state. As a result, the quantification of logic trees based on the combinatorial method yields a more exact result, and these are associated with a specific physical state of the system.

To illustrate the combinatorial approach, consider the fault tree in Figure 6.15. First, all possible combinations of success or failure events are generated. Because there are four events and two states (success or failure) for each event, there are $2^4 = 16$ possible system states. Some of these states constitute system operation (i.e., when top event T does not occur), and some states constitute failure (i.e., when top event T occurs). These 16 states are illustrated in Table 6.1. In Table 6.1, the subscript S refers to item success, and subscript F refers to item failure (i.e., the occurrence of the event in the fault tree). The probability of each combination is calculated directly by multiplying the probabilities of the corresponding elements of the combination.

Only combinations 14, 15, and 16 lead to the occurrence of the top event T. Summing the probability of these failure-contributing combinations results in system failure probability of $Pr(T) = 0.0018 + 0.0008 + 0.0002 = 0.0028$. This is the exact value of the top event probability (provided that the events are independent). This is consistent with the exact calculation by the Boolean reduction method. Note that the sum of the probabilities of all possible combinations (16 in this case) is one, because the combinations are all mutually exclusive and *collectively exhaustive* (they cover all event space in the universal set). Combinations 14, 15, and 16 are also mutually exclusive cut sets.

The Venn diagram technique helps visualize the difference between the results generated from the Boolean reduction and the combinatorial approach. Again, consider the simple system in Figures 6.14 and 6.15 with the minimal cut sets ABC and ABD. The left side of Figure 6.16 represents a Venn diagram for the two cut sets. Each cut set is represented by one shaded area. The two shaded overlap areas indicate that the cut sets are not mutually exclusive. Now consider how combinations 14, 15, and 16 are represented in the Venn diagram (the right side of Figure 6.16). Again, each shaded area corresponds to a combination. Here, there is no overlapping of the

TABLE 6.1
Combinatorial Method (Truth Table) of Evaluating Fault Tree

Combination Number, C_i	Combination (System States)	Probability of C_i	System Operation (T)
1	$\bar{A}\bar{B}\bar{C}\bar{D}$	0.5832	S
2	$\bar{A}\bar{B}\bar{C}D$	0.1458	S
3	$\bar{A}\bar{B}C\bar{D}$	0.0648	S
4	$\bar{A}\bar{B}CD$	0.0162	S
5	$\bar{A}B\bar{C}\bar{D}$	0.0648	S
6	$\bar{A}B\bar{C}D$	0.0162	S
7	$\bar{A}BC\bar{D}$	0.0072	S
8	$\bar{A}BCD$	0.0018	S
9	$A\bar{B}\bar{C}\bar{D}$	0.0648	S
10	$A\bar{B}\bar{C}D$	0.0162	S
11	$A\bar{B}C\bar{D}$	0.0072	S
12	$A\bar{B}CD$	0.0018	S
13	$AB\bar{C}\bar{D}$	0.0072	S
14	$AB\bar{C}D$	0.0018	F
15	$ABC\bar{D}$	0.0008	F
16	$ABCD$	0.0002	F
		$\sum Pr(C_i) = 1.00$	

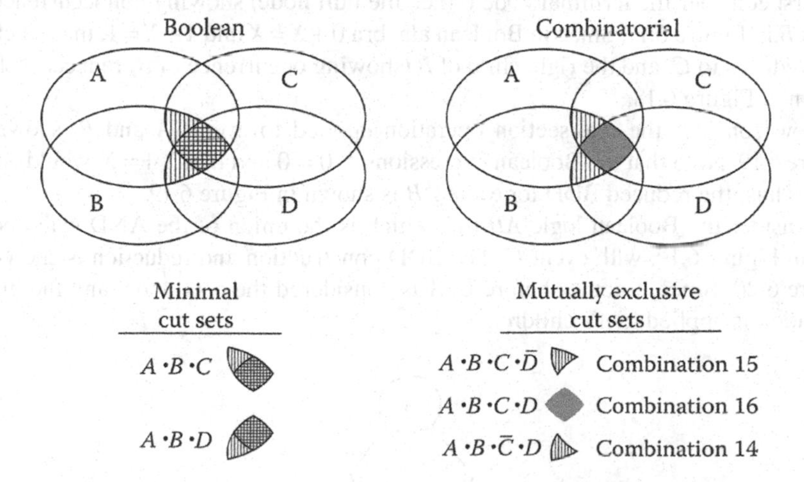

FIGURE 6.16 Venn diagrams illustrating Boolean and combinatorial methods.

shaded areas. The combinatorial approach generates mutually exclusive sets, and those sets that lead to system failure are called mutually exclusive cut sets. Therefore, when the rare event approximation is not used, the contributions generated by the combinatorial approach have no overlapping area and produce the exact probability. For sizable problems, usually the rare event approximation is the only practical choice. But, if the exact top event probabilities are desired, or failure probabilities are greater than 0.1, then the combinatorial approach is preferred.

A typical logic model may contain hundreds of events. For n binary events, there are 2^n combinations. Generating this many combinations is impractical for a large n (e.g., $n > 20$); a more efficient method would be needed. An algorithm to generate combinations whose probabilities exceed some cutoff limit (e.g., 10^{-7}), referred to as probabilistically significant combinations, has been proposed by Dezfuli et al. (1994).

6.4.2.3 Binary Decision Diagrams

The *binary decision diagram (BDD)* technique is the leading method used in modern fault tree software tools. The BDD logic manipulation process works directly with the logical expressions. A BDD is a graphical representation of the tree logic. The BDD used for logic tree analysis is called the *reduced ordered BDD* (i.e., the minimal form in which the events appear in the same order at each path). The BDD technique is described in more detail in (Stamatelatos et al., 2002). Bryant (2018) provides more information on the BDD approach to fault tree analysis.

The BDD is assembled recursively using the logic tree in a bottom-up fashion. Each basic event in the fault tree has an associated single-node BDD. For example, the BDD for a basic event B is shown in Figure 6.17. Starting at the bottom of the tree, a BDD is constructed for each basic event, and then combined according to the logic of the corresponding gate. The BDD for the OR gate "B or C" is constructed by applying the Boolean function for the union of two events to the BDD for event B and the BDD for event C. Since B is first in the relation, it is the *root node*. The union logic is expressed with the C BDD connected to each *child node* of B.

First consider the terminal node 0 (i.e., the null node) showing nonoccurrence of event B in Figure 6.17. Since in Boolean algebra $0+X=X$ and $1+X=1$, the left child of B reduces to C, and the right child of B (showing occurrence of B) reduces to 1, as shown in Figure 6.18.

Now consider the intersection operation applied to events A and B, shown in Figure 6.19. Note that the Boolean expressions $X \cdot 0 = 0$ and that $X \cdot 1 = X$ would apply here. Thus, the reduced BDD for event AB is shown in Figure 6.19.

Consider the Boolean logic $AB+C$, which is the union of the AND gate operation in Figure 6.19, with event C. The BDD construction and reduction is shown in Figure 6.20. Since A comes before C, A is considered the root node, and the union operation is applied to A's children.

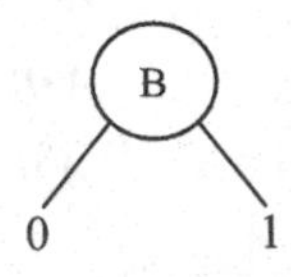

FIGURE 6.17 BDD for basic event B.

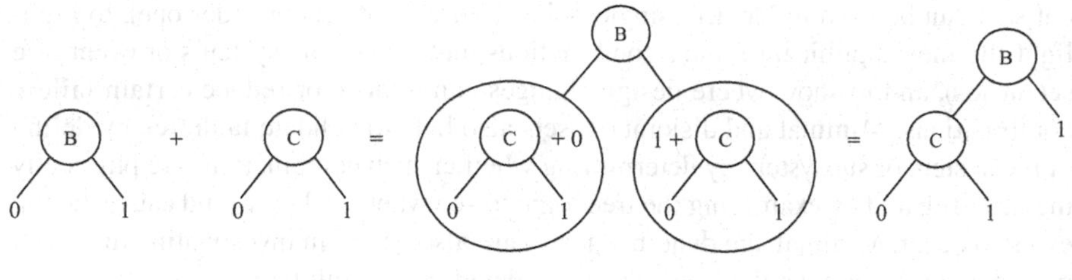

FIGURE 6.18 BDDs for the OR gate involving events B and C (step 1).

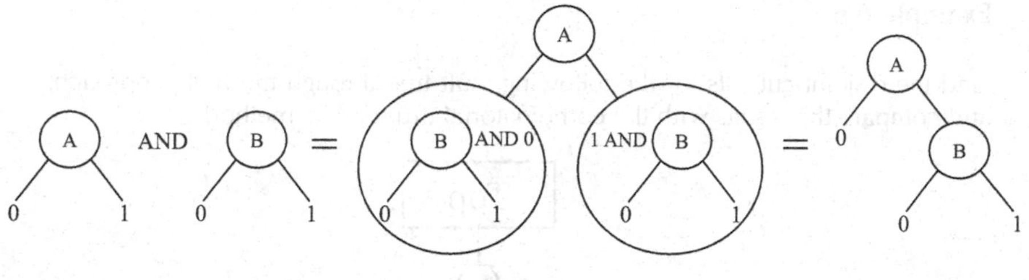

FIGURE 6.19 BDDs for the AND gate involving A and B (step 2).

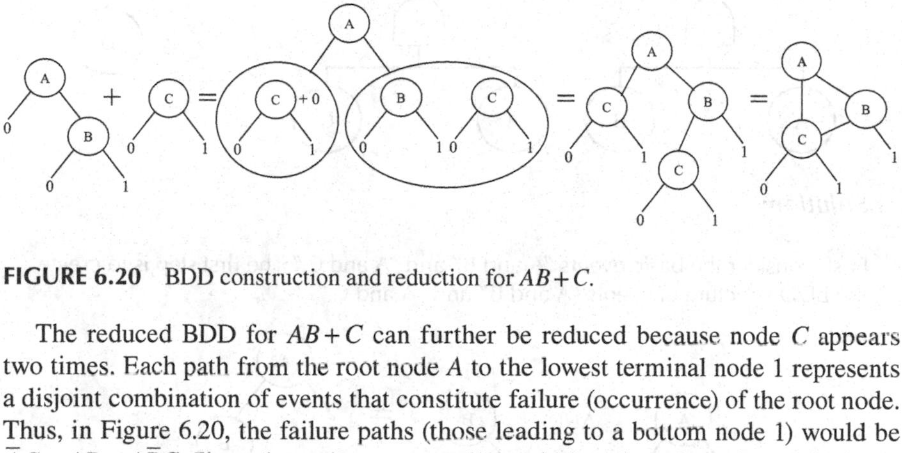

FIGURE 6.20 BDD construction and reduction for $AB+C$.

The reduced BDD for $AB+C$ can further be reduced because node C appears two times. Each path from the root node A to the lowest terminal node 1 represents a disjoint combination of events that constitute failure (occurrence) of the root node. Thus, in Figure 6.20, the failure paths (those leading to a bottom node 1) would be $\bar{A}C+AB+A\bar{B}C$. Since the paths are mutually exclusive, the calculation of the probability of failure is straightforward and similar to the truth table approach.

The failure paths leading to all end values of 1 can be identified and quantified by summing their probabilities. In developing the BDD for a fault tree, various reduction techniques are used to simplify the BDD.

Similar to the combinatorial (truth table) method, the BDD approach results in *disjoint cut sets* and yields an exact value of the top event probability. The exact probability is useful when many high probability events appear in the model. The BDD approach is also the most computationally efficient approach for logic tree evaluation. Because the cut sets and path sets generated in the BDD approach are disjoint, important measures and sensitivities can be calculated more efficiently. It is important to emphasize that generating the minimal and disjoint cut sets provides important qualitative information as well as quantitative information. For example, these cut sets can be used to identify single point failures (cut sets of order one), to highlight the most significant failure combinations (cut sets of the systems or event tree scenarios), and to show where design changes can remove or reduce certain failure combinations. Minimal and disjoint cut sets also help to validate fault tree modeling of the system or subsystem by determining whether such combinations are physically meaningful, and by examining the tree logic to see whether they would cause the top event to occur. Minimal and disjoint cut sets are also useful in investigating the effect or dependencies among the basic events modeled in the fault tree.

Example 6.6

Find the disjoint cut sets for the following fault tree through the BDD approach, and compare the results with the combinatorial (truth table) method.

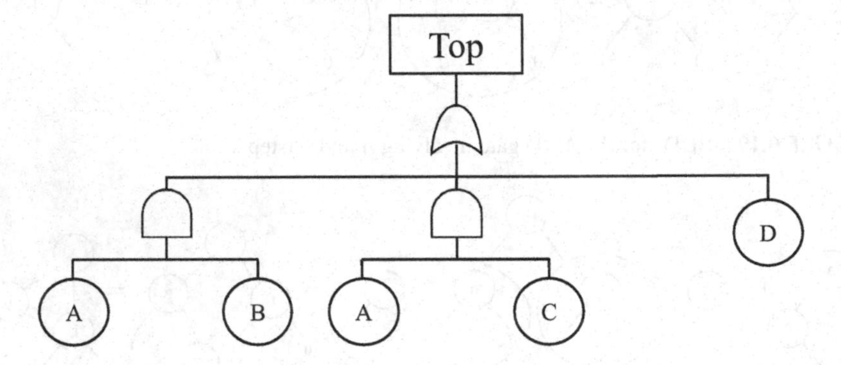

Solution:

First, consider the basic events "A and B" and "A and C." The first step is to create the BDD structure of events "A and B" and "A and C."

A AND B = A
0 1 0 1 0 B
0 1

A AND C = A
0 1 0 1 0 C
0 1

Now consider the union operation. The second step is to create the BDD structure for the union operation.

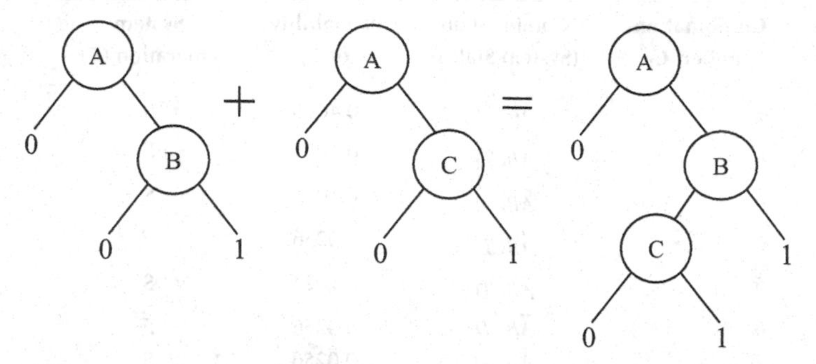

Finally, considering event *D*, the BDD structure for the fault tree is shown below. Each path from the root node, *A*, to a terminal node with value 1 represents a disjoint combination of events that causes system failure.

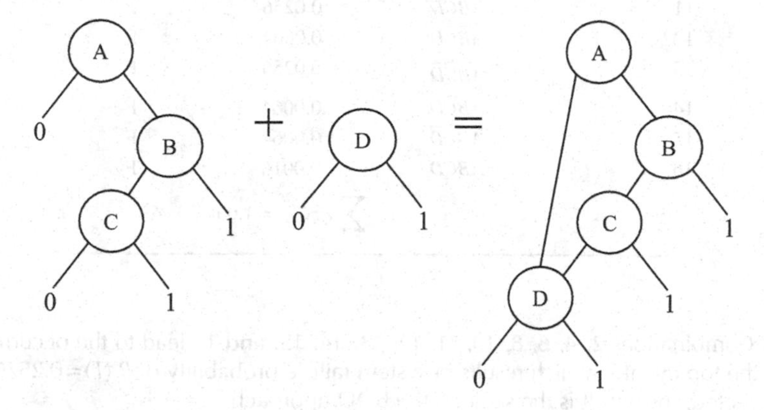

If the probability of failure for *A*, *B*, *C*, and *D* is 0.2,

Disjoint Path	Probability of Failure	
AB	$Pr(A)Pr(B)$	0.0400
$A\overline{B}C$	$Pr(A)(1-Pr(B))Pr(C)$	0.0320
$A\overline{B}\,\overline{C}D$	$Pr(A)(1-Pr(B))(1-Pr(C))Pr(D)$	0.0256
$\overline{A}D$	$(1-Pr(A))Pr(D)$	0.1600
Total	-	0.2576

The system failure probability for the top event is 0.2576.

The table below illustrates the results for the truth table method.

Combination Number, C_i	Combination (System States)	Probability of C_i	System Operation (T)
1	$\overline{A}\overline{B}\overline{C}\overline{D}$	0.4096	S
2	$\overline{A}\overline{B}\overline{C}D$	0.1024	F
3	$\overline{A}\overline{B}C\overline{D}$	0.1024	S
4	$\overline{A}\overline{B}CD$	0.0256	F
5	$\overline{A}B\overline{C}\overline{D}$	0.1024	S
6	$\overline{A}B\overline{C}D$	0.0256	F
7	$\overline{A}BC\overline{D}$	0.0256	S
8	$\overline{A}BCD$	0.0064	F
9	$A\overline{B}\overline{C}\overline{D}$	0.1024	S
10	$A\overline{B}\overline{C}D$	0.0256	F
11	$A\overline{B}C\overline{D}$	0.0256	F
12	$A\overline{B}CD$	0.0064	F
13	$AB\overline{C}\overline{D}$	0.0256	F
14	$AB\overline{C}D$	0.0064	F
15	$ABC\overline{D}$	0.0064	F
16	$ABCD$	0.0016	F
		$\sum Pr(C_i) = 1.00$	

Combinations 2, 4, 6, 8, 10, 11, 12, 13, 14, 15, and 16 lead to the occurrence of the top event T, which results in system failure probability of $Pr(T)=0.2576$. As expected, this result is the same as the BDD approach.

Example 6.7

Consider the system in the RBD below. Develop a fault tree representation of this system. Then compare the minimal cut sets of the fault tree obtained from the standard Boolean substitution method to the disjoint cut sets obtained from the BDD solution of the fault tree.

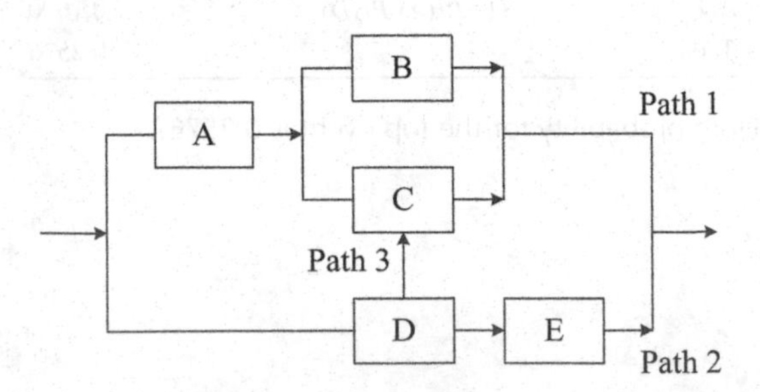

Solution:

The fault tree for this system is shown below.

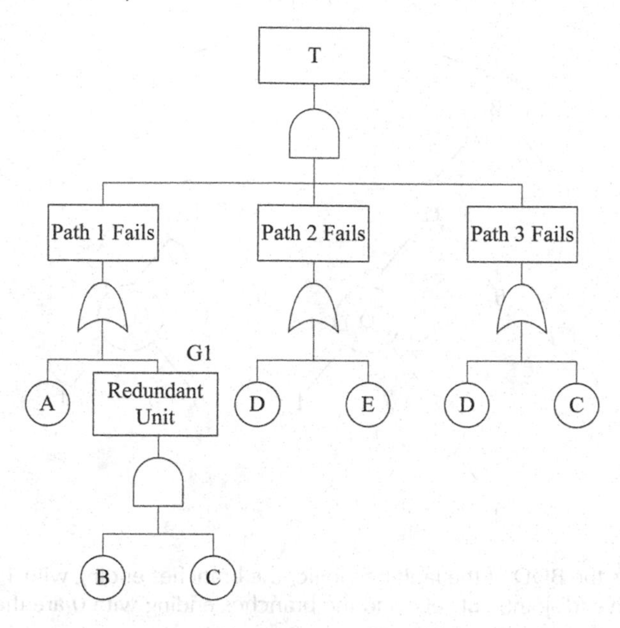

The minimal cut sets of the fault tree by the substitution method are:

$G_1 = BC$; Using the notation P_1F for "Path 1 Fails" $= G_1 + A$; $P_2F = D + E$; $P_3F = D + C$ and the cut sets are:

$$T = (BC + A)\cdot(D+E)\cdot(D+C) = AD + BCE + BCD + ACE.$$

The BDD solution of the fault tree for failure of each path set is shown below.

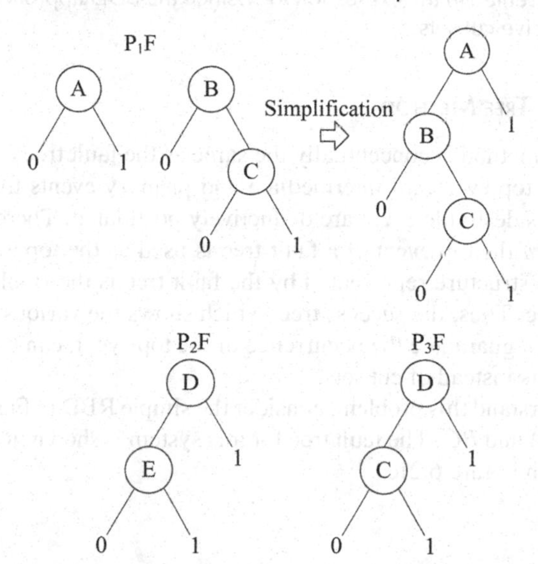

The final BDD structure is shown below right.

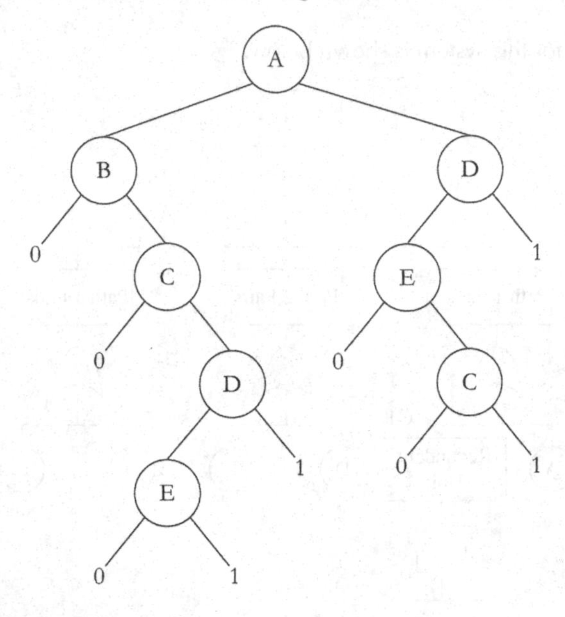

Since this is the BDD of the fault tree logic, the branches ending with 1 are mutually exclusive (disjoint) cut sets, and the branches ending with 0 are the mutually exclusive path sets. The cut sets are,

$$T = \bar{A}BC\bar{D}E + \bar{A}BCD + AC\bar{D}E + AD.$$

Note that the Boolean expression for T derived using the BDD approach, is the same as the one obtained using substitution method, except for components whose success is known (like $\bar{A}$ and $\bar{D}$ in $\bar{A}BC\bar{D}E$ term). Components guaranteed to work are presented in the expression for T since the BDD approach provides the mutually exclusive cut sets.

6.4.3 Success Tree Method

The success tree method is conceptually the same as the fault tree method. By defining the desirable top event, all intermediate and primary events that guarantee the occurrence of this desirable event are deductively postulated. Therefore, if the logical complement of the top event of a fault tree is used as the top event of a success tree, the Boolean structure represented by the fault tree is the Boolean complement of the success tree. Thus, the success tree, which shows the various combinations of success events that guarantee the occurrence of the top event, can be logically represented by path sets instead of cut sets.

To better understand this problem, consider the simple RBD in Figure 6.21a, which has two cut sets: A and BC. The fault tree for this system is shown in Figure 6.21b and the success tree in Figure 6.21c.

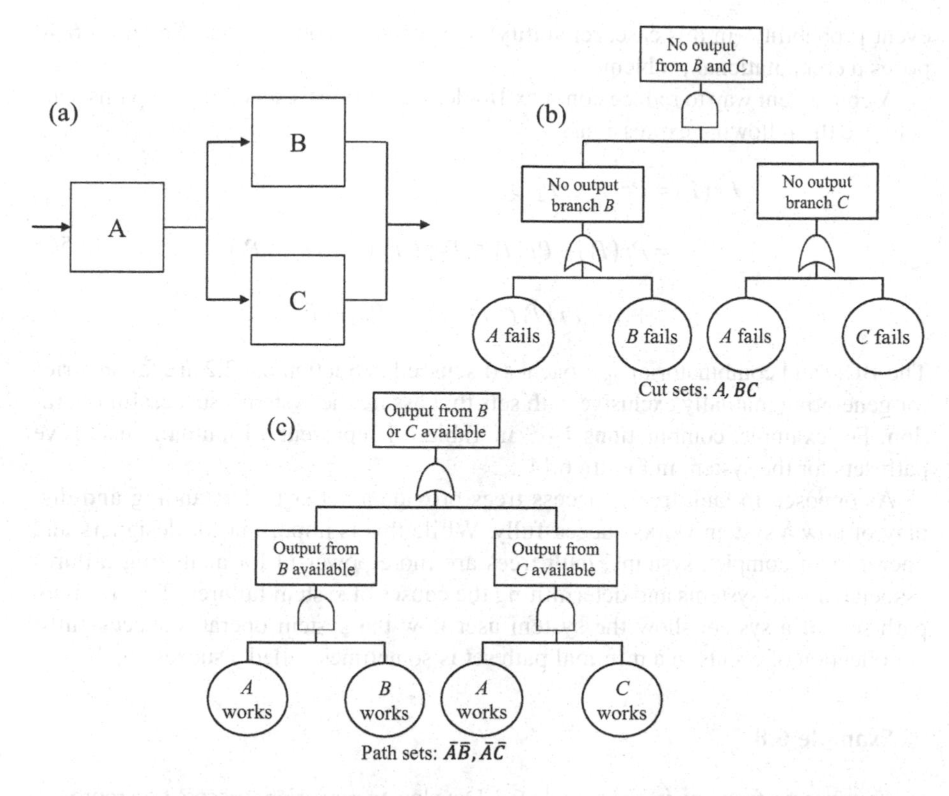

FIGURE 6.21 Correspondence between a fault tree and a success tree. For the RBD in (a), figure (b) is the corresponding fault tree and figure (c) the corresponding success tree.

From Figure 6.21b and c, it can be seen that changing the logic of one tree (changing AND gates to OR gates and vice versa) and changing all primary and intermediate events to their logical complements yield the other tree. This is also true for cut sets and path sets. The logical complement of the cut sets of the fault tree yields the path sets of the equivalent success tree. This can be seen in Figure 6.21. The complement of the cut sets is

$$\overline{A + B \cdot C} = (\text{apply de Morgan's theorem}),$$

$$\bar{A} \cdot \overline{B \cdot C} = (\text{apply de Morgan's theorem}), \tag{6.55}$$

$$\bar{A} \cdot (\bar{B} + \bar{B}) = \bar{A} \cdot \bar{B} + \bar{A} \cdot \bar{C},$$

which are the path sets.

Qualitative and quantitative evaluations of success trees are done the same way as fault trees. For example, the top-down successive substitution of the gates and reduction of the resulting Boolean expression yields the minimal path sets. Accordingly, Equation 6.40, or its lower bound Equation 6.41, can be used to determine the top event probability (in this case, reliability). However, as noted earlier, Equation 6.40 poses a computational problem.

A convenient way to reduce complex Boolean equations, especially the paths sets, is to use the following expressions:

$$
\begin{aligned}
Pr(T) &= Pr(P_1 \cup P_2 \cup \ldots \cup P_n) \\
&= Pr(P_1) + Pr(\bar{P}_1 \cap P_2) + Pr(\bar{P}_1 \cap \bar{P}_2 \cap P_3) \\
&\quad + \ldots + Pr(\bar{P}_1 \cap \bar{P}_2 \cap \ldots \cap \bar{P}_{n-1} \cap P_n).
\end{aligned}
\tag{6.56}
$$

The BDD and combinatorial approaches discussed in Section 6.4.2.2 are far superior for generating mutually exclusive path sets that assure the system's successful operation. For example, combinations 1–13 in Table 6.1 represent all mutually exclusive path sets for the system in Figure 6.14.

As opposed to fault trees, success trees provide a better understanding and display of how a system works successfully. While this is important for designers and operators of complex systems, fault trees are more powerful for analyzing failures associated with systems and determining the causes of system failures. The minimal path sets of a system show the system user how the system operates successfully. A collection of events in a minimal path set is sometimes called a success path.

Example 6.8

Consider the fault tree from Example 6.7. Develop an equivalent success tree representation of this system. Compare the minimal path sets of the success tree obtained from the standard Boolean substitution method compared to the same but mutually exclusive sets obtained from the BDD solution of the success tree.

Solution:

The success tree with the success criterion that at least one path is available is shown below.

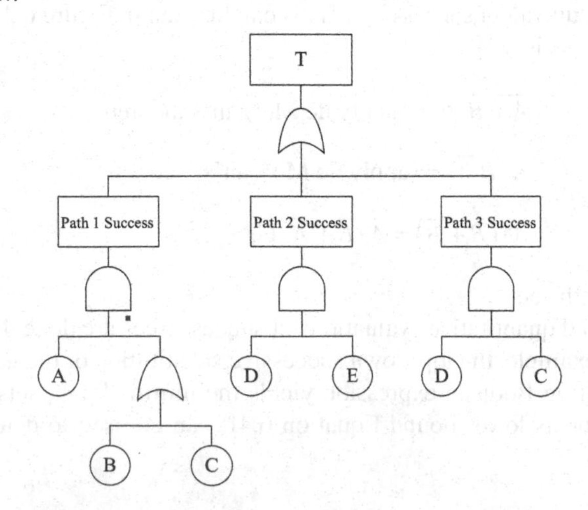

The path sets of the success tree by substitution are:

Path 1 *Available* = $\bar{A}\bar{B}+\bar{A}\bar{C}$;

Similarly for other two paths, *Path* 2 *Available* = $\bar{D}\bar{E}$; and *Path* 3 *Available* = $\bar{C}\bar{D}$. Therefore, $\bar{T}=\bar{A}\bar{B}+\bar{A}\bar{C}+\bar{D}\bar{E}+\bar{C}\bar{D}$.

Since this is the fault tree from Example 6.7, the resulting BDD is the same as in that example. The BDD branches ending with 0 are the mutually exclusive path sets. These are:

$$\bar{T}=\bar{A}\bar{B}+\bar{A}B\bar{C}+A\bar{D}\bar{E}+\bar{C}\bar{D}E.$$

6.5 EVENT TREE METHOD

Suppose the successful operation of a system depends on an approximately chronological but discrete operation of its items or subsystems (i.e., the items should work in a defined sequence for operational success). In that case, an *event tree* is appropriate way to model the reliability. This may not always be the case for a simple system, but it is often the case for complex systems, such as nuclear power plants, in which the subsystems must work according to a sequence of events to achieve a desirable outcome. Event trees, or a closely related model called the *event sequence diagram*, are particularly useful in these situations. Event trees are also used in probabilistic risk assessments to develop risk scenarios (the first question in the risk triplet discussed in Chapter 1). Development of event trees for risk analysis will be further discussed in Chapter 9.

6.5.1 Construction of Event Trees

Event trees are horizontally built structures that document a path (called a *scenario* or *sequence of events*) from the initiating event through a series of pivotal events that lead to the endpoint, outcome, or consequence. The event tree starts on the left, where the initiating event is modeled. An *initiating event* describes a situation when a legitimate demand for the operation of a system(s) occurs or an event moves the system out of its normal operating envelop. Development of the tree proceeds chronologically, with the demand on each item (or subsystem) being postulated and shown on the top of the event tree as an *event tree heading* or *pivotal event*. The first item demanded after the initiating event appears first, and so on.

An event tree for a nuclear power plant system is shown in Figure 6.22, with events:

- Initiating Event = Pipe break leading to loss of normal cooling;
- RP = (Failure of) Operation of the reactor-protection system to shut down the reactor;
- ECA = (Failure of) Injection of emergency coolant water by pump A;
- ECB = (Failure of) Injection of emergency coolant water by pump B;
- LHR = (Failure of) Long-term heat removal.

At each *branch point* in Figure 6.22, the upper branch shows the success of the pivotal event, and the lower branch shows its failure. Note: There is no single standard for the order of pivotal events and branching direction used in building event trees. Sometimes, the analyst puts the most consequential (worst consequence) scenario at

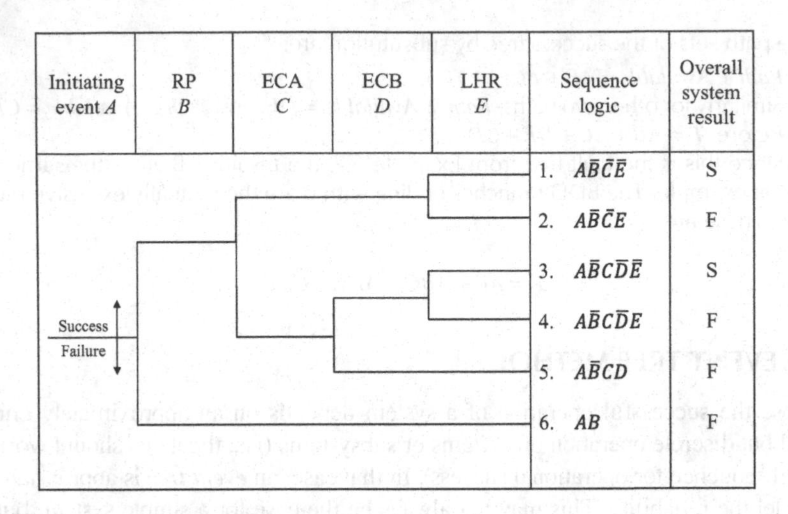

FIGURE 6.22 Example of an event tree.

the top but sometimes it goes at the bottom. The reader is advised to determine the most appropriate conventions for their application and to state the conventions at the beginning of model documentation.

The sequence logic is documented adjacent to each sequence as a Boolean expression. The final column of the event tree shows the system result for each scenario. This result describes the outcome of each sequence: whether the overall system succeeds, fails, initially succeeds but fails later, or vice versa.

In Figure 6.22, following the occurrence of the initiating event A, RP needs to operate (event $\bar{B}$). If RP does not operate (i.e., if failure event B occurs), the overall system will fail. This is shown by the lower branch of event B (sequence 6), which ends in system failure. If RP works, then the emergency cooling must operate. If either emergency cooling pump (ECA or ECB) functions properly, the emergency cooling works. Emergency cooling must be followed by proper long-term heat removal (LHR); failure of LHR (event E) leads to the failure of the overall system. If RP works, one emergency coolant pump operates, and the LHR succeeds, the result is a successful outcome for the system. Note that the operation of specific subsystems may not necessarily depend on the occurrence of some preceding events. For example, if ECA operates successfully, the failure or success of ECB does not matter for the overall system success.

The logical representation of each sequence is developed directly from the tree. For example, for sequence 5 in Figure 6.22, failure events A, C, and D occur, but event B was successful (indicated by the overbar). Clearly, these sequences are mutually exclusive. However, the notation on the sequence logic is Boolean notation—this does not imply the independence of these events.

Event trees are usually developed in a binary format; that is, the pivotal events are assumed to either occur or not occur. In cases where a spectrum of outcomes is possible, the branching process can proceed with more than two outcomes. In these cases, the qualitative representation of the event tree branches in a Boolean sense would not be possible.

Although inductive in principle, developing an event tree requires a good deal of deductive thinking by the analyst. To demonstrate this issue and further understand

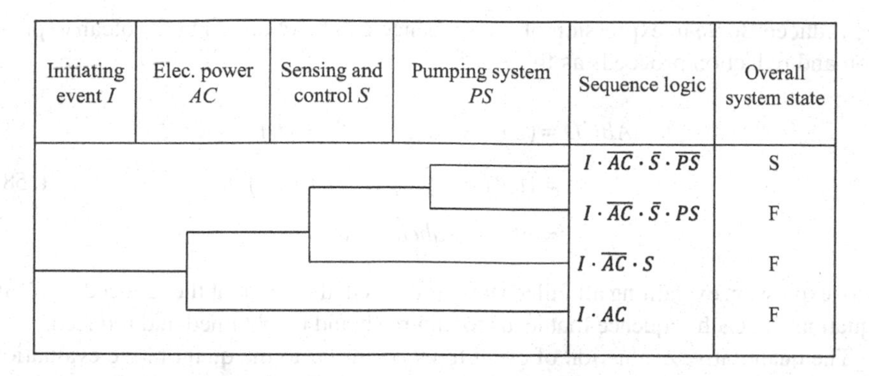

Initiating event I
Elec. power AC
Sensing and control S
Pumping system PS
Sequence logic
Overall system state
$I \cdot \overline{AC} \cdot \overline{S} \cdot \overline{PS}$
S
$I \cdot \overline{AC} \cdot \overline{S} \cdot PS$
F
$I \cdot \overline{AC} \cdot S$
F
$I \cdot AC$
F

FIGURE 6.23 Event tree for the pumping system in Figure 6.10.

the concept of event tree development, let us consider the pumping system in Figure 6.10. One can think of a situation where the sensing and control system device S initiates one of the two pumps, $P1$ and $P2$. However, the AC power source (AC) must exist to allow S and pumps $P1$ and $P2$ to operate. Thus, we can define three distinct events, AC, S, and pumping system PS for a sequence of events starting wherein the initiating event is a signal sent to the sensor S. An event tree that includes these three events can be constructed. Clearly, if AC fails, both PS and S fail; if S fails, only PS fails. This would place AC as the first pivotal event (heading) in the event tree, followed by S and PS. This event tree is illustrated in Figure 6.23.

Events represent discrete states of the items or systems. These items' failure probabilities can be modeled using the probability models for components or can be represented by fault trees. This way, the event tree sequences, and the logical combinations of events can be considered. This is a powerful aspect of the event tree technique. If the event tree pivotal events represent complex subsystems or items, building a fault tree for each pivotal event can conveniently model the failure logic in detail. Other system analysis models, such as RBDs, Bayesian networks, and logical representations in terms of cut sets or path sets, can also be used.

6.5.2 Evaluation of Event Trees

Qualitative evaluation of event trees is straightforward. The logical representation of each event tree heading, and ultimately each event tree sequence, is obtained and then reduced using Boolean logic rules.

For example, in sequence 5 of Figure 6.22, if events A, B, C, and D are represented by the following Boolean expressions (e.g., minimal cut sets),

$$\begin{aligned} A &= a, \\ B &= b + cd, \\ C &= e + d, \\ D &= c + eh, \end{aligned} \tag{6.57}$$

the reduced Boolean expression of the sequence can be obtained by Boolean expression and reduction proceeds as follows:

$$\begin{aligned} A\bar{B}CD &= (a)\left(\overline{b+cd}\right)(e+d)(c+eh) \\ &= (a)\left(\bar{b}\bar{c}+\bar{b}\bar{d}\right)(ec+eh+dc) \\ &= a\bar{b}\bar{c}eh + a\bar{b}c\bar{d}e + a\bar{b}\bar{d}eh. \end{aligned} \tag{6.58}$$

If an expression explaining all failed states is desired, the union of the reduced Boolean equations for each sequence that leads to failure should be obtained and reduced.

The quantitative evaluation of event trees is similar to the quantitative evaluation of fault trees. For example, to determine the probability associated with the sequence $A\bar{B}CD$:

$$\begin{aligned} Pr\left(A\bar{B}CD\right) &= Pr\left(a\bar{b}\bar{c}eh + a\bar{b}c\bar{d}e + a\bar{b}\bar{d}eh\right) \\ &= Pr\left(a\bar{b}\bar{c}eh\right) + Pr\left(a\bar{b}c\bar{d}e\right) + Pr\left(a\bar{b}\bar{d}eh\right) \\ &= Pr(a)\left[1-Pr(b)\right]\left[1-Pr(c)\right]Pr(e)Pr(h) \\ &\quad + Pr(a)\left[1-Pr(b)\right]Pr(c)\left[1-Pr(d)\right]Pr(e) \\ &\quad + Pr(a)\left[1-Pr(b)\right]\left[1-Pr(d)\right]Pr(e)Pr(h). \end{aligned} \tag{6.59}$$

Since the three terms are disjoint, the above probability is exact. However, if the terms are not disjoint, the rare event approximation can be used here.

Inherent in the structure of the events is that they are conditional on the previous event (i.e., the probabilities are conditional probabilities). However, conditional probability notation is rarely used explicitly because of the size of the logic expressions. The reader is reminded to consider this when quantifying.

6.6 EVENT SEQUENCE DIAGRAM METHOD

The Event sequence diagram (ESD) method is another scenario modeling technique, and has the potential to include dynamic nodes (Swaminathan and Smidts, 1999). The ESD starts with an initiating event and uses binary logic branching like the event tree method. However, ESDs use logic symbols directly in the model rather than listing events at the top of the diagram. Initiating events are represented with an oval, and pivotal events are represented with a rectangle. *End states* are represented with a diamond. In ESD logic, the branch coming out of the right side of a pivotal event denotes "yes," and a branch coming out the bottom of the pivotal event represents "no." An example of an ESD is given in Figure 6.24.

Each end state is quantified by multiplying the corresponding probabilities across the scenarios. The initiating and pivotal events can be quantified directly using data or various models, including parametric probability distributions, fault trees, and Bayesian networks.

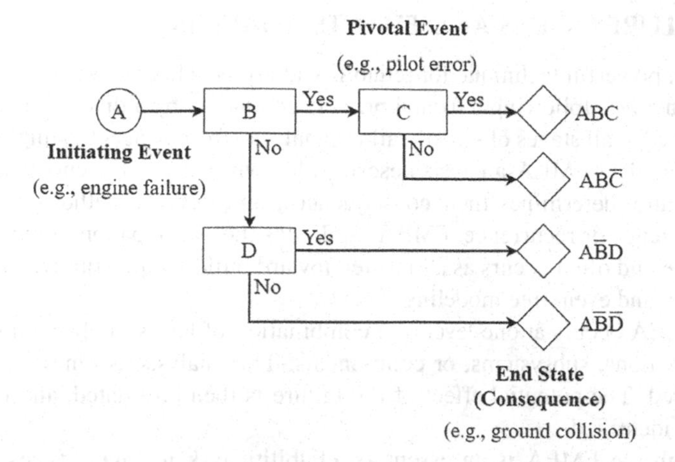

FIGURE 6.24 Example of an event sequence diagram.

Example 6.9

The ESD below (adapted from Groth and Hecht, 2017) shows a sequence of events that can happen after an unintended release of hydrogen gas from a system. Find the frequency of the end state "explosion." Assume that the hydrogen release (H2R) frequency is 1.0×10^{-4}/year, and the probability of successful leak detection (given a leak) is 0.10. If a leak is not detected and isolated, the probability of immediate ignition is 0.053. If the leak does not ignite immediately, it can accumulate, and the probability of delayed ignition of that accumulation is 0.027.

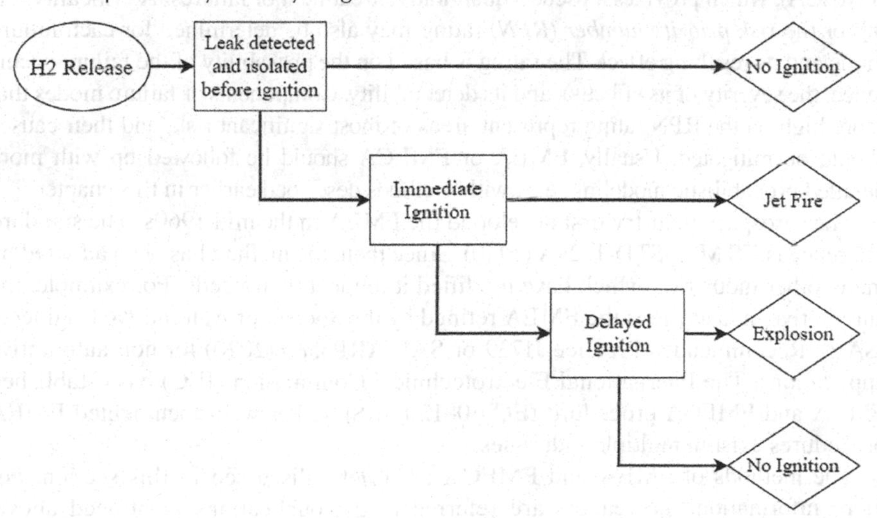

Solution:

$$f(\text{Explosion}) = f(\text{H2R})\cdot Pr(\text{LD}=\text{no})\cdot Pr(\text{II}=\text{no})\cdot Pr(\text{DI}=\text{yes})$$

$$=\left(1\times10^{-4}\right)\cdot(1-0.1)\cdot(1-0.53)\cdot 0.27 = 2.30\times10^{-6} \text{ explosions/year}$$

6.7 FAILURE MODES AND EFFECTS ANALYSIS

FMEA is a powerful technique for reliability analysis. This method is inductive and typically involves following a formal process conducted by a diverse team. In practice, it is used at all stages of system failure analysis, from concept to implementation to operation. The FMEA analysis describes known causes of events that lead to a system failure, determines their consequences, and devises methods to minimize their occurrence or recurrence. FMEA facilitates the participation of multiple types of expertise and often occurs as a first step toward building quantitative models such as fault tree and event tree modeling.

The FMEA occurs at one level or a combination of levels of abstraction, such as system functions, subsystems, or components. The analysis assumes that a failure has occurred. The potential effect of the failure is then postulated, and its possible causes are identified.

Although the FMEA is an essential reliability task for many types of system design and development processes, it is a more simplified technique. It provides limited insight into the probabilistic representation of system reliability. The quality of FMEA analyses can vary widely depending on the level of investment in the analysis. Another limitation is that FMEA can be performed for only one failure mode at a time. This single failure limitation means that FMEA may not be adequate for addressing systems in which multiple failures can coincide or systems with significant redundancy and diversity. (By contrast, deductive methods like fault trees are very powerful for identifying these failures.) However, FMEA provides valuable qualitative information about the system design and operation.

An extension of FMEA is called failure modes, effects and criticality analysis (FMECA), which provides a pseudo quantitative treatment of failures. A criticality rating or the *risk priority number (RPN)* rating may also be determined for each failure mode and its resulting effect. The rating is based on the probability of the failure occurrence, the severity of its effect(s), and its detectability. Components or failure modes that score high on the RPN rating represent areas of most significant risk, and their causes should be mitigated. Usually, FMEA or FMECA should be followed up with more detailed probabilistic modeling, e.g., with methods described earlier in this chapter.

The aerospace industry first developed the FMEA in the mid-1960s. The standard reference is US MIL-STD-1629A (1980). Since then, the method has been adopted by many other industries, which have modified it to meet their needs. For example, the automotive industry uses the FMEA refined by the Society of Automotive Engineers (SAE), Recommended Practice J1739 or SAE ARP5580 (2020) for non-automotive applications. The International Electrotechnical Commission (IEC) has established FMEA and FMECA procedure IEC 60812 (2018). Other well-documented FMEA procedures exist in multiple industries.

The methods of FMEA and FMECA are briefly discussed in this section. For more information, the readers are referred to the publications mentioned above. These and other documented FMEA standards provide a well-defined, documented starting point for conducting these analyses.

6.7.1 Objectives of FMEA

A team may perform different types of FMEA depending on the stage in product development and the reason for conducting the FMEA.

FMEA can be performed to address safety concerns or as part of a risk management process. FMEA can provide a basis for reliability engineering and reliability-centered maintenance programs. FMEA can be used to identify failure modes that result in system failure, hazardous conditions, or mission delays. FMEAs can support design decisions that improve system robustness and reliability. For example, SAE J1739 (2021) includes Design FMEA and Process FMEA methods.

Design FMEA is used to evaluate the failure modes and their effects for a product or system before it is released to production or operation. It is normally applied at the subsystem and the component abstraction levels. The major objectives of a Design FMEA are to:

1. Identify failure modes and rank them according to their effect on the system or product performance, thus establishing a priority system for design improvements.
2. Identify design actions to eliminate potential failure modes or reduce their rate of occurrence.
3. Document the rationale behind system or product design changes and provide a future reference for analyzing field concerns, evaluating new design changes, and developing next generation designs.

Process FMEA is used to analyze manufacturing, operations, and assembly processes. The major objectives of a process FMEA are to:

1. Identify failure modes that can be associated with manufacturing, operation errors, or assembly process deficiencies.
2. Identify highly critical process characteristics that may cause specific failure modes.
3. Identify the sources of manufacturing/assembly process variations (equipment performance, material, operator, and environment) and establish a strategy to reduce it.

The FMEA process should be carried out multiple times during a product, system, or process lifetime. A more detailed analysis should follow an analysis conducted during the design stage as the system or product matures and more information become available. An FMEA should be updated with field operating experience data as it becomes available.

6.7.2 FMEA/FMECA Procedure

Outlined below is a logical sequence of steps by which FMEA/FMECA is usually performed.

- Develop the FMEA plan, objectives, and approach and assemble the information basis and resources for the analysis.
- Define the system (or product) to be analyzed. Identify the system abstraction level (indenture level) that will be analyzed. Identify internal and interface system functions, restraints, and develop failure definitions.

- Construct a reliability or functional block diagram of the system, depending on the complexity of the system (functional diagrams are preferred for early design and very complicated systems).
- Identify all critical failure modes and define their effects on the immediate function of the system, and the mission to be performed.
- Evaluate each failure mode for the worst potential consequence and assign a severity classification category.
- Identify failure detection methods and compensating provision(s) for each failure mode.
- Identify corrective designs or other actions required to eliminate the failure or control the risk. Identify the problems that could not be corrected by design.
- Document the analysis.
- Revisit the analysis periodically to ensure that it reflects the current system, and to incorporate new data and information as it emerges.

6.7.3 FMEA Implementation

6.7.3.1 FMEA Using MIL-STD-1629A

The FMEA is usually performed using a tabular format. A worksheet implementation of a typical MIL-STD-1629A FMEA procedure is shown in Table 6.2. Once the FMEA plan is completed, the major steps of the analysis proceed as described below.

6.7.3.1.1 Information Basis

It is important to assemble information to use in the FMEA process to provide a strong technical basis for the analysis. While the process is inductive, having detailed information on the system, previous failures, and probabilities provides a stronger technical basis. The information basis can include system diagrams, system descriptions, failure databases, hazard checklists, failure mode models, probability models, operational experience, etc. Previous FMEAs on similar items or processes should also be consulted.

6.7.3.1.2 System Description and Block Diagrams

Next, we describe the system to allow the FMEA to be performed efficiently and to be understood by others. This description can be done at different levels of abstraction. For example, the system can be represented by a functional block diagram at the highest level (i.e., the functional level). The functional block diagram differs from the RBD discussed earlier in this chapter. Functional block diagrams illustrate the operation, interrelationship, and interdependence of the functional entities (e.g., subsystems) of a system. For example, the pumping system in Figure 6.10 can be represented by its functional block diagram, as shown in Figure 6.25. This figure also describes the components that support each system function.

6.7.3.1.3 Item/Functional Identification and Function

The first step in completing the FMEA worksheet (Table 6.2) is to provide the descriptive name and the nomenclature of the item under analysis. This step provides necessary information for the identification number, functional identification (nomenclature), and function columns in the FMEA. The item could be a function,

TABLE 6.2
FMEA Worksheet Format for MIL-STD-1629A

FAILURE MODE AND EFFECTS ANALYSIS

System ______________
Indenture level ________
Reference drawing ______
Mission ______________

Date ______________
Sheet ____ of ________
Compiled by __________
Approved by __________

IDENTIFICATION NUMBER	ITEM/FUNCTIONAL IDENTIFICATION (NOMENCLATURE)	FUNCTION	FAILURE MODES AND CAUSES	MISSION PHASE/ OPERATIONAL MODE	FAILURE EFFECTS			FAILURE DETECTION METHOD	COMPENSATING PROVISIONS	SEVERITY CLASS	REMARKS
					LOCAL EFFECTS	NEXT HIGHER LEVEL	END EFFECT				

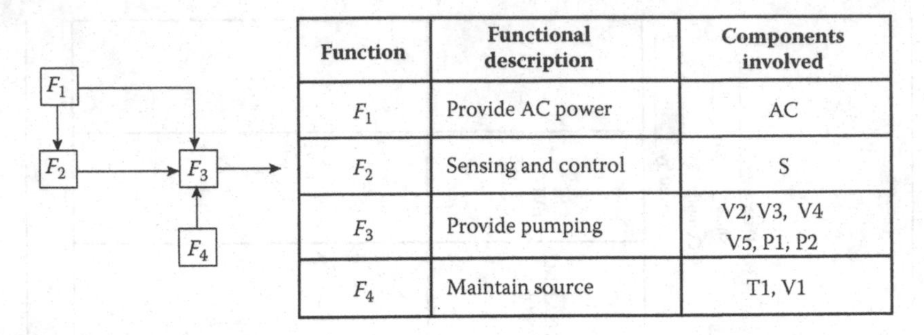

Function	Functional description	Components involved
F_1	Provide AC power	AC
F_2	Sensing and control	S
F_3	Provide pumping	V2, V3, V4 V5, P1, P2
F_4	Maintain source	T1, V1

FIGURE 6.25 Functional block diagram for the pumping system in Figure 6.10.

subsystem, component, or part. If the failures are postulated at a lower abstraction level, such levels should be shown. A fundamental item of the FMEA may be subject to a separate FMEA that further decomposes this item into more basic parts. More detailed levels of abstraction require greater levels of detail and effort. Therefore, one must carefully consider how far to decompose the FMEA events to meet the aims of the analysis within time and cost constraints.

6.7.3.1.4 Failure Modes and Causes and Mission Phase/Operational Mode

The next column of the FMEA worksheet contains the failure mode and causes. A failure mode can have more than one cause. Each failure mode and its associated cause(s) should be identified and listed in this column. The failure modes applicable to components and parts are often known *a priori*. Modern FMEA software tools have a library of components and associated failure modes. For example, typical failure modes for electronic components are open, short, corroded, drifting, misaligned, and so on, as described in Chapter 1. Some representative failure modes for mechanical components include deformed, cracked, fractured, sticking, leaking, and loosened, as described in Chapter 1. However, only specific failure modes may apply depending on the specific system under analysis, the environmental design, and other factors. These should be known and specified by the analyst.

6.7.3.1.5 Failure Effects

The consequences of each failure mode on the item's operation should be carefully examined and recorded in the column labeled "Failure Effects." In Table 6.2, the effects are distinguished at three levels: local, next higher abstraction level, and end effect. Local effects show the impact of the postulated failure mode on the operation and function of the item under consideration. Sometimes no local effects can be defined beyond the failure mode itself. However, the consequences of each postulated failure on the output of the item should be described along with second-order effects. The end effect analysis describes the effect of postulated failure on the operation, function, and status of the next higher abstraction level and ultimately on the system itself. The end effects in this column may result from multiple failures. For example, the failure of a supporting subsystem in a system can be catastrophic if it occurs along with another local failure. These cases should be clearly recognized and discussed in the end effect column.

6.7.3.1.6 Failure Detection Method

Failure detection features for each failure mode should be described. For example, based on the item's behavior pattern(s), previously known symptoms can indicate whether a failure has occurred. The related symptom can cover the operation of a component under consideration (logical symptom) or cover both the component and the overall system or equipment evidence of failure.

6.7.3.1.7 Compensating Provision

To maximize reliability, a detected failure should be corrected to eliminate its propagation to the whole system. Therefore, provisions that will alleviate the effect of a malfunction or failure should be identified at each abstraction level. These provisions include items such as (a) redundant elements for continued and safe operation, (b) safety devices, and (c) alternative modes of operation, such as backup and standby units. In addition, any action that may require operator action should be clearly described.

6.7.3.1.8 Severity

Severity classification provides a qualitative indicator of the worst potential effect resulting from the failure mode. The MIL-STD-1629A classifies severity levels as given in Table 6.3.

6.7.3.1.9 Remarks

Any additional information, clarifications, or notes should be entered in the "Remarks" column.

6.7.3.2 FMEA Using SAE J1739

The SAE J1739 FMEA procedure is conceptually similar to the MIL-STD-1629A. However, some definitions and ratings differ from those discussed so far. The key criterion for identifying and prioritizing potential design deficiencies here is the RPN, defined as the product of the severity, occurrence, and detection ratings. An example of an SAE J1739 FMEA format is shown in Table 6.4.

TABLE 6.3
Severity Levels Used in MIL-STD-1629A

Effect	Rating	Criteria
Catastrophic	I	A failure mode that may cause death or complete mission loss.
Critical	II	A failure mode that may cause severe injury, major system degradation or damage, or reduction in mission performance.
Marginal	III	A failure that may cause minor injury or degradation in system or mission performance.
Minor	IV	A failure that does not cause injury or system degradation but may result in system failure and unscheduled maintenance or repair.

Note: This table uses roman numerals for the rating number, because higher severity is given a lower rating number; in newer FMEA methods, this is reversed so that higher severity results in a higher RPN.

TABLE 6.4
FMEA Worksheet Format from SAE J1739

Potential Failure Mode and Effects Analysis (Design FMEA)

System: ____________ | Design Responsibility: ____________ | FMEA Number: ____________

Subsystem: ____________ | Key Date: ____________ | Page ___ of ___

Component: ____________ | Prepared by: ____________

Model Year/Vehicle(s): ____________ | FMEA Date (Orig): ____________ (Rev.) ________

Core Team:

Item/Function	Potential Failure Mode	Potential Effect(s) of Failure	Sever	Potential Cause(s)/ Failure Mechanism(s)	Occur	Current Design Controls	Detec	RPN	Recommended Actions	Responsibility and Target Completion Date	Action Results				
											Actions Taken	Sever	Occur	Detec	RPN

The contents of the item/function, potential failure mode, possible effect(s) of failure, likely cause(s)/failure mechanism(s), and the recommended actions steps of this FMEA procedure are the same as the MIL-STD-1629A FMEA discussed above. However, in SAE J1739, severity is evaluated on a ten-point scale, as shown in Table 6.5. In contrast to the MIL-STD-1629A, a higher rating in the SAE standard corresponds to higher severity (and, consequently, a higher RPN).

Also in contrast to MIL-STD-1629A, the SAE J1739 occurrence rate is expressed judgmentally by the expected frequency that a specific failure cause/mechanism will occur as shown in Table 6.6.

TABLE 6.5
Severity Scale Used in SAE J1739

Effect	Rating	Criteria
Hazardous	10	Safety-related failure modes causing noncompliance with government regulations without warning.
Serious	9	Safety-related failure modes causing noncompliance with government regulations with warning.
Very high	8	Failure modes resulting in loss of primary vehicle/system/component function.
High	7	Failure modes resulting in a reduced level of vehicle/system/component performance and customer dissatisfaction.
Moderate	6	Failure modes resulting in loss of function by comfort/convenience systems/components.
Low	5	Failure modes resulting in a reduced level of performance of comfort/convenience systems/components.
Very low	4	Failure modes resulting in loss of fit and finish, squeak, and rattle functions.
Minor	3	Failure modes resulting in partial loss of fit and finish, squeak, and rattle functions.
Very minor	2	Failure modes resulting in a minor loss of fit and finish, squeak, and rattle functions.
None	1	No effect.

TABLE 6.6
Occurrence Rate of Failure used in SAE J1739

Likelihood of Failure	Estimated or Expected Failure Frequency	Rating
Very high (failure is almost inevitable)	>1 in 2	10
	1 in 3	9
High (frequently repeated failures)	1 in 8	8
	1 in 20	7
Moderate (occasional failures)	1 in 80	6
	1 in 400	5
	1 in 2,000	4
Low (rare failures)	1 in 15,000	3
	1 in 150,000	2
Remote (failures are unlikely)	<1 in 150,000	1

TABLE 6.7
Detection Ratings Based on Design Control Criteria

Detection	Rating	Criteria
Uncertain	10	Design control will not and/or cannot detect a potential cause/mechanism and subsequent failure mode.
Very remote	9	Very remote chance that the design control will detect a potential cause/mechanism and subsequent failure mode.
Remote	8	Remote chance that the design control will detect a potential cause/mechanism and subsequent failure mode.
Very low	7	Very low chance the design control will detect a potential cause/mechanism and subsequent failure mode.
Low	6	Low chance that the design control will detect a potential cause/mechanism and subsequent failure mode.
Moderate	5	Moderate chance that the design control will detect a potential cause/mechanism and subsequent failure mode.
Moderately high	4	Moderately high chance that the design control will detect a potential cause/mechanism and subsequent failure mode.
High	3	High chance that the design control will detect a potential cause/mechanism and subsequent failure mode.
Very high	2	Very high chance that the design control will detect a potential cause/mechanism and subsequent failure mode.
Almost certain	1	The design control will almost certainly detect a potential cause/mechanism and subsequent failure mode.

Before the design is finalized, the engineering team has control over it in terms of possible design changes. Three types of design control are usually considered: those that (a) prevent the failure cause/mechanism or mode from occurring or reduce their rate of occurrence, (b) detect the cause/mechanism and lead to corrective actions, or (c) detect the failure mode (Table 6.7).

While in this standard the RPN is the primary measure of design risk in FMEAs, the designer should carefully review and propose corrective action remedies for the high-severity failure modes irrespective of the resultant RPN number. Action Results columns describe the implemented corrective actions along with the estimated reduction in severity, occurrence, and detection ratings and the resultant RPN.

6.7.4 FMECA Procedure: Criticality Analysis

Criticality analysis extends the FMEA table with the probability of failure mode occurrence and its impact on the system mission success. Table 6.8 shows an example of a criticality analysis worksheet format. The criticality analysis part of this worksheet is explained below.

6.7.4.1 Failure Probability Failure Rate Data Source

This column documents the source of information used for that item.

TABLE 6.8
FMECA Worksheet Format for MIL-STD-1629A

CRITICALITY ANALYSIS

System ______
Indenture level ______
Reference drawing ______
Mission ______

Date ______
Sheet ____ of ______
Compiled by ______
Approved by ______

Identification Number	Item/Functional Identification (Nomenclature)	Function	Failure Modes And Causes	Mission Phase/ Operational Mode	Severity Class	Failure Probability Failure Rate Data Source	Failure Effect Probability (β)	Failure Mode Ratio (α)	Failure Rate (λ_p)	Operating Time (T)	Failure Mode Crit # $C_m = \beta\alpha\lambda_p T$	Item Crit # $C_r = \Sigma(C_m)$	Remark

*NOTE: Both criticality number (C_r) and probability of occurrence level are shown for convenience

TABLE 6.9
Failure Effect Probabilities for Various Failure Effects (β)

Failure Effect	β Value
Actual loss	1.00
Probable loss	0.1 < 1.0
Possible loss	0.0 < 0.1
No effect	0

6.7.4.2 Failure Effect Probability β

The β value represents the conditional probability that the failure effect with the specified criticality classification will occur, given that the failure mode occurs. For complex systems, it is challenging to determine β unless a comprehensive logic model of the system (e.g., a fault tree) exists. Therefore, often, estimation of β becomes primarily a matter of judgment greatly driven by the analyst's experience or the underlying data use. The general guidelines in Table 6.9 can be used for determining β.

6.7.4.3 Failure Rate λ_p

The generic or specific failure rate for each failure mode of the item should be obtained and recorded in the failure rate (λ_p) column. The estimates of λ_p can be obtained from the test, field data, or generic sources of failure rates discussed in Chapter 5. Note this assumes a constant failure (hazard) rate, which is another limitation of FMECA.

6.7.4.4 Failure Mode Ratio α

The fraction of the item's failure rate, λ_p, related to the specific failure mode under consideration is evaluated and recorded in the failure mode ratio (α) column. The failure mode ratio is the probability that the item will fail in the identified failure mode. If all potential failure modes of an item are listed, the sum of their corresponding α values should be equal to 1. The values of α should typically be available from a data source (e.g., those described in Chapter 5). However, if not available, the values can be assessed based on the analyst's judgment.

6.7.4.5 Operating Time T

The operating time, in hours, or the number of operating cycles of the item should be listed in the corresponding column.

6.7.4.6 Failure Mode Criticality Number C_m

The *failure mode critical number* C_m is used to rank each potential failure mode of an item based on the failure mode's occurrence and the consequence of its effect.

$$C_m = \beta\alpha\lambda_p T. \tag{6.60}$$

6.7.4.7 Item Criticality Number (C_r)

The item criticality number is the criticality for the item under analysis. For a particular severity classification, the C_r of an item is the sum of the failure mode criticality numbers C_m with the same severity classification.

$$C_r = \sum_{i=1}^{n} (C_m)_i, \tag{6.61}$$

where $(C_m)_i$ is the criticality of an individual item's failure mode i, and n is the number of failure modes of an item with the same severity classification.

Based on the criticality number, a *criticality matrix* is usually developed to provide a visual way of identifying and comparing each failure mode to all other failures concerning severity. Figure 6.26 shows an example of such a matrix. This matrix can also be used for a qualitative criticality analysis in an FMEA. Along the vertical dimension of the matrix, the probability of occurrence level (subjectively estimated by the analyst in an FMEA study) or the criticality number C_r (calculated in an FMECA study) is entered. Along the horizontal dimension of the matrix, the severity classification of an effect is entered. The severity increases from left to right. Each item on the FMEA or FMECA could be represented by one or more points on this matrix. If the item's failure modes correspond to more than one severity effect, each failure mode will correspond to a different point in the matrix. Clearly, those severities that fall in the upper-right quadrant of the matrix require immediate attention for the reliability or design improvements.

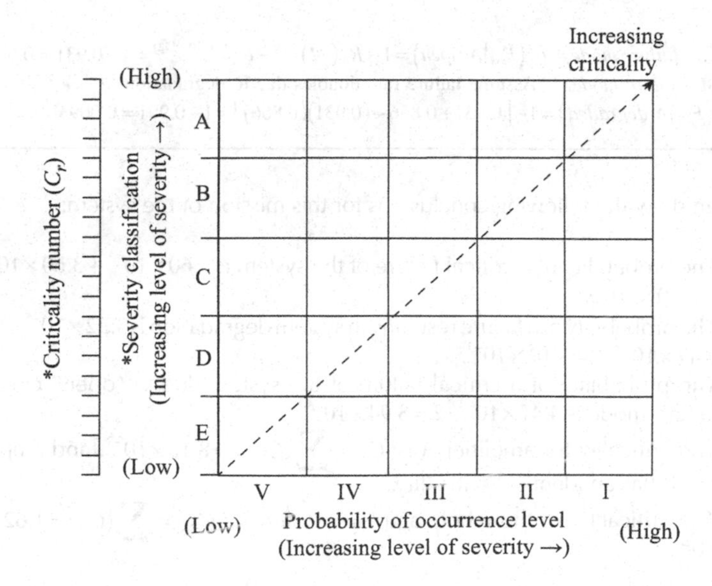

FIGURE 6.26 Example of criticality matrix.

There are several software tools that help the FMEA/FMECA developers. For example, the FMEA Tool (MSC-25379-1) developed by NASA is readily available. These tools help generate an FMEA model that include the block diagram and connecting the components on the diagram to a library of generic components and their failure modes. For more discussion on FMEA/FMECA and tool refer to Spreafico et al. (2017) or Stamatis (2003).

Example 6.10

Develop an FMECA for a system of two amplifiers, A and B, in a parallel configuration. In a given mission, this system should function for 72 hours.

Solution:

The FMEA Worksheet for this analysis is displayed below.

ID No.	Item	Failure Mode	Effects: Local	Effects: System	Severity Class	β	α	λ_p (hour⁻¹)	Time, T (hour)	Criticality No., C_m
1.	A	a. Open	Circuit A fails	Degraded	III	0.069	0.90			4.47×10^{-3}
		b. Short	Both A and B fail	Failed	II	1.000	0.05	1×10^{-3}	72	3.60×10^{-3}
		c. Other	Circuit A lost	Degraded	IV	0.009	0.05			3.32×10^{-5}
2.	B	a. Open	Circuit B fails	Degraded	III	0.069	0.90			4.47×10^{-3}
		b. Short	Both A and B fail	Failed	II	1.000	0.05	1×10^{-3}	72	3.60×10^{-3}
		c. Other	Circuit B lost	Degraded	IV	0.009	0.05			3.32×10^{-5}

Notes:

β *for "A open" failure mode* $= Pr\left(F_{sys}|A\ open\right) = 1 - R_A(72) = 1 - e^{-1.0\times10^{-3}(72)} = 1 - 0.931 = 0.069.$

β *for "A lost" failure mode:* **Assume failure rate doubles due to degradation: $R_A = e^{-2\times10^{-3}(72)} = 0.866$, then $\beta = Pr\left(F_{sys}|A\ degraded\right) = 1 - \left[0.931 + 0.866 - (0.931)(0.866)\right] = 1 - 0.991 = 0.009.$

One can draw the following conclusions for this mission of the system:

1. The probability of a critical failure of the system is $3.60\times10^{-3} + 3.60\times10^{-3} = 7.20\times10^{-3}$.
2. The probability of a failure resulting in system degradation is $3.32\times10^{-5}\cdot2 + 4.47\times10^{-3}\cdot2 = 9.01\times10^{-3}$.
3. The probability of a critical failure of the system due to "open" circuit failure mode is $4.47\times10^{-3}\cdot2 = 8.94\times10^{-3}$.
4. The criticality for amplifier A is: $C_{r_A} = \sum(C_m)_A = 8.10\times10^{-3}$, and amplifier B has an identical criticality.
5. The criticality for the whole amplifier system is: $C_{r_{Sys}} = \sum(C_m) = 1.62\times10^{-2}$.

The above approximate probabilities can only hold if the product of α, β, λ, and T are small (e.g., <0.1). Typically, criticality numbers are used as a crude measure of system reliability.

Therefore, the most effective design would allocate more engineering resources to the areas with high criticality numbers and minimize the Class I and Class II severity failure modes.

6.8 EXERCISES

6.1 Consider the RBD below. Assume the reliability of each item is $R(x_i) = e^{-\lambda_i t}$, and $\lambda_i = 2.0 \times 10^{-4}$ per hour for all *i*. Find the following:

a. Minimal path sets
b. Minimal cut sets
c. MTTF
d. Reliability of the system at 1,000 hours

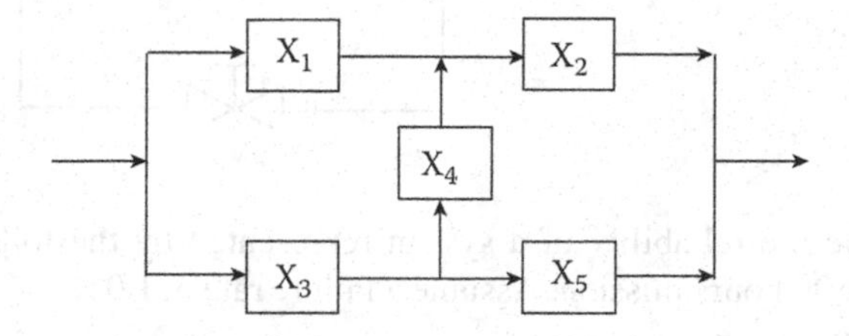

6.2 Consider the RBD below. Find the following if $\lambda_i = 1.0 \times 10^{-4}$ for all items.

a. Minimal path sets
b. Minimal cut sets
c. Reliability of the system at 1,000 hours
d. Probability of failure at 1,000 hours, using cut sets, to verify results from (c)
e. Accuracy of the results of (d) and/or (c), using an approximate method
f. MTTF of the system

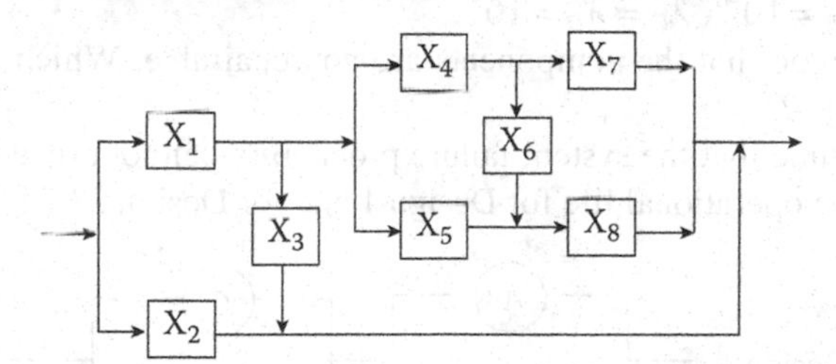

6.3 Calculate the reliability of the system shown in the figure below for a 1,000 hour mission. What is the MTTF for this system?

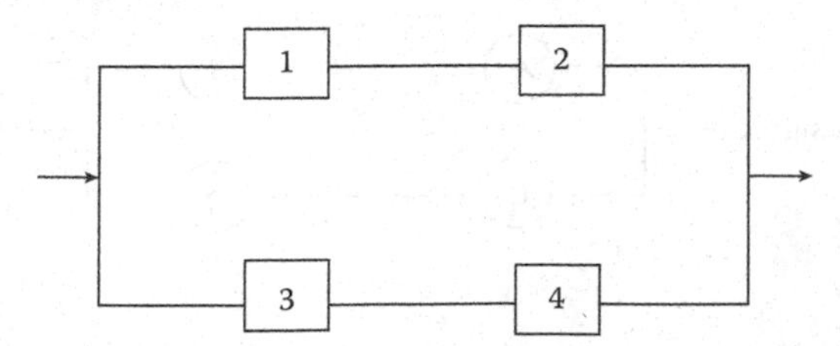

6.4 Consider the pumping system shown in the figure below. The purpose of the system is to pump water from point A to point B. The time to failure of all the valves and the pump can be represented by the exponential distributions with failure rates λ_v and λ_p, respectively.

a. Determine the reliability function of the system.

b. If $\lambda_v = 1.0 \times 10^{-3}$ and $\lambda_p = 2.0 \times 10^{-3}$ per hour, and the system has survived for 10 hours, what is the probability of surviving another 10 hours?

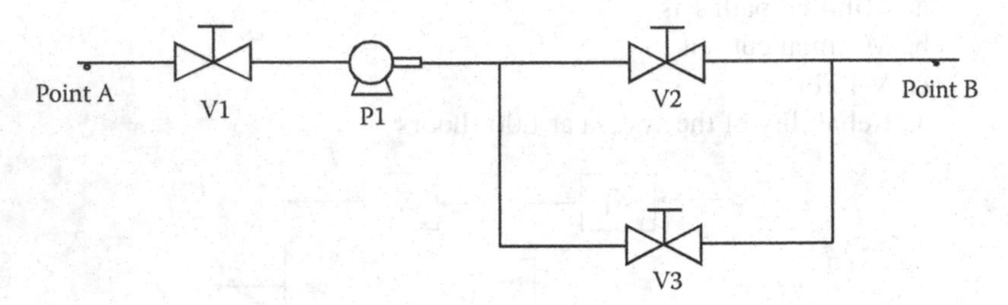

6.5 Estimate the reliability of a system represented by the following RBD for a 2,500 hours mission. Assume a failure rate of 1.0×10^{-6} per hour for each unit.

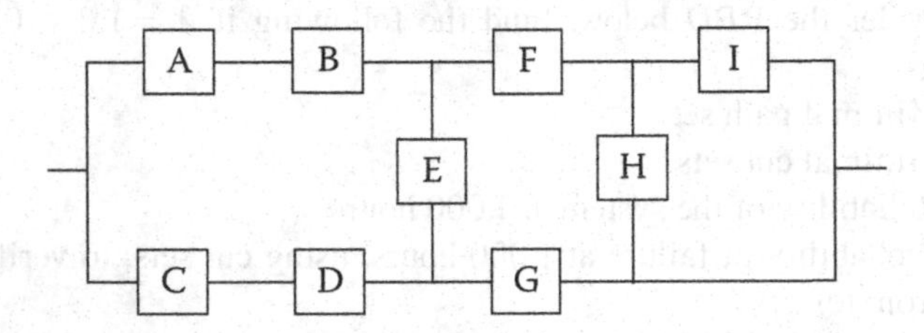

6.6 Compare Design 1 and Design 2 below. Use the failure rates (per hour) $\lambda_A = \lambda_B = 10^{-6}$, $\lambda_C = \lambda_D = 10^{-3}$.

a. Assume that the components are nonrepairable. Which is the better design?

b. Assume that the system failure probability cannot exceed 10^{-2}. What is the operational life for Design 1 and for Design 2?

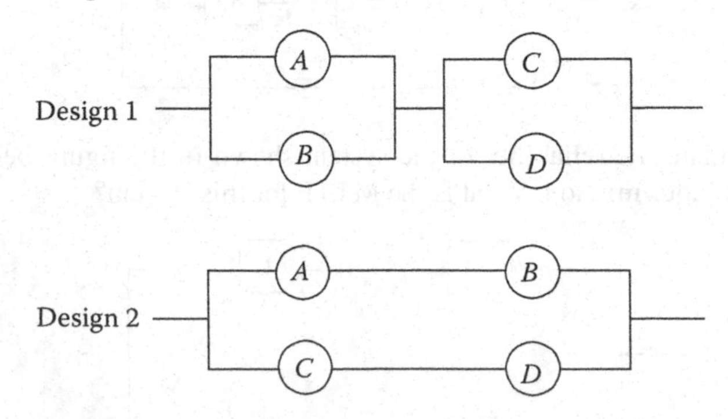

6.7 In the following system, which uses active redundancy, what is the probability that there will be no failures in the first year of operation? Assume constant failure rates given in the figure.

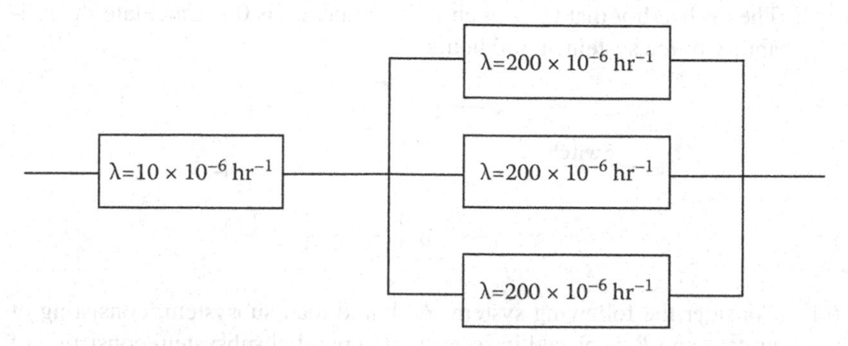

6.8 A filter system is composed of 30 elements, each with a failure rate of 2.0×10^{-4} per hour. The system will operate satisfactorily with two elements failed. What is the probability that the system will operate satisfactorily for 1,000 hours?

6.9 For the RBD below.

a. Find all minimal path sets.
b. Find all minimal cut sets.
c. Assuming each component has a reliability of 0.90 for a given mission time, compute the system reliability over mission time.

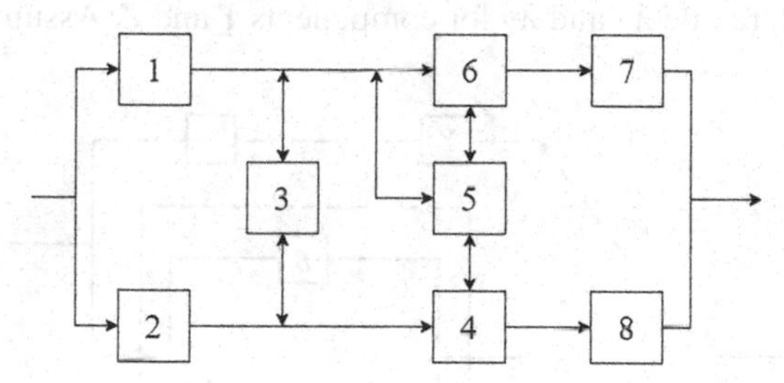

6.10 For the RBD below, the reliability of each component is: $A=0.90$, $B=0.95$, $C=0.77$, $D=0.90$, $E=0.8$. Find the reliability of the system.

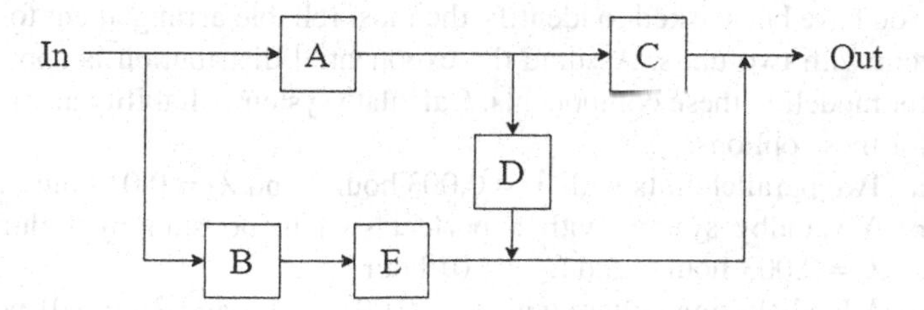

6.11 For the system below, the components A, B, and C have identical properties. Each has a constant failure rate of 2×10^{-3} per hour when operational and a constant failure rate of 1×10^{-4} per hour while in standby. The probability that the switch fails to operate is 0.1. Calculate the reliability of the system at 300 hours.

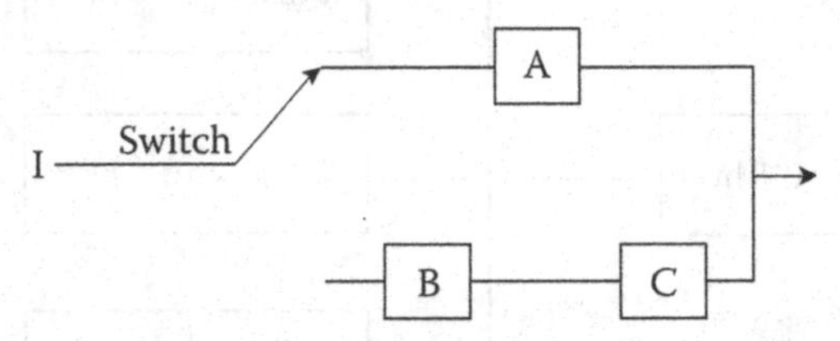

6.12 Consider the following system. A shared-load subsystem, consisting of units A and B, is placed in series with a parallel subsystem consisting of units C and D, with $\lambda_C = 0.001$ hour^{-1} and $\lambda_D = 0.002$ hour^{-1}. Times to failure for A and B are exponentially distributed, with the full-load failure rate being twice that of the half-load failure rate. Draw an RBD of this system. Find reliability at 20 hours.

6.13 For the following system, write the equation for determining the system reliability (solve the integrals in the expression). Upon failure of subsystems Y or Z, a perfect switch activates. Subsystem X comprises shared-load components B and C. B and C have identical constant failure rates λ_f and λ_h for full- and half-load operation, respectively. Assume a constant failure rate λ_Y and λ_Z for components Y and Z. Assume no standby failures.

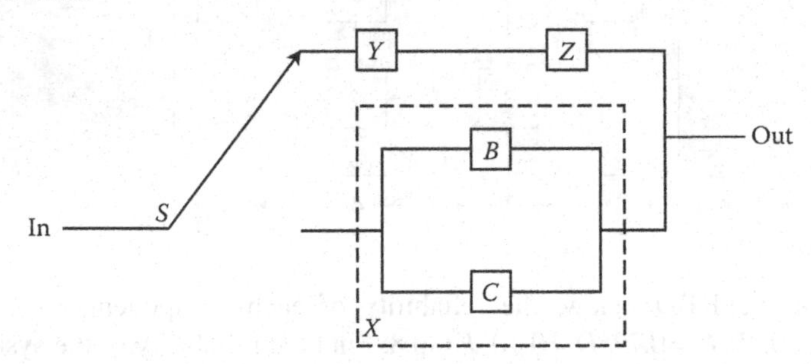

6.14 You have been asked to identify the most reliable arrangement for a system with two units. Assume the exponential distribution is appropriate for modeling these components. Calculate system reliability at 100 hours for these options.

a. Two parallel units with $\lambda_1 = 0.003$ hour^{-1} and $\lambda_2 = 0.01$ hour^{-1}.

b. A standby system with a perfect switch, no standby failure, and $\lambda_1 = 0.003$ hour^{-1} and $\lambda_2 = 0.01$ hour^{-1}.

c. A load-sharing system with $\lambda_h = 0.003$ hour^{-1} and $\lambda_f = 0.01$ hour^{-1}.

6.15 Tires used on construction trucks fail at a constant rate of one failure per 1,000 miles. In a truck with four tires, what is the probability that at least two tire failures occur in 1,500 miles?

6.16 An emergency diesel generator has three belts. The wear-out behavior of each belt can be modeled by a Weibull distribution with $\beta = 1.27$. However, each belt has a different scale parameter: 2,300, 7,700, and 4,400 operating hours. If the generator is new, determine the probability of a belt failure during an emergency where it is operated for 1 week.

6.17 Field observations show that a system with two identical components in parallel has an MTTF of 500 hours. Data from the manufacturer of the individual components indicate each component has a failure rate of about 0.02 per hour. Are these data consistent with each other?

6.18 Consider the fault tree below. Find the following:
   a. Minimal cut sets
   b. Minimal path sets
   c. Probability of the top event if the following probabilities apply: $Pr(A) = Pr(C) = Pr(E) = 0.01$; $Pr(B) = Pr(D) = 0.0092$.

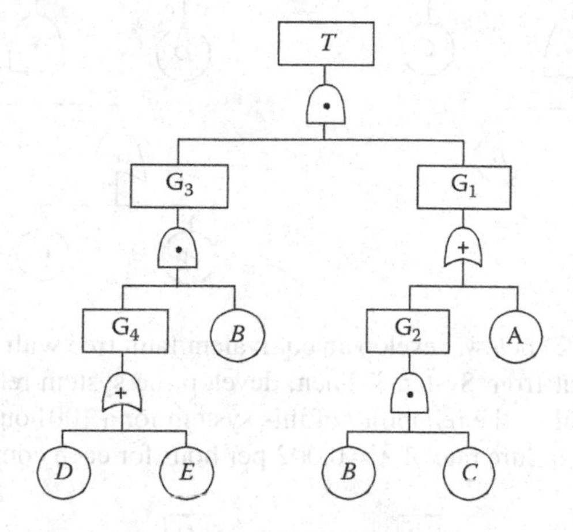

6.19 Use top-down substitution to find the cut sets and path sets of the fault tree below.

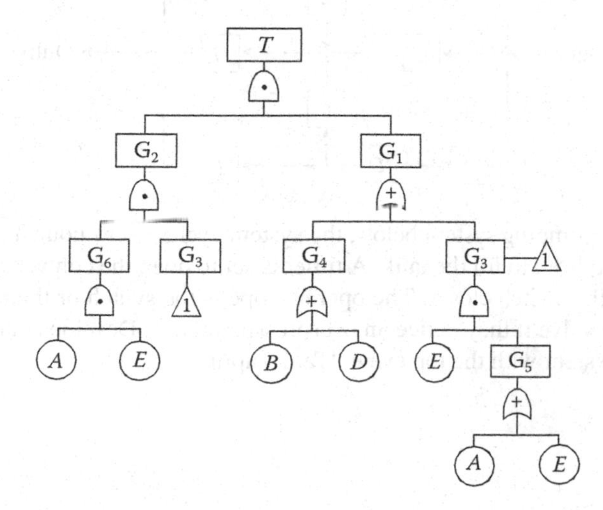

6.20 Consider the following fault tree.
   a. Find all minimal cut sets and path sets.
   b. Assuming all component failure probabilities are 0.01, find the top event probability.
   c. Generate a BDD for the FT and compare it with the minimal cut sets.
   d. Find the accuracy achieved by using the BDD vs. minimal cut sets.

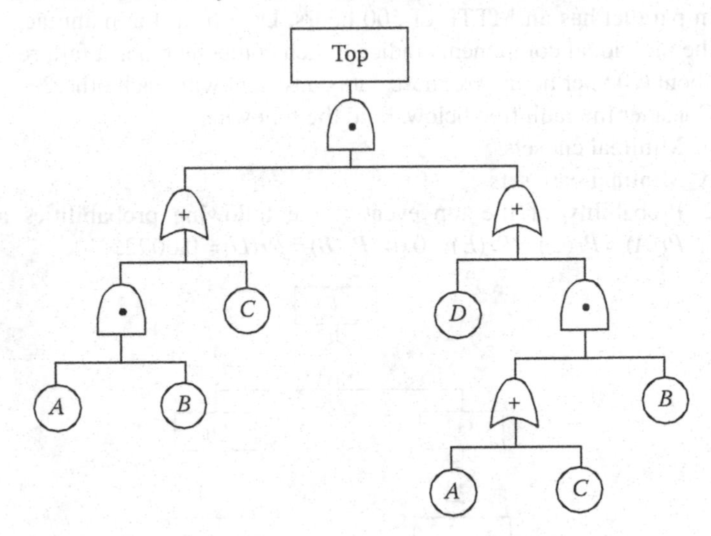

6.21 For the RBD below, develop an equivalent fault tree with the top event of "No Output from System." Then, develop the system reliability expression. Calculate the reliability of this system for a 100 hour mission using a constant failure rate of $\lambda = 0.002$ per hour for each component.

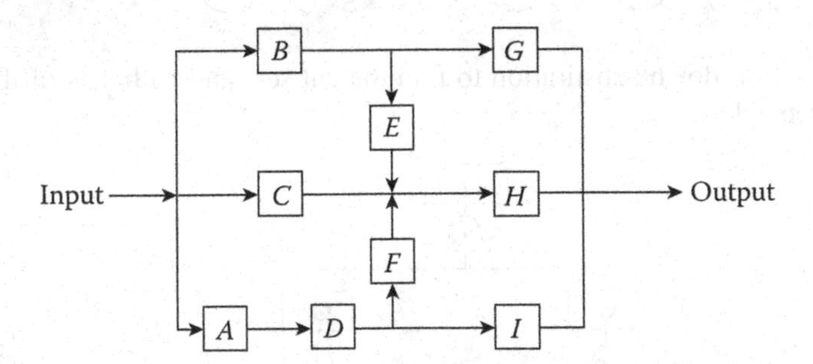

6.22 In the pumping system below, the system cycles every hour. Ten minutes are required to fill the tank. A timer is set to open the contact 10 minutes after the switch closes. The operator opens the switch or the tank emergency valve if they notice an overpressure alarm. Develop a fault tree for this system with the top event "Tank ruptures."

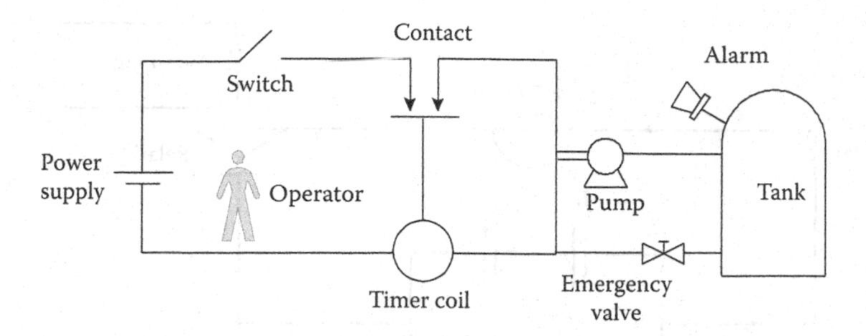

6.23 A containment spray system is used to scrub and cool the atmosphere around a nuclear reactor during an accident. Develop a fault tree with "No H_2O spray" as the top event. Assume the following:

- There are no secondary failures.
- There is no test and maintenance.
- There are no passive failures.
- Failures are independent.
- There is no human error.
- One of the two pumps and one of the two spray heads is sufficient to provide spray. (i.e., the operation of one train is enough for system success).
- One of the valves SV_1 or SV_2 is opened after demand. However, SV_3 and SV_4 are always normally open.
- Valve SV_5 is always in the closed position.
- SP_1, SP_2, SV_1, SV_2, SV_3, and SV_4 use the same power source P to operate.

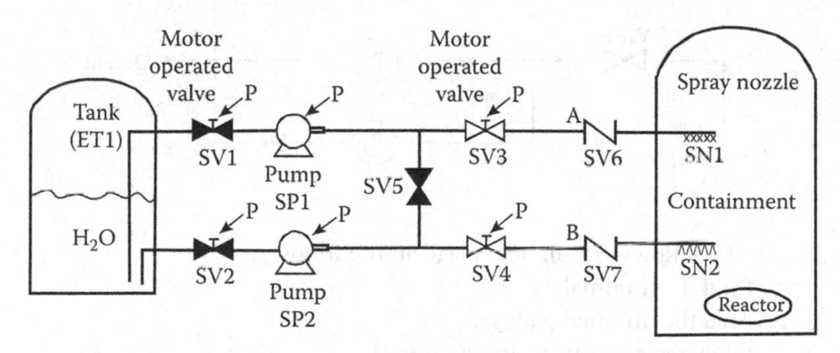

6.24 Consider the following electric circuit for providing emergency light during a blackout. In this circuit, the relay is held open as long as AC power is available, and any of the four batteries is capable of supplying light power.

a. Draw a fault tree with the top event "No Light When Needed."
b. Find the minimal cut sets of the system.
c. Find the minimal path sets of the system.

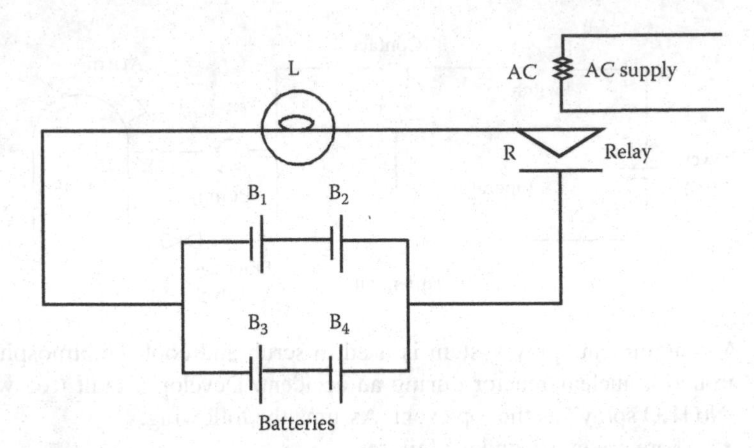

6.25 Consider the following pumping system consisting of three identical parallel pumps, and a valve in series. The pumps have a constant failure rate of λ_p (per hour) for the failure mode "fail off." The valve has a constant failure rate of λ_v (per hour) for the failure mode "spurious closure."

a. Develop an expression for the reliability of the system using the success tree method.
b. Find the average reliability of the system over time period t.
c. Repeat parts (a) and (b) for the case that $\lambda t < 0.1$, and then approximate the reliability functions. Find the average reliability of the system when $\lambda = 0.001$ hour^{-1} and $t = 10$ hours.

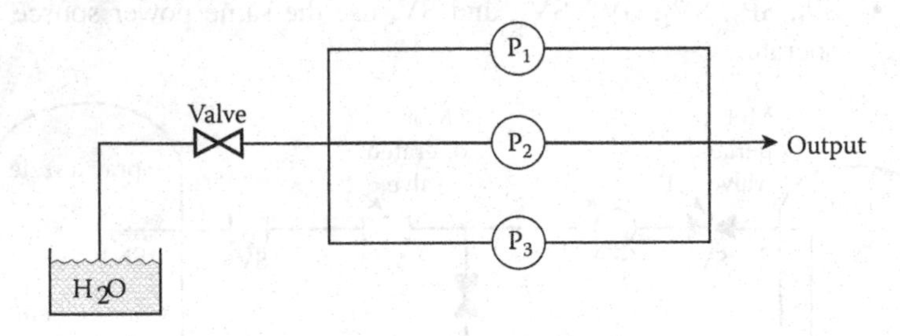

6.26 For the following fault tree, perform the following:

a. Find the minimal cut sets.
b. Find the minimal path sets.
c. Draw the equivalent success tree.
d. Calculate the unreliability of the system based on the constant failure rates below.

Component	A	B	Y	Z	K	G	V	P
Failure Rate	0.08	0.02	0.07	0.05	0.05	0.05	0.05	0.01

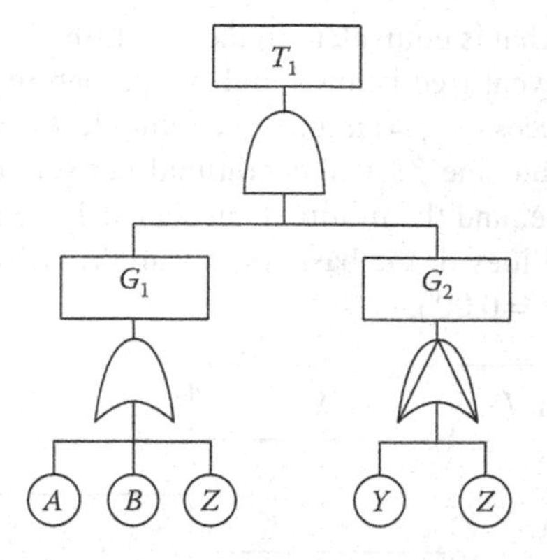

6.27 For the following fault tree, perform the following:

a. Find the minimal cut sets.
b. Find the minimal path sets.
c. Draw the equivalent success tree.
d. Calculate the unreliability of the system. Use the constant failure rate data from Exercise 6.26.

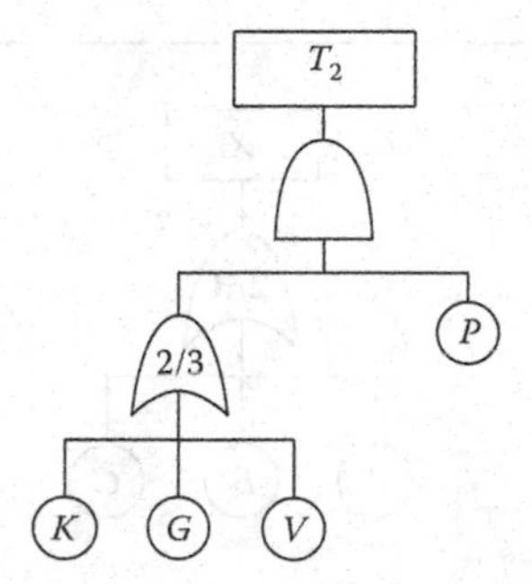

6.28 An event tree is used in reactor accident estimation as shown, where Sequence 1 is a system success and Sequence 2 and Sequence 3 are system failures. The minimal cut sets of subsystems X and Y are $X = AB + AC + D$ and $Y = BD + E + A$. Find the cut sets of Sequence 2 and Sequence 3.

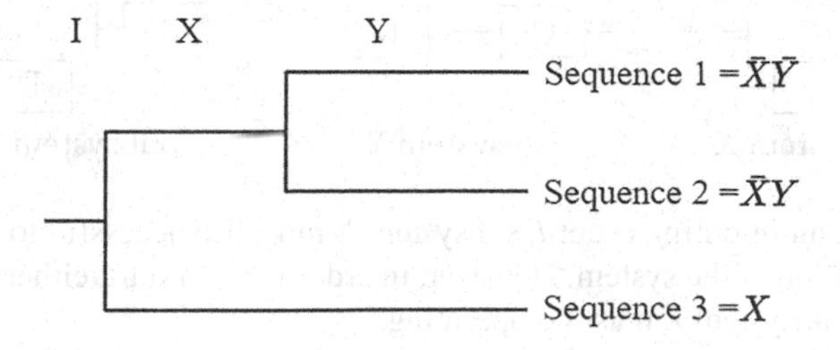

6.29 Build an ESD that is equivalent to the event tree in Exercise 6.28.

6.30 Consider the event tree below. Find the probability of each of the outcomes (S = success, F_1 = Failure Outcome 1, F_2 = Failure Outcome 2, F_3 = Failure Outcome 3), if the minimal cut sets of event X are given by the fault tree, and the minimal cut sets of Y are $BD + E$. Assume that failure probabilities of the basic events are: $A = 0.04$, $B = 0.03$, $C = 0.04$, $D = 0.02$, and $E = 0.003$.

Initiating event I	X	Y	Outcome
			S
			F_1
			F_2
			F_3

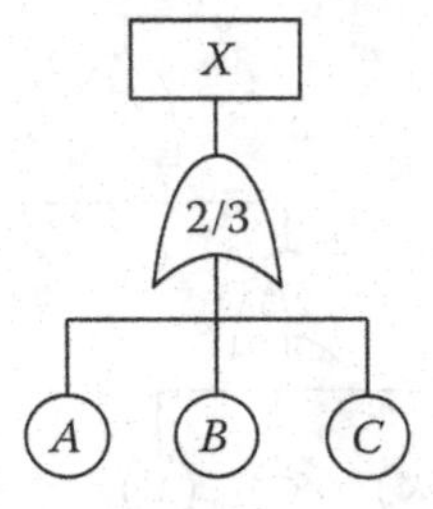

6.31 A mining crusher requires subsystem X and either subsystem Y or Z to function. Subsystem X, Y, and Z are configured as shown below. Draw an event tree for the mining crusher operation. Use the following component reliability values. $R_A = 0.99$, $R_B = 0.98$, $R_C = 0.999$, $R_D = 0.998$, $R_E = 0.99$.

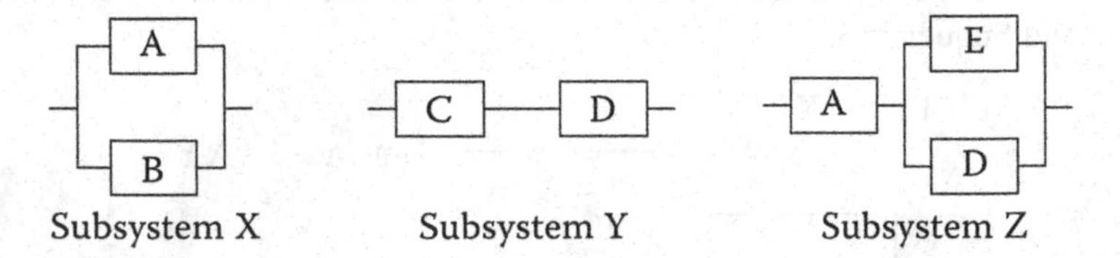

6.32 Upon an initiating event I, subsystem X must be successful to realize the operation of the system. However, in order for X to start, either subsystem Y or subsystem Z must be operating.

a. Draw an event tree for this system.

b. The minimal cut sets of the subsystems are: $X = ab + ac$; $Y = d + ef$; $Z = c + df$. Find the cut sets of all of the scenarios in the event tree.

6.33 For the system below, assume the following: One of the two product lines is sufficient for success. Control instruments receive the sensor values and transfer those values to the process control computer, which calculates the position of the control valves. The control instruments adjust the control valves as needed. The plant computer controls the process control computer.

a. Develop a fault tree for the top event "inadequate product feed."
b. Find the minimal cut sets of the fault tree.
c. Find the expression for quantifying the probability of the top event.
d. Which components are most critical to the design?

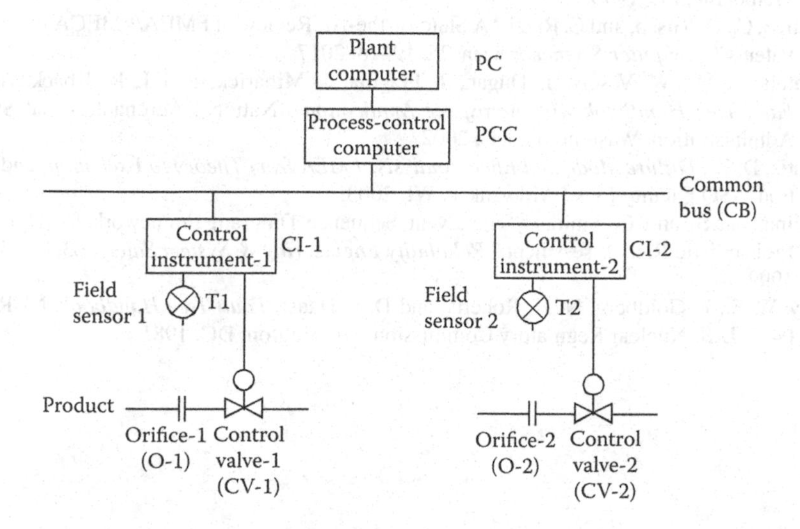

6.34 Perform a FMECA for the system described in Exercise 6.33. Compare the results with part (d) of Exercise 6.33.

REFERENCES

Bryant, R. E., "Binary Decision Diagrams." In *Handbook of Model Checking*, 191–217. Springer, Cham, 2018.

Crowder, M. J., A. C. Kimber, R. L. Smith, and T. J. Sweeting, *Statistical Analysis of Reliability Data*. Chapman & Hall, London, New York, 1991.

Daniels, H. E., "The Statistical Theory of the Strength of Bundles of Threads." *Proceedings of the Royal Society of London. Series A. Mathematical and Physical Sciences*, 183, 404–435, 1945.

Dezfuli, H., M. Modarres, and J. Meyer, "Application of REVEAL_W to Risk-Based Configuration Control". Reliability Engineering & System Safety, 44(3), 243–263, 1994.

Groth K. M. and E. S. Hecht, "HyRAM: A Methodology and Toolkit for Quantitative Risk Assessment of Hydrogen Systems." *International Journal of Hydrogen Energy*, 42, 7485–7493, 2017.

IEC 60812, *Failure Modes and Effects Analysis (FMEA and FMECA)*. International Electrochemical Commission, Geneva, Switzerland 2018.

Kapur, K. and L. Lamberson, *Reliability in Engineering Design*. Wiley & Sons, New York, 1977.

MIL-STD-1629A, *Procedure for Performing a Failure Mode, Effects, and Criticality Analysis*. Department of Defense, NTIS, Springfield, VA, 1980.

SAE ARP5580, *Recommended Failure Modes and Effects Analysis (FMEA) Practices for Non-Automobile Applications*, SAE International, Warrendale, PA 2020.

SAE J1739, *Potential Failure Mode and Effects Analysis (FMEA) Including Design FMEA, Supplemental FMEA-MSR, and Process FMEA*. SAE Internaional, Warrendale, PA, 2021.

Shen, J., M. Bensi, and M. Modarres, "Identification and Assessment of Current and Developing PRA Technologies for Risk-Informed-Decision-Making (RIDM)." In *Proceedings of the 16th Probabilistic Safety Assessment and Management Conference (PSAM16)*, Honolulu, HI. 2022.

Shooman, M. L., *Probabilistic Reliability: An Engineering Approach*. 2nd edition, Krieger, Melbourne, FL, 1990.

Spreafico, C., D. Russo, and C. Rizzi. "A State-of-the-Art Review of FMEA/FMECA Including Patents." *Computer Science Review* 25, 19–28. 2017.

Stamatelatos, M., W. Vesely, J. Dugan, J. Fragola, J. Minarick, and J. Railsback, *NASA Fault Tree Handbook with Aerospace Applications*. National Aeronautics and Space Administration, Washington, DC, 2002.

Stamatis, D. H., *Failure Mode and Effect Analysis: FMEA from Theory to Execution*, 2nd edition. ASQ Quality Press, Milwaukee, WI, 2003.

Swaminathan, S. and C. Smidts, "The Event Sequence Diagram Framework for Dynamic Probabilistic Risk Assessment." *Reliability Engineering & System Safety*, 63(1), 73–90, 1999.

Vesely, W. E., F. Goldberg, N. H. Roberts, and D. F. Haasl, *Fault Tree Handbook*. NUREG-0492. U.S. Nuclear Regulatory Commission, Washington, DC, 1981.

7 Reliability and Availability of Repairable Components and Systems

When we perform reliability studies, it is important to distinguish between repairable and nonrepairable items. The reliability analysis methods discussed in Chapters 3–5 largely apply to nonrepairable items. In this chapter, we examine the critical aspects of repairable systems and discuss methods used to determine the failure characteristics of these systems, and the methods for predicting their reliability and availability.

Nonrepairable components and systems are discarded or replaced (i.e., taken out of their sockets and replaced) with new ones when they fail. For example, light bulbs, transistors, contacts, satellites, and some small appliances are nonrepairable items. The reliability of a nonrepairable item is expressed in terms of its time to failure distribution, which can be represented by its respective cdf, pdf, or hazard (failure) rate function, as discussed in Chapter 3.

On the other hand, repairable items (components and systems) are not replaced following the occurrence of a failure; rather, they are repaired and put into operation again. However, if a nonrepairable item is a component of a repairable system, the distribution of the number of component replacements over a time interval is estimated within the framework of repairable systems reliability. In contrast to nonrepairable items, reliability problems associated with repairable items use different *random* (*stochastic*) *process* models that will be discussed.

Repair should not be confused with maintenance. *Maintenance* is a broader term used to describe all sorts of renewal processes and can be performed on an item that has not necessarily failed. Maintenance is carried out to prevent, predict, protect, or mitigate progression of the ongoing degradation processes (e.g., some failure mechanisms) that ultimately lead to failure. Examples of the types of maintenance include preventive maintenance, predictive maintenance associated with condition monitoring, and corrective maintenance. As a result of maintenance, repair may be necessary to correct incipient failures. A special kind of maintenance, corrective maintenance, reacts to failures or symptoms that point to the existence of a failure. Thus, repair is inevitable in corrective maintenance.

Both maintenance and repair lead to downtime of the components and systems, which reduce availability and thus increase unavailability. Items can be repaired, and repair activities take time. The probability that an item (system) is up (functioning) at a given random time can be measured by a probability value called the *pointwise availability*. Similarly, the expected fraction of time (or cycles) that an item is operational over a known period can be viewed as the *average availability*. Conversely, the probability that the item is nonoperational (i.e., down) is called *unavailability*.

DOI: 10.1201/9781003307495-7

We will begin this chapter with probabilistic models and statistical methods used to determine the failure characteristics of repairable items and their reliability. We will then define the concept of availability and explain availability evaluation methods for repairable items. Although the presentation of the material in this chapter focuses on system reliability and availability, the methods equally apply to components.

7.1 DEFINITION OF REPAIRABLE SYSTEM AND TYPES OF REPAIRS

In this book, the term *repairable item* is a synonym for any repairable or renewable component or system. The term *repair* applies to making a failed item (system, component, or part) operational. Repairable items generally experience several failures and repairs during their life. The repair may be perfect which restores the item to an *as good as new* condition. The repair may be imperfect which restores the item to a state less than perfect, for example *as bad as old*, or *better than old* (but worse than new). These repairs apply to restoration or fixes of failed items, incipient failures and to design/manufacturing changes.

7.2 VARIABLES OF INTEREST: AVAILABILITY, ROCOF, MTBF

Availability refers to the fraction of time that an item remains operational under normal conditions and maintains its intended function. A simple mathematical representation of availability is:

$$\text{Availability} = \frac{\text{Total elapsed time} - \text{Total downtime}}{\text{Total elapsed time}}. \tag{7.1}$$

As we discussed in Chapter 1, availability may be interpreted as the probability that a repairable item is functional at time t. That is,

$$A(t) = Pr\left[X(t) = 1\right], \tag{7.2}$$

where $Pr[X(t) = 1]$ denotes the probability that system X is in state 1 (functioning) at time t.

Note that the downtime is highly influenced by several factors, including the rate of at which failures occur, the length of time it takes to restore the item, and the condition of the item after repair (i.e., the degree of rejuvenation). These topics will be discussed in more detail later.

It is necessary to distinguish between several different times relevant to repairable system data collection and modeling.

- *Uptime, t_{up} or UT*, refers to times when the system is operating.
- *Time between failure T_f or TBF*: refers to the time between consecutive failures. Depending on the analysis, this may be operating time or it may be total time. It is important to verify which. In this text, we will use t_F as the age of the item when a failure occurs, and T_f to indicate operational time between consecutive failures.

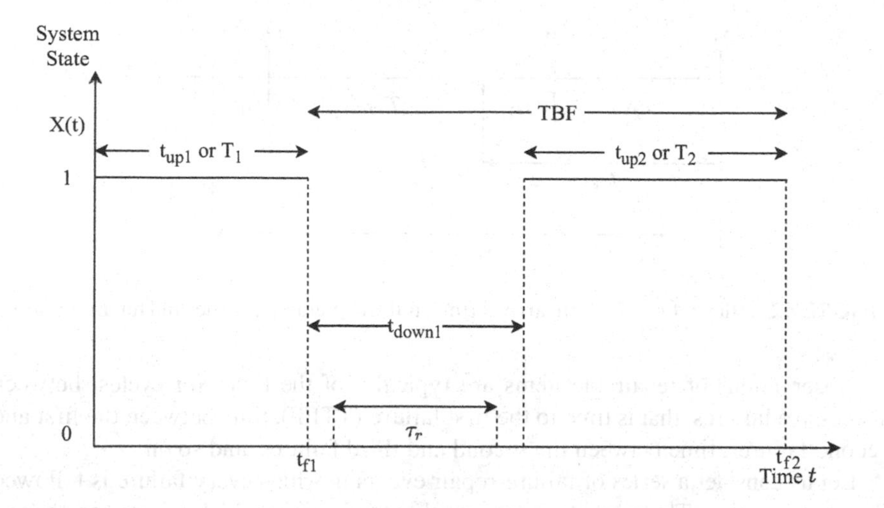

FIGURE 7.1 Illustration of different times used in repairable system modeling.

- *Time to repair, τ_r or TTR*, is the duration of the repair activity. We may also encounter t_r, the age of the item when the repair occurs.
- *Downtime, t_{down},* refers to the duration when the system is down. This includes downtime due to active repair, and includes time for failure detection, system diagnosis, logistic times to obtain parts and travel to the repair site, and other relevant activities (Figure 7.1).

Rate of occurrence of failures (ROCOF), $\lambda_f(t)$, is a measure of the rate that defines the trend (i.e., constant, increasing, decreasing) in the interarrival of failures. Note that the symbol λ is used for rate, but this is different than the λ parameter as the hazard rate (failure rate) for exponentially distributed variables used for nonrepairable items.

For a repairable item, it is desirable that the ROCOF values remain constant or decreasing (i.e., failure interarrival times increasing). It is critical to assess the rate of change in ROCOF, which will be discussed later in this chapter.

Rate of occurrence of repairs (ROCOR), $\mu(t)$ or $\lambda_r(t)$, is a measure of the rate that defines the trend (i.e., constant, increasing, decreasing) in the interarrival times (i.e., length) of repairs.

The mean time between successive failure events measured from the times of many interarrival of failures in repairable systems is known as the *mean time between failure (MTBF)*. We will denote this as MTBF, or $E(t_{BF})$ where t_{BF} is the time between failures. Therefore, the MTBF is related to a single item of interest. However, if times of a group of identical items' interarrival of failures are known, then the MTBF data can represent that group of items too.

7.3 REPAIRABLE SYSTEM DATA

For most repairable systems, data will come in the form of sequences of failures and repairs. Assume that we have a system which is put into operation at time $t=0$. The system operates until failure. Upon failure, the system is unavailable for a period of time; the system goes into maintenance.

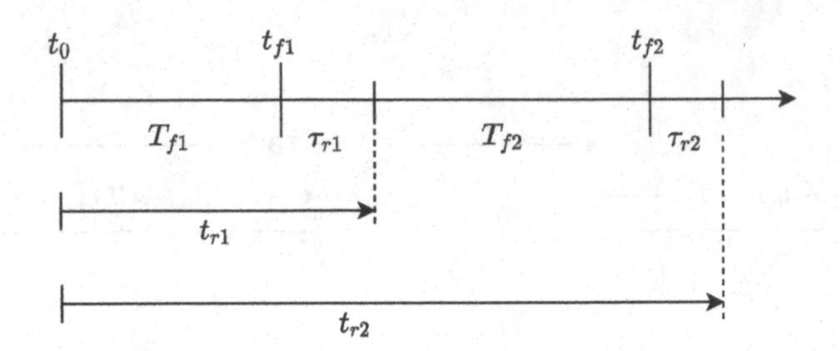

FIGURE 7.2 Illustration of failure arrival time, failure interarrival time, and repair duration.

Observations of repairable items are typically of the times (or cycles) between successive failures, that is time to the first failure (TTFF), time between the first and second failures, time between the second and third failures, and so on.

Let us consider a series of failure-repair events in which every failure is followed by a repair action. The respective data are the item's age at which failures and repair actions occur, which can be represented as

$$t_{f1}, t_{r1}, t_{f2}, t_{r1}, \ldots, t_{fk}, t_{rk}, \ldots, t_{fn}, t_{rn}. \tag{7.3}$$

Note these are calendar times. If we view this in terms of the duration of uptimes and the duration of repairs, we see this becomes:

$$T_{f1}, T_{r1}, \ldots, T_{fn}, T_{rn}, \tag{7.4}$$

where T_{fl} is the operational time to first failure (and is equivalent to t_{fl}), τ_i or T_{ri} is the duration of the ith repair following the occurrence of the ith failure. T_{f2} is the operational time between the end of the first repair and the occurrence of the second failure, and so on. See Figure 7.2 for illustration of these times.

Sometimes, we simplify Equations 7.3. and 7.4 by looking only at failure times. We call the sequence of times at which the system fails, $t_1, \ldots, t_n$ the *arrival times* of failure. We call a sequence of operational time between failures, $T_1, \ldots, T_n$ the *interarrival times* of failure.

This data set is typically needed for solving different reliability and risk analysis problems related to repairable systems, such as estimating whether the system is improving, deteriorating, or following a constant failure rate, predicting the number of failures observed in each time interval, developing the respective logistics, or analyzing reliability growth.

7.4 STOCHASTIC POINT PROCESSES (COUNTING PROCESSES)

The most basic availability model assumes the average time it takes to start and complete a repair, referred to as time to repair (or maintenance), is assumed to be negligible compared with the MTBF. This assumption makes it possible to apply different *stochastic point processes* as candidate models for real-life failure processes. For example, a minor system repair might take only a few hours to perform. Here, the assumption is rather realistic. However, an opposite example could be the time it takes to repair a very complex system, which might be comparable with its predicted

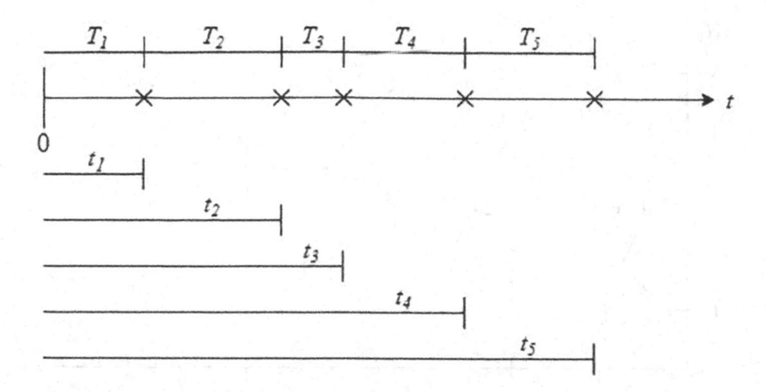

FIGURE 7.3 Stochastic point process data. *X* indicates occurrence of a failure at time t_i, (an arrival time) and T_i (an interarrival time).

MTBF; this more complex case will be discussed later. The assumption of negligible time to repair might be sufficient for some problems related to repairable systems, and thus point processes are a reasonable starting point.

By removing the repair times from the data set (Equations 7.3 and 7.4), one obtains the series of times between successive failures, including the TTFF, as seen in Figure 7.3.

These failure times can be analyzed using a point process model. Observations of a point process, a series of successive events observed during some time intervals, are called *realizations* or *trajectory* of a given point process. Note that in contrast to repairable systems, any typical reliability model of *nonrepairable* identical items is based on a positively defined r.v., such as time to the first (and the only) failure or the number of cycles to the first failure. Here, the failure time of the item is treated as the realization of an i.i.d. r.v. As such, the r.v. is a special case of the failure process, in which the process is terminated just after the first failure. However, for repairable items, the assumption that these successive failure times are i.i.d. is not realistic, and special point process models are considered an appropriate alternative approach to modeling the repairable system failure processes.

Similarly, taking only the repair times from the data set (Equation 7.3 and 7.4), forms repair arrival times and interarrival times. Based on this type of data, it is possible to determine whether the duration of repair actions of interest is decreasing or not. That is, one can perform a trend analysis of the repair times.

A *point process* can be informally defined as a mathematical model for highly localized events distributed randomly in time. The major r.v. of interest related to such processes is the number of events, $N(t)$, that occur in the time interval $(0, t]$. Therefore, such processes are also called *counting processes*. The point process $\{N(t), t \geq 0\}$ is formally defined as the one satisfying these conditions:

1. $N(t)$ is a nondecreasing integer,
2. $N(t) \geq 0$,
3. $N(0) = 0$,
4. If $t_2 > t_1$, then $N(t_2) \geq N(t_1)$,
5. If $t_2 > t_1$, then $N(t_2) - N(t_1)$ is the number of events occurring in the interval $(t_1, t_2]$.

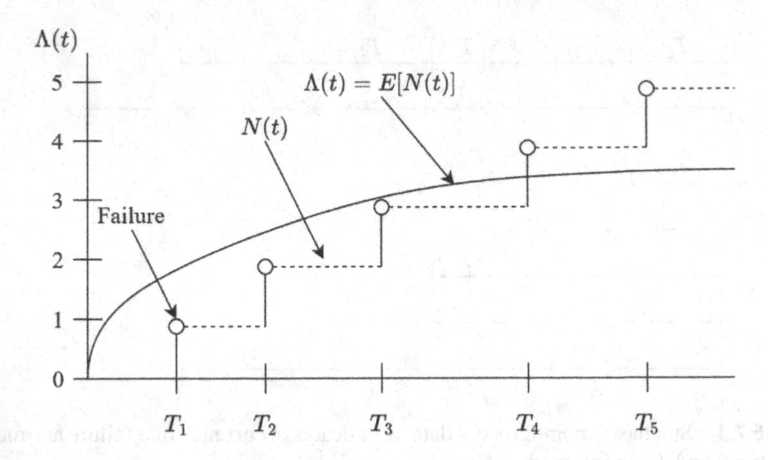

FIGURE 7.4 Graphical interpretation of $N(t)$ and $\Lambda(t)$ for a repairable system.

The integer-valued function $N(t)$ is the number of events observed in the time interval $(0, t]$, as illustrated in Figure 7.4. Therefore,

$$N(t) = max\left(n \mid T_n \le t\right). \tag{7.5}$$

The mean value $E(N(t))$ of the number of failures $N(t)$ observed in the time interval $(0, t]$ is called the *cumulative intensity function* (CIF), the *mean cumulative function* (MCF), or the *renewal function*. In the following, the term CIF is used. The CIF is usually denoted by $\Lambda(t)$; that is,

$$\Lambda(t) = E\left(N(t)\right). \tag{7.6}$$

Similar to the cdf for an r.v., from now on we will assume that the CIF, $\Lambda(t)$, is related to one item (e.g., a specific system). The item is supposed to be a member of a population (finite or infinite) consisting of identical (from reliability standpoint) items.

Another important characteristic of point processes is its corresponding ROCOF, which is the derivative of CIF with respect to time. That is,

$$\lambda(t) = \frac{\mathrm{d}\Lambda(t)}{\mathrm{d}t}. \tag{7.7}$$

Based on the ROCOF expression (Equation 7.7), the CIF is sometimes called the *cumulative* ROCOF.

Some point processes have constant ROCOF, indicating a system which is neither improving nor degrading. A point process with an increasing ROCOF represents an item that experiences *degradation* or *aging*. Analogously, the item modeled by a point process with a decreasing ROCOF is called an *improving* or *rejuvenating* system.

The distribution of TTFF of a point process is called *the underlying distribution*. For some point processes, this distribution coincides with the distribution of time between successive failures (the interarrival time); for others it does not. A point process is said to have *independent increments* if the numbers of failure events in mutually exclusive intervals are independent r.v.s. A point process is called *stationary* if the distribution of the number of failure events in any time interval depends on the length of the time interval only. Therefore, the expected number of failure events in equal intervals remains constant for stationary point processes.

7.4.1 Homogeneous Poisson Process

The homogeneous Poisson process (HPP) is the oldest and the simplest failure point process model. An HPP is a point process having independent increments, with parameter $\lambda > 0$, if the number of failure events N in any interval of length $\Delta t = (t_2 - t_1)$ has the Poisson distribution with the mean $E(N) = \lambda \Delta t$. That is,

$$Pr\left[N(t_2) - N(t_1) = n\right] = \frac{\left[\lambda \Delta t\right]^n}{n!} e^{[-\lambda \Delta t]}, \tag{7.8}$$

where $t_2 > t_1 \geq 0$. Note that the location of the time interval on the time axis does not matter.

It is easy to show that for the HPP, the times between successive failures, $T_{i+1} = t_{i+1} - t_i$, are i.i.d. r.v.s having the exponential distribution with parameter λ. In terms of failure processes, this parameter λ is the ROCOF.

The mean number of failures observed in interval $(0, t]$, $\Lambda(t)$, the CIF of this point process, is

$$\Lambda(t) = \int_0^t \lambda(x) \mathrm{d}x. \tag{7.9}$$

Since the ROCOF of the HPP is constant,

$$\Lambda(t) = \lambda t, \tag{7.10}$$

and the expected number of failures in an interval $\Delta t = (t_1, t_2]$ is:

$$\Lambda(t_2) - \Lambda(t_1) = E\left(N(t_2) - N(t_1)\right) = \lambda \Delta t. \tag{7.11}$$

In this case, the time of the nth failure has a gamma distribution with parameters (n, λ).

Due to the memoryless property of the exponential distribution, a repairable item's failure behavior modeled by HPP is independent of its age. Consequently, any preventive action makes little sense in the framework of the HPP model.

Often, it is necessary to model the failure behavior of different items *simultaneously*. Assume the items are put into service or on a reliability test at the same moment. Such situations can be modeled by the *superposition* of the respective point processes. The superposition of several point processes is the ordered sequence of all failures that occur in the individual point processes.

The superposition of k HPP processes with parameters $\lambda_1, \ldots, \lambda_k$ is an HPP with $\lambda = \lambda_1 + \cdots + \lambda_k$. An example of this is a series system, introduced in Chapter 6, with exponentially distributed elements.

7.4.2 Renewal Process

The renewal process (RP) can be considered as a generalization of the HPP for the case in which the time between failure distribution—that is, the *underlying distribution*—is an arbitrary continuous distribution of a positively defined r.v. In this case, the number of failures in any interval of length t no longer follows the Poisson distribution.

In an RP it is convenient to assume that the time origin (start of the process) is when the age of the item is zero (i.e., the item is new). Following the time origin, the RP model represents *as good as new* restorations (an equivalent term is *same as new*) when considering the *perfect repair assumption*. The term *as good as new* (and *perfect repair*) might be misleading. For example, for the decreasing failure rate (DFR) distribution, a restoration to the *as good as new* condition cannot be called a perfect repair because the item having the DFR lifetime distribution has its highest ROCOF when it is new (at zero age). As discussed, the notion of aging is not applicable for the exponential distribution, so an item having exponential time between failures is always as good as new.

The perfect repair assumption is not appropriate for a multicomponent system if only a few of the system components are repaired or replaced upon failure. But if a failed system (or a component) is replaced by an identical new one, the same-as-new repair assumption is appropriate for this system (or component).

In the following, it is assumed that both the pdf and the cdf of the underlying distribution are continuous. Xie (1989) describes the CIF, $\Lambda(t)$, for the RP from the *renewal function*,

$$\Lambda(t) = F(t) + \int_0^t F(t-x)\,\mathrm{d}\Lambda(x), \tag{7.12}$$

where $F(t)$ is the cdf. The respective ROCOF $\lambda(t)$ can be found as a solution of the derivatives of both sides of Equation 7.12 with respect to t. That is,

$$\lambda(t) = f(t) + \int_0^t f(t-x)\lambda(x)\,\mathrm{d}x. \tag{7.13}$$

Closed-form solutions of the above integral equations are known for the *exponential underlying distribution* (which is the HPP) and the *gamma underlying distribution*.

Several numeric solutions to Equation 7.12 have been developed. For example, Baxter et al. (1982) offered a numerical integration approach that covers the cases of the following underlying distributions: Weibull, gamma, lognormal, truncated normal, and inverse normal. Also, Blischke and Murthy (1994) used a Monte Carlo simulation, which provides a universal numerical approach to the problem.

7.4.3 Nonhomogeneous Poisson Process

A point process having independent increments is called the nonhomogeneous Poisson process (NHPP) with time-dependent ROCOF $\lambda(t) > 0$, if the probability that exactly n failure events occur in any interval (a, b) and the Poisson distribution with the mean equal to $\int_a^b \lambda(t)dt$ describe these events. That is,

$$Pr\left[N(b) - N(a) = n\right] = \frac{\left[\int_a^b \lambda(t)\,dt\right]^n \exp\left[-\int_a^b \lambda(t)\,dt\right]}{n!}, \tag{7.14}$$

for $n = 0, 1, \ldots, \infty$, and $N(0) = 0$. Unlike the RP, the times between successive events in the NHPP model are not i.i.d. because they are neither independent nor identically distributed. Similar to the RP, in NHPP applications the time origin (or zero age) is the time when a new (renewed) item has been put into operation.

Based on the above definition, the CIF and ROCOF of NHPP can be written as:

$$\Lambda(t) = \int_0^t \lambda(\tau)\mathrm{d}\tau. \tag{7.15}$$

The cdf of TTFF (i.e., the cdf of the underlying distribution) for the NHPP would be

$$F(t) = 1 - Pr\left[N(t) - N(0) = 0\right] = 1 - e^{-\Lambda(t)}. \tag{7.16}$$

Let us consider sequential failures occurring in an item according to the NHPP with the ROCOF $\lambda(t)$. Let t_k be the time to the kth failure. The probability that no failure occurs in interval $(t_k, t]$, where $t > t_k$, can be written as

$$Pr\left(N(t) - N(t_k) = 0\right) = R\left(t \mid t_k\right) = \mathrm{e}^{-\int_{t_k}^{t} \lambda(x)\mathrm{d}x} = \frac{\mathrm{e}^{-\int_0^t \lambda(x)\mathrm{d}x}}{\mathrm{e}^{-\int_0^{t_k} \lambda(x)\mathrm{d}x}} = \frac{R(t)}{R(t_k)}, \tag{7.17}$$

which is the conditional reliability function of a system having age t_k.

In other words, we can consider the NHPP as a process in which each failed item is instantaneously repaired (or replaced) by an identical, working one having the same age as the failed one. This type of restoration model is called the *same-as-old* (or *minimal repair*) condition. *Same-as-old* does not mean *bad* in the case of deteriorating systems, that is, systems having the increasing ROCOF.

If t_k is equal to zero, Equation 7.17 takes the following form:

$$R(t) = e^{-\int_0^t \lambda(x)\,dx}, \tag{7.18}$$

which means that the ROCOF of the NHPP coincides with the failure (hazard) rate function of the underlying TTFF distribution. Therefore, all future behavior of a repairable system is completely defined by this distribution. It also means that just after any repair/maintenance action carried out at time *t*, the ROCOF is equal to the failure rate of the TTFF distribution $\lambda(t)$. So, we can also consider the NHPP as a process in which each failed system is instantaneously replaced by an identical one having the same failure rate as the failed one.

Depending on the application, assumption of same-as-old may or may not be realistic. When applied to a single-component system, it is not a realistic assumption. However, for a complex system, composed of many components having close reliability functions, this assumption is more realistic.

An important case of the NHPP is when the ROCOF is a power function of time. That is,

$$\lambda(t) = \frac{\beta}{\alpha}\left(\frac{t}{\alpha}\right)^{\beta-1} \quad t \geq 0, \alpha, \beta > 0, \tag{7.19}$$

which results in the Weibull TTFF distribution. The NHPP process with the ROCOF given by Equation 7.19 is sometimes called the *Weibull NHPP, the power law NHPP process*, or the *Crow-AMSAA model*. Statistical procedures for this model were developed by Crow (1974, 1982), based on suggestions of Duane (1964). These procedures can also be found in MIL-HDBK-781A (1996) and IEC 61164 (2004). The main applications of the power law model are associated with reliability monitoring (called *reliability growth*) problems for both repairable items and nonrepairable items.

Another important case of NHPP is when the ROCOF is a simple log-linear function,

$$\lambda(t) = e^{(\beta_0 + \beta_1 t)} \quad t \geq 0. \tag{7.20}$$

This model was proposed by Cox and Lewis (1966) and is known as the *Cox-Lewis model* or the *log-linear NHPP model*. Like the power law model, the log-linear NHPP model has a monotonic ROCOF, which can be increasing, decreasing, or constant (if $\beta_1 = 0$). The respective underlying (TTFF) distribution is the *truncated Gumbel* (smallest extreme value) distribution. This distribution has the following cdf:

$$F(t) = 1 - e^{-\alpha\left(e^{\beta_1 t} - 1\right)}, \tag{7.21}$$

where $\alpha = e^{\frac{\beta_0}{\beta_1}}$ and $t \geq 0$.

The NHPP with the simple linear ROCOF, that is,

$$\lambda(t) = \beta_0 + \beta_1 t \quad t \geq 0, \tag{7.22}$$

is not very popular. This might be explained by the fact that for any time interval of interest $(0, t]$ the parameters of Equation 7.22 must satisfy the following inequality:

$$\lambda(t) = \beta_0 + \beta_1 t \geq 0. \tag{7.23}$$

Applications of the linear model in risk analysis are discussed by Vesely (1991) and Atwood (1992).

Statistical procedures for NHPP data analysis are well developed and can be found in many books on reliability data analysis (e.g., Rigdon and Basu, 2000).

7.4.4 Generalized Renewal Process

The minimal repair condition and the perfect repair condition are extremes and not realistic assumptions as noted by many authors (e.g., (Ascher and Feingold, 1984, Lindqvist, 1999, Thompson, 1981)). To make more realistic estimates, several generalizing models have been introduced (e.g., Lindqvist, 1999, Brown and Proschan, 1982, Kijima and Sumita, 1986, Zhang and Jardine, 1998, Wang and Pan 2021). Among these models, the *General Renewal Process* (GRP) introduced by Kijima and Sumita (1986) is the most popular, since it covers not only the RP and the NHPP, but also the intermediate *better-than-old-but-older-than-new* repair assumption. The GRP results in the G-renewal equation as a generalization of the ordinary renewal Equation 7.12.

The GRP model has been used in many applications, including the automobile industry (Kaminskiy and Krivtsov, 2000), the oil industry (Hurtado et al., 2005), and reliability studies of hydro-electric power plants (Kahle, 2005).

The GRP operates on the notion of *virtual age*. Let A_{v_n} be the virtual age of a system immediately after the nth repair. If $A_{v_n} = y$, then the system has time to the $(n+1)$th failure, t_{n+1}, which is distributed according to the following conditional cdf:

$$F\left(t \mid A_{v_n} = y\right) = \frac{F(t+y) - F(y)}{1 - F(y)}, \tag{7.24}$$

where $F(t)$ is the cdf of the TTFF distribution of a repairable item when it was new (i.e., the underlying distribution). Equation 7.24 is the conditional cdf of a working item at age y.

The following sum describes the *real age* of the system, A_{r_n} immediately after the nth repair

$$A_{r_n} = \sum_{i=1}^{n} t_i, \tag{7.25}$$

with $A_{r_0} = 0$.

In the framework of this GRP model, it is assumed that the nth repair can partially remove the damage incurred during the time between the (n–1)th and the nth failures, so that the respective virtual age after the nth repair is

$$A_{v_n} = A_{v\,n-1} + qt_n = qA_{r_n}, \quad n = 1, 2, \ldots, \tag{7.26}$$

where q is the *parameter of rejuvenation* (or *repair effectiveness parameter*) and the virtual age of a new item is $A_{v0} = 0$. Note that the virtual age expressed by Equation 7.26 can be interpreted as the virtual age at the (n–1)th failure plus the equivalent life taken away from the item between the (n–1)th and nth failures. The equivalent life is less than the time between the (n–1)th and nth failure but more than zero. For this case, the TTFF is distributed according to $F(t|0) = F(t)$, which is the underlying distribution.

The time between the first and second failures is distributed according to Equation 7.24, where $A_{v_1} = qt_1$. The time between the second and third failure is distributed according to according to the virtual age $A_{v2} = q(t_1 + t_2)$, and so on.

When $q = 0$, this process coincides with an ordinary RP, thus modeling the *same-as-new* repair assumption. When $q = 1$, the system is restored to the *same-as-old* repair assumption, which is the case of NHPP. The case of $0 < q < 1$ falls between these two repair assumptions. Finally, when $q > 1$, the virtual age $A_{v_n} > A_{r_n}$. In this case, the repair further damages the item beyond what it was just before the respective failure, leading to the *worse-than-old* repair assumption. Reversely, the repair may improve the item to a *better-than-new* condition, which is modeled by $q < 0$. The above interpretations are good for systems with monotonically increasing or decreasing ROCOF.

For the GRP, the expected number of failures in (0, t), that is, CIF $\Lambda(t)$, is given by a solution of the G-renewal equation (Kijima and Sumita, 1986),

$$\Lambda(t) = \int_0^t \left[g(z|0) + \int_0^z h(x) g\left[(z-x)|x\right] dx \right] dz, \tag{7.27}$$

where

$$g(t|x) = \frac{f(t+qx)}{1-F(qx)}, \quad t, x \geq 0, \tag{7.28}$$

is the conditional pdf function, such that $g(t|0) = f(t)$, $F(t)$ and $f(t)$ are the cdf and pdf of the TTFF underlying distribution. Since $\lambda(t) = d\left(\Lambda(t)\right)/dt$, one obtains the following equation for the ROCOF of the GRP

$$\lambda(t) = g(t|0) + \int_0^t \lambda(x) g\left[(z-x)|x\right] \mathrm{d}x. \tag{7.29}$$

Compared to the ordinary renewal Equation 7.11, Equation 7.28 has one additional parameter, which is the repair effectiveness parameter q.

Kijima and Sumita (1986) showed that the Volterra integral of Equation 7.29 has a unique solution. The closed-form solutions of Equations 7.28 and 7.29 are not trivial. However, MLE and Monte Carlo-based solutions of these equations are possible. For these solutions, readers are referred to (Hurtado et al., 2005) and Krivtsov (2000).

7.5 DATA ANALYSIS FOR POINT PROCESSES

7.5.1 Parameter Estimation for the HPP

Suppose that a failure process is observed for a predetermined time t_{end} during which n failures have been recorded at times $t_1,..., t_n$, where $t_n < t_{end}$. The process is assumed to follow an HPP. The corresponding likelihood function can be written as

$$L(\lambda|t_{\text{end}}, n) = \lambda^n e^{\lambda t_{\text{end}}}. \tag{7.30}$$

With t_{end} fixed, the number of events, n, is a *sufficient statistic* (note that one does not need to know $t_1, \ldots, t_n$ to construct the likelihood function). Thus, the statistical inference can be based on the Poisson distribution of the number of events. As a point estimate of λ, one usually takes $\hat{\lambda} = \frac{n}{t_{\text{end}}}$, which is the unique unbiased estimate for λ. One can find the confidence interval of λ using the Fisher information matrix in Equation 5.16.

7.5.2 Data Analysis for the NHPP

The NHPP can be used to model improving and deteriorating systems: if the intensity function (ROCOF) is decreasing, the system is improving, and if the intensity function is increasing, the system is deteriorating. ROCOF trend analysis is of great importance because the HPP, with the memoryless property of its exponential time to failure distribution, is inadequate and inconsistent with the true behavior of many repairable items.

Formally, we can test for trends in ROCOF by taking the null hypothesis of no trend (that is, that the events are an HPP) and applying a goodness of fit test for the exponential distribution representing the intervals between successive failures and the Poisson distribution representing the number of failures in the time intervals of constant length. Alternatively, a simple graphical procedure based on this property is to plot the cumulative number of failures versus the cumulative time. Deviation from linearity indicates the presence of a trend. However, this test is less sensitive to the NHPP alternatives, so it is better to apply the following methods (Cox and Lewis, 1966).

7.5.2.1 Maximum Likelihood Procedures

Consider successive failure times $t_1 < \cdots < t_n$ observed in the interval $(0, t_{end})$, $t_n < t_{end}$. The likelihood function for any ROCOF function $\lambda(t)$ may be written as

$$L(\lambda(t)|t_1\cdots t_n)=\prod_{i=1}^{n}\lambda(t_i)\mathrm{e}^{-\int_0^{t_1}\lambda(x)dx}\,\mathrm{e}^{-\int_{t_1}^{t_2}\lambda(x)dx}\cdots\mathrm{e}^{-\int_{t_{n-1}}^{t_n}\lambda(x)dx}\,\mathrm{e}^{-\int_{t_n}^{t_{\text{end}}}\lambda(x)dx}$$

$$=\prod_{i=1}^{n}\lambda(t_i)\mathrm{e}^{-\int_0^{t_n}\lambda(x)dx}\,\mathrm{e}^{-\int_{t_n}^{t_{\text{end}}}\lambda(x)dx}. \tag{7.31}$$

The corresponding log-likelihood function is given by

$$\Lambda=\sum_{i=1} n\ln\lambda(t_i)-\int_0^{t_{\text{end}}}\lambda(x)\mathrm{d}x. \tag{7.32}$$

Now, consider the special case when the ROCOF takes the simple form given by the log-linear NHPP model,

$$\lambda(t)=\mathrm{e}^{\alpha+\beta t}. \tag{7.33}$$

Note that the model above is more general than the linear one, $\lambda(t)=\alpha+\beta t$, which can be considered as a particular case of Equation 7.33, when $\beta t << 1$.

By plugging Equation 7.33 into Equations 7.31 and 7.32, one obtains

$$L=e\left[n\alpha+\beta\sum_{i=1}^{n}(t_i)-\frac{e^{\alpha}\left(e^{\beta t_{\text{end}}}-1\right)}{\beta}\right], \tag{7.34}$$

$$\Lambda=\ln[L]=n\alpha+\beta\sum_{i=1}^{n}(t_i)-\frac{e^{\alpha}\left(e^{\beta t_{\text{end}}}-1\right)}{\beta}. \tag{7.35}$$

As with other MLE procedures, the parameters α and β can be estimated by taking the derivative of this equation and setting it to zero.

Example 7.1

In a repairable system, eight failures have been observed at operational times 595, 905, 1,100, 1,250, 1,405, 1,595, 1,850, and 1,995 hours. Assume the observation ends at the time when the last failure is observed, and that the time to repair is negligible. Test whether these data exhibit a trend in a form of the log-linear NHPP model (Equation 7.33).

Solution:

Taking the derivative of Equation 7.35 with respect to β and the derivative with respect to α and equating them to zero results in the following system of equations for the MLEs of these parameters

$$\sum_{i=1}^{n}(t_i)+\frac{n}{\beta}-\frac{nt_n}{1-e^{-\beta t_n}}=0,$$

$$e^{\alpha}=\frac{n\beta}{e^{\beta t_n}-1}.$$

For the data given, $n=8$, $t_n=1{,}995$ hours, and $\Sigma t_i=10{,}695$ hours. Solving these equations numerically, one obtains $\hat{\alpha}=-6.8134$ and $\hat{\beta}=0.001103$, and the following trend model:

$$\lambda(t)=e^{(-6.8134+0.001103t)}.$$

Since $\beta=0.001103>0$, there is a very slight trend in the data. Since β is positive, this corresponds to a deteriorating system. The ROCOF is increasing with time.

Another form of ROCOF is

$$\lambda(t)=\lambda\beta t^{\beta-1}, \tag{7.36}$$

which has the same form as the hazard rate of the Weibull distribution (with $\lambda=1/\alpha^{\beta}$). For a repairable system having this ROCOF, the reliability function for an interval $(t,\,t+t_1)$ can be obtained as:

$$R\left[(t+t_1)|t\right]=e^{-\left[\lambda(t+t_1)^{\beta}-\lambda t^{\beta}\right]}. \tag{7.37}$$

Crow (1974) has shown that under the condition of a single system observed to its nth failure, when, $t_n=t_{\text{end}}$ the maximum likelihood estimators of β and λ can be obtained as

$$\hat{\beta}=\frac{n}{\sum_{i=1}^{n-1}\ln\left(\frac{t_n}{t_i}\right)}, \tag{7.38}$$

$$\hat{\lambda}=\frac{n}{t_n^{\hat{\beta}}}. \tag{7.39}$$

When $t_n<t_{\text{end}}$, $n-1$ is replaced with n in the summation in Equation 7.38 and t_n is replaced with t_{end} in Equations 7.38 and 7.39.

The confidence limits for inferences on β and λ have been developed and discussed by Bain (2017). Crow (1990) has expanded estimates in Equations 7.38 and 7.39 to include situations where data originate from multiunit repairable systems.

Example 7.2

Using the information in Example 7.1, calculate the MLEs of β and λ for the Weibull NHPP. Compare the result to the λ obtained for an HPP. Plot the results and discuss the difference.

Solution:

Using Equations 7.38 and 7.39, we can calculate $\hat{\beta}$ and $\hat{\lambda}$ as 2.155 and 6.209×10^{-7}, respectively. Using these values, the functional form of the demand ROCOF can be obtained by using the Weibull NHPP as

$$\lambda(t)=1.338\times10^{-6}t^{1.155},$$

where t represents the demand number (time in hours).

Using the HPP, the parameter λ is estimated as:

$$\hat{\lambda}=\frac{n}{t_n}=\frac{8}{1995}=4.0\times10^{-3}.$$

The plot of the demand ROCOF of NHPP as a function of calendar time for the system is shown below. As seen in the plot, the HPP produces a constant ROCOF, but the Weibull NHPP shows increasing ROCOF. This visual indicates that the HPP is inappropriate for this data.

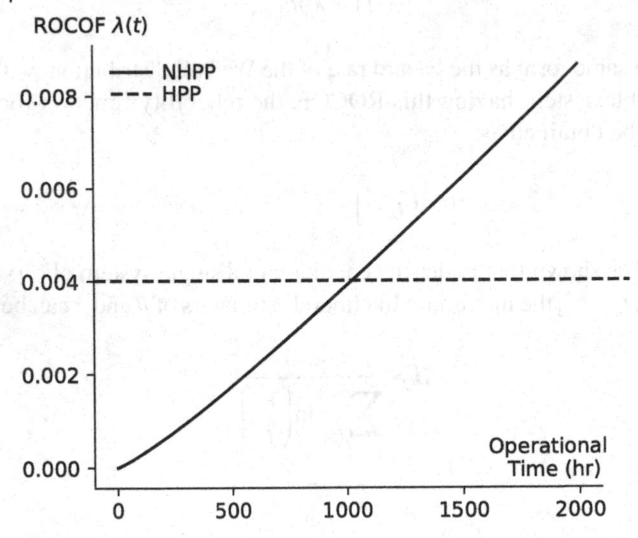

7.5.2.2 Laplace Test

Another approach to identify when to use an HPP vs. an NHPP is a hypothesis test known as the *Laplace test* (sometimes called the *Centroid test*). Consider successive failure times $0<t_1<\cdots<t_n<t_{\text{end}}$, the pdf of the truncated exponential distribution representing these failures is

$$f(t)=\frac{\beta}{e^{\beta t_{\text{end}}}-1}e^{\beta t},\quad 0\le t\le t_{\text{end}},\ \beta\ne0. \tag{7.40}$$

Thus, for any β the conditional pdf of $\sum t_i$ is the same as for the sum of n i.i.d. r.v.s having the pdf in Equation 7.40. Note that for $\beta=0$, the pdf becomes the uniform distribution over $(0, t_{\text{end}})$.

Now, we will use the conditional pdf (i.e., Equation 7.40) to test the null hypothesis, H_0: $\beta=0$ (meaning there is no trend in the times, i.e., the TBFs are exponentially distributed), against the alternative hypothesis H_1: $\beta\neq 0$ (meaning there is trend in the times, i.e., the TBFs are not exponentially distributed). Under the condition of $\beta=0$, the pdf (Equation 7.40) is reduced to the uniform distribution over $(0, t_{\text{end}})$, and $\sum t_i$ has the distribution of the sum of n independent, uniformly distributed r.v.s. Thus, one can use the distribution of the statistic given by Equation 7.41 in a hypothesis test:

$$U=\frac{\sum_{i=1}^{n}(t_i)-\left(\frac{nt_{\text{end}}}{2}\right)}{t_{\text{end}}\sqrt{\left(\frac{n}{12}\right)}}=\frac{\sum_{i=1}^{n}\left(\frac{t_i}{n}\right)-\left(\frac{t_{\text{end}}}{2}\right)}{t_{\text{end}}\sqrt{\left(\frac{1}{12n}\right)}}. \tag{7.41}$$

The test statistic U is approximately the standard normal distribution when the null hypothesis is true (Cox and Lewis, 1966).

If the alternative hypothesis is H_1: $\beta\neq 0$, then the large values of $|U|$ signify evidence against the null hypothesis. If the alternative hypothesis is H_1: $\beta>(<)\ 0$, then the large values of $U(-U)$ also provide evidence against the null hypothesis. If U is close to 0, there is no evidence of trend in the data, and the process is assumed to be stationary (i.e., an HPP). If $U<0$, the trend in $\lambda(t)$ is decreasing; that is, the intervals between successive failures (interarrival values) are increasing and the system is improving. If $U>0$, $\lambda(t)$ is increasing, meaning the intervals between failures are decreasing and the system is deteriorating. For the latter two situations, the process is not stationary (i.e., it is an NHPP).

If the data are failure terminated (Type II censored), Equation 7.41 should be replaced by

$$U=\frac{\dfrac{\sum_{i=1}^{n-1}t_i}{n-1}-\left(\dfrac{t_n}{2}\right)}{t_n\sqrt{\left(\dfrac{1}{12(n-1)}\right)}}. \tag{7.42}$$

Laplace's test is not very sensitive, and in practice, it would be nearly impossible to get a perfect zero for a process having no trend. So, as a matter practice, it is possible to assume values of the statistic U close to zero (for example, for $-0.3<U<0.3$) have no trend or are associated with an insignificant trend.

Example 7.3

Consider the failure interarrival times for a motor-operated rotovalve in a process system. Assume repair times are negligible. This valve is normally in standby mode and is demanded when overheating occurs in the process. The only major failure mode is "failure to start upon demand." The interarrivals of this failure mode are shown below in days. Determine whether an increasing ROCOF is justified.

Assume that a total of 5,256 demands occurred during this time. Assume a constant rate of occurrence for demands. Use a 5% significance level.

Failure Number	Interarrival Time (Days)	Failure Number	Interarrival Time (Days)
1	104	16	140
2	131	17	1
3	1,597	18	102
4	59	19	3
5	4	20	393
6	503	21	96
7	157	22	232
8	6	23	89
9	118	24	61
10	173	25	37
11	114	26	293
12	62	27	7
13	101	28	165
14	216	29	87
15	106	30	99

Solution:

There are $n=30$ failures in 5,256 demands. First, calculate the arrival times of the demands (the number of demands between two successive failures), and the demand arrival. These values are shown below.

Interarrival Time, T_i (Days)	Arrival Time, t_i (Days)	Interarrival Time, T_i (Days)	Arrival Time, t_i (Days)
104	104	140	3,591
131	235	1	3,592
1,597	1,832	102	3,694
59	1,891	3	3,697
4	1,895	393	4,090
503	2,398	96	4,186
157	2,555	232	4,418
6	2,561	89	4,507
118	2,679	61	4,568
173	2,852	37	4,605
114	2,966	293	4,898
62	3,028	7	4,905
101	3,129	165	5,070
216	3,345	87	5,157
106	3,451	99	5,256

Since the observation ends at the last failure, the following results are obtained using Equation 7.42. We use the null hypothesis that there is no trend in the ROCOF, $\beta = 0$,

$$\sum t_i = 95{,}899, \quad n-1 = 29,$$

$$\sum_{i=1}^{n-1} \frac{t_i}{n-1} = 3307, \quad \frac{t_n}{2} = 2628,$$

$$U = \frac{3307 - 2628}{5256\sqrt{1/12(29)}} = 2.41.$$

To test the null hypothesis that there is no trend in the data, and thus that the ROCOF, λ, of rotovalves is constant, we would use the standard normal distribution cdf (Table A.1) with $U = 2.41$. The acceptance region for the 5% significance level is (−1.96, +1.96). Since $2.41 > 1.96$, we reject the null hypothesis of no trend. Since $U > 0$, the trend is increasing, meaning that the system is deteriorating.

The existence of a trend in the data in this example indicates that the interarrivals of rotovalve failures are not i.i.d. r.v.s, and thus the stationary process for evaluating reliability of rotovalves is incorrect. Rather, these interarrival times can be described in terms of the NHPP.

Example 7.4

In a repairable system, six interarrival times 16, 32, 49, 60, 78, and 182 (in hours) between failures were observed. Assume the observation ends at the time when the last failure is observed. Test whether these data exhibit a trend at the 10% significance level. If there is a trend, estimate the trend model parameters for the Weibull NHPP. Find the probability that the interarrival time for the seventh failure will be greater than 200 hours.

Solution:

Use Laplace's test to test the null hypothesis that there is no trend in the data at the 10% significance level. The respective acceptance region is (−1.645, +1.645). From Equation 7.42, find the test statistic:

$$U = \frac{\frac{[16 + (16+32) + \cdots]}{6-1} - \left(\frac{417}{2}\right)}{417\sqrt{1/12(5)}} = -1.82.$$

Notice that $t_n = 417$. The value of U obtained indicates that H_0 is rejected, and the sign of U shows that the trend is decreasing and the system is improving.

Using Equations 7.38 and 7.39, we can find

$$\hat{\beta} = \frac{6}{\ln\left(\frac{417}{16}\right) + \ln\left(\frac{417}{16+32}\right) + \cdots} = 0.712,$$

$$\hat{\lambda} = \frac{6}{(417)^{0.712}} = 0.0818 \text{ hour}^{-1}.$$

Thus, $\lambda(t) = 0.0583t^{-0.288}$.

From Equation 7.37, with $t_i = 200$ hours, $R\left((t_{\text{end}} + 200) | t_{\text{end}}\right) = e^{-\lambda(t_{\text{end}} + t_1)^{\beta} - \lambda(t_{\text{end}})^{\beta}}$ $= 0.145$. So the probability that the seventh failure occurs within 200 hours is: $Pr\left[T_7 > 200 = 1 - R\left((t_{\text{end}} + 200) | t_{\text{end}}\right)\right] = 0.855.$

7.6 AVAILABILITY OF REPAIRABLE SYSTEMS

The notion of availability is related to repairable items only. We define availability as the probability that a repairable item will function (or is in an up state) properly at a given time t. Conversely, the unavailability of a repairable item, $q(t)$, is defined as the probability that the item is in a failed state (down) at a given time t.

There are several mathematical representations of availability; the most common are as follows.

1. *Instantaneous (point) availability* of a repairable item at time t, $a(t)$, is the probability that the item is up (functioning properly) at time t.
2. *Limiting availability*, a, is defined as the horizontal asymptote of the instantaneous availability, $a(t)$:

$$a = \lim_{t \to \infty} a(t). \tag{7.43}$$

3. *Average availability*, $\bar{a}$, is defined for a fixed time interval, T, as

$$\bar{a} = \frac{1}{T} \int_0^T a(t) \mathrm{d}t. \tag{7.44}$$

4. *Limiting average availability*, $\bar{a}_l$, is defined as

$$\bar{a}_l = \lim_{T \to \infty} \frac{1}{T} \int_0^T a(t) \mathrm{d}t. \tag{7.45}$$

The limiting average availability has few applications in reliability engineering. We only elaborate on the first three representations of availability in the remainder of this section.

If an item is nonrepairable, mathematically speaking, its instantaneous availability coincides with its reliability function, $R(t)$. That is,

$$a(t) = R(t) = e^{-\int_0^t \lambda(x)\,dx}, \tag{7.46}$$

where $\lambda(t)$ is the hazard rate function. The unavailability, $q(t)$, is related to $a(t)$ as

$$q(t) = 1 - a(t). \tag{7.47}$$

From the modeling point of view, repairable items can be divided into two groups:

- Repairable items for which failure is immediately detected and fixed (*revealed faults*). If the duration of repair is negligible, the downtime contribution to availability due to the item's failure detection can be practically neglected compared to the repair and operating times.
- Repairable items for which failure is detected upon inspection and repaired subsequently. Here, the downtime due to repair is not negligible. A subclass of this group of repairable items undergoes preventive maintenance over fixed intervals of inspection (testing), which are called *periodically-inspected* (or *periodically-tested)* items.

The two groups discussed above will be further examined in the next two subsections.

7.6.1 Instantaneous (Point) Availability of Revealed-Fault Items

For the first group systems, it can be shown that $a(t)$ and $q(t)$ are obtained from the following ordinary differential equations:

$$\begin{aligned} \frac{da(t)}{dt} &= -\lambda(t)a(t) + \mu(t)q(t), \\ \frac{dq(t)}{dt} &= \lambda(t)a(t) - \mu(t)q(t), \end{aligned} \tag{7.48}$$

where $\lambda(t)$ is the ROCOF and $\mu(t)$ is the ROCOR or repair rate. Repair rate describes the rate (per unit time) with which a repair action is performed. Note that as the number of repairs over an interval of time increases, the repair rate also increases. However, as the lengths of times to repair increase, the repair rate decreases. Therefore, contrary to the poor reliability performance when an item's failure rate is high, larger values of repair rate are associated with good repair performance.

Solutions to the Equation 7.48 are trivial when no trends in the ROCOF and repair rate exist (i.e., λ and μ are constants independent of time). The point availability and point unavailability of the item are given by

$$a(t)=\frac{\mu}{\lambda+\mu}+\frac{\lambda}{\lambda+\mu}e^{-(\lambda+\mu)t},$$
$$q(t)=\frac{\lambda}{\lambda+\mu}-\frac{\lambda}{\lambda+\mu}e^{-(\lambda+\mu)t}. \tag{7.49}$$

Note that in Equation 7.49, $\mu = 1/E(\tau)$, where $E(\tau)$ is the mean time interval per repair, also called *mean time to repair* (MTTR). Therefore, larger MTTR values are associated with poor repair performance. Clearly, MTBF $= 1/\lambda$ in this case.

For simplicity, the pointwise availability and unavailability functions in Equation 7.49 can be represented in an approximate form if repair downtimes can be neglected (i.e., assuming they are very few or extremely short in length), thus $\mu \approx 0$. This simplifies availability calculations significantly. In this case, for small λt values, the availability and unavailability in each interval are

$$a(t)\approx 1-\lambda t,$$
$$q(t)\approx \lambda t. \tag{7.50}$$

7.6.2 Instantaneous (Point) Availability of Periodically-Tested Items

The pointwise unavailability of periodically-tested items is of major interest in preventive maintenance applications. The periodic tests occur at a fixed interval, every T_{pt} hours (or other units of time). For the special case of preventive maintenance where inspections (tests) are performed over equal periods, but assuming inspections and repair durations are negligible, Equation 7.50 can be used by replacing t with the inspection duration T_{pt}. In this case, Figure 7.5 shows the unavailability as a function of time, using Equation 7.50. If the inspection is periodic with no trends in ROCOF or repair rate, but the inspection duration and repair length are non-negligible compared to the inspection duration T_{pt}, one must include the effect of the downtimes due to inspection and repair.

A practical, but approximate method of computing the availability and unavailability of periodically-tested items in a preventive maintenance regime is the application of the average values of availability during operation, testing, and repair. Vesely

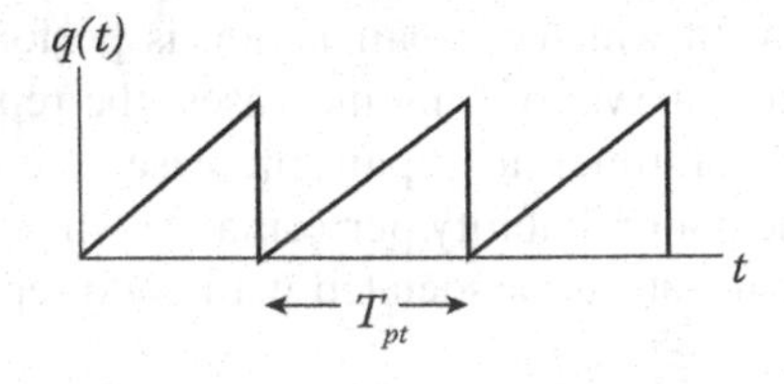

FIGURE 7.5 Approximate pointwise unavailability for a periodically-tested item.

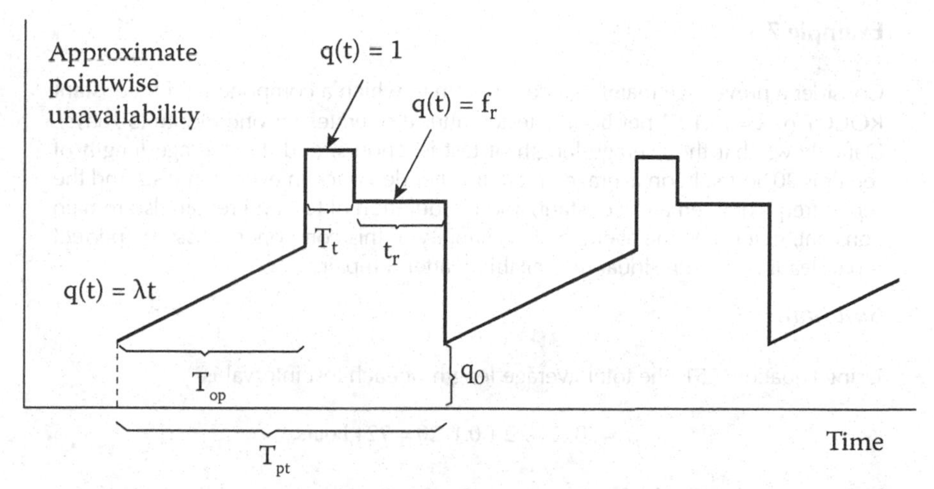

T_{pt} = Test interval, t_r = Average repair time,
T_t = Average test duration, f_r = Frequency of repair (fraction of tests leading to repair), q_0 = Residual unavailability.

FIGURE 7.6 Pointwise unavailability for a periodically-tested item including test and repair outages.

and Goldberg (1977) have proposed the approximate average pointwise unavailability functions for this case. Consider Figure 7.6, where two cycles of a periodically-tested item are shown. In this case the item operates for a fixed length of time, T_{op}, after which the item is tested or inspected to for any faults or incipient failures. The average test duration (time) is T_t, and if a repair is needed then it is repaired for an average length of time, T_r. Clearly, not all intervals (cycles) of the test or inspection identify any need for repair; only a fraction, f_r, of the intervals require repair. Therefore, the expected total length of a downtime interval, T_{pt}, would be,

$$T_{pt} = T_{op} + T_t + f_r T_r. \tag{7.51}$$

The values of T_{op}, T_t and T_r would be slightly different in each test cycle or interval. In this case the average values of these parameters and the approximate average values of the unavailability, $\overline{q}$, during the operation, testing and inspection, and repair for an item experiencing constant ROCOF may be expressed by

$$\overline{q} = \frac{1}{2}\lambda T_{op} + f_r \frac{T_r}{T_{pt}} + \frac{T_t}{T_{pt}}. \tag{7.52}$$

It should be noted that, due to imperfection in test and repair activities, it is possible that a mean residual unavailability, q_0, remains following a test and/or repair. The average residual unavailability represents a constant contribution to the total average unavailability from imperfect repair.

Example 7.5

Consider a preventive maintenance program in which a component with constant ROCOF of $\lambda = 1 \times 10^{-4}$ per hour is tested after it operates for one month (30 days). Data shows that the average length of test is 2 hours, and the average length of repair is 20 hours. If, on average, a repair is needed once in every ten tests and the repair frequency remains constant, and the duration of test and repair also remain constant, determine the average unavailability of this component. Assume perfect repair leading to no residual unavailability after a repair.

Solution:

Using Equation 7.51, the total average length of each test interval is

$$T_{pt} = 30 \cdot 24 + 2 + 0.1 \cdot 20 = 724 \text{ hours.}$$

Using Equation 7.52, $\bar{q} = 0.042$, or $\bar{a} = 1 - \bar{q} = 0.958$.

7.6.3 Limiting Point Availability

The constant ROCOF and repair rate pointwise availability equations discussed in Sections 7.6.1 and 7.6.2 are useful in many practical applications and have asymptotic values. For example, the limit of Equation 7.49 for availability can be expressed as:

$$a = \lim_{t \to \infty} a(t) = \frac{\mu}{\lambda + \mu}, \tag{7.53}$$

or its equivalent

$$a = \frac{\text{MTBF}}{\text{MTTR} + \text{MTBF}}. \tag{7.54}$$

Equation 7.54 is referred to as the asymptotic availability of a repairable item with constant failure and repair rates.

7.6.4 Average Availability

According to its definition, average availability is a constant measure of availability over a period T. For non-inspected items, T can take on any value (preferably, it should be about the mission length, T_m). For inspected items, T is normally the inspection (or periodic test) interval (T_{pt}) or mission length T_m. Thus, for a repairable item, the approximate expression for point availability with constant λ can be used. If we assume the approximate unavailability from Equation 7.50 (which may be applicable for $\lambda t < 0.1$), then

$$\bar{a} = \frac{1}{T}\int_0^T (1 - \lambda t)\, dt = 1 - \frac{\lambda T}{2}. \tag{7.55}$$

TABLE 7.1
Average Availability Functions

Type of Item	Average Unavailability	Average Availability
Nonrepairable	$\frac{1}{2}\lambda T_m$	$1-\frac{1}{2}\lambda T_m$
Repairable revealed fault	$\frac{\lambda\tau}{1+\lambda\tau}$	$\frac{1}{1+\lambda\tau}$
Repairable periodically tested	$\frac{1}{2}\lambda T_{op}+f_r\frac{T_r}{T_{pt}}+\frac{T_t}{T_{pt}}$	$1-\left(\frac{1}{2}\lambda T_{op}+f_r\frac{T_r}{T_{pt}}+\frac{T_t}{T_{pt}}\right)$

λ, constant failure rate; τ, average downtime or MTTR; f_r, frequency of repair per test interval; T_m, mission length; T_{op}, operating time (up time) $= T_{pt}-f_r T_r-T_t$; T_{pt}, periodic test interval; T_r, average repair time; T_t, average test duration.

Accordingly, for all types of systems, one can obtain such approximations for average availabilities. Vesely et al. (1981) have discussed the average unavailability for various types of systems.

Table 7.1 summarizes the average unavailability of nonrepairable, and repairable revealed-fault and periodically-tested items. The equations in Table 7.1 can also be applied to standby equipment, with λ representing the standby (or demand) ROCOF, and the mission length or operating time being replaced by the time between two tests. Barroeta and Modarres (2006) have discussed equations and examples for availability cases with NHPP, including situations in which there are cost considerations associated with items that are periodically tested and maintained.

7.6.5 Other Average Measures of Availability

There are other measures of availability used in industry which specifically consider different types of average unavailability (or downtime) to measure the availability of an item. These measures are used to categorize the different types of unavailability measures and understand the cause(s) of an unavailability. Examples of these measures are *inherent availability*, *achieved availability*, and *operational availability*.

7.6.5.1 Inherent Availability

Inherent availability is defined as

$$A_i = \frac{\text{MTBF}}{\text{MTBF}+\text{MCMT}}, \tag{7.56}$$

where MCMT is the Mean Corrective Maintenance Time. The MCMT considers only the repair actions resulting from *corrective maintenance*, which are conducted when unexpected failures occur. Inherent availability provides a metric to measure the proportion of the unavailability attributed to failures and may allow improvements to the overall design of the engineering systems to reduce the impact of failures on availability.

7.6.5.2 Achieved Availability

Achieved availability is defined as

$$A_a = \frac{\text{MTBM}}{\text{MTBM} + \bar{M}}, \tag{7.57}$$

where MTBM is the Mean Time Between Maintenance, and $\bar{M}$ is the mean maintenance time. The mean maintenance time is inclusive of both *corrective maintenance* and *preventive maintenance* actions to consider both the unscheduled and scheduled maintenance actions. Achieved availability provides a metric to understand the total maintenance burden of the item. This is commonly used in engineering system specifications as the designer of the system has control of all the factors which influence achieved availability.

7.6.5.3 Operational Availability

Operational availability is defined as:

$$A_O = \frac{\text{MTBM}}{\text{MTBM} + \text{MDT}}, \tag{7.58}$$

where MDT is the Mean Downtime and includes administrative and logistics downtimes. Operational availability is typically what an end user ultimately is concerned with. However, it needs to be considered from the perspective of not just the operation of the system, but also, the supporting elements such as spares availability, personnel availability and the logistics chain which may delay the maintenance action from being completed.

7.7 MARKOV PROCESSES FOR SYSTEM AVAILABILITY

A *Markov process* is a random (stochastic) process shown by a sequence of possible states. In Markov processes, the probability of each state depends only on the immediate past states. A *Markov chain* is a type of Markov process having a discrete state space or a discrete time of an item. Markov chains are useful tools for evaluating availability of items that have multiple states (e.g., up, down, and degraded). For example, consider a system with the states shown in Figure 7.7. As the item moves from one state to another, the process is called a *transition*. A transition matrix describes the probabilities of transitions across the states and an initial state (or initial distribution).

In the framework of Markovian models, the transition probabilities between various states are characterized by *transition rates*. The transition rates may not necessarily be constant, leading to a time-dependent Markov process. But, in most

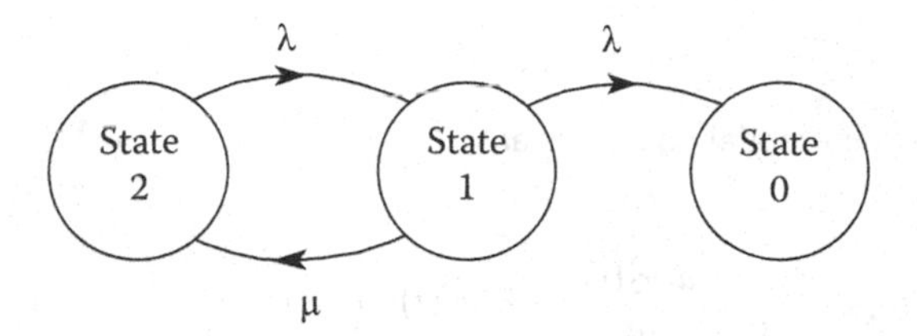

FIGURE 7.7 Markovian model for a system with discrete states.

applications Markov models use constant transition rates. Consider an item with n discrete states. If $Pr_i(t) = Pr\left(\text{the item is in state } i \text{ at time } t\right)$, then $\sum_{i=1}^{n} Pr_i(t) = 1$. Let ρ_{ij} = transition rate from state i to state j, ($i, j = 1, 2, \ldots, n$). Because ρ_{ij} is constant, the random time the item is in state i until the transition to state j follows the exponential distribution with rate $\lambda = \rho_{ij}$. If $Pr_i(t)$ is differentiable,

$$\frac{dPr_i(t)}{dt} = -Pr_i(t)\left(\sum_{j(j\neq i)} \rho_{ij}\right) + \left(\sum_{j(j\neq i)} \rho_{ji} Pr_j(t)\right). \tag{7.59}$$

If a differential equation like Equation 7.59 is written for each state, and the resulting set of differential equations is solved, one can obtain the time-dependent probability of each state.

Example 7.6

A system that consists of two cooling units has the three states shown in the Markovian model in the figure below. When one unit fails, the other unit takes over, and repair on the first starts immediately. When both units are down, there are two repair crews to simultaneously repair the two units. The three states are as follows:

- State 0, when both units are down.
- State 1, when one unit is operating, and the other is down.
- State 2, when the first unit is operating and the second is in standby (in an operating-ready condition).

Determine the probability of each state Determine the availability of the entire system.

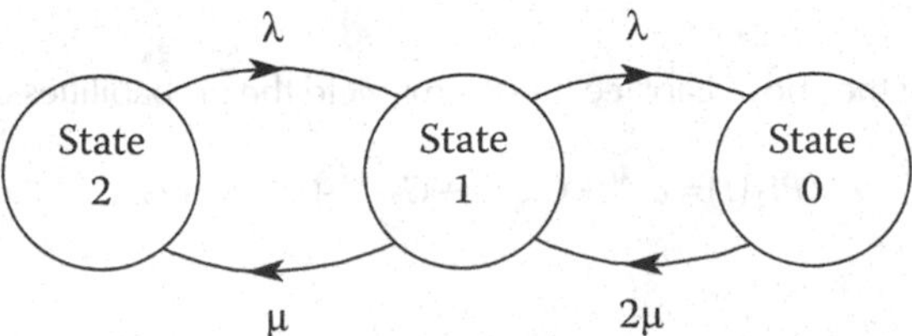

Solution:

The governing differential equations are

$$\frac{dPr_2(t)}{dt} = -\lambda Pr_2(t) + \mu Pr_1(t),$$

$$\frac{dPr_1(t)}{dt} = \lambda Pr_2(t) - (\mu + \lambda) Pr_1(t) + 2\mu Pr_0(t),$$

$$\frac{dPr_0(t)}{dt} = \lambda Pr_1(t) - 2\mu Pr_0(t).$$

Taking the Laplace transform of both sides of the equations yields

$$sPr_2(s) - Pr_2(0) = -\lambda Pr_2(s) + \mu Pr_1(s),$$

$$sPr_1(s) - Pr_1(0) = \lambda Pr_2(s) - (\mu + \lambda) Pr_1(s) + 2\mu Pr_0(s),$$

$$sPr_0(s) - Pr_0(0) = \lambda Pr_1(s) - 2\mu Pr_0(s).$$

Assuming the initial conditions $Pr_2(0) = 1$ and $Pr_1(0) = Pr_0(0) = 0$. Solving the above set of equations, $Pr_i(s)$ can be calculated as

$$Pr_2(s) = \frac{1}{\lambda - s} + \frac{\mu\lambda(2\mu + s)}{s(s+\lambda)(s-k_1)(s-k_2)},$$

$$Pr_1(s) = \frac{\lambda(2\mu + s)}{s(s-k_1)(s-k_2)},$$

$$Pr_0(s) = \frac{\lambda^2}{s(s-k_1)(s-k_2)},$$

where,

$$k_1 = \frac{-2\lambda - 3\mu - \sqrt{4\lambda\mu + \mu^2}}{2},$$

$$k_2 = \frac{2\mu\lambda + \lambda^2 + 2\mu^2}{k_1}.$$

The inverses of the above Laplace transforms yield the probabilities of each state as

$$Pr_2(t) = e^{-\lambda t} + G_1 e^{-\lambda t} + G_2 e^{k_1 t} + G_3 e^{k_2 t} + G_4,$$

where,

$$G_1 = \frac{\mu(\lambda+2\mu)}{(\lambda+k_1)(\lambda+k_2)}, \quad G_2 = \frac{\mu\lambda(2\mu+k_1)}{(k_1+\lambda)(k_1-k_2)(k_1)},$$
$$G_3 = \frac{\mu\lambda(2\mu+k_1)}{k_2(k_2-k_1)(k_2+\lambda)}, \quad G_4 = \frac{2\mu^2}{2\mu\lambda+\lambda^2+2\mu^2},$$

and

$$Pr_1(t) = A_1 e^{k_1 t} + A_2 e^{k_2 t} + A_3,$$

where,

$$A_1 = \frac{2\mu\lambda}{(k_1-k_2)(k_1)} + \frac{\lambda}{k_1-k_2},$$
$$A_2 = \frac{\lambda(2\mu+k_2)}{(k_1-k_2)(k_2)},$$
$$A_3 = \frac{2\mu\lambda}{2\mu\lambda+\lambda^2+2\mu^2},$$

and

$$Pr_0(t) = B_1 e^{k_1 t} + B_2 e^{k_2 t} + B_3,$$

where,

$$B_1 = \frac{\lambda^2}{(k_1-k_2)(k_1)},$$
$$B_2 = \frac{\lambda^2}{(k_2-k_1)(k_1)},$$
$$B_3 = \frac{\lambda^2}{2\mu\lambda+\lambda^2+2\mu^2}.$$

The availability of the two units in the system is $a(t) = Pr_2(t) + Pr_1(t)$, and the unavailability of the entire system is $q(t) = Pr_0(t)$.

If a trend exists in the parameters that characterize system availability (i.e., in ROCOF and repair rate), one cannot use the Markovian method; only solutions of Equation 7.59 with time-dependent ρ can be used. Solving such equations is more difficult, especially in systems with many states. However, there are numerical algorithms available to solve these equations.

Markov chains have been used for generating sequences of random numbers to depict very complicated probability distributions, via a process called Markov Chain Monte Carlo (MCMC) simulation. The MCMC simulation methods have radically improved the practicability of Bayesian inference methods, allowing numerical generation of multivariate posterior distributions. For this topic refer to Gamerman and Lopes (2006).

7.8 AVAILABILITY OF COMPLEX SYSTEMS

In Chapter 6, we discussed methods for estimating the reliability of a system as a function of the reliability of its individual components. The same methods also apply if some or all the components described in the system are repairable. In these cases, one can use the availability (or unavailability) functions for each component of a system and use, for example, system cut sets to obtain system availability (or unavailability). The method and process of determining system availability is the same as the system reliability estimation methods discussed in Chapter 6.

Example 7.7

Assume all components of the system shown in Figure 6.5 are repairable (revealed fault) with a constant ROCOF of 10^{-3} (per hour) and a mean downtime of 15 hours, except component 7. Component 7 has a constant ROCOF of 10^{-5} per hour, with a mean downtime of 10 hours. Calculate the average system unavailability.

Solution:

The cut sets are (C_7), (C_1, C_2), (C_1, C_5, C_6), (C_2, C_3, C_4), and (C_3, C_4, C_5, C_6). The unavailability of components 1–6, according to Table 7.1 is

$$q_{C1\text{-}C6} = \frac{\lambda\tau}{1+\lambda\tau} = \frac{10^{-3}\cdot 15}{1+10^{-3}\cdot 15} = 9.85\times 10^{-3}.$$

Similarly,

$$q_{C7} = \frac{10^{-5}\cdot 10}{1+10^{-5}\cdot 10} = 9.99\times 10^{-5}.$$

Using the rare event approximation,

$$q_{sys} = q(\text{cut sets}) = q_{C7} + q_{C1}q_{C2} + q_{C1}q_{C5}q_{C6} + q_{C2}q_{C3}q_{C4} + q_{C3}q_{C4}q_{C5}q_{C6}.$$

Thus,

$$q_{sys} = 9.99\times 10^{-5} + 9.70\times 10^{-5} + 9.56\times 10^{-7} + 9.56\times 10^{-7} + 9.41\times 10^{-9} = 1.99\times 10^{-4}.$$

Example 7.8

The auxiliary feedwater system in a pressurized water reactor (PWR) plant is used for emergency cooling of steam generators. The simplified piping and instrument diagram (P & ID) of a typical system like this is shown in form of a reliability block diagram below.

Calculate the system unavailability. Assume all components are in standby mode and are periodically tested with the characteristics described below.

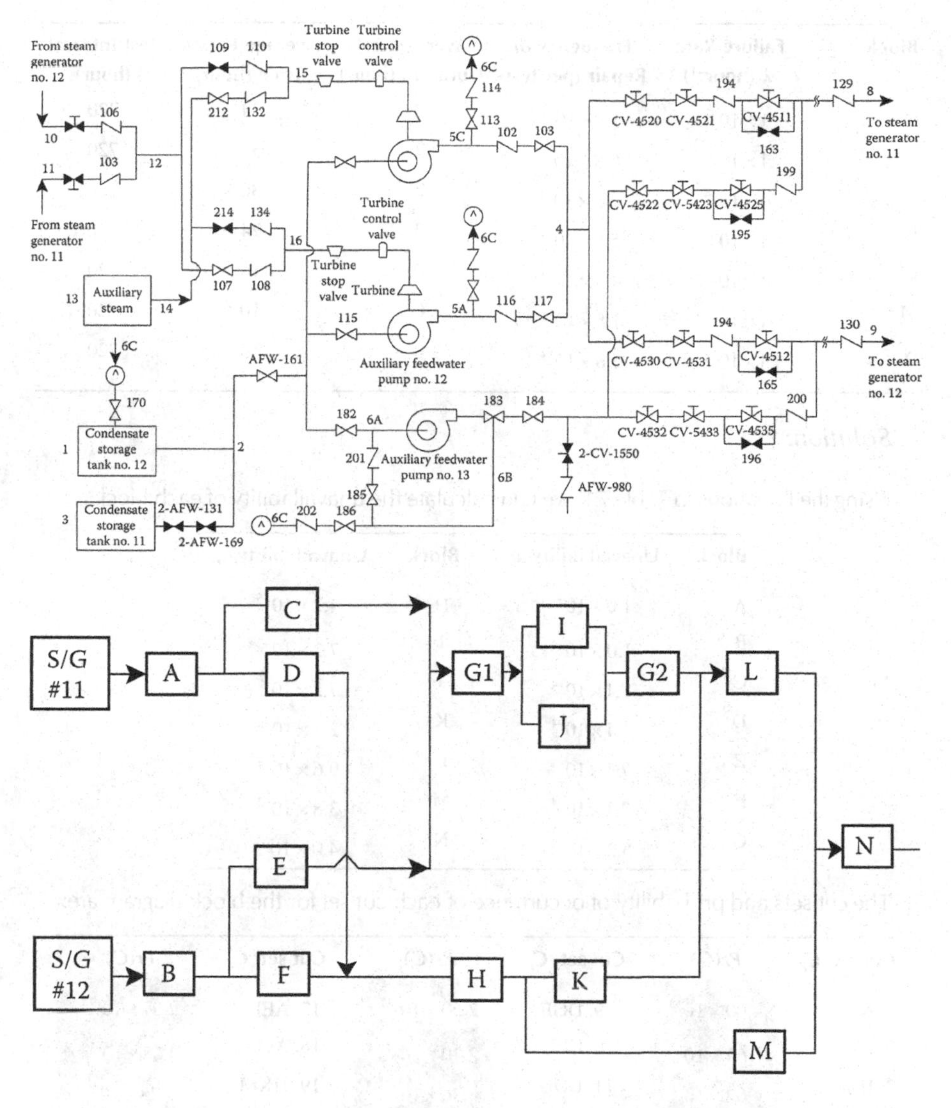

Block	Failure Rate, λ (hour^{-1})	Frequency of Repair (per test)	Average Test Duration (hour)	Average Repair Time (hour)	Test Interval (hour)
A	1×10^{-7}	9.2×10^{-3}	0	5	720
B	1×10^{-7}	9.2×10^{-3}	0	5	720
C	1×10^{-6}	2.5×10^{-2}	0	10	720
D	1×10^{-6}	2.5×10^{-2}	0	10	720
E	1×10^{-6}	2.5×10^{-2}	0	10	720
F	1×10^{-6}	2.5×10^{-2}	0	10	720
$G=G_1=G_2$	1×10^{-7}	7.7×10^{-4}	0	15	720

(Continued)

Block	Failure Rate, λ (hour^{-1})	Frequency of Repair (per test)	Average Test Duration (hour)	Average Repair Time (hour)	Test Interval (hour)
H	1×10^{-7}	1.8×10^{-4}	0	24	720
I	1×10^{-4}	6.8×10^{-1}	2	36	720
J	1×10^{-4}	6.8×10^{-1}	2	36	720
K	1×10^{-5}	5.5×10^{-1}	2	24	720
L	1×10^{-7}	4.3×10^{-3}	0	10	720
M	1×10^{-4}	1.5×10^{-1}	0	10	720
N	1×10^{-7}	5.8×10^{-4}	0	5	720

Solution:

Using the Equations in Table 7.1, we can calculate the unavailability of each block as:

Block	Unavailability, q_i	Block	Unavailability, q_i
A	1.0×10^{-4}	H	4.2×10^{-5}
B	1.0×10^{-4}	I	7.2×10^{-2}
C	7.1×10^{-4}	J	7.2×10^{-2}
D	7.1×10^{-4}	K	2.5×10^{-2}
E	7.1×10^{-4}	L	9.6×10^{-5}
F	7.1×10^{-4}	M	3.8×10^{-2}
G	5.2×10^{-5}	N	4.0×10^{-5}

The cut sets and probability of occurrence of each cut set for the block diagram are:

Cut set, C_i	$Pr(C_i)$	Cut set, C_i	$Pr(C_i)$	Cut set, C_i	$Pr(C_i)$
1. N	4.00×10^{-5}	9. DGF	2.60×10^{-11}	17. AEH	2.97×10^{-12}
2. LM	3.64×10^{-6}	10. CEH	2.10×10^{-11}	18. AFG	3.68×10^{-12}
3. HL	4.02×10^{-9}	11. BDL	6.76×10^{-12}	19. IJKM	4.78×10^{-6}
4. GH	2.19×10^{-9}	12. BDG	3.68×10^{-12}	20. DFIJ	2.55×10^{-9}
5. AB	9.98×10^{-9}	13. BCH	2.97×10^{-12}	21. CEKM	4.68×10^{-10}
6. HJI	2.14×10^{-7}	14. BCD	4.99×10^{-11}	22. BDIJ	3.61×10^{-10}
7. GKM	4.87×10^{-8}	15. AFL	6.76×10^{-12}	23. BCKM	6.61×10^{-11}
8. DFL	4.79×10^{-11}	16. AEF	4.24×10^{-7}	24. AFIJ	3.61×10^{-10}
				25. AEKM	6.61×10^{-11}

Using the rare event approximation, we can compute the average system unavailability as $q_{sys} = 4.9\times10^{-5}$.

In estimating the availability of redundant systems with periodically-tested components, it is important to recognize that components whose simultaneous failures cause the system to fail (i.e., sets of components in each cut set of the system) should be tested in a *staggered* manner. This way the system will not become totally unavailable during the testing and repair of its redundant components. For example, consider a system of two parallel units, each of which is periodically tested and has a pointwise unavailability behavior that can be approximated by the RBD model at the top of Figure 7.8. If the components are not tested in a staggered manner, the system's pointwise unavailability exhibits the shape in Figure 7.8. On the other hand, if the components are tested in a staggered manner, the system unavailability would exhibit the shape illustrated in Figure 7.9. Clearly, the average unavailability

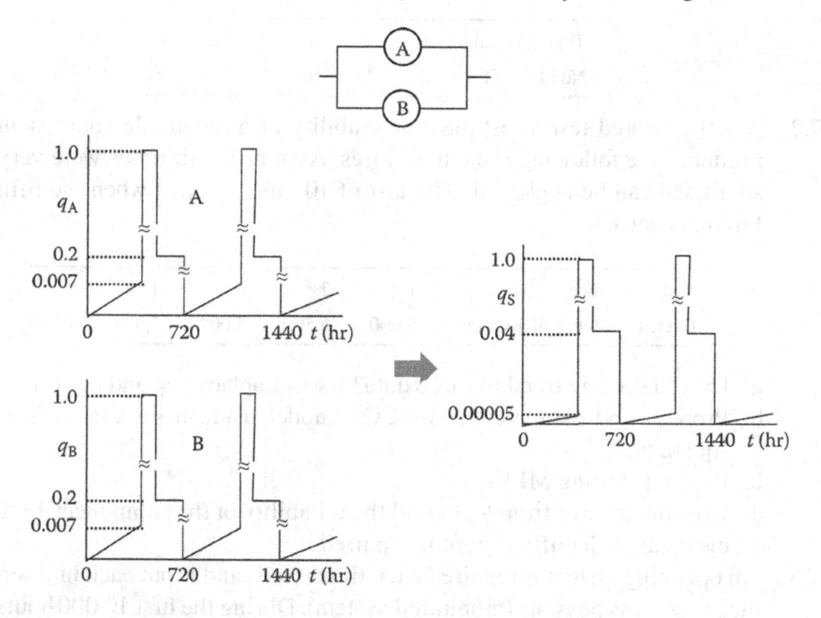

FIGURE 7.8 Unavailability of a two-unit parallel system using non-staggered testing.

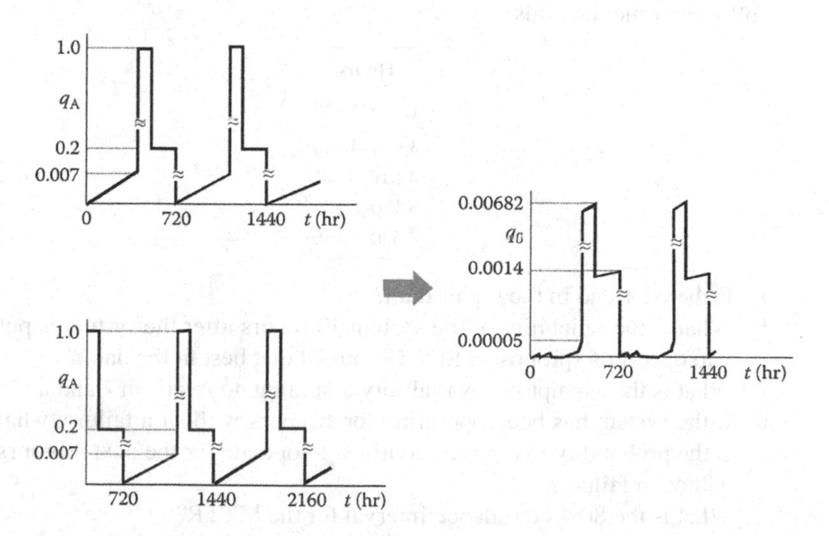

FIGURE 7.9 Unavailability of a two-unit parallel system using staggered testing.

with staggered testing is lower. Vesely et al. (1981) and Ginzburg and Vesely (1990) discuss this subject in more detail. Also, to minimize unavailability, one can find an optimum value for test intervals and the optimum degree of staggering.

7.9 EXERCISES

7.1 The following table shows fire incidents during six equal time intervals of 22 chemical plants. Are the fire incident times dependent? Prove your answer.

Time interval	1	2	3	4	5	6
Number of fires	6	8	16	6	11	11

7.2 A left-censored test to estimate availability of a repairable component produced the following cycle to failures. Assume repair times were very small and can be neglected. The test of 10 units stopped when the fifth failure occurred.

Repair number	1	2	3	4	5
Cycle to failure (interarrivals)	5,000	6,500	4,000	4,000	4,200

a. Do you see any trend in these data? Use a Laplace test and plotting.
b. Propose and estimate the ROCOF model using a regression least square fit.
c. Repeat (b) using MLE.
d. Use your answer from (c) to find the reliability of the component 1000 cycles after the fifth failure is repaired.

7.3 An operating system is repaired each time it fails and is put back into service as soon as possible (monitored system). During the first 10,000 hours of service, it fails five times and is out of service for repair during the following time intervals:

Hours
1,000–1,050
3,960–4,000
4,510–4540
6,130–6,170
8,520–8,560

a. Is there a trend in the repair data?
b. What is the reliability of the system 100 hours after the system is put into operation (propose a ROCOF model that best fit the data)?
c. What is the asymptotic availability assuming no trends in λ and μ?
d. If the system has been operating for 10 hours without a failure, what is the probability that it will continue to operate for the next 10 hours without a failure?
e. What is the 80% confidence interval for the MTTR?

7.4 Consider the following data showing failure times and repair duration of a component.

Age in Days	Repair Duration (hour)
120	3.2
210	3.8
465	6.5
600	3.5
780	7.7

If the component has been in service in the past 900 days, determine:

a. If there is a trend in the ROCOF and ROCOR. Find the model parameters for each.
b. The instantaneous and limiting point availabilities of the component at present and after additional 100 days of operation.

7.5 Consider a periodically-tested system with a constant ROCOF of 2×10^{-4} hour^{-1}. Field data show that the average constant repair duration is 8 hours, the average test duration is 1.5 hours, and the average constant frequency of repair is 4% per test. Find the optimum inspection interval such that the average availability is maximized.

7.6 Field data for a component are shown below.

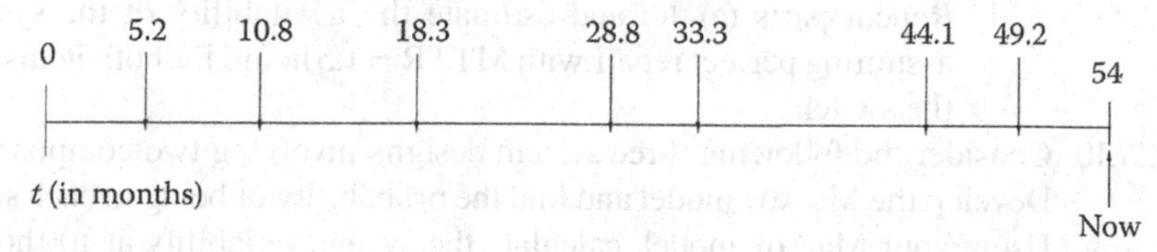

a. Is there a trend in ROCOF? Find the appropriate HPP or NHPP model.
b. Plot the conditional probability of no failure for the next year (plot in monthly intervals). The current age of the component is 5 years.
c. Assuming the average downtime for each repair is 10 hours and ROCOF has no trend, determine the average conditional unavailability 1 year from now.
d. In a periodic testing regime that reduces the failure rate by a factor of 5 (e.g., by using a preventive maintenance program) with an average test duration of 5.5 minutes, average repair duration of 1.5 hours and average frequency of repair of 2%, what inspection interval we should adopt to maintain the same level of unavailability as in part (c)?

7.7 Data from a periodically-tested component shows five durations of repair (in hours) as: 6, 8.5, 9.2, 9.5, and 9.7. There were a total of 55 tests. The average length of each test is reported as 2 hours. The total accumulated component operating time is 1,000 hours.

a. Use a Laplace test to determine if there is any trend in the repair rate. Test whether the data exhibit a trend with a 5% significance level.
b. Find the parameters of the appropriate point process model for repair rate and the 90% confidence interval on those parameters.

c. Calculate average unavailability of the component assuming a constant rate of occurrence of failure and constant duration of repair.

7.8 The following tables shows times of failure of a repairable component. The current age of the component coincides with the occurrence of last failure.

Failure number	1	2	3	4	5	6
Interarrival of failures (hours)	800	695	812	731	911	834

a. Estimate the parameters of a Weibull NHPP ROCOF model.
b. For the model in part (a) determine the MTBF of the future repairs.

7.9 Consider two identical, independent components with failure rate $\lambda = 0.01$ hour^{-1} operating for a mission time of 24 hours. Consider two systems that could be made if these components are placed in:
i. Standby configuration with perfect switching and no standby failures,
ii. Standby configuration with imperfect switching and standby failure rates of $\lambda_{ss} = 1\times10^{-6}$ hour^{-1} and $\lambda' = 1\times10^{-5}$ hour^{-1}, respectively.
   a. Develop a Markov model representing the states of the system.
   b. Write and solve the differential equations representing the probability of being in each possible system state.
   c. Compare your results with those given in parts (c) and (d) of Example 6.4.
   d. Repeat parts (a)–(c) and estimate the availability of the system assuming perfect repair with MTTR = 1.5 hours for both items and the switch.

7.10 Consider the following three system designs involving two components. Develop the Markov model and find the probability of being in each state. Using your Markov model, calculate the system reliability at 100 hours. Assume an exponential distribution for time to failure of the components with $\lambda_1 = 0.003$ hour^{-1} and $\lambda_2 = 0.01$ hour^{-1}.
i. Two parallel units.
ii. A standby system assuming perfect switching and no standby failures.
iii. A load-sharing system with $\lambda_h = 0.003$ hour^{-1} and $\lambda_f = 0.01$ hour^{-1}.

7.11 Consider the system represented by the RBD below. All components are identical and are repaired immediately after a failure. If each component has a ROCOF of $\lambda = 0.002$ (per hour) and an MTTR of 15 hours, find the limiting pointwise unavailability of the system.

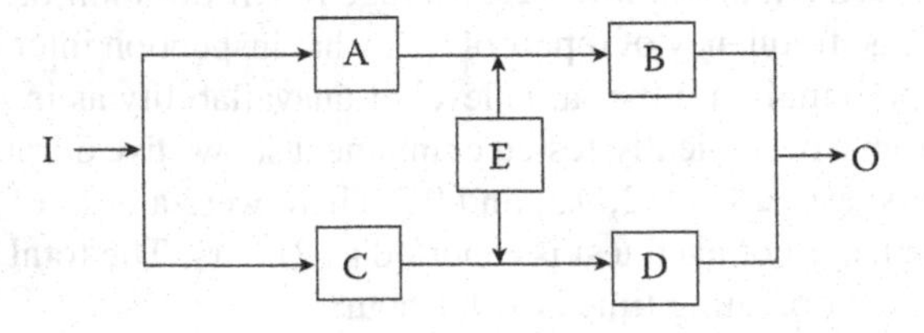

7.12 Consider the fault tree below. If all the components are periodically tested with a test interval of 1 month and average test duration of 1 hour, determine the average unavailability of the system.

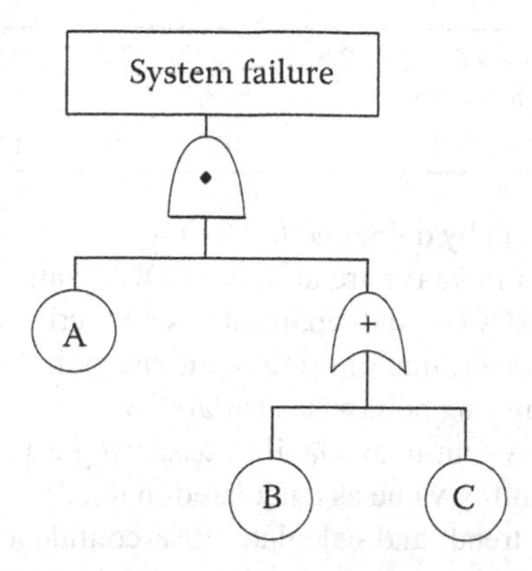

Component	Failure Rate (per hour)	Repair Rate (per hour)	Frequency of Repair
A	2×10^{-3}	1×10^{-1}	0.1
B	5×10^{-4}	2×10^{-2}	0.1
C	5×10^{-4}	2×10^{-2}	0.2

7.13 Consider the following set of data obtained for one component of the system shown in the fault tree below. The current age of the component is 13,500 hours. Assume all components have the same availability.

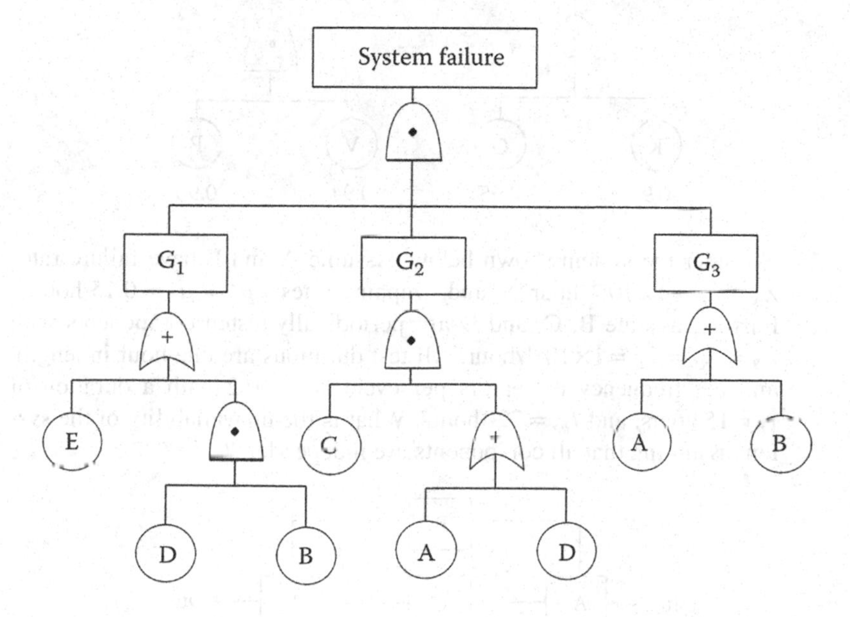

Failure time from when component was new (hour)	2,820	3,990	7,224	9,823	12,543
Repair duration (hour)	15	10	30	12	34

Assess the system by doing the following:

a. Determine if there is a trend in the ROCOF and repair rate.
b. Determine ROCOF and repair rate as a function of time (or, if there is if no trend, determine limiting pointwise point estimates).
c. Calculate limiting pointwise availability.
d. Determine system unavailability assuming all components have the same availability value as calculated in (c).
e. Assume no trend, and calculate 90% confidence limits for ROCOF using data in (a).

7.14 Consider the success tree for system S shown below. Develop all mutually exclusive cut sets and path sets and use them to determine the system availability.

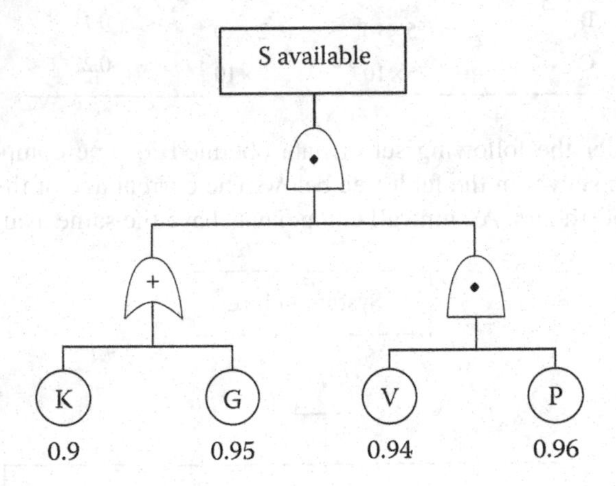

7.15 Consider the system shown below. Assume A and E have failure rates $\lambda_A = \lambda_E = 1 \times 10^{-4}$ hour^{-1} and repair rates $\mu_A = \mu_E = 0.15$ hour^{-1}. Further, assume B, C, and D are periodically-tested components with $\lambda_B = \lambda_C = \lambda_D = 1 \times 10^{-4}$/hour. All test durations are one hour in length, and the frequency of repairs per cycle is $f = 0.2$ with a duration of $\tau_r = 15$ hours, and $T_{op} = 720$ hours. What is the unavailability of the system assuming that all components are independent?

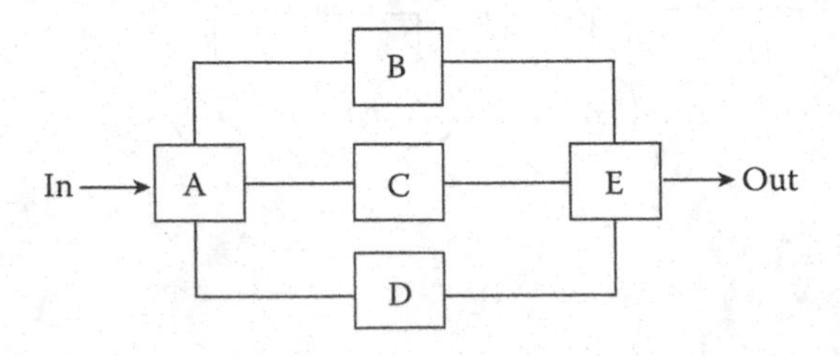

7.16 A simplified schematic of the electric power system at a nuclear power plant is shown in the figure below.

a. Draw a fault tree with the top event "Loss of Electric Power from Both Safety Load Buses."
b. Determine the unavailability of each event in the fault tree for 24 hours of operation.
c. Determine the top event probability.

Assume the following:

- Either the main generator or one of the two diesel generators is sufficient.
- One battery is required to start the corresponding diesel generator.
- Normally, the main generator is used. If that is lost, one of the diesel generators provides the electric power on demand.

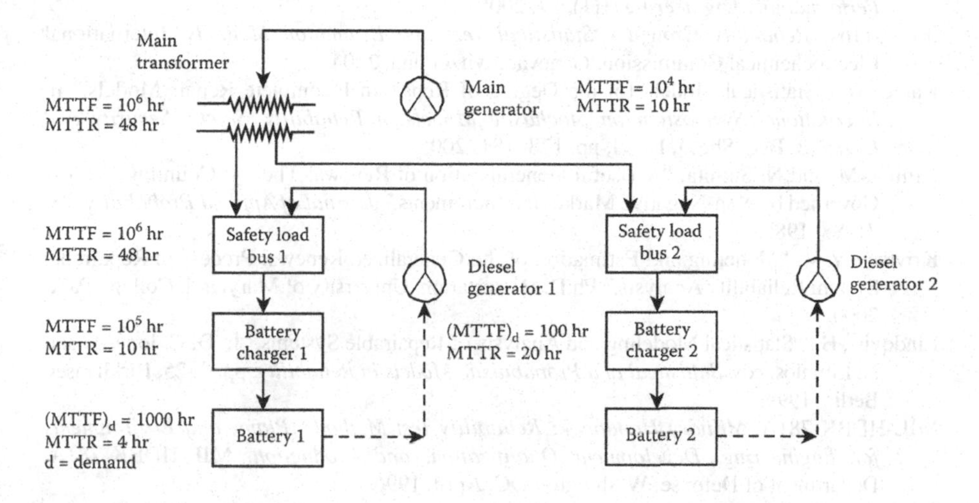

REFERENCES

Ascher, H. and H. Feingold, *Repairable Systems Reliability: Modeling and Inference, Misconception and Their Causes*, Marcel Dekker, New York, 1984.

Atwood, C., "Parametric Estimation of Time-Dependent Failure Rates for Probabilistic Risk Assessment." *Reliability Engineering & System Safety*, 37, 181–194, 1992.

Bain, L. J. *Statistical Analysis of Reliability and Life-Testing Models; Theory and Methods*, 2nd Edition, Rutledge/CRC, New York, 2017.

Barroeta, C. and M. Modarres, "Risk and Economic Estimation of Inspection Policy for Periodically Tested Repairable Components." In *Proceedings of PSAM-8*, New Orleans, 2006.

Baxter, L. A., E. M. Scheuer, D. J. McConalogue and W. R. Blischke. "On the Tabulation of the Renewal Function." *Technometrics*, 24, 151–156, 1982.

Blischke, W. R. and D. N. P. Murthy, *Warranty Cost Analysis*, Marcel Dekker, New York, 1994.

Brown, M. and F. Proschan, "Imperfect Maintenance." In J. Crowley and R. Johnson, eds, *Survival Analysis*, pp. 179–188, IMS Lecture Notes—Monograph Series, Hayward, Vol. 2, 1982.

Cox, D. R. and P. A. W. Lewis, *The Statistical Analysis of Series of Events*, Chapman & Hall, London, 1966.

Crow, L. H., "Reliability Analysis for Complex Repairable Systems." In F. Proschan and R. J. Serfling, eds, *Reliability and Biometry*, pp. 379–410, SIAM, Philadelphia, 1974.

Crow, L. H., "Confidence Interval Procedures for the Weibull Process with Applications to Reliability Growth." *Technometrics*, 24, 67–72, 1982.

Crow, L. H., "Evaluating the Reliability of Repairable Systems." In *Proceedings of the Annual Reliability and Maintenance Symposium*, IEEE, Orlando, FL, 1990.

Duane, J. T., "Learning Curve Approach to Reliability Monitoring." *IEEE Transactions on Aerospace*, 2, 563–566, 1964.

Gamerman, D. and H. F. Lopes, *Markov Chain Monte Carlo: Stochastic Simulation for Bayesian Inference*, 2nd edition, Chapman and Hall/CRC, New York, 2006.

Ginzburg, T. and W. E. Vesely, *FRANTIC-ABC User's Manual: Time-Dependent Reliability Analysis and Risk Based Evaluation of Technical Specifications*, Applied Biomathematics, Inc., Setauket, NY, 1990.

Hurtado, J. L., F. Joglar, and M. Modarres, "Generalized Renewal Process: Models, Parameter Estimation and Applications to Maintenance Problems." *International Journal of Performability Engineering*, 1(1), 37, 2005.

IEC 61164. *Reliability Growth - Statistical Test and Estimation Methods*. International Electrochemical Commission, Geneva, Switzerland, 2004.

Kahle, W., "Statistical Models for the Degree of Repair in Incomplete Repair Models." In *International Symposium on Stochastic Models in Reliability, Safety, Security and Logistics*, Beer Sheva, Israel, pp. 178–181, 2005.

Kijima, M. and N. Sumita, "A Useful Generalization of Renewal Theory: Counting Process Governed by Non-Negative Markovian Increments." *Journal of Applied Probability*, 23, 71–88, 1986.

Krivtsov, V. V., "Modeling and Estimation of the Generalized Renewal Process in Repairable System Reliability Analysis." Ph.D. Dissertation. University of Maryland, College Park, 2000.

Lindqvist, H., "Statistical Modeling and Analysis of Repairable Systems." In D. C. Ionescu and N. Limnios, eds, *Statistical and Probabilistic Models in Reliability*, pp. 3–25, Birkhauser, Berlin, 1999.

MIL-HDBK-781A, *Military Handbook: Reliability Test Methods, Plans, and Environments for Engineering, Development Qualification, and Production*, MIL-HDBK-781A, Department of Defense, Washington DC, April, 1996.

Rigdon, S. E. and A. P. Basu, *Statistical Methods for the Reliability of Repairable Systems*, Wiley, New York, 2000.

Thompson, W. A. Jr., "On the Foundation of Reliability." *Technometrics*, 23, 1–13, 1981.

Vesely, W., "Incorporating Aging Effects into Probabilistic Risk Analysis Using a Taylor Expansion Approach." *Reliability Engineering & System Safety*, 32, 315–337, 1991.

Vesely, W. E. and F. F. Goldberg, *FRANTIC—A Computer Code for Time-Dependent Unavailability Analysis*, NUREG-0193, 1977.

Vesely, W. E., F. F. Goldberg, J. T. Powers, J. M. Dickey, J. M. Smith, and E. Hall, *FRANTIC II—A Computer Code for Time-Dependent Unavailability Analysis*, U.S. Nuclear Regulatory Commission, NUREG/CR-1924, Washington, DC, 1981.

Wang, Z. and R. Pan, "Point and Interval Estimators of Reliability Indices for Repairable Systems Using the Weibull Generalized Renewal Process." *IEEE Access*, 9, 6981–6989, 2021.

Xie, M., "On the Solution of Renewal-Type Integral Equations." *Communications in Statistics. Simulation and Computation*, 18, 281–293, 1989.

Zhang, F. and. A. K. S. Jardine, "Optimal Maintenance Models with Minimal Repair, Periodic Overhaul and Complete Renewals." *IIE Transactions*, 30, 1109–1119, 1998.

8 Selected Advanced Topics in Reliability and Risk Analysis

In this chapter, we discuss several complementary topics to support the methods presented in the other chapters of this book. These topics are intended to introduce more advanced methods required to model complex systems. While the topics are not covered in complete depth, it is important to be aware of these topics and resources for advanced studies. The topics covered in this chapter are uncertainty analysis, dependent failure analysis, importance measures, and human reliability analysis. These four topics are critical to a comprehensive analysis, particularly for analyzing complex engineering system reliability and risk assessment problems.

8.1 UNCERTAINTY ANALYSIS

Uncertainty arises from a lack of, or insufficient knowledge and information about, events, system states, processes, and phenomena that factor into the analysis. Reliability and risk model output uncertainties are influenced by many factors, including: (a) Input to the deterministic reliability and risk models, (b) Model parameters and physical constants, (c) Model form or structure, and (d) Quality and availability of information and data. For further discussion on sources of these uncertainties, see the references (e.g., Modarres et al., 2021, Bensi et al., 2022, and Drouin, 2017).

These uncertainties should be characterized and propagated through the model to find the output uncertainties. In general, the uncertainty analysis in reliability and risk tries to measure the "goodness" of the model outputs with regard to an estimation problem. Without such a measure, it would be difficult to judge how closely the estimated values relate to or represent the reality (i.e., the expected or true value of an outcome of interest) and provide a basis for making decisions. Critical elements of the uncertainty analysis include characterizing data and measurement uncertainties, and assigning uncertainties (usually in the form of probability distributions) to the model parameters, model error, and model inputs. Then, these uncertainties are propagated to assess their compounded contribution to the output uncertainty. It is beneficial for the decision-maker to know and appreciate the aleatory and epistemic parts of the reliability and risk results.

In a classical (frequentist) probabilistic estimation, uncertainty about the output of a reliability or risk model is described by a point estimate along with a confidence interval. In Bayesian estimation, the reliability and risk outputs are represented by a probability distribution describing the credible interval. Typically, a description of the source of uncertainty and its effects should accompany the analysis to facilitate

DOI: 10.1201/9781003307495-8

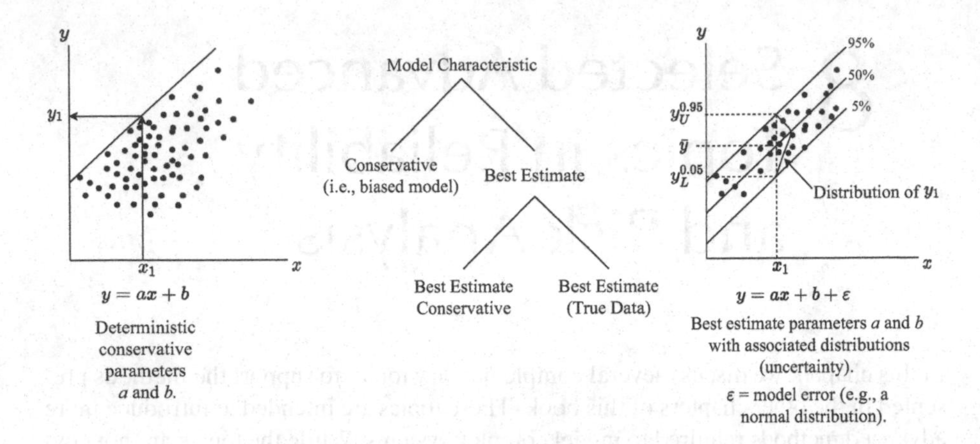

FIGURE 8.1 Illustration of the deterministic approaches: Conservative vs. Best Estimate Plus Uncertainty Representation.

understanding the nature of the uncertainties and whether they are reducible with further study.

A reliability or risk model uncertainty analysis comprises one of two possible methods. First, the parameters and the independent input variables are described probabilistically (for example, by treating them as random variables and representing them using continuous or discrete distributions, or by describing their statistics and moments such as the mean and standard deviation). Alternatively, the best estimate deterministic model is represented by the best estimate single value parameters, but the model output is treated as random by adding or multiplying it by an *error distribution*. A normal distribution is often used for the additive error distribution and a lognormal distribution is used for multiplicative error. The uncertainties and errors associated with the best estimate models are found from uncertain and limited data and information. In some reliability and risk analyses, it may be necessary by choice or through the data analysis used in the modeling process to bias the best estimate model outputs. For example, some non-destructive flaw or damage detection tools may bias their measurements consistently to larger or smaller than the true values. In these cases, the amount of bias should be measured and addressed when calculating the reliability and risk outputs. Figure 8.1 graphically depicts the discussions above. For more readings on conservative and best estimate models, refer to Bucalossi et al. (2010) and Modarres et al. (2021).

8.1.1 Steps in Uncertainty Analysis for Risk and Reliability Models

The substeps below summarize the necessary parts of comprehensive uncertainty analysis.

a. **Identify the structure and parameters of the model and uncertainties**
   The deterministic part of a reliability or risk model is a mathematical form, and a reliability or risk model or its implementation in a computer code may

contain a collection of submodels. It is essential to know the confidence in the form (structure) of the model. Further, the parameters and input to those models may also be uncertain. These uncertainties often arise from using limited data to estimate the parameters and model inputs. It is also helpful to list the limitations of the models and possible impacts on confidence to the reliability and risk model output. The effects of those limitations can be analyzed later through a sensitivity analysis. Finally, in complex reliability and risk codes containing a collection of submodels, sometimes computational uncertainties are significant.

b. **Characterize the underlying data used in model development and analysis**
The uncertainties in the model and parameters identified in Step (a) should be characterized. This includes a description of the experimental or field data used to build the models (for example, through a regression analysis). Typically, the analysts in this part identify limitations of the database's size, the accuracy of the measured data points, type of data (sensor-based, human measured or observed events, expert judgment), and missing data such as the probability of not detecting possible data points.

c. **Quantify and describe uncertainties, model parameters, model inputs, and model form**
In this step, the analyst should assign qualitative levels of uncertainty (in the absence of information and data) through engineering judgment or compute uncertainty metrics for the parameters, model input, and the model itself. The data described in Step (b) should be used to support the uncertainty metrics. For example, the corresponding model parameters can be estimated and described using a classical maximum likelihood estimation or Bayesian estimation. Uncertainties associated with the model itself are usually described qualitatively, and their impact is measured quantitatively through sensitivity analysis. However, a more formal approach would be using the underlying data in Step (b) and estimating the model error (that is, the data scatter around the model) using the classical maximum likelihood or Bayesian estimation. Quantitatively, uncertainties are presented in the form of a distribution function (in the Bayesian treatment of uncertainty, credible intervals are shown) or through a confidence interval and associated confidence level (e.g., 90%, in the classical estimation methodology). In any reliability and risk analysis, the analysts must describe and be consistent in their quantitative treatments. That is, either consistently use a Bayesian analysis (i.e., treat variables as random and express them through probability distributions) or use a classical method (i.e., treat the variable as an unknown metric described by a best estimate and confidence interval conditioned on the underlying data). If multiple competing models exist, performing a model selection analysis (Kuhn and Johnson, 2013) or using a model averaging method is advised.

d. **Describe the types of uncertainties**
Uncertainties in Step (c) for the model parameters (or hyperparameters), model inputs, and the model error should be divided into those that are reducible (epistemic) or irreducible (aleatory). This is a judgmental but

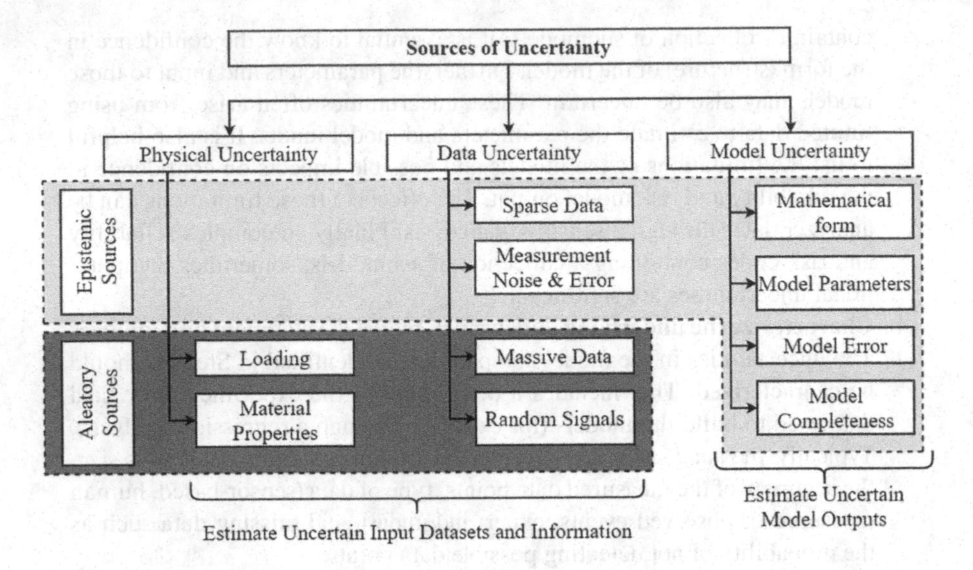

FIGURE 8.2 Types and sources of uncertainties.

critical step for appropriately interpreting the reliability and risk analysis results. Experience shows that after some team deliberations, analysts can decide on reasonably reducible uncertainties that are expected to contribute critically to the final reliability and risk result and designate them into one of these two types. Figure 8.2 illustrates an approximate breakdown of aleatory and epistemic uncertainties. Table 8.1 lists specific examples of the types and sources of uncertainties.

e. **Propagate uncertainties**

Once the uncertainties about the reliability and risk models in the form of parameter, model error, and input uncertainties are described, they should be propagated through the analysis so the uncertainties about the reliability and risk model (or code) output uncertainties can be computed. The output uncertainties show the compounded value of these uncertainties for specific reliability or risk analyses. The propagation of uncertainty, especially in complex models and codes, is computer-based. There are two types of methods used for propagation: (a) Computing the cumulative effect of the statistical moments (e.g., the mean and standard deviation) of the parameters and input on the statistical moments of the output; (b) Using a Monte Carlo simulation to generate samples of the output values from random samples of the model parameters, input, or model error. In propagating uncertainties, it is highly advisable to propagate aleatory and epistemic uncertainties separately so the decision-makers can recognize the extent to which the uncertainties are reducible in the reliability and risk outputs. The propagation method and the approach for separate propagation of aleatory and epistemic uncertainties are further described in Section 8.1.2.

TABLE 8.1
A Categorization of the Uncertainties in Reliability and Risk Analysis

Type	Epistemic				Aleatory				
Sources	Different physical/mathematical models Limited/missing data Lack of knowledge Incomplete information Missing knowledge Systematic bias in measurements				Inherent variations Random data Stochastic behaviors Stochastic properties Noise in measurement				
Treatment	Reducible				Irreducible				
Reliability & Risk-related Categories	Model Parameters		Model Structures		Unpredictable Events	Random Phenomena		Physical Properties	
Subcategories	Some physical properties and parameters	Mathematical model parameters	Empirical predictive models	Probability distribution models	Random environmental conditions	Random changes in model input variables	Random measurement and detection noise	Material properties	Material structures
Examples	• Crack densities • Yield strength • Fatigue endurance • Fracture toughness	• Parameters of regression models for physical degradation • Shape, scale, and location parameters of pdfs	• Form of regression models • Empirical model errors (errors in prediction of accelerated life models)	• Pdf chosen to represent deterministic model input and parameters • Choice of prior distribution in Bayesian Inference	• Weather changes • Ambient conditions • Hostile environments	• Applied stresses • Operating regimes • Residual stresses • Material Density	• Failure event detection • Flaw detection • Sensor output • Sampling	• Alloys metal fractions • Modulus	• Geometric parameters • Presence of notches and inclusions

f. **Describe and graph the output uncertainties**
Visualization aids understanding, and therefore, the output uncertainties should be represented by one or more of these methods:
- Histogram of the output values, including the mean, median, and quantiles
- A box plot of the output values
- A continuous pdf or cdf of the output variables, including the mean, median, and quantiles
- A cumulative complementary distribution plot showing the likelihood that the output variable exceeds a given damage level or time of failure

The results should also include a qualitative description of the significance and sources of the output uncertainties.

When aleatory and epistemic uncertainties are separated, the above representations would be used to express the aleatory part of the uncertainty, each representing a given realization of the epistemic uncertainties (whereby, for example, multiple cdfs of aleatory distribution are plotted, the ranges of which show the epistemic component of the uncertainty).

g. **Perform sensitivity analysis**
For qualitatively described uncertainties, uncertainties described through expert elicitation and judgment, and uncertain assumptions used in the reliability and risk model development, perform a sensitivity analysis to highlight and measure their impact. To do so, quantify the risk and reliability model using different values of the input and document the different outputs to illuminate the effect of the epistemic uncertainties.

h. **Report uncertainty results**
In the final report, reliability and risk output uncertainties and sensitivity analyses are documented, reported, and used in decision-making.

8.1.2 Uncertainty Propagation Methods

Two primary uncertainty propagation methods may be used in risk and reliability analyses: the method of moments and the probabilistic methods (mostly Monte Carlo).

8.1.2.1 Method of Moments for Propagation of Uncertainties

The method of moments is the simplest and most commonly used method for propagating uncertainties. It is a suitable method when a deterministic model representing physics or a reliability and risk model are simple and at least two of the first moments (i.e., of the mean and variance) of the model input and parameters are known. A good illustration of this method can be found in the work by Wu (1994) and Apostolakis and Lee (1977). This method is computationally cheap and scalable. In this method, the first-order moments (i.e., mean and variance) of the model's independent input variables and model parameters are used to derive the same moments associated with the model output using the Taylor expansion of a mathematical model. The method's accuracy may be improved by using higher-order Taylor expansion terms. However, calculating higher-order derivatives is computationally expensive, and usually unnecessary.

To implement this approach, consider a deterministic model in the following form:

$$y = f(x_1, \ldots, x_n, s_1, \ldots, s_n) \; for \; (i = 1, \ldots, n), \tag{8.1}$$

where x_i are the point estimates of the model parameters (e.g., MTTF, failure rate, and probability of failure on demand) of a system component, and s_i are the respective standard deviations (errors) for the n model outputs.

Assume that

- $f(x_1, \ldots, x_n, s_1, \ldots, s_n) = f(x, s)$ satisfies the conditions of Taylor's theorem, and
- The estimates x_i are independent and unbiased with expectations (true values) μ_i.

Using Taylor's series expansion about μ_i, and denoting $(x_1, \ldots, x_n)$ by x and $(s_1, \ldots, s_n)$ by s, we can write

$$\begin{aligned} y = f(x, s) = f(\mu_1, \ldots, \mu_n, s) &+ \sum_{i=1}^{n} \left[\frac{\partial f(x)}{\partial x}\right]_{x_i = \mu_i} \times (x_i - \mu_i) \\ &+ \frac{1}{2!} \sum_{j=1}^{n} \sum_{i=1}^{n} \left[\frac{\partial^2 f(x)}{\partial x_i \partial x_j}\right]_{x_i = \mu_i,\, x_j = \mu_j} \times (x_i - \mu_i)(x_j - \mu_j) + R, \end{aligned} \tag{8.2}$$

where R represents the residual terms.

Taking the expectation of Equation 8.2 (using the algebra of expectations in Table 2.2), we obtain

$$\begin{aligned} E(y) = f(\mu_1 \ldots, \mu_n, s) &+ \sum_{i=1}^{n} \left[\frac{\partial f(x)}{\partial x}\right]_{x_i = \mu_i} \times E(x_i - \mu_i) \\ &+ \frac{1}{2!} \sum_{j=1}^{n} \sum_{i=1}^{n} \left[\frac{\partial^2 f(x)}{\partial fx_i \partial fx_j}\right]_{x_i = \mu_i,\, x_j = \mu_j} \times E\left[(x_i - \mu_i)(x_j - \mu_j)\right] + E(R). \end{aligned} \tag{8.3}$$

Because the estimates x_i are unbiased with expectations (true values) μ_i, the second term in the above equation cancels out. Dropping the residual term, $E(R)$, and assuming that the estimates x_i are independent, one obtains the following approximation:

$$E(y) \approx f(\mu_1 \ldots, \mu_n, s) + \frac{1}{2} \sum_{i=1}^{n} \left[\frac{\partial^2 f(x)}{\partial x_i^2}\right]_{x_i = \mu_i} \times s^2(x_i). \tag{8.4}$$

For the more general and practical applications of the method of moments, we need to obtain the point estimate $\hat{y}$ and its variance $var(\hat{y})$. Replacing μ_i with x_i, we obtain

$$\hat{y} \approx f(x_1, \ldots, x_n, s) + \frac{1}{2}\sum_{i=1}^{n}\left[\frac{\partial^2 f(x)}{\partial x_i^2}\right]_{x=x_i} \times s^2(x_i). \tag{8.5}$$

If, for a given uncertainty analysis problem, the second term can be neglected, the estimate (Equation 8.4) is reduced to these simple forms:

$$\hat{y} \approx f(x_1, x_2, \ldots, x_n). \tag{8.6}$$

Taking the variance and treating the first term as constant, one obtains

$$var(\hat{y}) = var\left\{\sum_{i=1}^{n}\left[\frac{\partial f(x)}{\partial x_i}\right]_{x_i=\mu_i} \times (x_i - \mu_i)\right\} = \sum_{i=1}^{n}\left[\frac{\partial f(x)}{\partial x_i}\right]_{x_i=\mu_i}^2 \times s^2(x_i). \tag{8.7}$$

Example 8.1

For the system shown below, each component has a constant failure rate with a mean value of 5×10^{-3} per hour. If the failure rate can be represented by an r.v. that follows a lognormal distribution with a coefficient of variation of 2, calculate the mean and standard deviation of the system unreliability at $t=1$, 10, and 100 hours.

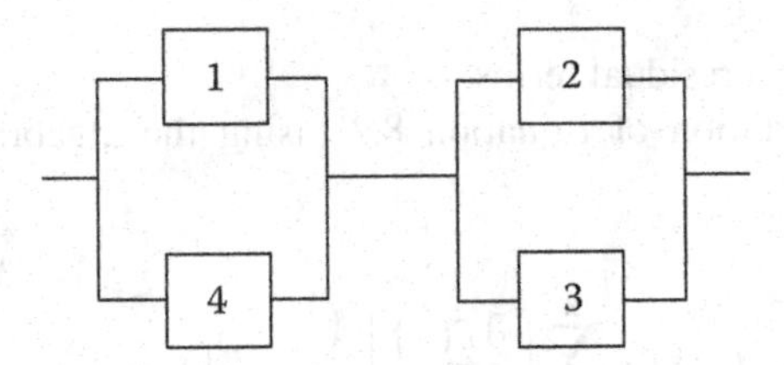

Solution:

System unreliability can be obtained from the following expression:

$$Q = q_1 \cdot q_4 + q_2 \cdot q_3 - q_1 \cdot q_2 \cdot q_3 \cdot q_4$$

since $q_i = 1 - e^{-\lambda t}$, then

$$Q_{sys} = \left(1-e^{-\lambda_1 t}\right)\left(1-e^{-\lambda_4 t}\right) + \left(1-e^{-\lambda_2 t}\right)\left(1-e^{-\lambda_3 t}\right) - \left(1-e^{-\lambda_1 t}\right)\left(1-e^{-\lambda_2 t}\right)\left(1-e^{-\lambda_3 t}\right)\left(1-e^{-\lambda_4 t}\right).$$

Note that $\hat{\lambda}_1 = \hat{\lambda}_2 = \hat{\lambda}_3 = \hat{\lambda}_4 = \hat{\lambda} = 5\times10^{-3}$ / hour. Using Equation 8.6 (i.e., neglecting the second term of Equation 8.5 due to its insignificance),

$$\hat{Q}=\left(1-e^{-\hat{\lambda}_1 t}\right)\left(1-e^{-\hat{\lambda}_4 t}\right)+\left(1-e^{-\hat{\lambda}_2 t}\right)\left(1-e^{-\hat{\lambda}_3 t}\right)-\left(1-e^{-\hat{\lambda}_1 t}\right)\left(1-e^{-\hat{\lambda}_2 t}\right)\left(1-e^{-\hat{\lambda}_3 t}\right)\left(1-e^{-\hat{\lambda}_4 t}\right)$$

$$=2\left(1-e^{-\hat{\lambda} t}\right)^2-\left(1-e^{-\hat{\lambda} t}\right)^4$$

$$=1-4e^{-2\hat{\lambda}t}+4e^{-3\hat{\lambda}t}-e^{-4\hat{\lambda}t}$$

$$\hat{Q}(t=1\text{ hour})=4.98\times10^{-5}$$

$$\hat{Q}(10)=4.75\times10^{-3}$$

$$\hat{Q}(100)=0.286$$

The partial derivatives are:

$$\frac{\partial Q}{\partial \lambda_1}=te^{-\lambda_1 t}\left(1-e^{-\lambda_4 t}\right)-te^{-\lambda_1 t}\left(1-e^{-\lambda_2 t}\right)\left(1-e^{-\lambda_3 t}\right)\left(1-e^{-\lambda_4 t}\right).$$

Repeating for other partial derivatives of Q for λ_2, λ_3, and λ_4 yields

$$\frac{\partial Q}{\partial \lambda_i}=te^{-\lambda_i t}\left[\left(1-e^{-\lambda t}\right)-\left(1-e^{-\lambda t}\right)^3\right],$$

and from Equation 8.7,

$$s^2(Q)=4\,var\left(\hat{\lambda}_i\right)\left\{te^{-\hat{\lambda}_i t}\left[\left(1-e^{-\hat{\lambda} t}\right)-\left(1-e^{-\hat{\lambda} t}\right)^3\right]\right\}^2$$

$$=4\left(s\left(\hat{\lambda}_i\right)\right)^2\left\{te^{-\hat{\lambda}_i t}\left[\left(1-e^{-\hat{\lambda} t}\right)-\left(1-e^{-\hat{\lambda} t}\right)^3\right]\right\}^2.$$

Using $c_v=2$, we find $s(\lambda_i)=2E(\lambda)=2\hat{\lambda}=2\left(5\times10^{-3}\right)=0.01$. Therefore, $var(\lambda_i)=10^{-4}$. It is now possible to calculate the variance and standard deviation for system unreliability.

$$s^2(Q)_{1\text{ hour}}=9.85\times10^{-9},\ s(Q)_{1\text{ hour}}=9.93\times10^{-5},$$

$$s^2(Q)_{10\text{ hours}}=8.57\times10^{-5},\ s(Q)_{10\text{ hours}}=9.26\times10^{-3},$$

$$s^2(Q)_{100\text{ hours}}=0.163,\ s(Q)_{100\text{ hours}}=0.403.$$

For a special case when $y=\sum_{i=1}^{n}x_i$ (e.g., a series composed of components having the exponential time to failure distributions with failure rates x_i), and dependent x_i's, the variance of y is given by

$$var(y)=\sum_{i=1}^{n}\text{var}(x_i)+2\sum_{i=1}^{n-1}\sum_{j=i+1}^{n}\text{cov}(x_i,\,x_j). \tag{8.8}$$

In the case where $y=\prod_{i=1}^{n}x_i$, and independent x_i's (e.g., a parallel system composed of components having reliability functions, x_i for $i=1,\ldots,n$),

$$E(y)=\prod_{i=1}^{n}E(x_i), \tag{8.9}$$

and

$$var(y)\approx\left[\sum_{i=1}^{n}\frac{\text{var}(x_i)}{E^2(x_i)}\right]\times E^2(y). \tag{8.10}$$

The method of moments provides a quick and accurate estimation of low-order moments of Y based on the moments of x_i, and the process is simple. However, for highly non-linear expressions of Y, using only low-order moments can lead to significant inaccuracies, and using higher moments is complex.

8.1.2.2 Monte Carlo Methods for Propagation of Uncertainties

In the Monte Carlo simulation, a value is randomly selected (sampled) from the pdf of each uncertain model input and model parameter. Each sample is called a realization of model input. The mathematical model is then used with each realization of the model input to calculate the associated output(s) by assessing the model deterministically. This is repeated many times (e.g., 10,000 times or more) to obtain many realizations of the model input, parameters, and associated output values. The outputs for each realization are then ordered from the lowest to highest, and either converted to a probability distribution or more simply the statistical moments and intervals are directly derived from the output values. For more readings on the convergence, accuracy, and the adequate number of Monte Carlo iterations, see Oberle (2015). The output values, distributions are represented in histograms but are occasionally fitted to a known pdf model. For more readings on the procedure and codes for Monte Carlo simulation, see Thomopoulos (2013).

In Monte Carlo sampling, the user can assume the model inputs and model parameters are independent or not. A value is independently sampled from each model input and model parameter pdfs in independent sampling. In dependent cases, the values sampled would be conditional. For example, if two model parameters are dependent, a random value is taken from the marginal pdf of the first parameter. Then for the second parameter, the sample is taken from the conditional pdf of the model parameter (conditioned on the first parameter value).

There are various Monte Carlo type techniques for uncertainty propagation, but five methods are noteworthy: Classical Monte Carlo simulation, Latin hypercube sampling (LHS), importance sampling, discrete probability distribution (DPD), and Wilks tolerance.

8.1.2.2.1 Classical Monte Carlo Simulation

The following illustrates the general steps for constructing the model output's uncertainty using this method. These steps are:

1. For each model input and model parameter, select a random value using a random number generator.
2. Calculate the model output value corresponding to the randomly selected input.
3. Repeat Steps 1–2 many times (usually several thousand times) to get many of the output values that include values in the extremes of the model output.
4. Using the output values from Step 3, choose a confidence level and construct the respective confidence limit for the output values or fit a pdf to the data and find the corresponding credible intervals.

Considering the reliability or risk model, then $y = f(\theta, x_i)$, $i = 1,\ldots, m$ in which there are m input variables (model inputs, parameters, and model error) denoted by a set or vector of random variables x_i, each described by a pdf. The Monte Carlo simulation process to find the model output y is conceptually illustrated in Figure 8.3. Each iteration of the Monte Carlo sampling would yield a realization of the model inputs and output, y^i, where the simulation process is repeated n times. When the y^i values are ranked from the lowest to the highest and binned into small intervals of y, the histogram or pdf of y is found, from which the probability intervals and other statistics of the model output are calculated. For a more detailed description of Monte Carlo simulation and example coding scripts, refer to Robert and Casella (2010).

See Sankararaman et al. (2011) for an example of Monte Carlo-based uncertainty quantification and model validation in fatigue crack growth analysis. For an application of Monte Carlo simulation, see Chatterjee and Modarres (2012) and Beardsmore et al. (2010). It is important to note that concern over the computational cost of carrying out a classical Monte Carlo simulation may not always justify using the alternative sampling techniques described in the following sections. In many analyses, the human cost of developing the model using alternative sampling techniques, performing the analysis, and documenting and defending the analysis may far exceed the computational cost of the traditional Monte Carlo iterative approach.

To quantify the contribution of the two types of uncertainties, the uncertainties need to be separated using a double-loop Monte Carlo simulation, where the epistemic uncertainties of the model input (mainly parameters) are sampled to create an instance or realization of the risk and reliability model in the outer loop. Given the epistemic sample, the aleatory variables are repeatedly simulated as part of the inner loop to generate an instance of the aleatory uncertainty of the risk and reliability output (e.g., probability and timing of a leak). The process is continued with additional realizations of the models by taking more epistemic uncertainty samples and output aleatory distributions. When epistemic sampling is exhausted, the results will include a family of aleatory distributions. When ranked, the epistemic uncertainty quantiles of the family of the output aleatory distributions can be established. Figure 8.4 depicts the two-loop sampling process (adapted from Jyrkama and Pandey, 2016). For an example

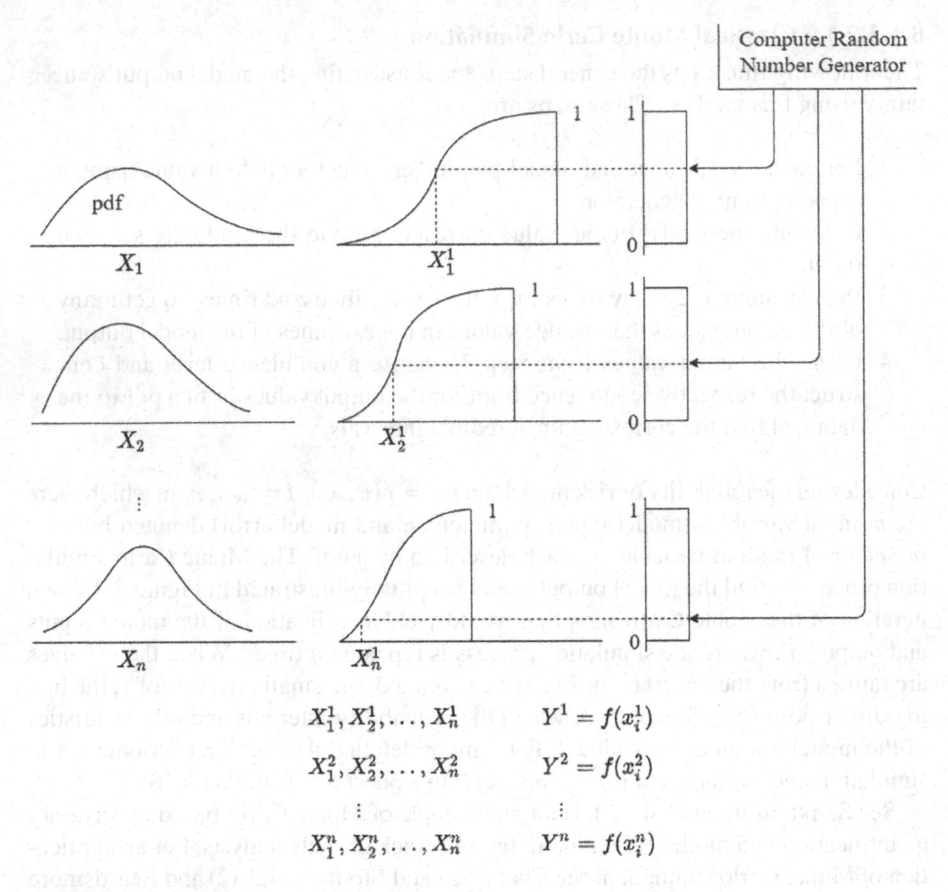

FIGURE 8.3 Illustration of the classical Monte Carlo simulation process.

of a two-loop Monte Carlo application that separately simulates the model input's aleatory and epistemic uncertainties in reliability and risk, see Duan et al. (2015). Further, Jyrkama and Pandey (2016) show through an example, that a two-staged simulation approach is preferable. In this approach, model parameters are represented by pdfs of their own, which allows the impact of both aleatory and epistemic uncertainties of the results to be displayed distinctly, including the estimation of the confidence bounds.

The uncertainty partitioning process may lead to misinterpretation of the results, particularly about the confidence (i.e., lower and upper bounds) in the estimated probabilities. Care must be taken to assure that the partitioned uncertainties show the uncertainty part that is reducible (i.e., epistemic). The total uncertainty about the output is still a combination of epistemic and aleatory components. Such a total uncertainty may be computed from a single-loop Monte Carlo simulation.

The output of partitioning aleatory and epistemic uncertainty is typically shown through the family of the aleatory model output variable. For example, the probability of detecting damage in a component can be expressed through a lognormal distribution whose cdf shows the aleatory component, and a family of the cdf curves represents the epistemic part. An example of this is shown in Figure 8.4.

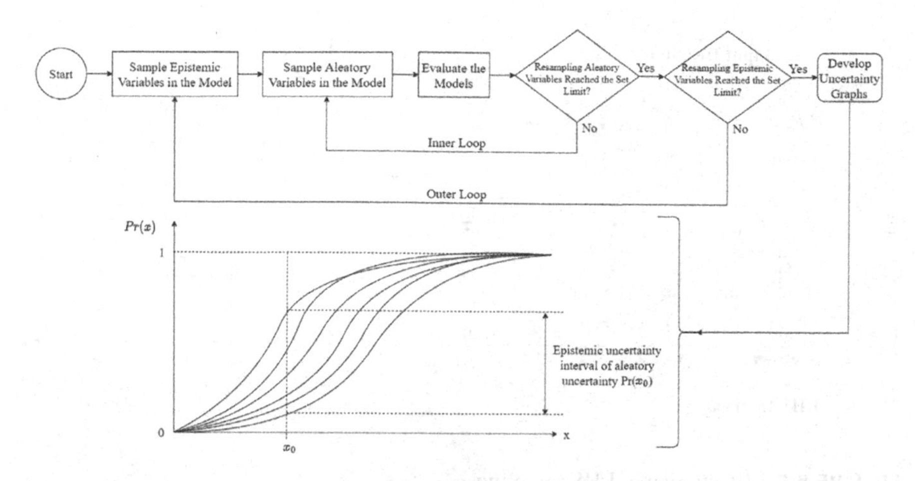

FIGURE 8.4 The two-loop Monte Carlo simulation process.

8.1.2.2.2 Latin Hypercube Sampling (LHS)

Sometimes when the reliability model involves a complex array of submodels with many inputs and parameters and requires a large amount of computational effort for its evaluation, it would be time-consuming to sample thousands of unbiased samples. To remedy this issue, LHS is used to reduce the number of samplings necessary while assuring that the entire range of the input distributions is sampled.

In the LHS method, the range of probable values for each uncertain model input and parameter, n, is divided into several segments, m, of equal probability. Thus, the whole input space is partitioned into m^n cells, each having equal probability. For example, for the case of four model inputs and parameters, each divided into five equal probability segments, we will end up with 5^4 or 625 cells. The cell number indicates a specific combination of segment numbers of the n model input and parameters that form a sample. For example, the cell number (2, 5, 2, 3, 1) indicates that the sample lies in segment 2 concerning the first uncertain model input (or parameter), segment 5 concerning the second model input, and so on. A random value for each cell segment is then taken and used as input in the reliability or risk model to find the corresponding output. Typically, two to several hundred samples would be sufficient to evaluate the uncertainties of the model output. The advantage of the LHS approach is that the random samples are generated from all the ranges of possible values, thus giving insight into important regions of the pdfs, particularly the distributions' tails. The LHS process is illustrated in Figure 8.5. On the left side of Figure 8.5, four intervals of an input random variable's pdf are found using four equal probability intervals from its cdf. The right side of Figure 8.5 illustrates two possible independent random regions based on the LHS intervals defined for each pdf. For more readings and coverage of this topic, refer to the references (e.g., Olsson et al., 2003). Akramin et al. (2017) compares Monte Carlo to LHS in fatigue analysis.

8.1.2.2.3 Importance Sampling

In importance sampling, the sampling space is divided into many nonoverlapping subregions called strata. The number of samples from each stratum depends on that subregion's importance, but a simple random sampling is performed within each stratum.

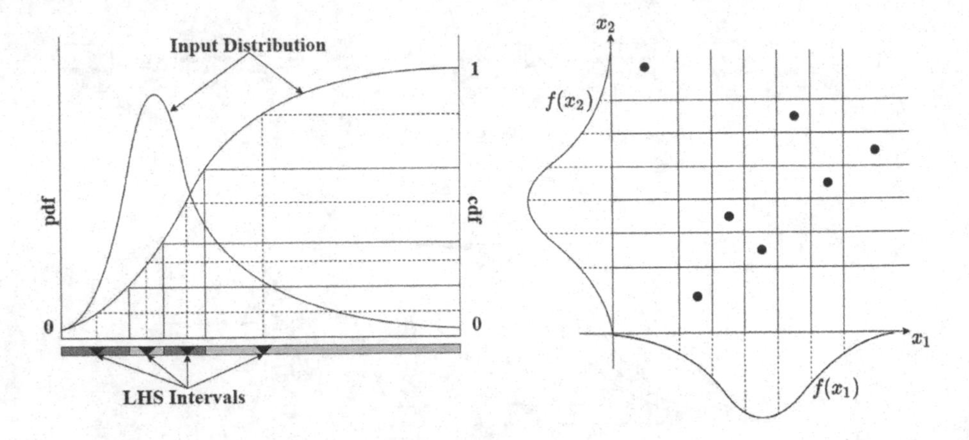

FIGURE 8.5 Illustration of LHS sampling.

Importance sampling is used to ensure the inclusion of regions with low probability and high impact on the model output values. When importance sampling is used, the probability assigned and the number of samples from each stratum must be folded back into the analysis to present and characterize the model output uncertainties meaningfully. See Srinivasan (2002) for more discussions on importance sampling.

8.1.2.2.4 Discrete Probability Distributions Sampling

Discrete Probability Distributions (DPD) sampling is a probabilistic sampling and propagation technique in which the input distributions are discretized, and the discrete representation of the output distribution is calculated. First, a distribution is determined for each model input and parameters (similar to the Monte Carlo Method), then these distributions are divided into discrete intervals. The number of discrete intervals can be different for each input distribution. The range of each interval's values is chosen subjectively based on the importance or significance of that interval. Finally, the probability that the input variable occurs in each interval is calculated. Thus, the distribution of each model input or parameter variable is discretized into m intervals; each interval has a corresponding probability and a corresponding central value. The model output variable is evaluated $m \times k$ times, where k is the number of independent model input and parameter variables (assuming that the pdf of each input variable is divided into m discrete intervals). Each evaluation's result has an associated probability equal to the product of the independent input variables' probabilities. Thus, a model output DPD can be constructed from these intervals. The DPD technique is a valid method but becomes quickly impractical for substantial problems with many variables. For more discussions on this topic, see Kaplan (1981).

8.1.2.2.5 Wilks Tolerance Limit

For very complex models (mostly complex codes) containing multiple submodels and many uncertain parameters (each represented by a probability or probability density), the Monte Carlo sampling methods discussed in the preceding sections become complex. In complex computer-based models, calculating output values often requires significant time and effort. Wilks's tolerance limit is used in these cases.

TABLE 8.2
(a) Minimum Sample Size (One-Sided). (b) Minimum Sample Size in Tolerance Limit (Two-Sided)

(a)

	β			
γ	0.50	0.90	0.95	0.99
0.50	3	17	34	163
0.80	5	29	59	299
0.90	7	38	77	388
0.95	8	46	93	473
0.99	11	64	130	663

(b)

	β		
γ	0.90	0.95	0.99
0.90	22	45	239
0.95	29	59	299
0.99	44	90	459

A *tolerance interval* is an interval of the model input and parameter variable that contains probability (or confidence) β that at least a fraction γ of the model input and parameter values are represented in that interval—the probability β and fraction γ are selected by the analyst. This area's pioneering work is attributed to Wilks (1941) and later to Wald (1943). Depending on the values of β and the fraction γ, a fixed number of Monte Carlo simulations is used (this number is a lookup value, usually less than 100). The number of iterations does not depend on the number of uncertain parameters in the model, making it very appealing and scalable for application to complex models such as computer codes.

There are two kinds of tolerance limits: *One-sided tolerance limits* and *two-sided tolerance limits*. In the one-sided, the tolerance limit would correspond to the upper limit of the model output with the probability β (say 95%) that at least the fraction γ of the true output values are represented. For example, according to Wilks, a Monte Carlo sample size of just 45 iterations generates sufficient realizations of the model input and model parameter values to assure, with a probability (confidence level) of $\beta = 0.95$ and a fraction $\gamma = 0.9$, that the range of the 45 model calculated outputs represent 90% of the true output values (i.e., 90% of the true values fall below the highest output value among the 45 output values). To obtain the *two-sided tolerance limit*, the fixed Wilks' sample size of 93 Monte Carlo integrations, corresponding to the β and γ values (e.g., the same 0.95 and 0.9, respectively), will result in a range of 93 outputs that represent 90% of the true model output with 95% confidence. For example, in the 93 samples taken from the model output (i.e., by using the standard Monte Carlo sampling), the 93 model outputs may be used to find the highest and lowest values corresponding to the two tolerance limits. Table 8.2a and b show the number of samples for one-sided and two-sided tolerance limits, respectively, for a few commonly used values. For comparing Wilks' tolerance limits with the classical Monte Carlo methods, see Lee et al. (2014).

8.2 ANALYSIS OF DEPENDENT FAILURES

Dependent failures are critical in reliability analysis and must be adequately treated to minimize significantly overestimating reliability. In general, dependent failures are defined as events in which the probability of each failure depends on the occurrence of other failures. According to the chain rule of probability (Equation 2.22), if a set of dependent events $\{E_1, \ldots, E_n\}$ exists, the probability of each failure in the set depends on the occurrence of other failures in the set.

The probabilities of dependent events on the right-hand side of Equation 2.22 are usually, but not always, greater than the corresponding independent probabilities. Determining the conditional probabilities in Equation 2.22 is generally difficult. However, parametric methods can consider the conditionality and generate the probabilities directly. These methods are discussed later in this section.

Generally, dependence among various events, for example, failure events of two items, is due to either the internal environment of these systems or the external environment (or events). The internal aspects can be divided into internal challenges, intersystem dependencies, and intercomponent dependencies. The external aspects are natural or human-made environmental events that make failures dependent. For example, the failure rates for items exposed to extreme heat, earthquakes, moisture, and flood will increase. The intersystem and intercomponent dependencies can be categorized into four broad categories: Functional, shared equipment, physical, and human-caused dependencies. These are described in Table 8.3.

TABLE 8.3
Categories and Examples of Dependencies

Dependent Event Type	Dependent Event Category	Dependent Event Subcategory	Example
Internal	1. Challenge	—	1. Internal transients or deviations from the normal operating envelope introduce a challenge to several items
	2. Intersystem (failure between two or more systems)	1. Functional 2. Shared equipment 3. Physical 4. Human	1. Power to several independent systems is from the same source 2. The same equipment, e.g., a valve, is shared between otherwise independent systems 3. The extreme environment, e.g., high temperature causes dependencies between independent systems 4. Operator error causes failure of two or more independent systems
	3. Intercomponent	1. Functional 2. Shared equipment 3. Physical 4. Human	1. A component in a system provides multiple functions 2. Two independent trains in a hydraulic system share the same common header 3. Same as system interdependency above 4. Design errors in redundant pump controls introduces a dependency in the system
External	—	—	Earthquake or fire fails several independent systems or components

The significant causes of dependence among a set of systems or components, as described in Table 8.3, can be explicitly identified and modeled, for example, by system reliability analysis models, such as fault trees or Bayesian networks. However, modeling the rest of the causes (e.g., those with less significance) in detail can result in combinatorial explosion of items to model. To simplify this, groups of causes can be collectively modeled using the concept of *common cause failures* (CCF). CCFs are the collection of dependent failures described in Table 8.3 (especially between components) where the dependence is implicit and is hard to explicitly model in the system or component reliability analysis. Typically, functional and shared equipment dependencies are explicitly modeled in the system analysis, but other dependencies are mostly implicit and considered collectively using CCF.

Many reliability studies have shown CCFs to contribute significantly to complex systems' overall unavailability or unreliability. There is no unique and universal definition for CCFs. However, Mosleh et al. (1988) define a CCF as "a subset of dependent events in which two or more component fault states exist at the same time, or in a short time interval, and are direct results of a shared cause."

To better understand CCFs, consider a system with three redundant components A, B, and C. The total failure probability of A can be expressed in terms of its independent failure A_I and dependent failures by defining three mutually exclusive failure states:

- C_{AB} is the failure of components A and B (and not C) from common causes.
- C_{AC} is the failure of components A and C (and not B) from common causes.
- C_{ABC} is the failure of components A, B, and C from common causes.

Component A fails if any of the above states exist. The equivalent Boolean representation of the total failure of component A, A_t, is $A_t = A_I + C_{AB} + C_{AC} + C_{ABC}$. Similar expressions can be developed for components B and C.

Now, suppose that the success criteria for the system is two-out-of-three for components A, B, and C. Accordingly, the failure of the system can be represented by these events (cut sets): $(A_I\ B_I)$, $(A_I\ C_I)$, $(B_I\ C_I)$, C_{AB}, C_{AC}, C_{BC}, and C_{ABC}. Thus, the Boolean representation of the system failure is

$$S = (A_I B_I) + (A_I C_I) + (B_I C_I) + C_{AB} + C_{AC} + C_{BC} + C_{ABC}. \tag{8.11}$$

If independence is assumed, the first three terms of the above Boolean expression are used, and the remaining terms are zero. Assuming that the CCF terms incorporate all of the necessary dependencies, and assuming the rare event approximation applies, the system failure probability Q_S is given by

$$\begin{aligned} Q_S \approx\ & Pr(A_I)Pr(B_I) + Pr(A_I)Pr(C_I) + Pr(B_I)Pr(C_I) \\ & + Pr(C_{AB}) + Pr(C_{AC}) + Pr(C_{BC}) + Pr(C_{ABC}). \end{aligned} \tag{8.12}$$

If components A, B, and C are identical (which is often the case since common causes among different components have a much lower probability), then

$$
\begin{aligned}
&Pr(A_I) = Pr(B_I) = Pr(C_I) = Q_1, \\
&Pr(C_{AB}) = Pr(C_{AC}) = Pr(C_{BC}) = Q_2, \\
&Pr(C_{ABC}) = Q_3.
\end{aligned}
\tag{8.13}
$$

Therefore,

$$
Q_s = 3(Q_1)^2 + 3Q_2 + Q_3. \tag{8.14}
$$

One can introduce the probability Q_k representing the probability of CCF among k specific components in a component group of size m, such that $1 \leq k \leq m$. In these models, Q_t is the total probability of failure accounting both for the common cause and independent failures. The remainder of this section discusses two types of CCF models in more detail and elaborates on the parameter estimation of the CCF models.

8.2.1 Single-Parameter Models

Single-parameter models use one parameter in addition to the total component failure probability to calculate the CCF probabilities. One of the most used single-parameter models defined by Fleming (1975) is the β-factor model. It is the first parametric model applied to CCF events in risk and reliability analysis. The sole parameter of the model, β, is defined as the fraction of the component failure rate attributed to the CCFs divided by the total failure rate due to all causes. That is,

$$
\beta = \frac{\lambda_c}{\lambda_c + \lambda_I} = \frac{\lambda_c}{\lambda_t}, \tag{8.15}
$$

where λ_c is the CCFs failure rate, λ_I is a failure rate due to independent failures, and the total failure rate is $\lambda_t = \lambda_c + \lambda_I$.

An essential assumption of this model is that whenever a CCF event occurs, all components of a redundant set of components fail simultaneously. In other words, a CCF can be viewed as a fatal shock striking a set of redundant components.

Based on the β-factor model, for a system of m redundant (parallel) components, the probabilities of basic events involving k specific components (Q_k), where $1 < k < m$, are equal to zero, except Q_1 and Q_m. These quantities are given as

$$
\begin{aligned}
&Q_1 = (1-\beta)Q_t \\
&Q_2 = Q_3 = \ldots = Q_{m-1} = 0 \\
&Q_m = \beta Q_t.
\end{aligned}
\tag{8.16}
$$

Generally, the total component failure rate estimate is generated from generic sources of failure data or specific datasets. The estimators of the corresponding β-factor do not explicitly depend on generic failure data but instead rely on specific assumptions about observed data interpretation. The point estimator of β is discussed in Section 8.2.3. In addition, some recommended values of β are given in Mosleh et al. (1988). Although this model can be used with a certain degree of accuracy for two-component redundancy, the results tend to be conservative for a higher level of redundancy. However, this model has been widely used in risk and reliability studies due to its simplicity.

Example 8.2

Consider the following system with two redundant trains. Suppose each train comprises a valve and a pump (each driven by a motor). The pump failure modes are "failure to start" (PS) and "failure to run following a successful start" (PR). The valve failure mode is "failure to open" (VO). Assume both pumps can fail simultaneously due to common causes. Assume the same for the valves. Assume you have data for each failure modes given as Q (the total probability of failure on demand), λ (the failure rate for failure to run), and t (the mission time) and that the rare event approximation applies. Develop an expression for the probability of system failure using a fault tree and the β-factor method.

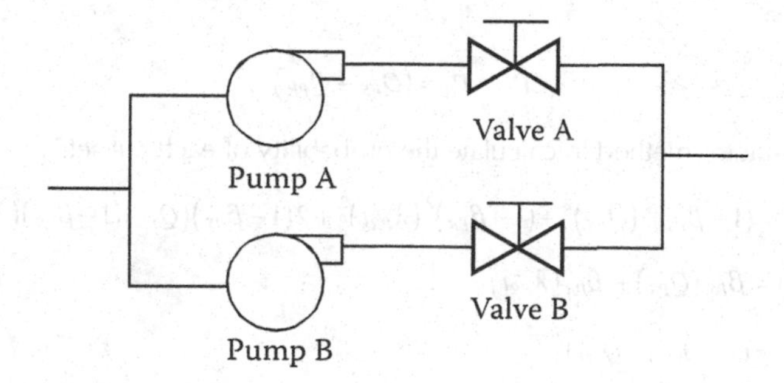

Solution:

Develop a system fault tree to include both independent and CCFs of the components, where

P_A is the independent failure of pump A
P_B is the independent failure of pump B
P_{AB} is the dependent failure of pumps A and B
V_A is the independent failure of valve A
V_B is the independent failure of valve B
V_{AB} is the dependent failure of valves A and B.

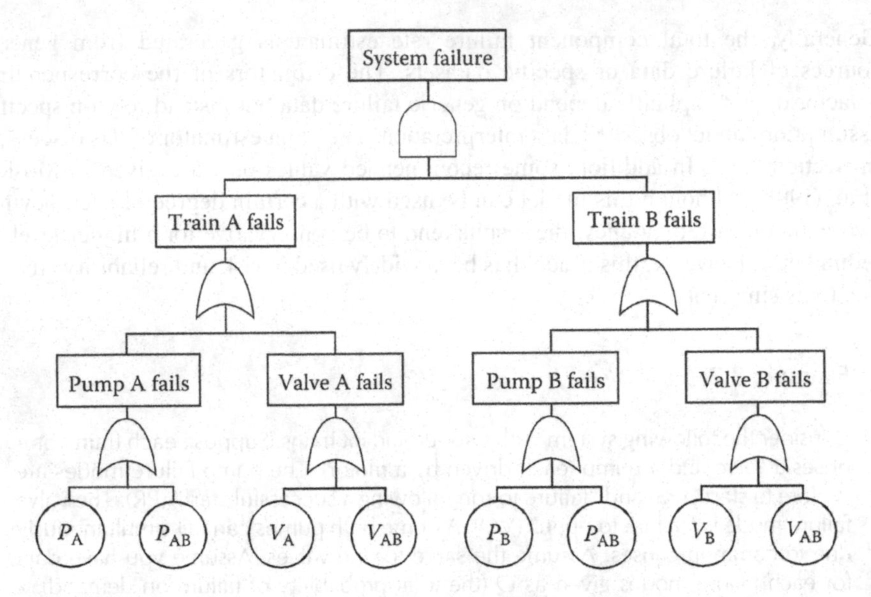

By solving the fault tree, these cut sets can be identified:

$$C_1 = (P_A, P_B),\ C_2 = (P_{AB}),\ C_3 = (V_A, V_B),\ C_4 = (V_{AB}),\ C_5 = (P_A, V_B),\ C_6 = (P_B, V_A),$$

where,

$$P_A = P_B = (Q_{PS} + Q_{PR}).$$

Use the β-factor method to calculate the probability of each cut set:

$$Pr(C_1) = (1-\beta_{PS})^2 (Q_{PS})^2 + (1-\beta_{PR})^2 (\lambda_{PR}t)^2 + 2(1-\beta_{PS})(Q_{PS})(1-\beta_{PR})(\lambda_{PR}t),$$

$$Pr(C_2) = \beta_{PS}(Q_{PS}) + \beta_{PR}(\lambda_{PR}t),$$

$$Pr(C_3) = (1-\beta_{VO})^2 (Q_{VO})^2,$$

$$Pr(C_4) = \beta_{VO}(Q_{VO}),$$

$$Pr(C_5) = Pr(C_6) = \left[(1-\beta_{PS})(Q_{PS}) + (1-\beta_{PR})(\lambda_{PR}t)\right]\left[(1-\beta_{VO})(Q_{VO})\right].$$

System failure probability is calculated using rare event approximation:

$$Q_{sys} = \sum_{i=1}^{6} Pr(C_i).$$

8.2.2 Multiple-Parameter Models

Multiple-parameter models are used to obtain a more accurate assessment of CCF probabilities in systems with a higher level of redundancy. These models have several parameters that are usually associated with different event characteristics. This category of models can be further divided into two subcategories, shock and non-shock models. The α-factor model and the multiple Greek letter (MGL) model are non-shock models, whereas the binomial failure rate (BFR) model is a shock model.

TABLE 8.4
Generic Values of the α-Factor Parameters

Number of Items (m)	α-Factor			
	α_1	α_2	α_3	α_4
2	0.95	0.050	—	—
3	0.95	0.040	0.01	—
4	0.95	0.035	0.01	0.005

We will only present the α-factor model in this section. The readers are referred to Mosleh et al. (1998), Mosleh and Siu (1987), and Hokstad and Rausand (2008) to learn more about the other CCF models.

The α-factor model discussed by Mosleh and Siu (1987) develops CCF failure probabilities from a set of failure ratios and the total component failure rate. The parameters of the model are the fractions of the total probability of failure in the system that involves the failure of k components due to a common cause, a_k.

The probability of a common cause basic event involving the failure of k components in a system of m components is calculated according to these equations.

$$Q_k = \frac{k}{\binom{m-1}{k-1}} \frac{\alpha_k}{\alpha_t} Q_t,\ k = 1,\ldots m, \tag{8.17}$$

$$\alpha_t = \sum_{k=1}^{m} k\alpha_k, \tag{8.18}$$

where Q_t is the total probability of failure for a component accounting both for the common cause and independent failures and Q_k is the probability of CCF among k specific components. Table 8.4 (Mosleh, 1991) provides generic values of α factors.

For example, the probabilities of the basic events of the three-component system described earlier will be

$$\begin{aligned} Q_1 &= (\alpha_1/\alpha_t)Q_t, \\ Q_2 &= (\alpha_2/\alpha_t)Q_t, \\ Q_3 &= (3\alpha_3/\alpha_t)Q_t, \end{aligned} \tag{8.19}$$

where $\alpha_t = \alpha_1 + 2\alpha_2 + 3\alpha_3$.

Therefore, the system failure probability for the two-out-of-three system discussed earlier can now be written as

$$Q_s = 3\left(\frac{\alpha_1}{\alpha_t}Q_t\right)^2 + 3\left(\frac{\alpha_2}{\alpha_t}Q_t\right) + 3\left(\frac{\alpha_3}{\alpha_t}Q_t\right). \tag{8.20}$$

Using the generic values for the 2-out-of-3 success $\alpha_t = 0.95 + 0.08 + 0.03 = 1.06$. If we assume $Q_t = 8 \times 10^{-3}$ for each component, then

$$Q_{sys} = 3\left[\frac{0.95}{1.06}\left(8\times10^{-3}\right)\right]^2 + 3\left[\frac{0.04}{1.06}\left(8\times10^{-3}\right)\right] + 3\left[\frac{0.01}{1.06}\left(8\times10^{-3}\right)\right] = 1.28\times10^{-3}. \tag{8.21}$$

8.2.3 Data Analysis for CCFs

Despite the difference among the models described in Sections 8.2.1 and 8.2.2, they all have similar data requirements in parameter estimation. Therefore, one should not expect major differences between the numerical results provided by these models. Most of the difference in the results may be attributed to the statistical aspects of the parameter estimation, which has to do with the assumptions made in developing a parameter estimator and the dependencies assumed in CCF probability quantification.

The most important steps in quantifying CCFs are collecting information from the raw data and selecting a model that can use most of this information. Statistical estimation procedures discussed in Chapters 4, 5, and 7 can be applied to estimate the CCF model parameters. If separate models rely on the same type of information in estimating the CCF probabilities and similar assumptions regarding the mechanism of CCFs are used, comparable numerical results can be expected. Table 8.5 summarizes simple point estimators for parameters of the CCF models discussed above. In this table, n_k is the total number of observed failure events involving failure of k similar components due to a common cause; m is the total number of redundant items considered; and d is the total number of system demands. If an item is normally operating (not on standby), then d can be replaced by the total test (operation)

TABLE 8.5
Point Estimates of α-Factor and β-Factor Models

Model	Point Estimator
β-factor	$\hat{Q}_t = \frac{1}{md}\sum_{k=1}^{m} kn_k$
	$\hat{\beta} = \sum_{k=2}^{m} kn_k \Big/ \sum_{k=1}^{m} kn_k$
α-factor	$\hat{Q}_t = \frac{1}{md}\sum_{k=1}^{m} kn_k$
	$\hat{\alpha}_k = n_k \Big/ \sum_{k=1}^{m} kn_k,\ k = 1,\ldots,m$

time *t (or TTT) of the item.* The estimators in Table 8.5 assume that in every system demand, all components and possible combinations of components are challenged. Therefore, the estimators apply to systems whose tests are non-staggered.

Example 8.3

For the system described in Example 8.2, estimate the β parameters, λ and Q, for the valves and pumps based on the following failure data:

	Event Statistic		
Failure Mode	n_1	n_2	**Total (hour or day)**
Pump fails to start (PS)	10	1	500 (demands)
Pump fails to run (PR)	50	2	10,000 (hours)
Valve fails to open (VO)	15	1	10,000 (demands)

In the data table, n_1 is the number of observed independent failures, and n_2 is the number of observed events involving two-unit CCF. Calculate the system unreliability for a 10 hour mission.

Solution:

From Table 8.5,

$$\hat{\beta} = \frac{2n_2}{n_1 + 2n_2}.$$

Apply this formula to β_{PR}, β_{PS}, and β_{VO} using appropriate values for n_1 and n_2:

$$n_{PS} = n_1 + 2n_2 = 12,$$

$$n_{PR} = n_1 + 2n_2 = 54,$$

$$n_{VO} = n_1 + 2n_2 = 17.$$

Use Equations 5.24 and 5.59 for estimating λ and Q, respectively,

$$Q_{PS} = \frac{12}{500} = 2.4 \times 10^{-2}\text{/demand}; \ \beta_{PS} = \frac{2}{12} = 0.17,$$

$$\lambda_{PR} = \frac{54}{10{,}000} = 5.4 \times 10^{-3}\text{/hour}; \ \beta_{PR} = \frac{4}{54} = 0.07,$$

$$Q_{VO} = \frac{17}{10{,}000} = 1.7 \times 10^{-3}\text{/demand}; \ \beta_{VO} = \frac{2}{17} = 0.12.$$

Therefore, using the cut set probability equations developed in Example 8.2, the estimates of the failure probabilities at 10 hours of operation for each cut set are

$$Pr(C_1)=(1-0.17)^2(2.4\times10^{-2})^2+(1-0.07)^2(5.4\times10^{-3}\times10)^2$$
$$+2(1-0.17)(2.4\times10^{-2})(1-0.07)(5.4\times10^{-3}\times10)=4.9\times10^{-3},$$

$$Pr(C_2)=0.17(2.4\times10^{-2})+0.07(5.4\times10^{-3})(10)=7.9\times10^{-3},$$

$$Pr(C_3)=(1-0.12)^2(1.7\times10^{-3})^2=2.2\times10^{-6},$$

$$Pr(C_4)=0.12(1.7\times10^{-3})=2.0\times10^{-4},$$

$$Pr(C_5)=\left[(1-0.17)(2.4\times10^{-2})+(1-0.07)(5.4\times10^{-3}\times10)\right]\left[(1-0.12)(1.7\times10^{-3})\right]$$
$$=1.1\times10^{-4},$$

$$Pr(C_6)=Pr(C_5)=1.1\times10^{-4}.$$

Thus, the system failure probability is

$$Q_{sys}\cong\sum_{i=1}^{6}Pr(C_i)=1.3\times10^{-2}.$$

8.3 IMPORTANCE MEASURES

During the reliability analysis or risk assessment of a system, some components may become more critical than others in terms of system functionality, risk, and reliability because of the system's structural arrangement. For example, a set of components arranged in series is much more critical to system reliability than the same components in the parallel arrangement within the system. There are several formal methods to gauge the importance of each component or event within a system from the perspective of the system's reliability and risk. These methods are called *importance measures*. In this section, we will describe five popular importance measures: Birnbaum, criticality, Fussell-Vesely, risk reduction worth (RRW), and risk achievement worth (RAW). Usually, importance measures are used in the failure space (e.g., for risk, unavailability and unreliability evaluations); however, this book also discusses their application in the success space (e.g., for reliability and design evaluations).

8.3.1 Birnbaum Importance

Introduced by Birnbaum (1969), this measure of component importance, $I_i^B(t)$, for component i in success (reliability) space is defined as the rate of change in the reliability of the system with respect to a change in the reliability of component i.

$$I_i^B(t)=\frac{\partial R_S\left[R(t)\right]}{\partial R_i(t)},\qquad(8.22)$$

where $R_S\left[R(t)\right] = R_S(t)$ is the reliability of the system as a function of the reliability of its components $R(t)$ which can be further expanded to a function of time, t, alone, i.e., $R_S(t)$. $R_i(t)$ is the reliability of the specific component i at time t.

If, for the component i, $I_i^B(t)$ is large, it means that a slight change in the reliability of component i, $R_i(t)$, will significantly change the system reliability $R_S(t)$.

If system components are assumed to be independent, the Birnbaum measure of importance can be simplified to

$$I_i^B(t) = R_S\left[R(t)|R_i(t)=1\right] - R_S\left[R(t)|R_i(t)=0\right], \tag{8.23}$$

where $R_S\left[R(t)|R_i(t)=1\right]$ and $R_S\left[R(t)|R_i(t)=0\right]$ are the values of the reliability function of the system given the reliability of component i is 1 (working) and 0 (failed), respectively.

Equations 8.22 and 8.23 are often used in conjunction with the system's unreliability, unavailability, or risk function, $F_S\left[Q(t)\right]$. The Birnbaum's importance measure of component i would be the rate of change in $F_S\left[Q(t)\right]$ with respect to the rate of change of individual component unreliability or unavailability $Q_i(t)$. In this case, the Birnbaum importance in the failure space is written in form of Equation 8.24 where failure of component i is $Q_i(t)=1$, and success (non-failure) of component i is $Q_i(t)=0$.

$$I_i^B(t) = \frac{\partial F_S\left[Q(t)\right]}{\partial Q_i(t)} = F_S\left[Q(t)|Q_i(t)=1\right] - F_S\left[Q(t)|Q_i(t)=0\right]. \tag{8.24}$$

Example 8.4

Consider the system shown below. Determine the Birnbaum importance of each component at $t=720$ hours with respect to the system's reliability (success space). Assume an exponential time to failure for each component.

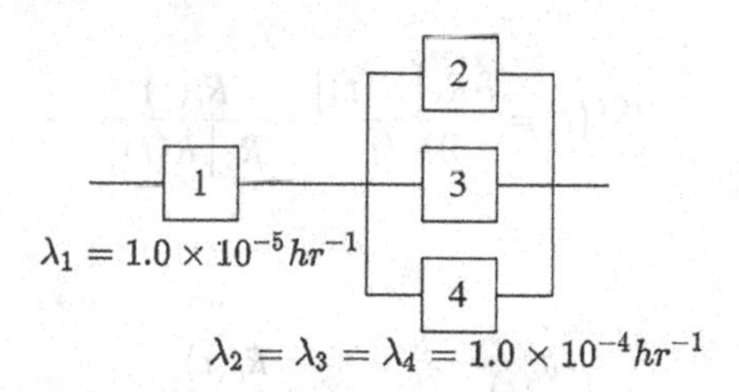

Solution:

$$R_1(t=720) = 0.993,\ R_2(t=720) = R_3(t=720) = R_4(t=720) = 0.9305 /$$

The reliability function of the system is

$$R_S\left[R(t)\right] = R_1(t)\left\{1-\left[1-R_2(t)\right]\left[1-R_3(t)\right]\left[1-R_4(t)\right]\right\} = 0.9925.$$

Using Equation 8.23,

$$I_i^B(t) = R_S\left[R(t)|R_i(t)=1\right] - R_S\left[R(t)|R_i(t)=0\right]$$

$$I_1^B(t) = 1\cdot\{1-[1-R_2(t)]\}[1-R_3(t)][1-R_4(t)] - 0.$$

Therefore, $I_1^B(t=720) \approx 0.99967$.

For component 2,

$$I_2^B(t) = R_1(t)\left[(1-R_3(t))(1-R_4(t))\right]$$

$$I_2^B(t=720) \approx 0.00479$$

Similarly,

$$I_3^B(t=720) = I_4^B(t=720) \approx 0.00479.$$

It can be concluded that the rate of improvement in component 1 has far more importance (impact) on system reliability than components 2, 3, and 4. For example, if the reliability of the parallel units increases by an order of magnitude, clearly, the importance of components 2, 3, and 4 decreases (e.g., for $\lambda_2 = \lambda_3 = \lambda_4 = 10^{-4}\,hr^{-1}$, $I_2^B = I_3^B = I_4^B \approx 0$, and $I_1^B = 1$). Similarly, the importance measures change if identical units are in parallel with component 1.

8.3.2 Criticality Importance

Birnbaum's importance for component i is independent of the reliability of component i itself. Therefore, I_i^B is not a function of $R_i(t)$. It would be more difficult and costly to improve the more reliable components further than to improve the less reliable ones. Therefore, the criticality importance of component i to further remedy this shortcoming is defined as

$$I_i^{CR}(t) = \frac{\partial R_S\left[R(t)\right]}{\partial R_i(t)} \times \frac{R_i(t)}{R_S\left[R(t)\right]}, \tag{8.25}$$

or

$$I_i^{CR}(t) = I_i^B(t) \times \frac{R_i(t)}{R_S\left[R(t)\right]}. \tag{8.26}$$

The criticality importance improves the Birnbaum importance to account for the reliability of the individual components relative to the reliability of the whole system. Therefore, if the Birnbaum importance of a component is high, but its reliability is low with respect to the system's reliability, then the criticality importance assigns a low importance value to this component. Similarly, Equation 8.26 can be represented in the failure space using the unreliability or unavailability function:

$$I_i^{CR}(t) = I_i^B(t) \times \frac{Q_i(t)}{F_S[Q(t)]}. \tag{8.27}$$

Using the criticality importance for component 1 in Example 8.4, $I_1^{CR}(t) = 0.99967 \times \frac{0.993}{0.9925} \approx 1$. Since component 1 is more reliable, its contribution to the system's reliability (i.e., its criticality importance) increases, whereas components 2, 3, and 4 will have a less important contribution to the overall system reliability.

A subset of the criticality importance measure in the failure space is the *inspection importance* measure (I_i^W). This measure is defined as the product of the Birnbaum importance and the component's failure probability (unreliability or unavailability). Accordingly,

$$I_i^W(t) = I_i^B(t) \times Q_i(t). \tag{8.28}$$

This measure prioritizes operability test activities to ensure high component readiness and performance.

8.3.3 Fussell-Vesely Importance

When component i contributes to system reliability but is not necessarily critical, the Fussell-Vesely importance measure can be used. This measure, introduced by W.E. Vesely and later applied by Fussell (1975), is in the form of

$$I_i^{FV}(t) = \frac{R_i[R(t)]}{R_S[R(t)]}, \tag{8.29}$$

where $R_i[R(t)]$ is the contribution of component i to the reliability of the system. $R_i[R(t)]$ is obtained by only retaining terms involving $R_i(t)$ in the system reliability expression $R_S(t)$. Similarly, using unreliability or unavailability functions Fussell-Vesely importance in the failure space is,

$$I_i^{FV}(t) = \frac{F_i[Q(t)]}{F_S[Q(t)]}, \tag{8.30}$$

where $F_i[Q(t)]$ denotes the contributions from component i to system failure (unreliability) or system risk and $F_S[Q(t)]$ is the system unreliability or unavailability.

The Fussell-Vesely importance measure has been frequently applied to system cut sets (e.g., obtained from fault trees or event tree scenarios) to determine the importance of individual cut sets to the failure probability of the whole system. For example, consider the importance of the cut set k, I_k, obtained from the cut sets. In that case, Equation 8.30 changes to

$$I_k^{FV}(t) = \frac{Q_k(t)}{Q_S(t)}, \tag{8.31}$$

where $Q_k(t)$ is the time-dependent probability or frequency that mutually exclusive (disjoint) cut set or minimal cut set that contains component k occurs, and $Q_S(t)$ is the total time-dependent probability or frequency that the system fails, or risk outcomes occur (due to all cut sets).

Generally, the minimal cut sets with the largest values of I_k are the most important ones. Equation 8.31 is equally applicable to mutually exclusive cut sets. Consequently, system improvements should initially be directed toward the minimal cut sets with the most significant importance values.

If the probability of all minimal cut sets or mutually exclusive cut sets is known, then the following approximate expression can be used to find the importance of individual components:

$$I_i^{FV}(t) = \frac{\sum_{j=1}^{m} \left[Q_j(t) | i = \text{true}\right]}{Q_S(t)}, \tag{8.32}$$

where $\left[Q_j(t) | i = \text{true}\right]$ shows the probability of the jth cut set that contains failure event or component i failure and m is the number of minimal cut sets containing component i.

Expression 8.32 is an approximation; the situation of two minimal cut sets containing component i failing simultaneously is neglected since its probability in practice is remote. This is not an issue if mutually exclusive cut sets are used to calculate I_i^{FV}.

Example 8.5

Consider the pumping system below. Determine the Birnbaum, criticality, and Fussell-Vesely importance measures of the valve (V), pump 1 ($P1$), and pump 2 ($P2$) using both reliability and unreliability versions of the importance measures.

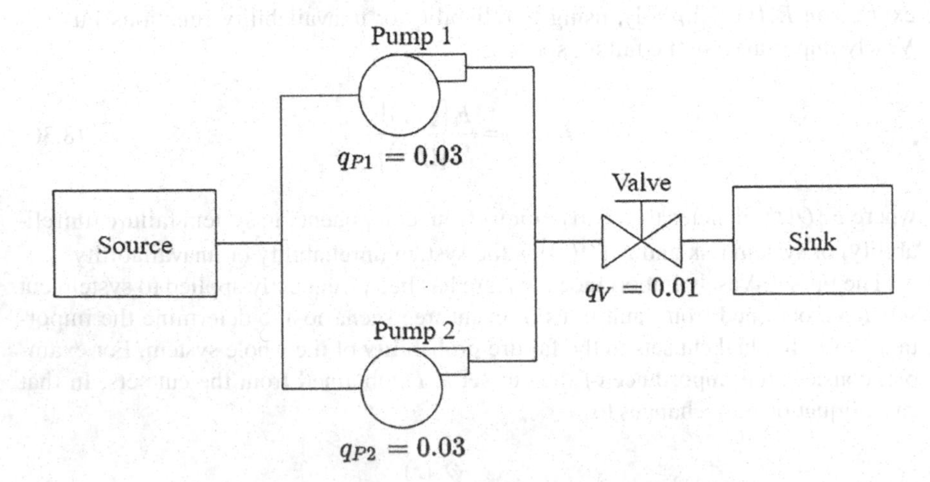

Solution:

Because of the individual component reliabilities, $R_{P1} = R_{P2} = 0.97$, $R_V = 0.99$, the reliability function is

$$R_S\left[R(t)\right] = R_V\left[R_{P1} + R_{P2} - R_{P1}R_{P2}\right] = 0.989.$$

Using the rare event approximation, the unreliability function is

$$F_S\left[Q(t)\right] = Q_{P1}Q_{P2} + Q_V = 0.011.$$

1. Birnbaum's importance:

$$I_V^B = R_{P1} + R_{P2} - R_{P1}R_{P2} \approx 0.999,$$

$$I_{P1}^B = R_V - R_V R_{P2} \approx 0.03,$$

$$I_{P2}^B = R_V - R_V R_{P1} \approx 0.03.$$

Using the unreliability function,

$$I_V^B \approx 0.999,$$

$$I_{P1}^B = Q_{P2} \approx 0.03,$$

$$I_{P2}^B = Q_{P1} \approx 0.03.$$

2. Criticality importance:

$$I_V^{CR}(t) = 0.999 \times \frac{0.99}{0.989} \approx 1,$$

$$I_{P1}^{CR}(t) = I_{P2}^{CR}(t) = 0.03 \times \frac{0.97}{0.989} \approx 0.029.$$

The same criticality importance values are expected for the unreliability function.

3. Fussell-Vesely importance:
$R_i\left[R(t)\right]$ is obtained by retaining terms in $R_S(t)$ involving $R_i(t)$.

$$R_V\left[R(t)\right] = R_S\left[R(t)\right] \approx 0.989,$$

$$R_{P1}\left[R(t)\right] = R_V \times R_{P1} - R_V \times R_{P1} \times R_{P2} \approx 0.029,$$

$$R_{P2}\left[R(t)\right] = R_V \times R_{P2} - R_V \times R_{P1} \times R_{P2} \approx 0.029,$$

$$I_V^{FV} = \frac{0.989}{0.989} = 1,$$

$$I_{P1}^{FV} = I_{P2}^{FV} = \frac{0.029}{0.989} \approx 0.029.$$

Using the unreliability function,

$$F_V[Q(t)] = Q_V = 0.01,$$

$$F_{P1}[Q(t)] = Q_{P1} \times Q_{P2} = 0.0009,$$

$$F_{P2}[Q(t) = Q_{P2} \times Q_{P1} = 0.0009.$$

Then,

$$I_V^{FV} = \frac{0.01}{0.011} = 0.91,$$

$$I_{P1}^{FV} = I_{P2}^{FV} = \frac{0.0009}{0.011} \approx 0.0820.$$

8.3.4 Risk Reduction Worth Importance

The RRW importance is defined in the failure space only. It is a measure of the change in unreliability (unavailability or risk) when an input variable, such as the unavailability of a component, is set to zero—that is, by assuming that a component is perfect (or having a failure probability of zero) and thus eliminating any postulated failure. This importance measure shows how much better the system can become as its components are improved.

The RRW importance can be expressed either in the form of a ratio or a difference. Accordingly, RRW importance in the ratio form is

$$I_i^{RRW}(t) = \frac{F_S[Q(t)]}{F_S[Q(t)|Q_i(t)=0]}, \tag{8.33}$$

and in difference form is

$$I_i^{RRW}(t) = F_S[Q(t)] - F_S[Q(t)|Q_i(t)=0], \tag{8.34}$$

where $F_S[Q(t)\,|Q_i(t)=0]$ is the system unreliability (unavailability or risk) when the unreliability (or unavailability) of component i is set to zero. Note that in the ratio form I_i^{RRW} is larger than one. So, it shows the factors by which improving reliability (i.e., reducing unreliability or unavailability) will improve system reliability (or reduce system unreliability, unavailability, or risk). Conversely, in the difference form, I_i^{RRW} would be a number between zero and one. The ratio form is the preferred way of measuring I_i^{RRW} in practice because the significance of important components is more apparent.

In practice, this measure is used in design and operations to identify elements of the system (such as a component failure mode or frequency of event) that are the best candidates to lower or eliminate for reducing system unreliability (risk or unavailability).

8.3.5 Risk Achievement Worth Importance

The RAW importance is the counter of the RRW measure. In this measure, the input variable (e.g., component unavailability) is set to one, and the effect of this change on system unreliability (unavailability or risk) is measured. Similar to RRW, the calculation may be done as a ratio or a difference form. By setting component failure probability to one, RAW measures the increase in system failure probability assuming the worst case of the component performance. As a ratio, the RAW measure is

$$I_i^{RAW}(t) = \frac{F_S\left[Q(t)|Q_i(t)=1\right]}{F_S\left[Q(t)\right]}, \tag{8.35}$$

and as a difference,

$$I_i^{RAW}(t) = F_S\left[Q(t)|Q_i(t)=1\right] - F_S\left[Q(t)\right], \tag{8.36}$$

where $F_S\left[Q(t)|\ Q_i(t)=1\right]$ is the system unreliability (unavailability or risk) when the unreliability (or unavailability) of component i is set to one. Similar to RRW, the preferred method for using RAW is the ratio form in Equation 8.35.

The RAW measure is helpful for identifying elements of the system that are the most crucial for making the system unreliable (unavailable or increasing the risk). Therefore, components with high I^{RAW} are those that will have the most impact should their failure probability unexpectedly rise. Consequently, it is best used to identify components that are good candidates for scheduling maintenance, performing surveillance, or applying prognosis and health management.

Example 8.6

Repeat Example 8.5 and calculate I^{RRW} and I^{RAW} for all components. Compare the results with I^B, I^{CR}, and I^{FV}.

Solution:

From Example 8.5, the unreliability function is $F_S\left[Q(t)\right] = Q_{P1}Q_{P2} + Q_V = 0.011$.
For RRW,

$$F_S\left[Q|Q_V=0\right] = Q_{P1} \times Q_{P2} = 0.03 \times 0.03 = 0.0009.$$

Therefore, for the ratio measure,

$$I_V^{RRW} = \frac{0.011}{0.0009} = 12.2.$$

For the difference measure,

$$I_V^{RRW} = 0.011 - 0.0009 = 0.01.$$

Similarly, for pumps as a ratio,

$$I_{P1}^{RRW} = I_{P2}^{RRW} = \frac{0.011}{0.01} = 1.1.$$

As difference,

$$I_{P1}^{RRW} = I_{P2}^{RRW} = 0.011 - 0.01 = 0.001.$$

In the ratio and difference methods, the larger numbers indicate increasing importance. This is only an index for identifying components whose assured performance will most highly affect system operation.

Similarly, for RAW, the ratio method yields

$$I_V^{RAW} = \frac{1}{0.011} = 90.91.$$

For the difference method,

$$I_V^{RAW} \approx 0.989.$$

For the pumps, using the ratio method,

$$I_{P1}^{RAW} = I_{P2}^{RAW} = \frac{1 \times 0.03 + 0.01}{0.011} = 3.64.$$

For the difference method,

$$I_{P1}^{RAW} = I_{P2}^{RAW} = (1 \times 0.03 + 0.01) - 0.011 = 0.029.$$

The I_i^{RAW} shows the importance of component *i* with respect to system unreliability when component *i* fails. Clearly, by comparing the results to I^B, I^{CR}, and I^{FV} with I^{RAW} and I^{RRW}, the relative importance value measured by I_i^{RAW} is consistent. This is expected since all other measures are related to the degradation of the component. I^{RAW} is related to the worth of improvement in component reliability. For a more in-depth discussion of importance measures used in reliability engineering and risk assessment, see Modarres, (2006).

Several other measures of importance have been introduced, as well as computer program importance calculations. For more information, the readers are referred to NUREG/CR-4550, Bertucio and Julius (1990), and Zio and Podofillini (2006). An extension of the importance measures is the uncertainty importance. Uncertainty importance is used to identify contributors to the uncertainty associated with system reliability, unreliability, or risk. Such uncertainties are attributed to the inputs of a model or a system, as discussed in Section 8.1. For more information on this subject, see Modarres (2006), Bier (1983), and NUREG-1150 (1990).

8.4 HUMAN RELIABILITY ANALYSIS

Humans play many roles in engineering systems and thus it is essential to appropriately model human reliability as part of reliability and risk analysis. *Human reliability analysis* (HRA) is the subset of reliability engineering that provides the methodologies to represent and quantify human-machine failures alongside hardware and software failures. It is widely stated that human errors and failures are a root cause of between 60% and 80% of engineering failures. If human elements are omitted from a system reliability analysis, the results are incomplete and inaccurate. The human or crew—or more precisely the human-machine team (Groth et al., 2019)—is another item to be modeled as part of system reliability analysis.

As mentioned in Chapter 1, HRA must be an embedded part of the reliability engineering process, not a separate process to be conducted independently and separately from hardware. Humans play a role in all engineered systems, and in most complex systems, human-machine teams are critical to accomplishing the mission or achieving a multitude of functions. Most HRA methods are focused on crews because complex engineering systems typically involve many people working together. However, most HRA concepts could be applied to a single human if analyzing a system involving only one human for a mission or function, such as driving a truck.

HRA is typically more difficult to model than hardware reliability. Literature shows there is no one best HRA method for many reasons. Humans play a diverse role in systems, fulfilling many functions at multiple levels of abstraction, and adapting to situations in a way that hardware cannot. Additionally, human performance can be affected by a wide range of influencing factors that are difficult to control, and human actions rarely have discrete, clearly defined binary success and failure states. Finally, it is difficult to obtain performance data on human behavior in extreme situations. Therefore, it is often necessary to use HRA models that inherently consider the context of the activity.

There are many facets of human reliability. To understand human reliability, one needs to understand both the engineering system and human behavior. HRA includes several core activities: Defining the critical human actions and corresponding failures, modeling the factors that influence the outcomes, and assessing the probability of those factors and outcomes. We will introduce each of these in this section. Then, we will shift to a discussion of specific methods. Numerous HRA methods have been developed for a range of applications. We will introduce a few of the most important methods, starting with the first HRA method, Technique for Human Error Rate Prediction (THERP). We will then discuss one of the most widely used methods, SPAR-H, and then we will introduce the IDHEAS model, developed to blend the simplicity of older methods like SPAR-H with a deeper cognitive basis. Finally, we will introduce more advanced HRA methods that are both systems-oriented and cognitively based: Information, Decision, Action in Crew Context (IDAC) and Phoenix. HRA is covered in more depth in many references (e.g., Kirwan, 1994).

8.4.1 The HRA Process

At its core, HRA involves these four major classes of activities:

1. *Identify:* Define human failure events to include in risk and reliability modeling.
2. *Represent:* Model the factors that contribute to human failure events.

3. *Quantify:* Assign human error probabilities (HEPs) to these events.
4. *Documentation:* Document the analysis to ensure that is it understandable, reproducible, and traceable.

8.4.2 Identifying Human Failure Events

The first step of HRA is to define the relevant human actions and select those which are significant to system reliability. The basic unit of analysis is an *event*, which is modeled as either a basic event in a fault tree or a pivotal event in an event tree or event sequence diagram. For HRA, the events of interest are often called *human failure events* (HFE). Sometimes the term human error is used, but this term is avoided in modern HRA since it oversimplifies a complex process involving both humans and machines. *The HFE is a failure by the crew (or human-machine team) to meet a system requirement of a complex engineering system.*

The core activity of the HFE identification process is to determine what human actions (or inactions) could contribute to the loss of a critical function in each scenario. The resources to use in this process include:

- A diverse team, including participants with HRA experience, risk assessment and reliability analysis experience, human factors/psychology backgrounds, workers, trainers, and personnel involved in system design, operations, and maintenance
- System procedures and training manuals
- Existing risk and reliability assessment models and results
- Crew interviews and work observations
- Formal task analyses
- Taxonomies of error types (e.g., unsafe acts).

In the nuclear industry, HFEs are grouped into three types of categories:

- *Preinitiator events*: The term used for potential errors in maintenance that occur before an event. This includes errors like miscalibration and failure to restore equipment correctly.
- *Initiators (or initiating events)*: The term used when a human action initiates a challenge to the physical system which causes it to deviate from its normal operating envelope.
- *Postinitiator events*: The term used for errors in diagnosing and manipulating equipment in response to various off-normal situations. This type of event is usually for an action in the control room. An example would be a failure to respond to an alarm in the control room or failure to initiate a manual action in response to a transient.

In HRA it is often necessary to further subdivide HFEs into *tasks* or fundamental actions taken as part of the event. While the terminology around task decomposition for HRA can vary widely, recent efforts by Paglioni and Groth (2022) have created the following definitions.

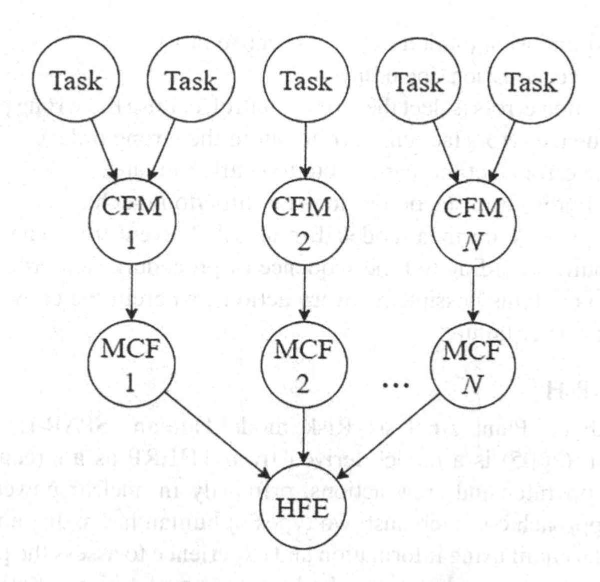

FIGURE 8.6 Elements involved in HFE decomposition. (Adapted from Paglioni and Groth, 2022.)

The top level of abstraction is the *objective*, which refers to the ultimate goal of crew actions with respect to the system need in a given scenario. It is clear that the failure of an objective would be an HFE. The next level of abstraction below an objective is a function. The term *function* refers to the high-level purpose(s) that the crew must perform to achieve an objective. For more specificity about functions, the term *Major Crew Function (MCF)* is used to distinguish human-machine team functions from purely hardware functions. Each MCF is associated with several failure modes, or *Crew Failure Modes (CFMs)*, that define *how* the function could be failed. Finally, at the lowest level of decomposition are *tasks* or *crew activity primitives*, which are the most fundamental actions taken to achieve the MCF. The relationship between the elements involved in breaking down an HFE is shown in Figure 8.6.

In HRA, the variable of interest is most often the occurrence (or nonoccurrence) of the HFE (i.e., the failure or non-failure of the MCF or series of MCFs that comprise an objective). It is useful to understand the variety of functions in which human-machine teams may be engaged, and the ways they can fail. Each HRA method offers guidance on how to identify functions and key activities and defines the taxonomies of HFEs or CFMs.

8.4.2.1 THERP

The oldest HRA technique is the THERP method, developed by Swain and Guttman (1983). THERP requires analysts to develop a detailed task analysis. The task analysis delineates the necessary steps and required human performance to achieve the system objective. Note that these largely assume the humans are following detailed procedures to conduct the actions. Then, the analyst determines the errors that could occur by identifying human error categories with each task. The following human error categories are defined by THERP.

a. Errors of omission (omit a step or the entire task)
b. Errors of commission, including:
    a. Selection errors (select the wrong control, choose the wrong procedures),
    b. Sequence errors (actions carried out in the wrong order),
    c. Time errors (actions carried out too early/too late),
    d. Qualitative error (action is done too little/too much).

Task sequences are then modeled in an HRA event tree. The HRA event tree is built according to time sequence or procedure step order. The event tree also contains possible recovery actions, wherein the crew can recover from an earlier failure.

8.4.2.2 SPAR-H

The Standardized Plant Analysis Risk model-Human (SPAR-H) method from Gertman et al. (2005) is a model derived from THERP as a streamlined method for assessing operator and crew actions, primarily in nuclear power plant control rooms. The approach considers just two types of human tasks: diagnosis and action. Diagnosis tasks entail using information and experience to assess the plant condition, or to plan, prioritize, and determine the best course of actions. Action tasks entail physical manipulation of the plant. SPAR-H decomposes the HFE into contributions from diagnosis failures and action failures.

8.4.2.3 IDAC and Phoenix

Over the past several decades, HRA researchers have focused on developing more rigorous methods for identifying the major crew functions and articulating how they can fail. Newer methods are being developed to expand the technical basis of HRA, integrating elements from cognitive science and systems thinking. One model with considerable technical depth is the IDAC model (Chang and Mosleh, 2007). The IDAC model leverages an expanded cognitive basis for evaluating human-machine team performance, whereas earlier HRA methods such as THERP and SPAR-H focus on behaviorist models of human behavior.

The IDAC model considers human/crew activities to be based on three major types of responses (Figure 8.7): Information processing (I), Diagnosis and decision-making (D), and Action taking (A), which take place in the context of coordination and communication among the crew of human operators. IDAC is focused on the context of control room activities but has the potential to be used more broadly because it is rooted in a deeper modeling basis that pulls from both engineering and cognitive psychology. IDAC is one of the most flexible HRA methods available, and it can be applied to a wide range of engineering systems.

- *Information processing (I)*: I-phase functions involve the detection and processing (i.e., understanding) of input signals from the human-system interfaces (HSIs) in a control room. HSIs are those systems with which the human crew interacts directly—in a "traditional" control room setting, these are mainly analog or digital display boards and input devices. I-phase functions require the crew to find, observe, and understand the relevant information from the display boards.

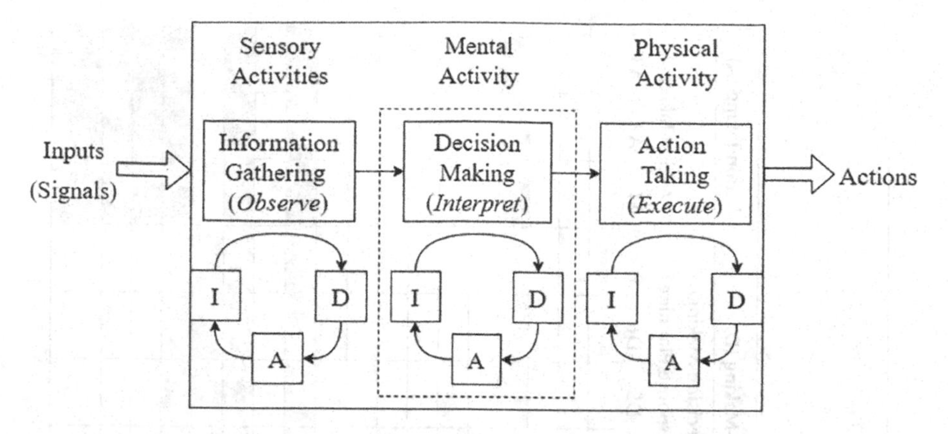

FIGURE 8.7 IDAC model Information-Decision-Action structure. (Adapted from Chang and Mosleh, 2007, and Ekanem et al., 2016.)

- *Diagnosis and decision-making (D)*: D-phase functions take the information gathered in the I-phase and require the crew to make sense of the information. This often involves the comparison of the I-phase data against a mental model (i.e., impressions, inferences, and/or assumptions about the state of the system) and/or procedure(s) to diagnose and/or make a decision regarding their response. Diagnosis entails evaluating the salience and importance of the information to create an updated mental model of the scenario. Decision-making evaluates this mental model against known scenarios and available procedures to determine a path forward for response.
- *Action (A)*: A-phase functions involve acting upon the information gathered in the I-phase and the corresponding diagnosis and/or decision made in the D-phase. Actions include communication and coordination between the crew and the execution of manipulations on the input-HSIs. Manipulations of the HSIs can occur as "regulating" or "maintaining" actions, wherein system parameters are kept within tolerances and processes are controlled to meet regulatory and safety requirements. Alternatively, manipulations can begin, end, or alter system safety processes.

Table 8.6 shows how the IDAC model is used to delineate the possible human-machine team activities into the appropriate I, D, or A phase.

Because of its general nature, IDAC is applicable to a wide range of engineering systems and has influenced the development of other methods. IDAC has been adapted for Phoenix (Ekanem et al., 2016) and within the NRC's IDHEAS method.

The Phoenix method builds beyond IDAC and associates each IDAC activity with relevant CFMs through which the human-machine team can fail to perform the function. Table 8.7 provides the CFMs for each phase of human response. Phoenix also provides detailed guidance for generating Crew Response Trees (CRTs), which are event trees that document the human-team response to the scenario, and for pruning the CRTs to avoid combinatorial explosion.

TABLE 8.6
IDAC Model of Human-Machine Team Activities (Ekanem et al., 2016)

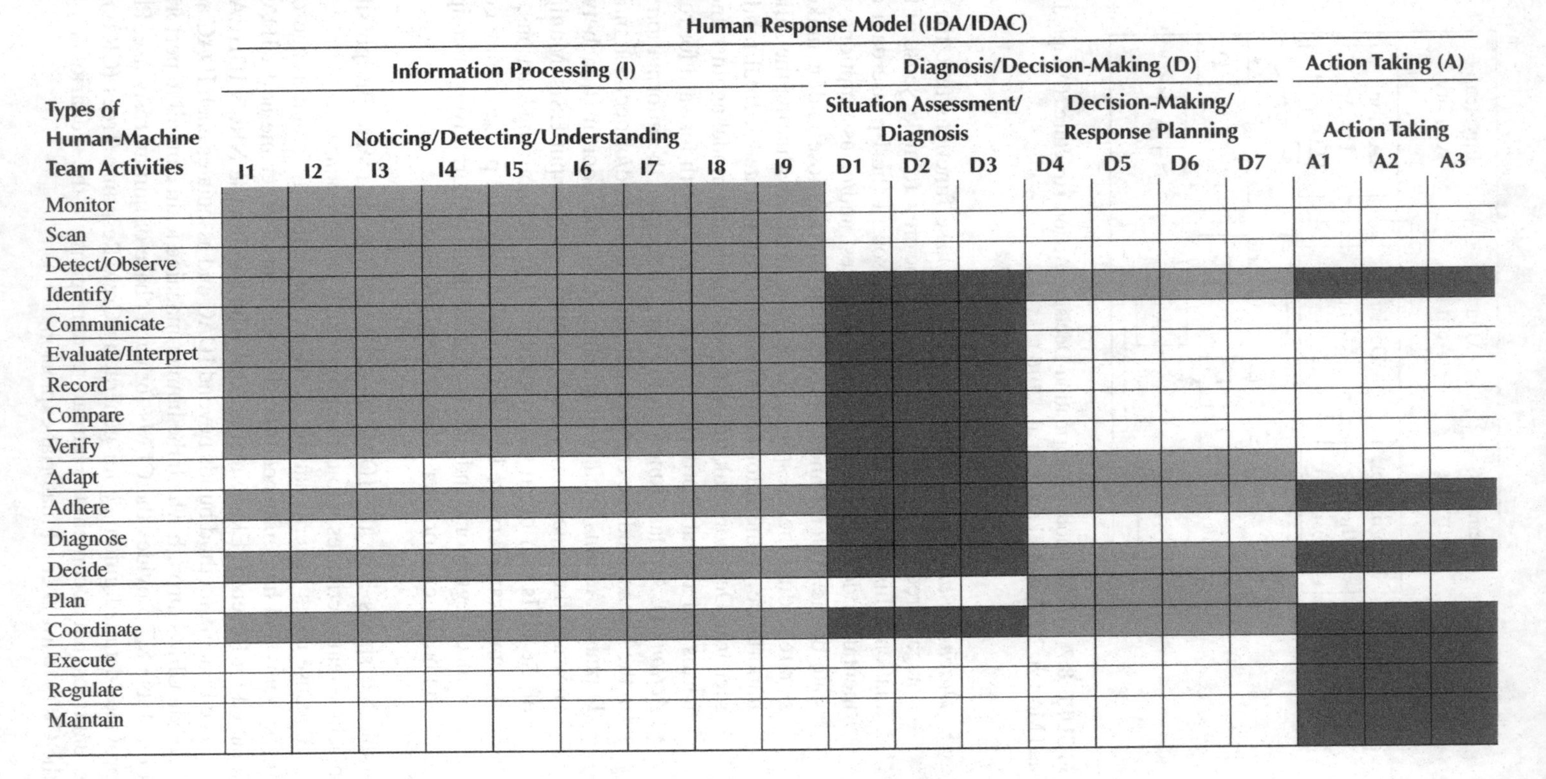

Types of Human-Machine Team Activities	Human Response Model (IDA/IDAC)																		
	Information Processing (I)									Diagnosis/Decision-Making (D)							Action Taking (A)		
	Noticing/Detecting/Understanding									Situation Assessment/ Diagnosis			Decision-Making/ Response Planning				Action Taking		
	I1	I2	I3	I4	I5	I6	I7	I8	I9	D1	D2	D3	D4	D5	D6	D7	A1	A2	A3
Monitor																			
Scan																			
Detect/Observe																			
Identify																			
Communicate																			
Evaluate/Interpret																			
Record																			
Compare																			
Verify																			
Adapt																			
Adhere																			
Diagnose																			
Decide																			
Plan																			
Coordinate																			
Execute																			
Regulate																			
Maintain																			

TABLE 8.7
CFMs Defined in the Phoenix Method (Ekanem et al., 2016)

#	CFMs in "I" Phase	#	CFMs in "D" Phase	#	CFMs in "A" Phase
I1	Key Alarm not Responded to (Intentional and Unintentional)	D1	Plant/System State Misdiagnosed	A1	Incorrect Timing of Action
I2	Data Not Obtained (Intentional)	D2	Procedure Misinterpreted	A2	Incorrect Operation of Component/Object
I3	Data Discounted	D3	Failure to Adapt Procedure to the Situation	A3	Action on Wrong Component/Object
I4	Decision to Stop Gathering Data	D4	Procedure Step Omitted (Intentional)		
I5	Data Incorrectly Processed	D5	Inappropriate Transfer to a Different Procedure		
I6	Reading Error	D6	Decision to Delay Action		
I7	Information Miscommunicated	D7	Inappropriate Strategy Chosen		
I8	Wrong Data Source Attended to				
I9	Data Not Checked with Appropriate Frequency				

8.4.2.4 IDHEAS

The NRC's IDHEAS method, described in NUREG/CR-2199 (Xing et al., 2017), leverages concepts from the IDAC method, situational awareness models and macrocognition models (Whaley et al., 2016). The IDHEAS method draws on five major macrocognitive functions (abbreviated in IDHEAS as "MCF"): Detection, Understanding (Status Assessment), Decision-Making (Response Planning), Action Execution, and Coordination. IDHEAS reiterates the need to model the "important human actions"—and that a failure is "failure of any macrocognitive function demanded by the task" (Xing et al., 2017). IDHEAS contains a set of CFMs corresponding to each macrocognitive function, developed for human response to internal events during at-power operations in nuclear power plants. IDHEAS includes one decision tree for each of the 15 CFMs for at-power operations, which assumes proceduralized actions:

- Key Alarm Not Attended To
- Data Misleading or Not Available
- Wrong Data Source Attended To
- Critical Data Misperceived
- Critical Data Dismissed/Discounted
- Premature Termination of Critical Data Collection
- Misinterpret Procedure
- Choose Inappropriate Strategy
- Delay Implementation

- Critical Data Not Checked with Appropriate Frequency
- Fail to Initiate Response
- Fail to Correctly Execute Response (Simple Task)
- Fail to Correctly Execute Response (Complex Task)
- Misread or Skip Step in Procedure
- Critical Data Miscommunicated

A more general extension of the IDHEAS method is still under development by the U.S. NRC. Readers are encouraged to visit the primary sources for more information on the method.

8.4.3 Representing and Quantifying HFEs

The second facet of HRA involves representing and modeling the causal factors that contribute to the occurrence of HFEs. At this step, the focus is on determining what factors and circumstances can enhance or degrade human performance (and thus change the probability of an HFE) and how those factors are related to each other. These Performance Influencing Factors (PIFs) are used within the third facet of HRA, which is to quantify the HEP for each HFE.

The PIFs (also called Performance Shaping Factors (PSFs)) are akin to the "agents of failure" in hardware reliability. PIFs are environmental, crew, personal, situational, or task-oriented characteristics that describe the context in which the CFM or HFE occurs. These PIFs may lead to the degradation of performance, and eventually a human-machine failure process occurs. Each HRA method uses a different set of PIFs—some methods use as few as 3, some use over 50.

The PIFs are causal elements that affect human-machine team performance and change the probability of occurrence of CFMs and failures of MCFs. The most comprehensive and rigorously defined set of PIFs comes from Groth and Mosleh (2012). The Groth and Mosleh taxonomy, shown in Table 8.8, provides a consistent vocabulary, clear definitions, and defined structure for combining PIFs and contextual factors at multiple levels of detail in a transparent and repeatable way. This PIF taxonomy considers human, organizational, machine, situational and environmental factors (rather than human or cognitive factors alone), recognizing the need to characterize a broad range of factors that contribute to human-machine failure. Using such a taxonomy is important because it provides a comprehensive list of factors that addresses the full spectrum of HRA contexts and includes factors relevant to both HRA modeling and human reliability data collection at multiple levels of abstraction. Furthermore, the clear definitions make it possible to map variables from multiple sources or HRA methods onto a single set of variables for use in modeling activities.

Within any of the HRA methods, the PIFs are then used as the foundation of a quantitative model. The quantitative model will elicit information about the PIFs (or context) for a particular HFE, and then that information is passed through an HRA model to obtain the corresponding HEP. A PIF providing a positive influence can reduce the HEP, whereas a negative influence increases the HEP.

TABLE 8.8
Data-Informed PIF Taxonomy

Organization-Based	Team-Based	Person-Based	Situation/Stressor-Based	Machine-Based
Training program	Communication	Attention	External environment	HSI
Availability	Availability	To task	Conditioning events	Input
Quality	Quality	To surroundings	Task load	Output
Corrective action program	Direct supervision	Physical and physiological abilities	Time load	System response
Availability	Leadership	Alertness	Other loads	
Quality	Team coordination	Fatigue	Non-task	
Other programs	Team cohesion	Impairment	Passive information	
Availability	Role awareness	Sensory limits	Task complexity	
Quality		Physical attributes	Cognitive	
Safety culture		Other	Execution	
Management activities		Knowledge/experience	Stress	
Staffing		Skills	Perceived situation	
Scheduling		Bias	Severity	
Workplace adequacy		Familiarity with situation	Urgency	
Resources		Morale/motivation/attitude	Perceived decision	
Procedures			Responsibility	
Availability			Impact	
Quality			Personal	
Tools			Plant	
Availability			Society	
Quality				
Necessary info.				
Availability				
Quality				

Source: Adapted from Groth and Mosleh (2012).

8.4.3.1 THERP

The THERP method contains dozens of data tables containing nominal HEPs for specific actions. Within each table, a handful of PIFs are used to modify the nominal HEPs in each table. See Swain and Guttmann (1983) for the data tables and further discussion about the dependency between HFEs.

8.4.3.2 SPAR-H

In SPAR-H, eight PIFs are used to modify the nominal HEPs for the diagnosis and action tasks. The PIFs are:

1. Available time
2. Stress
3. Experience and training
4. Complexity
5. Human-machine Interface
6. Procedures
7. Fitness for Duty
8. Work Practices

In SPAR-H, the PIFs are treated as if they are independent of each other, and each directly influences the HEP. Each PIF is given a multiplier for its effect on HEP. See Table 8.9 for the SPAR-H states and multipliers for action.

The formula for calculating HEP is:

$$\mathrm{HEP} = \mathrm{NHEP} \cdot \prod_{i=1}^{8} \mathrm{PIF}_i, \tag{8.37}$$

where the nominal or baseline HEP (NHEP) is NHEP = 0.01 for diagnosis tasks and 0.001 for action tasks. Note that HEP should never exceed 1. If more than three PIFs are rated as having a negative influence (i.e., the SPAR-H multiplier is >1.0), a correction factor is applied to ensure the resulting HEP does not exceed 1.0:

$$\mathrm{HEP} = \frac{\mathrm{NHEP} * \prod_{i=1}^{8} \mathrm{PIF}_i}{\mathrm{NHEP} * \left(\prod_{i=1}^{8} \mathrm{PIF}_i - 1 \right) + 1}. \tag{8.38}$$

The task failure probability is then calculated as the Diagnosis HEP plus the Action HEP. SPAR-H also includes a simple method, based on THERP, for calculating dependency between HFEs. See Gertman et al. (2005) for additional information.

8.4.3.3 IDAC and Phoenix

IDAC was built as a dynamic HRA method and contains approximately 100 PIFs, many of which are also incorporated into Groth and Mosleh (2012). The IDAC method is intended to be used as a dynamic simulation platform. See Chang and Mosleh (2007) for more details.

TABLE 8.9
Multiplier Tables for SPAR-H Diagnosis and Action Tasks (Gertman et al., 2005)

PIFs	PIF Levels	Multiplier for Diagnosis	PIFs	PIF Levels	Multiplier for Action
Available time	Inadequate time	*Pr*(failure) = 1.0	Available time	Inadequate time	*Pr*(failure) = 1.0
	Barely adequate time (≈2/3 × nominal)	10		Time available ≈ time required	10
	Nominal time	1		Nominal time	1
	Extra time (1–2 × nominal and >30 minutes)	0.1		Time available ≥5× time required	0.1
	Expansive time (>2× nominal and >30 minutes)	0.01		Time available ≥50× time required	0.01
	Insufficient information	1		Insufficient information	1
Stress/stressors	Extreme	5	Stress/ stressors	Extreme	5
	High	2		High	2
	Nominal	1		Nominal	1
	Insufficient information	1		Insufficient information	1
Complexity	Highly complex	5	Complexity	Highly complex	5
	Moderately complex	2		Moderately complex	2
	Nominal	1		Nominal	1
	Obvious diagnosis	0.1		Insufficient information	1
	Insufficient information	1			
Experience/ training	Low	10	Experience/ training	Low	3
	Nominal	1		Nominal	1
	High	0.5		High	0.5
	Insufficient information	1		Insufficient information	1

(*Continued*)

TABLE 8.9 *(Continued)*

Multiplier Tables for SPAR-H Diagnosis and Action Tasks (Gertman et al., 2005)

PIFs	PIF Levels	Multiplier for Diagnosis
Procedures	Not available	50
	Incomplete	20
	Available, but poor	5
	Nominal	1
	Diagnostic/symptom oriented	0.5
	Insufficient information	1
Ergonomics/ HMI	Missing/misleading	50
	Poor	10
	Nominal	1
	Good	0.5
	Insufficient information	1
Fitness for duty	Unfit	*Pr*(failure) = 1.0
	Degraded fitness	5
	Nominal	1
	Insufficient information	1
Work processes	Poor	2
	Nominal	1
	Good	0.8
	Insufficient information	1

PIFs	PIF Levels	Multiplier for Action
Procedures	Not available	50
	Incomplete	20
	Available, but poor	5
	Nominal	1
	Insufficient Information	1
Ergonomics/ HMI	Missing/misleading	50
	Poor	10
	Nominal	1
	Good	0.5
	Insufficient Information	1
Fitness for duty	Unfit	*Pr*(failure) = 1.0
	Degraded Fitness	5
	Nominal	1
	Insufficient Information	1
Work processes	Poor	5
	Nominal	1
	Good	0.5
	Insufficient Information	1

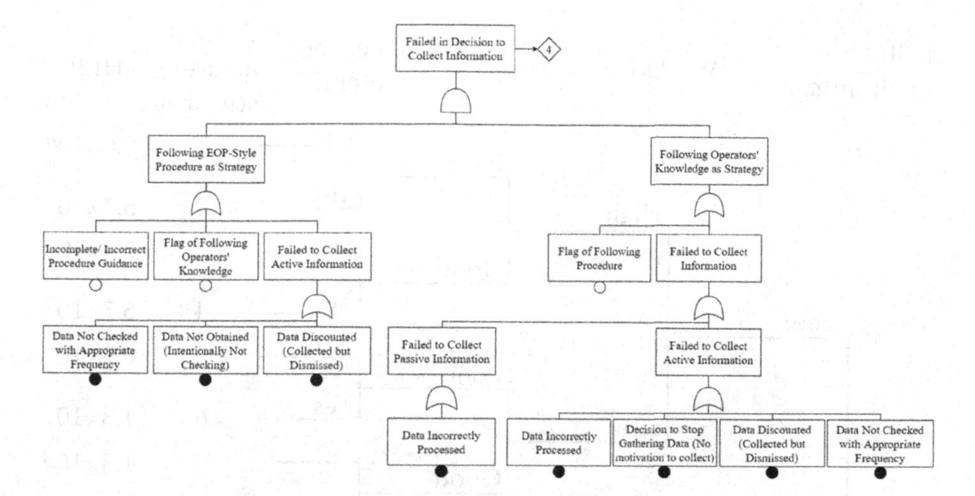

FIGURE 8.8 Example of CFM fault tree from Phoenix. (Adapted from Ekanem et al., 2016.)

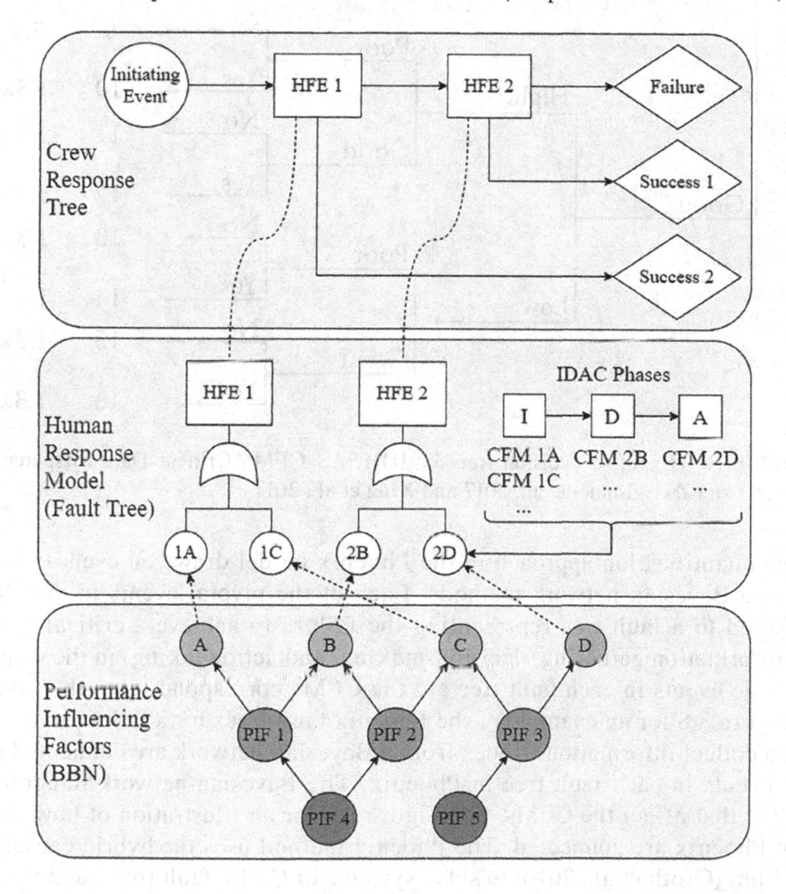

FIGURE 8.9 Connections between different modeling layers in the Phoenix method. (Adapted from Ramos et al., 2021.)

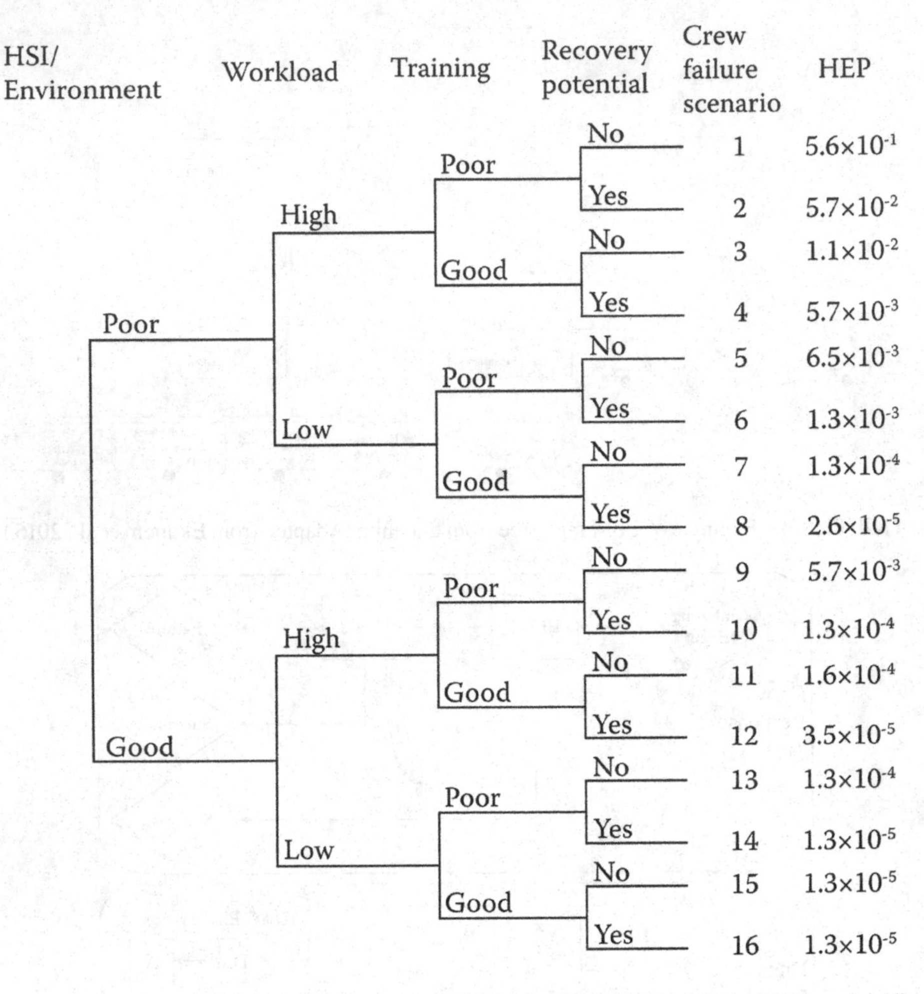

FIGURE 8.10 Example decision tree for IDHEAS CFM "Critical Data Misperceived." (Adapted from Zwirglmaier et al., 2017 and Xing et al., 2017.)

The quantification approach in the Phoenix model draws on event tree, fault tree, and Bayesian network methods. Each of the pivotal events in the CRT is connected to a fault tree representing the failure to achieve a critical function (i.e., information gathering, decision-making, and action taking) in the response. The basic events in each fault tree are the CFMs corresponding to the function. See Figure 8.8 for an example of the Phoenix fault trees for a failure in the decision to collect information. Nodes from a Bayesian network are connected to the basic events in each fault tree in Phoenix. The Bayesian network integrates all the PIFs that affect the CFMs. See Figure 8.9 for an illustration of how the layers of Phoenix are connected. The Phoenix method uses the hybrid causal logic algorithm (Groth et al., 2010) to solve systems of CRTs, fault trees, and Bayesian networks.

8.4.3.4 IDHEAS

IDHEAS represents each of its CFMs using a decision tree. Each decision tree contains approximately 4 PIFs, which are the branching points. An analyst follows the branches, answers a series of questions about the PIFs, and chooses an HEP from the end state of the decision tree. IDHEAS includes one decision tree for each CFM. Figure 8.10 provides an example of a decision tree from IDHEAS for the CFM Critical Data Misperceived.

8.5 EXERCISES

8.1 The minimal cut sets of a simple system are expressed by: $F = A + BC$. The following data describe components A, B, and C.

Components	**A**	**B**	**C**
Number of failures	2	9	11
Total test time (hour)	1,470	3,315	4,012

a. Use the standard Monte Carlo simulation to calculate the mean and 90% credible interval of the unreliability of this system.
b. Compare the results with the 90% confidence interval using methods of moments.
c. Discuss the meaning of these two intervals regarding the interpretation of the unknown reliability function.

8.2 Consider the risk expression $R = f \times C$, where f is the frequency of the scenario and C is the consequence of the scenario. The mean frequency is 1×10^{-5} per year with a standard deviation of 1×10^{-5} per year, and the mean consequence is 100 injuries with a standard deviation of 10.
a. Assuming the random variable R follows a lognormal distribution, determine R's 90% confidence interval.
b. Now assume both f and C are lognormally distributed and determine the 90% confidence interval for R. Compare these results to the results from (a).

8.3 The stress-life fatigue model is expressed by $NS^m = c$, where S = stress amplitude, N = number of cycles to failure, and m and c are constants. Suppose the mean and variance of applied stress are known, and the stress is described by a lognormal distribution. The parameter m is known with certainty. Further, the parameter c is described by a normal distribution with known mean and variance. Determine the mean and variance of N using the LHS method.

8.4 Consider the scenario:

$$F = IXY + IXZ + IW,$$

where F is the frequency of failure for the system, and I is the frequency of an undesirable event. The probability of occurrence of events X and Y can be obtained from the cdf: $F_x(t) = 1 - e^{-\lambda t}$. For both events Z and W, the probability of occurrence is given by the Weibull cdf: $F_z(t) = 1 - e^{-\left(\frac{t}{\alpha}\right)^{\beta}}$. The mean values of the probability distribution model parameters are $\lambda_X = \lambda_Y = 2\times10^{-4}$ / hour, $\alpha_Z = \alpha_W = 1{,}800$ hours, and $\beta_Z = \beta_W = 1.7$. Assume the mean value of event I is 0.1/year.

a. What is the mean value of F as a function of t?
b. If the events I, X, Y, Z, and W each have a coefficient of variation of 5%, what is the coefficient of variation of F?

8.5 Assume the Weibull distribution describes the time to failure of a device. Bayesian analysis of the failure data from the field has shown that the Weibull's shape parameter can be described by a normal distribution with mean of 1.2 and coefficient of variation of 20%. Similarly, the analysis shows the scale parameter follows a lognormal distribution with a mean of 1,380 hours and an error factor of 6.7.

a. Assign aleatory or epistemic uncertainties to each distribution.
b. Propagate the distributions and plot and label the aleatory and epistemic uncertainties in a single graph.
c. Plot the total time to failure cdf, using mean parameter values.
d. Assume this is a device with which you are familiar. Discuss three sources of aleatory and two sources of epistemic uncertainties that are present.

8.6 Repeat Exercise 6.23 and assume CCF between the valves and the pumps exist. Using the generic data in Appendix B, calculate the probability that the top event occurs. Use a β-factor method with β=0.08 for valves and pumps. Discuss whether the selection of β is sensitive to the result.

8.7 Consider a 3-out-of-4 failure system. If all components are identical and subject to common cause failure, calculate the failure probability of the system using the α-factor model for parametric CCF probability assessment (use generic data discussed in this chapter for α). The total failure probability of one unit is 0.001.

8.8 Consider the system shown below and assume that the reliability values of the individual components (A, B, C, D, and S) are known. Assume all components have only one failure mode, leading to the component failure that interrupts the flow through the component.

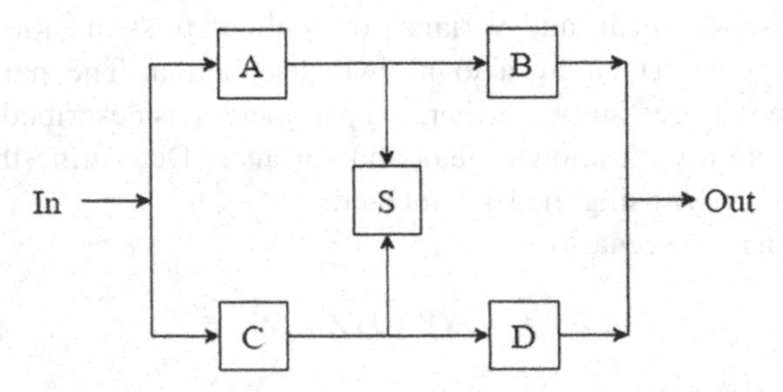

a. Develop a fault tree with top event "No Flow out of the System."
b. Find the minimal cut sets of the top event using the substitution method.

c. If the probability of failure of the components for a known mission are $Pr(A)=Pr(B)=Pr(C)=Pr(D)=0.015$ and $Pr(S)=0.02$, find the probability of the top event occurrence.
d. If two of these systems are made in parallel configuration into a new system, what would the reliability of the new system.
e. Repeat (c) assuming CCF between the components A & C, and B & D. Use the α-factor method with the parametric values listed in Table 8.4.

8.9 In a system of three identical parallel units, the failures are dependent. It has been observed that over 12 years of continuous operation, the following failures have occurred: one dependent failure of all three units; five dependent failures of two units; and fifteen single independent failures.
a. Estimate the probability of a single total independent failure per hour.
b. Estimate the probability of two dependent failures per hour.
c. Estimate the probability of three dependent failures per hour.
d. Find the probability of system failure (assuming two out of the three units need to function for the system to work).

8.10 The system diagram below describes the control of a NO_2 supply system for anesthetic use in a dental operating room. For this problem, assume that system failure corresponds to an excessive supply of gas at the patient end.

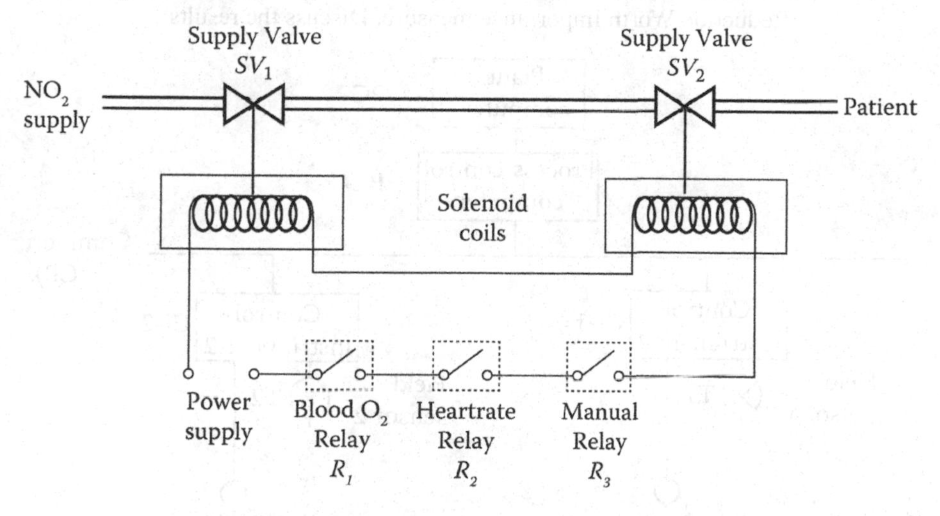

Assume the following failure modes and probabilities:

- Valve fails open (probability $F_1 = 10^{-3}$).
- Leakage past valve seat (probability $F_2 = 10^{-4}$).
- The three relays are identical and arranged in series; each relay has three associated failure modes:
- Fails closed (probability $F_3 = 2.5\times10^{-3}$).
- Short circuit (probability $F_4 = 2.5\times10^{-3}$).
- Ground fault (probability $F_5 = 1.5\times10^{-3}$).

If current is flowing through the circuit, the solenoids are energized and supply valves SV_1 and SV_2 are both open.

a. Draw a fault tree for the top event "patient asphyxiation—excessive gas supply."
b. Use the data provided to compute the probability of this event.

8.11 For the system below, assume the following: One of the two product lines is sufficient for success. Control instruments receive the sensor values and transfer those values to the process control computer, which calculates the position of the control valves. The control instruments adjust the control valves as needed. The plant computer controls the process control computer.

The probability of failure of PC, PCC, and CB are all equal to 2×10^{-3}. The probability of failure of each unit O-1, O-2, CV-1, CV-2, T1, T2, CI-1, and CI-2 is 1×10^{-2}, but there is a common cause failure between CI-1 and CI-2 described by a β-factor model with parameter $\beta = 0.08$. Failure of these units is the only failure that should be considered, and they represent total failure due to all failure modes.

a. Develop a fault tree for the top event, "inadequate product feed."
b. Find the minimal cut sets of the FT.
c. Find the probability of the top event.
d. Determine which components are critical to the design using the Risk Reduction Worth Importance measure. Discuss the results.

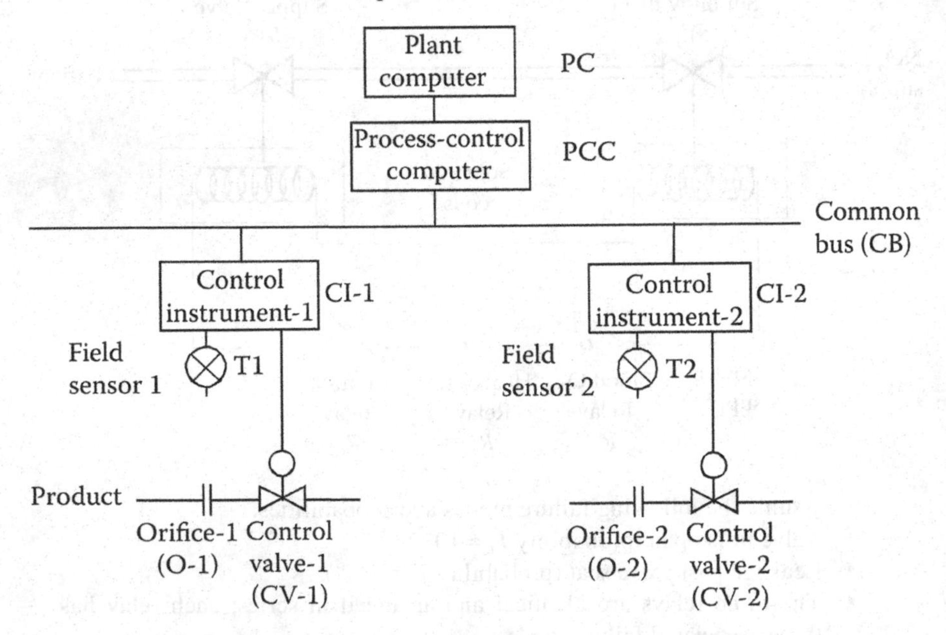

8.12 Analyze the system diagram given. Do not consider any other failure mode or human actions except the total failure of the components shown in the diagram.

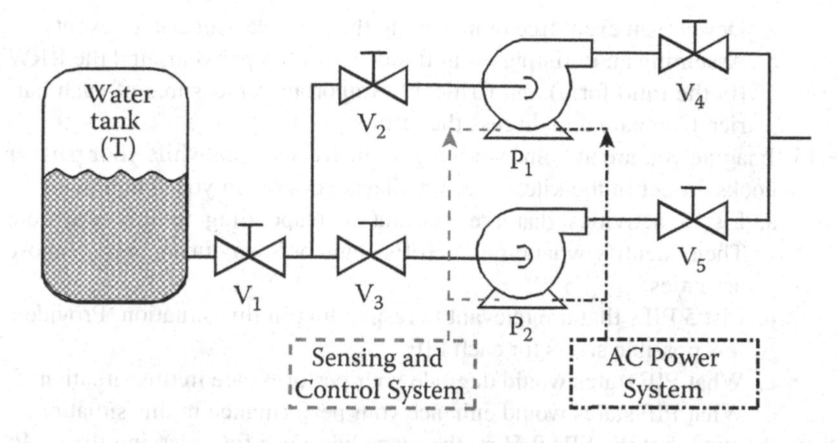

Component	Failure Probability
T	0.0001
V_1	0.0001
V_2, V_3, V_4, V_5	0.0005
P_1, P_2	0.0007
AC Power Source, Control System	0.0002

a. Draw a fault tree top event "No flow out of the system."
b. Find the minimal cut sets and, using the data in the table, find the probability of the top event assuming common cause failure between components P-1 and P-2 with a β-factor model assuming $\beta = 0.075$.
c. Determine how uncertainty about the top event is propagated to establish uncertainties in the top event probability, assuming factors of 10 for the lower and upper 90% confidence limits with respect to the point estimates given in the table.

8.13 The following describes the failure logic of a system: $F = XY + XZ + T$. The probability of X and Y can be obtained from an exponential model such as $Pr(t) = 1 - e^{-\lambda t}$. The probability of Z and T can be obtained from a Weibull model such as $Pr(t) = 1 - e^{-\left(\frac{t}{\alpha}\right)^{\beta}}$. If $\lambda_X = \lambda_Y = 5 \times 10^{-5}$/hour, $\alpha_Z = \alpha_T = 3{,}150$ hours, and $\beta_Z = \beta_T = 1.5$, determine the importance of parameters α, β, and λ at a given time t with respect to the unreliability of the system using Birnbaum, Criticality, Fussell-Vesely, RAW, and RRW. Compare the results from each importance measure and discuss the interpretations of each.

8.14 Three independent and diverse preventive barriers exist in a design to avert exposure to a hazard. Assume the probability of failure of each barrier (given an initiating event) is 0.001 for the first barrier, 0.0025 for the second barrier, and 0.004 for the third barrier, and the consequence of the hazard exposure is 100 units of losses.

a. Develop an event tree representing the possible scenario of events.
b. Assuming an initiating event frequency of 0.5 per year, find the RRW (in the ratio form) and Criticality importance measures of each barrier. Compare and discuss the results.

8.15 Imagine you are at home studying in the living room while your partner cooks dinner in the kitchen. A fire alarm goes off in your kitchen.
a. List 3 activities that are relevant to responding to this situation. Then, identify what type of HFEs could be seen during each of those activities.
b. List 5 PIFs that are relevant to responding in this situation. Provide a few possible states for each PIF.
c. What PIF states would degrade your performance in this situation?
d. What PIF states would enhance your performance in this situation?

8.16 Assume that the SPAR-H method is valid to use for assessing the HEPs of students taking exams for this course. If the baseline probability for exam failure (HEP) is 0.05, use the SPAR-H method to calculate your probability of failing the final exam for this course. Include both the calculations and justification of the selection of the states for the PIFs.

8.17 Assume that the SPAR-H method is validated for HRA of semi-truck drivers. Calculate the HEP for a semi-truck driver for the task "change lanes on the highway" under the following conditions. Justify your choice of PIFs.
a. The weather and traffic conditions you observed on the way to class today.
b. Snowy weather, assuming a new driver who has just completed training.

8.18 An important concern for the storage of low-level nuclear waste is accidental damage to the canister. If the outer shell of the storage canister is damaged, an alarm sounds, alerting the operators. After acknowledging the alarm, the operators should implement emergency procedure EP-01 to lower a sealed isolation dome around the canister. You have the following data: Impacts occur on average once every ten container-years. The container is damaged in 5% of impacts. The alarm fails on 0.1% of demands. Operators fail to acknowledge the alarm 3% of the time. Operators implement procedure EP-01 correctly 97% of the time.
a. Draw an event tree for this situation and calculate the annual probability of each scenario.
b. Which CFMs from the Phoenix method are relevant to the operator failing to implement procedure EP-01 correctly?

8.19 Discuss why PIFs are necessary factors in HRA but are not typically used in hardware component reliability modeling.

8.20 Your team has been asked to provide insight into human reliability for a nuclear power plant. The plant owner monitors the workers extremely closely, especially John and Jane. The owner estimates that the amount of time it takes Jane to find the right procedure when working alone under nominal conditions is lognormally distributed with a median of

3 minutes and shape parameter of 0.25. However, the owner also analyzed the working conditions at the plant and found several PIFs negatively affecting operator performance. In particular, the presence of John results in high stress for his co-workers.

a. Determine the probability that Jane finds the procedure in under 2 minutes on a normal day working alone.
b. What are two HRA frameworks the owner should consider for modeling the reliability of the workers? Which one would you suggest implementing and why?
c. Determine the probability that Jane finds the procedure in under 2 minutes on a day when John is working.

REFERENCES

Akramin, M. R. M., A. K. Ariffin, M. Kikuchi, and S. Abdullah, "Sampling method in probabilistic S-version finite element analysis for initial flaw size." *Journal of the Brazilian Society of Mechanical Sciences and Engineering*, 39(1), 357–365, 2017.

Apostolakis, G. and Y. T. Lee, "Methods for the Estimation of Confidence Bounds for the Top-Event Unavailability of Fault Trees." *Nuclear Engineering and Design*, 41(3), 411–419, 1977.

Beardsmore, D. W., K. Stone, and H. Teng, "Advanced Probabilistic Fracture Mechanics Using the R6 Procedure." In *Pressure Vessels and Piping Conference*, Vol. 49224, pp. 735–742. 2010.

Bensi, M. et al., "Uncertainty in External Hazard Probabilistic Risk Assessment (PRA): A Structured Taxonomy." In *Proceedings of the 16th Probabilistic Safety Assessment and Management Conference (PSAM16)*, Honolulu, HI, June 26–July 1, 2022.

Bertucio, R.C., and J. A. Julius, *Analysis of Core Damage Frequency: Sequoyah, Unit 1 Internal Events*. NUREG/CR-4550, Vol. 5. U.S. Nuclear Regulatory Commission, Washington, DC, 1990.

Bier, V. M., "A Measure of Uncertainty Importance for Components in Fault Trees." In *Transactions of the 1983 Winter Meeting of the American Nuclear Society*, San Francisco, CA, 1983.

Birnbaum, Z. W., "On the Importance of Different Components in a Multicomponent System." In P. R. Krishnaiah, ed., *Multivariate Analysis-II*, Academic Press, New York, 1969.

Bucalossi, A., A. Petruzzi, M. Kristof, and F. D'Auria, "Comparison between Best-Estimate-Plus-Uncertainty Methods and Conservative Tools for Nuclear Power Plant Licensing." *Nuclear Technology*, 172(1), 29–47, 2010.

Chang, Y. H. J., and A. Mosleh, "Cognitive Modeling and Dynamic Probabilistic Simulation of Operating Crew Response to Complex System Accidents: Part 1: Overview of the IDAC Model." *Reliability Engineering & System Safety*, 92(8), 997–1013, 2007.

Chatterjee, K., and M. Modarres, "A Probabilistic Physics-of-Failure Approach to Prediction of Steam Generator Tube Rupture Frequency." *Nuclear Science and Engineering*, 170(2), 136–150, 2012.

Drouin, M., *Treatment of Uncertainties Associated with PRAs in Risk-Informed Decision Making*, NUREG-1855, U.S. Nuclear Regulatory Commission, 2017.

Duan, X., M. Wang, and M. Kozluk, "Acceptance Criterion for Probabilistic Structural Integrity Assessment: Prediction of the Failure Pressure of Steam Generator Tubing with Fretting Flaws." *Nuclear Engineering and Design* 281, 154–162, 2015.

Ekanem, N. J., A. Mosleh, and S-H. Shen, "Phoenix–a Model-based Human Reliability Analysis Methodology: Qualitative Analysis Procedure." *Reliability Engineering & System Safety*, 145, 301–315, 2016.

Fleming, K. N., "A Reliability Model for Common Mode Failures in Redundant Safety Systems." In *Proceedings of the 6th Annual Pittsburgh Conference on Modeling and Simulations*, Instrument Society of America, Pittsburgh, PA, 1975.

Fussell, J. B. "How to Hand-Calculate System Reliability and Safety Characteristics." *IEEE Transactions on Reliability*, 24(3), 169–174, 1975.

Gertman, D., H. Blackman, J. Marble, J. Byers, and C. Smith. *The SPAR-H Human Reliability Analysis Method*, NUREG/CR-6883. US Nuclear Regulatory Commission, Washington, DC, 2005.

Groth, K. M. and A. Mosleh. "A Data-Informed PIF Hierarchy for Model-based Human Reliability Analysis." *Reliability Engineering & System Safety*, 108, 154–174, 2012.

Groth, K. M., R. Smith, and R. Moradi. "A Hybrid Algorithm for Developing Third Generation HRA Methods using Simulator Data, Causal Models, and Cognitive Science." *Reliability Engineering & System Safety*, 191, 106507, 2019.

Groth, K., C. Wang and A. Mosleh, "Hybrid Causal Methodology and Software Platform for Probabilistic Risk Assessment and Safety Monitoring of Socio-Technical Systems." *Reliability Engineering & System Safety*, 95, 1276–1285, 2010.

Hokstad, P., and M. Rausand, "Common Cause Failure Modeling: Status and Trends." In K.B. Misra, ed., *Handbook of Performability Engineering*, pp. 621–640. Springer, London, 2008.

Jyrkama, M. I., and M. D. Pandey, "On the Separation of Aleatory and Epistemic Uncertainties in Probabilistic Assessments." *Nuclear Engineering and Design*, 303, 68–74, 2016.

Kaplan, S., "On the Method of Discrete Probability Distributions in Risk and Reliability Calculations–Application to Seismic Risk Assessment." *Risk Analysis*, 1(3), 189–196, 1981.

Kirwan, B., A *Guide* to *Practical Human Reliability Assessment*. CRC Press, London, 1994.

Kuhn, M., and K. Johnson, *Applied Predictive Modeling*. Vol. 26. Springer, London, 2013.

Modarres, M., *Risk Analysis in Engineering: Techniques, Tools, and Trends*. CRC Press, Boca Raton, 2006.

Modarres, M., S. Krahn, and J. O'Brien, "Understanding and Effectively Managing Conservatisms in Safety Analysis of Nonreactor Nuclear Facilities." *Nuclear Technology*, 207(3), 424–440, 2021.

Mosleh, A., "Common Cause Failures: An Analysis Methodology and Examples." *Reliability Engineering & System Safety*, 34, 249–292, 1991.

Mosleh, A., D. M. Rasmuson, and F. M. Marshall, *Guidelines on Modeling Common-Cause Failures in Probabilistic Risk Assessment*, NUREG/CR-5485. U.S. Nuclear Regulatory Commission, Washington, DC, 1998.

Mosleh, A., K. N. Fleming, G. W. Parry, H. M. Paula, D. H. Worledge, and D. M. Rasmuson, *Procedures for Treating Common Cause Failures in Safety and Reliability Studies*, NUREG/CR-4780. U.S. Nuclear Regulatory Commission, Washington, DC, 1988.

Mosleh, A. and N. O. Siu, "A Multi-Parameter, Event-Based Common-Cause Failure Model." In *Proceedings of the 9th International Conference on Structural Mechanics in Reactor Technology*, Lausanne, Switzerland, 1987.

Lee, S. W., B. D. Chung, Y-S. Bang, and S. W. Bae, "Analysis of Uncertainty Quantification Method by Comparing Monte-Carlo Method and Wilks' Formula." *Nuclear Engineering and Technology*, 46(4), 481–488, 2014.

Oberle, W., *Monte Carlo Simulations: Number of Iterations and Accuracy*, ARL-TN-0684. Army Research Lab Aberdeen Proving Ground MD Weapons and Materials Research Directorate, 2015.

Olsson, A., G. Sandberg, and O. Dahlblom. "On Latin Hypercube Sampling for Structural Reliability Analysis." *Structural Safety* 25(1), 47–68, 2003.

Paglioni, V. P., and K. M. Groth. "Dependency Definitions for Quantitative Human Reliability Analysis." *Reliability Engineering & System Safety* 220, 108274, 2022.

Robert, C. P., and G. Casella. *Introducing Monte Carlo Methods with R*. Springer, New York, Vol. 18, 2010.

Sankararaman, S., Y. Ling, and S. Mahadevan. "Uncertainty Quantification and Model Validation of Fatigue Crack Growth Prediction." *Engineering Fracture Mechanics* 78(7), 1487–1504, 2011.

Srinivasan, R., *Importance Sampling – Applications in Communications and Detection*. Springer-Verlag, Berlin, 2002.

Swain, A.D., and H. E. Guttmann, *Handbook of Human Reliability Analysis with Emphasis on Nuclear Power Plant Operations*, NUREG/CR-1278. US Nuclear Regulatory Commission, Washington, DC, 1983.

Thomopoulos, N. T., *Essentials of Monte Carlo Simulation: Statistical Methods for Building Simulation Models*. Springer, London, 2013.

U.S. Nuclear Regulatory Commission. *Severe Accident Risks: An Assessment for Five U.S. Nuclear Power Plants*, NUREG-1150. Washington, DC. 1990.

Wald, A., "An Extension of Wilks' Method for Setting Tolerance Limits." *The Annals of Mathematical Statistics*, 14(1), 45–55, 1943.

Whaley, A. M., et al. *Cognitive Basis for Human Reliability Analysis*. NUREG-2114. U.S. Nuclear Regulatory Commission, Washington, DC, 2016.

Wilks, S. S., "Determination of Sample Sizes for Setting Tolerance Limits." *The Annals of Mathematical Statistics*, 12(1), 91–96, 1941.

Wu, Y.-T., "Computational Methods for Efficient Structural Reliability and Reliability Sensitivity Analysis." *AIAA Journal*, 32(8), 1717–1723, 1994.

Xing, J., G. Parry, M. Presley, J. Forester, S. Hendrickson, and V. Dang, *An Integrated Human Event Analysis System (IDHEAS) for Nuclear Power Plant Internal Events At-Power Application*, NUREG-2199, Vol. 1. U.S. Nuclear Regulatory Commission. Washington, DC, 2017.

Zio, E., and L. Podofillini, "Accounting for Components Interactions in the Differential Importance Measure." *Reliability Engineering & System Safety*, 91(10–11), 1163–1174, 2006.

Zwirglmaier, K., D. Straub, and K. M. Groth, "Capturing Cognitive Causal Paths in Human Reliability Analysis with Bayesian Network Models." *Reliability Engineering & System Safety*, 158, 117–129, 2017.

9 Risk Analysis

It is of considerable interest to measure potential losses from failure in terms of, for example, loss of human life, adverse health effects, loss of property, loss of mission, and damage to the environment. This falls into the field of study known as *risk analysis*. Within risk analysis, one element is *risk assessment* which is defined as *a systematic process to comprehend the nature of risk and express risk, with the available knowledge* (Aven et al., 2018, p. 8). This definition is often also used for risk analysis, but typically the term risk analysis is understood as a broader term than risk assessment, where *the field of risk analysis is defined to include risk assessment, risk characterization, risk communication, risk management, and policy relating to risk, in the context of risks of concern to individuals, to public and private sector organizations, and to society at a local, regional, national, or global level* (Aven et al. 2018, p. 8).

Thus, risk analysis is the process of characterizing, managing, and communicating the existence, causes, magnitude, prevalence, contributing factors, and uncertainties of the potential losses (consequences). Characterization of risk is typically achieved through the formal process of risk assessment. *Risk management* provides inputs to decision makers about the sources of risk, priorities, and strategies to manage risk and reduce risk to acceptable levels. *Risk communication* provides independent information to decision makers and stakeholders about the results of risk assessment and risk management including any uncertainties.

In this chapter we will discuss Quantitative Risk Assessment (QRA) focused on complex engineering systems. From an engineering viewpoint, the risk or potential loss for a system composed of hardware, software, humans, and organizations is the exposure to a hazard by some recipients or targets (e.g., people, organization, economic assets, ecosystem, and environment). The risk can be quantitatively expressed as the frequency of a loss interpreted as the probability or amount of loss (consequence) over a fixed intervals of time or space, within available knowledge.

If there are adequate historical data on such losses, the frequency of losses can be directly measured from the statistics of the actual loss. The loss may be external to the system, caused by the system to one or more end recipients. This approach is often used for cases in which data on such losses are readily available. For example, ample field data are usually available for events such as car accidents, cancer occurrence, or for severe weather events such as storm-induced floods. When there are rare events, including those that have not occurred or have not caused the actual losses, QRA methods are used to model the possible event scenarios, probabilities, and associated consequences. In this analysis, the potential loss is predicted. There are few cases, especially for highly reliable complex engineering systems, for which data on losses are available. Therefore, often we must model and quantitatively predict the risk.

Generally, there are three classes of risk assessment methods: Quantitative, qualitative, and a mix of the two (sometimes the latter is called semi-quantitative). Table 9.1 gives some examples of each. Each of these comprises widely used methods

DOI: 10.1201/9781003307495-9

TABLE 9.1
Example Risk Assessment Methods

Type	Example Methods
Qualitative to semi-quantitative	• FMEA (Failure Modes and Effects Analysis) • FMECA (Failure Modes, Effects, and Criticality Analysis) • PHA (Process Hazard Analysis) • HAZOP (Hazards and Operability Analysis)

Quantitative	• QRA (Quantitative Risk Assessment) or PRA (Probabilistic Risk Assessment)
	  • Fault trees
	  • Event trees
	  • Bayesian networks
	  • Simulation
	  • Hazard models

that have different purposes, strengths, and weaknesses. In this book, we advocate for the use of quantitative forms, typically called QRA in oil and gas, chemical processes, and similar domains, and a sophisticated form of QRA called Probabilistic Risk Assessment (PRA) or Probabilistic Safety Assessment (PSA) in the nuclear and aviation domains.

Categories of risk analysis include these applications:

- *Safety risk analysis* involves estimating potential harms caused by accidents occurring due to natural events (climatic conditions, earthquakes, brush fires, etc.) or human-made products, technologies, and systems (i.e., aircraft crashes, chemical plant explosions, nuclear plant accidents, technology obsolescence, or failure).
- *Security risk analysis* involves estimating access and harm caused due to war, terrorism, riot, crime (vandalism, theft, etc.), and misappropriation of information (national security information, intellectual property, etc.) in physical or digital systems (e.g., cybersecurity).
- *Health risk analysis* estimating potential diseases, injuries, and losses of life affecting humans, animals, and plants.
- *Financial risk analysis* involves estimating potential individual, institutional, and societal monetary losses such as project losses, bankruptcy, market loss, property damage, misappropriation of funds, and currency fluctuations, interest rates, market share.
- *Environmental risk analysis* involves estimating losses due to noise, contamination, and pollution in ecosystem (water, land, air, and atmosphere) and in space (e.g., space debris).

Risk analysis is widely used by private companies and government agencies to support regulatory and resource-allocation decisions. Risk analysis can be used in all stages of design, development, construction, and operation of engineering systems. For example, possible applications in different stages of design include:

- *Conceptual design:* Compare risks of alternative design options; Assess product viability.
- *Design:* Provide barriers to prevent, minimize, or eliminate harm; Minimize life cycle cost; Apportion risk limits and performance goals.
- *Development:* Identify systems or subsystems that contribute most to safety and risk; Test safety and risk-significant elements of the design; Support warranty development.
- *Regulation:* Develop risk-based or risk-informed regulations, codes, and standards; Determine the significance of the elements of the system and/or dominant contributors to the consequences; Set monitoring and performance criteria; Determine what, where, and how frequently the regulator should conduct inspections; Demonstrate regulatory compliance.
- *Operation:* Optimize cost and schedule of maintenance and other operational activities; Define surveillance requirements and schedules; Define replacement policies and decisions; Support aging considerations and management decisions; Develop security measures.
- *Decommissioning:* Assess the safety of possible decommissioning activities; Select the most appropriate disposal method; Assess long-term liability issues.
- *Accident investigation and forensic engineering:* Support root cause identification; Identify the most likely scenarios.

Risk assessment, the first element of risk analysis, consists of two aspects: A qualitative aspect that includes identifying, characterizing, and ranking hazards; and a quantitative aspect of risk evaluation that includes estimating the frequencies and consequences of hazard exposure. The estimation of the frequency of hazard exposure and the consequences of such exposures depends greatly on the reliability of the system's components, typically involving human, software, and hardware failure events and their interactions. These topics have been extensively addressed in previous chapters of this book.

After risk assessment, the second element of risk analysis is risk management to eliminate or minimize the occurrence of major failures and accidents by reducing the frequency of their occurrence (e.g., minimizing hazard occurrence frequency); to reduce the impacts of uncontrollable failure events and accidents (e.g., implement risk mitigations; install barriers; prepare and adopt emergency response strategies); and to transfer risk (e.g., via insurance coverage). In the remainder of this chapter, we discuss how the reliability evaluation methods addressed in the preceding chapters are used, collectively, in a risk assessment.

9.1 QRA DEFINED

As explained in Chapter 1, according to Kaplan and Garrick (1981) QRA is defined as a set of triplets seeking answers to three basic questions:

1. What can go wrong?
2. How likely is it to happen?
3. If it does happen, what are the consequences?

TABLE 9.2
Risk Triplets

Scenario	Probability or Frequency	Consequence
S_1	P_1	C_1
S_2	P_2	C_2
S_3	P_3	C_3
.	.	.
.	.	.
S_n	P_n	C_n

Answering these questions for engineering systems often involves significant analysis and modeling. Answers to the first question can be expressed in terms of scenarios of events leading to the outcome of some losses (consequences). Answers to the second question can be expressed quantitatively by the probability or frequency that each scenario occurs. Answers to the third question provide a quantitative value of the losses. Therefore, for multiple scenarios of events that describe things that go wrong, Table 9.2 depicts the mathematical representation of the answers to the set of triplets.

Based on the above definition, there are three major parts of a QRA:

- Identifying the scenarios, S_i, that comprise the potential sequences of events, corresponding root causes, and outcomes resulting from a challenge to the system.
- Determining probability, P_i, or frequency of occurrence, F_i, for each event, E_i, in each scenario S_i. In some cases, estimates of the probability or frequency are computed through a detailed analysis of experience and available historical data; sometimes they are judgmental estimates based on an expert's belief of the situation; and sometimes they are simply a best guess based on the available information. The confidence in such estimates depends on the quality and quantity of the data available. The most realistic way to estimate the probability and frequency is by detailed logical modeling of the systems and generating the values of interest using fault trees and event trees discussed in Chapter 6.
- Evaluating the consequence, C_i, of when each scenario S_i occurs. Consequence, C_i, is a quantitative measure of the outcome of scenario S_i. The choice of the type of consequence may affect the acceptability threshold and the tolerance level for the risk. Examples of consequence or loss measures include the amount of human, economic, environmental, and safety losses such as the mission loss, payload damage, damage to property, number of injuries, number of fatalities, or monetary loss.

The results of the risk assessment are then used to interpret the various contributors to risk, which are compared, ranked, and placed in perspective. This process consists of calculating a risk profile based on an engineering system.

A simple representation of risk can be the total the expected risk value R expressed as

$$R = \sum_{S_i} P_i \times C_i. \tag{9.1}$$

Naturally, calculations of P_i and C_i involve some uncertainties, approximations, and assumptions. Therefore, uncertainties must be considered explicitly in risk assessments, as discussed in Chapter 8. Using expected losses, scenario elements, and the risk profile, one can evaluate the reasonable risk management strategies to control, mitigate, or avoid risks.

According to Equation 9.1, it is possible to interpret the total risk value, R, as the expected consequence values summed over the possible scenarios. Thus, if a scenario i occurs with a frequency of 0.01 per year, and if the associated loss is \$1 million, then the expected loss (or risk value) associated with that scenario is $R_i = 0.01 \times \$1{,}000{,}000 = \$10{,}000$/year. Conversely, if the frequency of scenario occurrence is 1 per year, but the loss is \$10,000, the risk value is still $R_i = 1 \times \$10{,}000 = \$10{,}000$/year. Thus, the expected loss (risk value) for these two situations is the same; that is, both events are equally risky from a rational viewpoint. Occasionally, for risk decisions, value factors (weighting factors) are assigned to each event contributing to risk.

Another method for interpreting the results is to construct a risk profile. With this method, the probability or frequency values are plotted against the consequence values. Figure 9.1a illustrates these methods. Figure 9.1a shows the use of logarithmic scales, which are usually used because they can cover a wide range of values. The error brackets denote epistemic uncertainties in the probability estimate (vertical) and the consequences (horizontal). This approach provides a means of illustrating events with low probability, high consequences, and high uncertainty. It is useful when discrete probabilities and consequences are known. Figure 9.1b shows the construction of the complementary cumulative probability risk profile (sometimes known as a *Farmer's curve* (Farmer, 1961)). In this case, the logarithm of the probability that the total consequence C exceeds C_i is plotted against the logarithm of C_i. The most

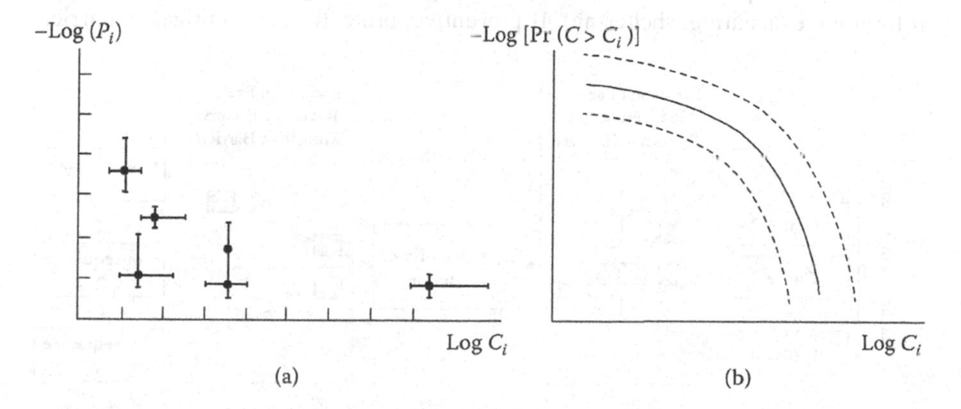

FIGURE 9.1 Two constructions of a risk profile including uncertainty.

notable application of this method was in the landmark Reactor Safety Study (U.S. NRC, WASH-1400, 1975). With this method, the low-probability/high-consequence risk values and high-probability/low-consequence risk values can be easily differentiated, which was not the case in the expected risk methods described earlier. That is, the extreme values of the estimated risk can be easily displayed. To represent aleatory and epistemic uncertainty in the consequence values in frequency-consequence or Farmer's curve, either a family of such curves is generated (e.g., through Monte Carlo simulation) and directly plotted, or the 5% and 95% the credible or confidence intervals of the frequency-consequence and Farmer's risk curves are plotted.

9.2 QRA PROCESS

Now let us turn to the process of QRA which, as discussed in the previous section, is a function of scenarios, probabilities or frequencies, and consequences. The modeling process, therefore, entails scenario development and assessing the frequency and quantifying the risk contribution from each possible scenario that leads to the end consequence. The QRA process attempts to identify all possible scenarios that lead to losses, the probabilities or frequencies of each event in the scenario (including root causes), and a description and amount of the consequences that result from each scenario.

A convenient way to depict the key elements of QRA is by the bowtie construct depicted in Figure 9.2. In Figure 9.2, the word hazard is the source of potential harm or loss—typically used in unintentional incidents. The word threat is related to intentional exposure of harm or loss and is often used in a security risk analysis instead of the word hazard.

In this construct, a scenario starts with an initiating event that puts the engineering system in an off-normal condition. Protective and preventive features (called barriers) in form of passive barriers (walls, pipes, etc.), protective and preventive systems (e.g., cooling, scrubbing, extinguishing fire systems, human response) can stop or mitigate the exposure of any hazard, accident progression, or unwanted events. However, escalation factors (e.g., wind, barrier failures) can exacerbate the situation. If exposure happens, other barriers can mitigate the degree of exposure (e.g., by informing, evacuating, sheltering). If preventive, protective, and mitigative barriers

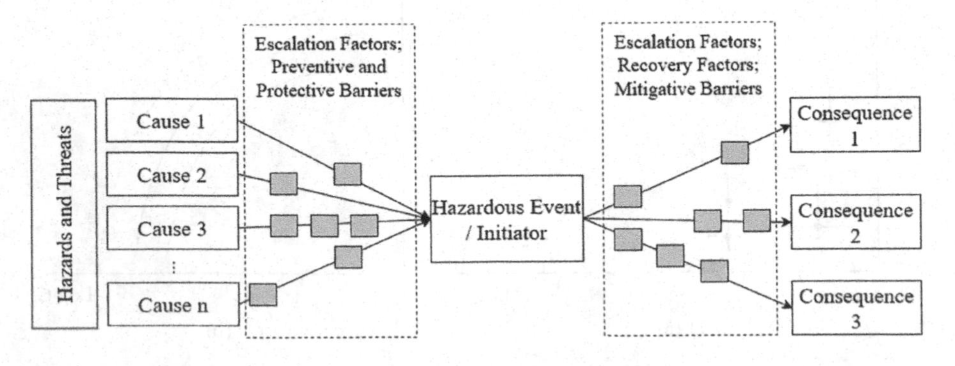

FIGURE 9.2 The bowtie model construct.

fail or work insufficiently, different degrees of exposure (outcome) can happen, and each exposure will be associated with a consequence.

9.2.1 Initiating Event

The initiating event can be caused by one of several root causes, spanning hardware, human, software, and other types of failures. There will also be preventive and protective barriers that reduce the probability of the root cause turning into a hazard exposure event. Likewise, there can be escalation factors that exacerbate the situation. We will discuss how to identify elements of these scenarios and how to quantify them later in this chapter.

9.2.2 Hazard Exposure Scenarios

Now let us turn to the formation of the chain of events represented by the scenario S_i, which leads to exposure of a hazard and, if not mitigated, an outcome or consequence, C_i. Hazard exposure scenario development requires a set of descriptions of how a barrier confining a hazard is threatened, the manner of barrier failure, and the effects on an end recipient when exposed to the uncontained hazard. In QRA all this information is used to find the frequency of each scenario S_i.

9.2.3 Identification of Hazards

In a risk assessment it is critical to perform a survey of the process or system under study to identify the hazards of concern. These hazards typically belong to one of these categories:

- Thermal hazard (e.g., heat flux from a fire from chemical reactor; freezing from exposure to cryogenic fluids)
- Mechanical hazard (e.g., kinetic or potential energy from a moving object; fragments or pressure from mechanical explosions or ruptures of containers)
- Chemical hazard (e.g., toxic materials or by-products released from a chemical process; chemical reactions; smoke exceeding tenability limits)
- Electrical hazard (e.g., potential difference, electrical and magnetic fields, and electrical shock)
- Ionizing radiation (e.g., radiation from a nuclear reactor core)
- Nonionizing radiation (e.g., radiation from a microwave or the sun)
- Biological hazard (e.g., pathogens and diseases; insects in imported products).

Presumably, each of these hazards will be part of the process, and normal process boundaries will be used as their barrier or containment. However, there will also be challenges to the system and its barrier, including potential component and system failures, and human actions, external events, and conditions that challenge the system. In a scenario, one postulates the challenges to the process or system and tries to estimate the probability or frequency of these challenges.

9.2.4 Identification of Barriers

Each identified hazard must be examined to determine all the physical barriers that contain it or that can intervene to prevent or minimize exposure to the hazard. These barriers may be *protective* by physically surrounding the hazard (e.g., walls, pipes, valves, fuel clad, and structures); they can be *mitigative* based on a specified distance from a hazard source to minimize exposure to the hazard (e.g., to minimize exposure to radioactive materials); or they may provide direct shielding of the subject from the hazard (e.g., protective clothing and bunkers). Barriers may also be *preventive* to stop or minimize the progression of the hazard (e.g., fire suppression systems).

9.2.5 Identification of Challenges to Barriers

After the identification of the individual barriers, a concise definition of the requirements for maintaining each one follows. Hierarchical analytical models provide a means to express these requirements. One can also simply identify what is needed to maintain the integrity and performance of each barrier. Barrier failure is due to the degradation of strength of the barrier and the presence of high stress impinged on the barrier.

Barrier strength degrades because of:

- Reduced thickness (due to deformation, erosion, or corrosion)
- Changes in material properties (e.g., toughness and yield strength). Material properties may be affected by the local environment (e.g., temperature).

Stress on the barrier increases with:

- Internal forces or pressure
- Penetration or distortion by external objects or forces.

The above causes of degradation are often the result of one or more of these conditions:

- Malfunction of process equipment (e.g., the emergency cooling in a nuclear plant)
- Problems with human-machine interface
- Poor design or maintenance
- Adverse natural phenomena
- Adverse human-made environment.

9.3 PRA PROCESS

We now focus on the most well-known QRA methods, PRA. PRA involves the estimation of the frequency of loss or probability of loss over an interval of time or space, assesses the most significant contributors to the risks of the system, and determines the value of the risk. As a type of QRA, the PRA procedure formally follows the risk triplet concept in which scenarios of events and probabilities (or frequencies) of

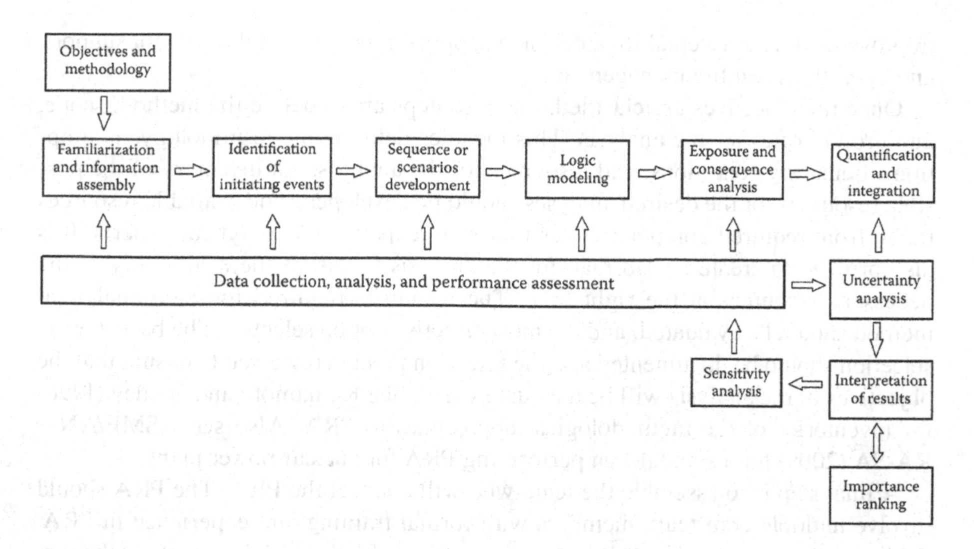

FIGURE 9.3 Components of the overall PRA process. (Adapted from Stamatelatos et al., 2011.)

those scenarios leading to exposure of hazards are estimated, and the corresponding magnitudes of health, safety, environmental, and economic consequences for each scenario are predicted. The risk value (i.e., expected loss) of each scenario is often measured as the product of the scenario frequency and its consequences (Equation 9.1). The most significant result of the PRA is not the actual value of the risk computed (the so-called bottom-line number); rather, it is the determination of the elements of the system that substantially contribute to the risks of that system, the uncertainties associated with such estimates, and the effectiveness of various risk reduction strategies available. *The primary value of a PRA is to highlight the system design and operational deficiencies and support subsequent risk management efforts to identify and optimize resources that can be invested on improving the design and operation of the system.*

The following subsections discuss the essential components of PRA and the steps that must be performed in a PRA. The major components of a PRA are shown in Figure 9.3 (adapted from *The NASA PRA Guide* by Stamatelatos et al. (2011)). Notice that there are feedback loops: this indicates that the PRA process is typically iterative. The initial results are often used to inform and refine the analysis. Furthermore, documentation is critical to ensure the quality of the PRA. Therefore, a good PRA should involve detailed and consistent documentation at each step. For further discussion on the elements of PRA, see Modarres (2006). For the connection of PRA to prognostics and health monitoring, see Moradi and Groth (2020).

9.3.1 Objectives and Methodology

Preparing for a PRA begins with defining or reviewing the objectives of the analysis. Among the many objectives that are possible, the most common ones include design improvement, risk acceptability, decision support, regulatory and oversight support, and operations and life management.

Once the objectives are clarified, the next steps are to define the method, scope, and ground rules for the analysis. This includes defining the terminology, assumptions, boundary conditions, and resources for the analysis. An inventory of the possible resources for the desired analyses should be developed. The available resources range from required computer codes to system experts and analytical experts. It is also prudent to create a road map for the analysis to ensure the availability of the necessary resources at the right time. The resources required for each analytical method should be evaluated, and the most effective option selected. The basis for the selection should be documented and the selection process reviewed to ensure that the objectives of the analysis will be adequately met. See Kumamoto and Henley (1996) for inventories of the methodological approaches to PRA. Also see ASME/ANS-RA-SA (2009) for a standard on performing PRA for nuclear power plants.

A final step is to assemble the team who will conduct the PRA. The PRA should involve multiple core team members with formal training and experience in PRA. Additionally, the team should have access to experts from multiple aspects of the system, including design, operation, and maintenance. The team will also need access to information and data experts. Finally, the team should contain experienced members, newer members, and diverse perspectives.

One important caution that applies when assembling the team is that experts should have the requisite experience with PRA. While systems experts are an important part of the team, they may lack the formal training in the PRA modeling techniques, the probabilistic methods, and the reliability engineering methods necessary to conduct an accurate, repeatable, well-documented PRA.

9.3.2 Familiarization and Information Assembly

This step involves understanding the system and assembling the necessary information to conduct a rigorous analysis. A general knowledge of the physical layout of the overall system (e.g., facility, design, process, plant, aircraft, or spacecraft), administrative controls, maintenance and test procedures, as well as barriers and subsystems that protect, prevent, or mitigate hazard exposure conditions, is necessary to begin the PRA. All systems, subsystems, components, structures, locations, and activities expected to play a role in the initiation, propagation, or arrest of a hazard exposure condition must be understood in sufficient detail to construct the models necessary to capture all possible scenarios. A detailed inspection of the overall system must be performed in the areas expected to be of interest to the analysis. The following items should be performed in this step:

1. Identify major critical barriers, structures, emergency safety systems, and human interventions for the system.
2. Identify and explicitly describe the interactions among all major subsystems (or parts of the system). Summarize the results in a dependency matrix.
3. Study and document past major failures, dependent events, near-misses, and abnormal or initiating events that have been observed in the facility.

Such information would help ensure the inclusion of important applicable scenarios.
4. Create a consistent computer-based filing system at the outset and maintain it throughout the study.

With the help of the designers, operators, and the system owners, the analysts should determine the configuration and phases of the operation of the overall system to be analyzed. The analysts should also determine the faults and conditions to be included or excluded, the operating modes of concern, and the hardware configuration on the design freeze date (i.e., the date after which no additional changes in the overall system design and configuration will be modeled). The results of the PRA only apply to the overall system at the freeze date.

9.3.3 Identification of Initiating Events

This task involves identifying those events (failures, abnormal events, conditions internal to the system, or conditions external to the system such as natural events) that could, if not correctly and promptly responded to, result in hazard exposure. As in Section 9.2.3, the first step in risk assessment involves identifying sources of hazard and barriers around these hazards. The next step involves identifying events that can lead to a direct threat to the integrity of the barriers.

A system may have one or more operational modes. In each operational mode, specific functions are achieved. Each function is directly realized by one or more systems by making certain actions and behaviors. These systems, in turn, comprise more basic units (e.g., subsystems, components, assemblies, software, and human actions) that accomplish the objective of the system. If a system is operating within its design parameter tolerances, there is little chance of challenging the system boundaries in such a way that hazards will escape those boundaries. These operational modes are called normal operation modes.

During normal operation modes, the loss of certain functions or systems will cause the process to enter an off-normal (transient) state. Once in this transition, there are two possibilities. First, the state of the system could be such that no other function is required to maintain the process in a safe condition. (Safe refers to a mode where the chance of exposing hazards beyond the system boundaries is tolerable.) The second possibility is a state wherein other functions (and associated items) are required to prevent exposing hazards beyond the system boundaries. For this second possibility, the loss of the function or the system is considered an initiating event. Since such an event is related to the normally operating equipment, it is called an *operational initiating event.*

Operational initiating events can also apply to various modes of the system (if they exist). The terminology remains the same since, for each mode, certain equipment, people, or software must be functioning. For example, an operational initiating event found during the PRA of a nuclear reactor could be Low Primary Coolant System Flow. Flow is required to transfer heat produced in the reactor to heat exchangers and ultimately to the cooling towers and the outside environment. If this coolant flow function is reduced to the point where an insufficient amount of heat is transferred, it could initiate an off-normal scenario which could lead to core damage (and thus the possibility of exposing radioactive materials—the main source of hazard in this case). Therefore, another system must operate to remove the heat produced by the reactor (i.e., a protective barrier). Thus, Low Primary Coolant System Flow is an operational initiating event.

One method for determining the operational initiating events begins with first drawing a functional block diagram of the system. From the functional block diagram, a hierarchical relationship is produced, with the process objective being successful completion of the desired system. Each function can then be decomposed into its subsystems and components and can be combined in a logical manner to represent operations needed for the success of that function.

Potential initiating events are events that result in failures of functions, subsystems, or components, the occurrence of which causes the overall system to fail. These potential initiating events are "grouped" such that members of a group require the same subsystem responses to cope with the initiating event. These groupings are the operational initiator categories.

An alternative to using functional hierarchy for identifying initiating events is the use of Failure Modes and Effects Analysis (FMEA) (see Chapter 6). The difference between these two methods is that the functional hierarchies are deductively and systematically constructed, whereas FMEA is an inductive and experiential technique. Using FMEA for identifying initiating events consists of identifying failure events (modes of failures of equipment, software, and human) whose effect is a threat to the integrity and availability of the hazard barriers of the system.

Both methods can be used together, and one can always supplement the set of initiating events with generic initiating events (if known). For example, see Sattison and Hall (1990) for these initiating events for nuclear reactors, and the *NASA Guide* (Stamatelatos et al., 2011) for space vehicles.

To simplify the process, after identifying all initiating events, it is necessary to combine those initiating events that pose the same threat to hazard barriers and require the same mitigating functions of the process to prevent hazard exposure.

The following procedure should be followed when grouping initiating events:

1. Combine the initiating events that directly break all needed hazard barriers.
2. Combine the initiating events that break the same hazard barriers (not necessarily all the barriers).
3. Combine the initiating events that require the same group of mitigating human or automatic actions following their occurrence.
4. Combine the initiating events that simultaneously disable the normal operation as well as some of the available mitigating human, software, or automatic actions.

Events that cause off-normal operation of the overall system and require other systems to operate to maintain hazards within their desired boundaries, but that are not directly related to a hazard mitigation, protection, or prevention function, are called *nonoperational initiating events*. Nonoperational initiating events are identified with the same methods used to identify operational events. One class of nonoperational initiating events of interest is those that are primarily external to the overall system or facility.

There are two kinds of external initiating events. The first kind refers to events that originate from within the facility or the overall system boundary but which are not part of the facility or system operation; these events can adversely affect the facility or overall system and are called *internal events external to the system*. Examples of these internal events external to the system are fires from fuel or oil tanks stored within the system boundary or facility, or floods inside the facility or a system compartment caused by the rupture of a water tank that is part of the overall system. The effects of these events should be modeled with event trees or event sequence diagrams to show all possible scenarios.

The second kind of external events are those that originate outside of the overall system's physical boundary. These are called *external events*. Examples of external initiating events are fires and floods that originate from outside of the system and loss of off-site power. Additional examples include seismic events, extreme heat, extreme drought, transportation events, volcanic events, high-wind events, terrorism, hacking, and sabotage. Again, this classification can be used in developing and grouping the event tree scenarios.

The following procedure should be followed in this step of the PRA:

1. Implement a method for identifying specific operational and nonoperational initiating events. Two representative methods are functional hierarchy and FMEA. If a generic list of initiating events is available, it can be used as a supplement.
2. Using the method selected, identify a set of initiating events.
3. Group the initiating events having the same effect on the system. The initiating events in each group are those requiring the same mitigating functions to prevent or mitigate hazard exposures.

9.3.4 Scenario Development

The goal of scenario development is to delineate a complete set of scenarios that encompasses all the potential exposure propagation paths that following the occurrence of an initiating event. To describe the cause-and-effect relationship between initiating events and subsequent event progression, it is necessary to identify those functions (e.g., safety functions) that must be maintained to prevent the loss of hazard barriers. Event trees or event sequence diagrams are used to display the scenarios that describe the functional response of the system or process to the initiating events. Event tree development techniques are discussed in Chapter 6.

Event trees order and depict, in an approximately chronological manner, the success or failure of key systems or hazard barriers (e.g., human protective or mitigative actions or hardware failure) that are required to act in response to the initiating event. In PRA, two types of event trees are typically developed: Functional and systemic. The *functional event tree* uses protective and mitigating functions in its pivotal events. The main purpose of the functional tree is to better understand the scenario of events at an abstract level. The functional tree also guides the PRA analyst in developing a more detailed systemic event tree. The *systemic event tree* reflects the sequences of specific hardware, software, and human failures and errors that realize the functions described in the functional events. Therefore, systemic event tree pivotal events are specific failures, human actions, or protective or mitigative subsystem operations. A systemic event tree fully delineates the overall system response to an initiating event and serves as the main tool for further analyses in the PRA. For a detailed discussion on specific tools and techniques used for this purpose, see Modarres (2006).

The following procedure should be followed in this step of the PRA:

1. Identify the mitigating functions for each initiating event (or group of events).
2. Identify the corresponding systems, human actions, or hardware operations associated with each function, along with their necessary conditions for success.
3. Develop a functional event tree for each initiating event (or group of events). Although not strictly required, functional event trees are important during a design-stage PRA, and in other cases they help avoid errors and missing the system's critical protective, preventive, and mitigative capabilities when developing the systemic event trees. They are helpful for communication among the team members and other stakeholders.
4. Develop a systemic event tree for each initiating event, delineating the success conditions, the initiating event progression phenomena, and the end effect of each scenario.

   For specific examples of scenario development refer to Chapter 6. For further readings see Stamatelatos et al. (2011) and Kumamoto and Henley (1996).

9.3.5 Causal Logic Modeling

Event trees commonly involve branch points at the pivotal events describing the state of a barrier such as a subsystem or event. The pivotal event describes the barrier either working (or an event happening) or not working. If sufficient failure data are available to directly estimate the probability or frequency of any of the pivotal events, they can be used to quantify that aspect of the event tree.

However, often failure of these barriers or events is rare, and there may not be an adequate record of observed failure events to provide a historical basis for estimating the probability or frequency of their failure. In such cases, other logic-based analysis methods, such as fault trees, are used. The most common method used in PRA to calculate the probability of a pivotal event is fault tree analysis, although other methods have been developed, including methods incorporating Bayesian networks (Groth et al., 2010). As described in Chapter 6, fault tree analysis involves developing a logic model in which the pivotal event is broken down into its more elementary constituents (components, subsystems, or segments) for which adequate reliability data exist. For more discussions on methods and techniques used for logic modeling, see Chapter 6.

Different event tree modeling approaches imply variations in the complexity of the logic models that may be required. If the event tree pivotal events describe main protective or mitigative functions (rather than system or component failures), the corresponding fault trees become more complex. If support functions (or systems) are explicitly included as pivotal events, event trees will be more complex (i.e., having many pivotal events), but leads to smaller fault trees. In either case, the models must also accommodate all dependencies among the main and support functions (or subsystems). Which approach is preferred is a stylistic choice by the PRA team.

It is important to identify dependencies and to explicitly include dependencies in both the event trees and fault trees in the PRA. As the reliability of individual pivotal events increases due to redundancy, the contribution from dependent failures becomes more important; often, dependent failures dominate the overall reliability of pivotal events. Including the effects of dependent failures in the reliability models and fault trees used in the PRA is a difficult process and requires sophisticated, fully integrated models to account for unique failure combinations. The treatment of dependent failures is not a single step performed during the PRA; it must be considered throughout all models in the analysis.

The following procedure should be followed to develop the fault trees or similar causal logic models:

1. Develop a fault tree for each pivotal event for which observed historical failure data is inadequate.
2. Determine the level of detail for the fault tree based on the availability of reliability data or models to assess probabilities of failure modes of hardware components, software failures, unavailability due to test and maintenance, and human errors.
3. Identify the hardware, software, and human elements that are identical and could cause dependent or common cause failure (CCFs). Typically, in chemical processes and nuclear power plants, identical redundant pumps, motor-operated valves, air-operated valves, generators, and batteries, including similar human actions, software, and maintenance events are sources of CCFs.
4. Explicitly model dependencies between fault tree basic events and/or ensure the dependencies are captured in the event tree pivotal events. These dependencies include subsystems and intercomponent dependencies (e.g., CCFs). For more discussion, see Chapter 8.
5. Explicitly model both functional dependencies and items that are potentially susceptible to CCFs into the corresponding fault trees and event trees.

9.3.6 Failure Data Collection, Analysis, and Assessment

A critical building block in assessing the reliability and availability of events modeled in the event trees and fault trees is the collection and analysis of reliability data of the items, barriers, and events. The best resources for predicting future reliability and availability are data from closely related past field experiences and tests. Hardware, software, and human reliability data are inputs to assess the performance of hazard barriers. For example, see Moradi and Groth (2020) for an outline of the types of data necessary for conducting risk analysis on hydrogen storage systems. The validity of the results depends highly on the quality of the input information. It must be recognized, however, that historical data and experiments have predictive value only to the extent that the conditions under which the data were generated remain applicable.

Collecting the various failure data consists of the following activities: Collecting generic data, assessing generic data, statistically evaluating facility- or overall system-specific data, and developing failure probability distributions using test and/or facility- and system-specific data. Three types of events identified during the risk scenario definition and system modeling must be quantified for the event trees and fault trees to estimate the frequency of occurrence of sequences: Initiating events, component failures, and human error.

Quantifying the probability or frequency of initiating events and failure of hazard barriers and components involves two activities. First, the probabilistic failure model for each barrier or component failure event must be established, and then the parameters of the model must be estimated. Typically the necessary data include times of failures, repair times, test frequencies, test downtimes, and CCF events; these data and methods for their analysis have been discussed in Chapters 5, 7, and 8. Further, uncertainties associated with such data must also be characterized as discussed in Chapters 5 and 8. For addressing software reliability, see Smidts (1996).

To attain the very low levels of risk, the systems and hardware that comprise the barriers to hazard exposure must have very high levels of reliability. This high reliability is typically achieved through well-designed systems with adequate margins of safety in terms of uncertainties and redundancy and/or diversity in hardware, which provides multiple success paths. The problem then becomes one of ensuring the independence of the paths, since there is always some degree of coupling between agents of failures, such as those activated by failure mechanisms, either through the operating environment (events external to the system) or through functional and spatial dependencies.

The following procedure should be followed as part of the data analysis task:

1. Determine generic values of material properties, strength or endurance, load or damage agents, failure times, failure occurrence rate, and failures on demand for each item (hardware, human action, or software) identified in the PRA models. This can be obtained either from facility- or system-specific experiences, from generic sources of data, or both.
2. Gather data on hazard barrier tests, repair, and maintenance data primarily from experience, if available. Otherwise use generic performance data.
3. Assess the frequency of initiating events and other probability of failure events from experience, expert judgment, or generic sources. For each data type, ensure that the conditions under which the data were generated remain applicable.
4. Determine the dependent or CCF probabilities for identical or similar items. However, when substantial facility- or system-specific data are available, they should be primarily used. Otherwise, use generic values.

9.3.7 Consequence Analysis

The range of effects produced by exposure to the hazard may encompass harm to people, damage to equipment, and contamination of land or facilities. It is necessary to characterize two aspects of consequences: The physical hazards and the effect of those hazards on the population, system, or asset exposed to the hazard. These effects are evaluated using the knowledge of the behavior of the hazardous material(s) and the specific outcomes of the scenarios considered. In the case of the dispersal of hazardous materials, the size of the release is combined with the potential dispersion mechanisms to characterize the physical outcomes. Then, the effects of those outcomes on the exposed population and/or anticipated loss are calculated.

For example, for a release of hydrogen, one can use physical simulation models to calculate physical behaviors associated with the release, dispersion, accumulation, and ignition of hydrogen using validated simulation models such as those in HyRAM (Groth and Hecht, 2017). For hydrogen, physical effects can include damaging levels of heat flux, pressure, or impact damage from fires and explosions.

Then, the physical effects of those models are translated into probabilities or expected values of harm or losses. For example, probit regression models can be used to determine the probability of injury, damage, or fatality from a given physical effect. For example, LaChance et al. (2011) document a variety of probit models relevant to hydrogen infrastructure. Alternatively, the consequences could also be quantified in terms of financial or environmental losses associated with different levels of damage to the system. Such approaches are typically used in actuarial and insurance analysis or land contamination studies.

More detailed discussion of consequence analysis is available in many references. See Vinnem (2007), CCPS (1998), and Mannan (2012) for a starting point.

The following procedure should be followed in this step of the PRA:

1. Evaluate the physical hazard exposure (i.e., calculate the hazard exposure outcome for each scenario).
2. Evaluate the effects of those exposures on the population, system, or asset considering exposure paths, meteorological factors, impact of evacuations or other mitigative actions, and loss or harm criteria (i.e., calculate losses).

9.3.8 Quantification and Integration

At this stage, fault trees and event trees are integrated into a single logic model and the minimal cut sets and frequencies of each event tree scenario are calculated. The relevant methods are described in Chapter 6. This integration depends somewhat upon the way system dependencies have been modeled. We will describe the more complex situation, in which two or more fault trees representing the pivotal events of an event tree are dependent. The case in which all fault trees are independent is a straightforward simplification of this.

First, as discussed in Section 9.3.6, data and models are used to calculate the failure probabilities of each basic event in the fault trees (including assigning probabilities to CCF basic events), the initiating events, and any directly quantified pivotal events in the event tree.

Then, we use a Boolean reduction process to arrive at a Boolean logic expression (e.g., minimal cut sets or scenario logic) for each scenario. The Boolean logic is first generated for each fault tree representing a pivotal event in the event tree. Next, the Boolean logic from fault trees corresponding to pivotal events, e.g., for the main subsystems and support units, is merged into the corresponding event trees. For this merged logic model, the Boolean expressions are then generated and reduced to obtain the minimal cut sets or disjoint cut sets for each pivotal event. Then, the minimal cut sets for the event tree pivotal events are combined based on the logic of each event tree scenario to determine the cut sets of each scenario.

Finally, the cut sets are evaluated probabilistically to generate scenario probabilities. If possible, all minimal cut sets must be generated and retained during this process; unfortunately, in complex systems and facilities sometimes this leads to an unmanageably large number of cut sets. To manage this problem, the collection of cut sets generated for a scenario is often truncated (i.e., cut sets are discarded based on the number of terms in a cut set or on the probability or frequency of the cut set). This is usually a practical necessity because of the overwhelming number of cut sets that result from the analysis of a complex system; the number of combinations will be large, and the probability or frequency of many combinations will be vanishingly small. The truncation process rarely affects the determination of the dominant scenarios, since the contribution of the discarded events to the final risk values would be extremely small. Even though the discarded cut sets may individually be several orders of magnitude less probable than the ones retained, in rare situations a large number of them may add up to a reasonable contribution to the risk. The actual risk might thus be larger than what the PRA results indicate. However, a detailed examination of a few PRA studies of very complex systems, such as nuclear power plants, shows that cut set truncation will not introduce any significant error in the total risk assessment results (Dezfuli and Modarres, 1984).

Other methods for evaluating scenarios also exist, which directly estimate the frequency of the scenario without specifying cut sets. These methods are often used in highly dynamic systems whose configuration changes as a function of time, leading to dynamic event trees and fault trees. For more discussion on these systems, see the references (e.g., Stamatelatos et al., 2011, Aldemir, 2013, Villa et al., 2016, Swaminathan and Smidts, 1999, Parhizkar et al., 2021).

Once the consequences have been evaluated for each scenario, the total risk is calculated for each scenario, and the total risk for all scenarios is calculated using Equation 9.1.

The following procedure should be followed as part of the quantification and integration step in the PRA:

1. Use failure data and probability models to calculate the probabilities of each basic event in the fault trees (including assigning probabilities to CCF basic events).
2. Obtain Boolean logic expressions (minimal cut sets) of each fault tree.
3. Merge the corresponding fault trees associated with each pivotal event modeled in the event tree scenarios (i.e., combine them in a Boolean form). Then, obtain the reduced Boolean function for each scenario (i.e., truncated minimal cut sets).
4. Calculate the total frequency of each scenario, using the frequency of initiating events, the probability of each basic or pivotal event in the scenario

(including barrier failures, unavailability from test and maintenance outages, CCF probability, and human error probability). Use the minimal cut sets of each sequence for the quantification process. If needed, simplify the process by truncating based on the cut sets or probability.
5. Calculate the total frequency of each scenario.
6. Calculate the total frequency of all scenarios of all event trees.
7. Calculate the total risk by combining the scenario frequencies with scenario consequences using Equation 9.1.

9.3.9 Uncertainty Analysis

Uncertainties are part of any assessment, model, or estimation. In engineering calculations, we routinely overlook the estimation of uncertainties associated with failure models and parameters, either because the uncertainties are very small, the risk analysis is based on inputs that indirectly represent uncertainties, or the analyses are done conservatively (e.g., by using high safety factors or design margins). Since PRAs are primarily used for decision making and management of risk, it is critical to incorporate uncertainties in all facets of the PRA and the risk management decisions that use it.

In PRAs, uncertainties are primarily shown by probability distributions. For example, the probability of failure of a subsystem (e.g., a hazard barrier) may be represented by a probability distribution showing the range and likelihood of risk values. Similarly, probability distributions may be developed to show the range of possible consequences.

Other sources of uncertainties are in the models used, such as fault tree and event tree models, stress-strength and damage-endurance models used to estimate failure probability of barriers, time to failure distribution models of hardware, frequency of initiating events, software faults and failures and human errors, correlations between the amount of hazard exposure and the consequence, exposure models and pathways, and models to treat inter- and intra-barrier failure dependencies. Another important source of uncertainty is the incompleteness of the risk models and other failure models used in the PRAs, such as the level of detail used in decomposing subsystems using fault tree models, the scope of the PRA, and omission of certain scenarios in the event tree simply because they were not known, or known but with no historical failures or reliability data.

Once the uncertainties associated with hazard barriers have been estimated and assigned to models and parameters, they must be propagated through the PRA logic model to find the uncertainties associated with the end results of the PRA, primarily with the bottom-line risk calculations, and with the list of risk-significant elements of the system. Propagation is done using one of several techniques, but the most popular method used is Monte Carlo simulation. The results are then shown and plotted in form of probability distributions. See Chapter 8 for more technical discussion on uncertainty analysis.

Steps in uncertainty analysis include:

1. Identify models and parameters that are uncertain and the method of uncertainty estimation to be used for each.
2. Describe the scope of the PRA and the significance and contribution of uncertainties that are not modeled or considered explicitly.

3. Estimate and assign probability distributions depicting model and parameter uncertainties in the PRA.
4. Propagate uncertainties associated with the hazard barrier models and parameters to find the uncertainty associated with the risk value.
5. Present the uncertainties associated with risks and contributors to risk in an easy to understand and visually straightforward manner.

9.3.10 Sensitivity Analysis

Sensitivity analysis is the method for determining the significance of the choice of a model or its parameters, the inclusion (or omission) of a barrier or failure dataset, assumptions about phenomena or hazards, the performance of specific barriers, the intensity of hazards, and any highly uncertain input parameters or variables to the final risk value calculated. The process of sensitivity analysis is straightforward. The effects of the input variables and assumptions are measured by modifying them one at a time by several folds, factors, or even one or more orders of magnitude, and measuring relative changes observed in the PRA's risk results. Those models, variables, and assumptions whose change leads to the highest variation in the final risk values are determined as "sensitive." In such a case, revised assumptions, models, additional failure data, and more mechanisms of failure may be needed to reduce the uncertainties associated with sensitive elements of the PRA.

Sensitivity analysis helps focus resources and attention on those elements of the PRA that need better characterization, data, or modeling. A good sensitivity analysis strengthens the quality and validity of the PRA results. Elements of the PRA that could exhibit multiple impacts on the final results, such as certain phenomena (e.g., pitting corrosion, fatigue cracking, and CCF) and uncertain assumptions, are usually good candidates for sensitivity analysis. The steps involved in the sensitivity analysis are:

1. Identify the elements of the PRA (including assumptions, failure probabilities, models, and parameters) that analysts believe might be sensitive to the final risk results.
2. Change the contribution or value of each sensitive item in either direction (i.e., increase or decrease) by multiples of 2–100. Note that certain changes in the assumptions may require multiple changes of the input variables. For example, assessing a change in the failure rate of similar equipment requires changing the failure rates of all identical equipment modeled in the PRA.
3. Calculate the impact of the changes in Step 2 one at a time and list the elements that are most sensitive.
4. For the most sensitive elements, propose additional data, any changes in the assumptions, the use of alternative models, or modification of the scope of the PRA analysis.

9.3.11 Risk Ranking and Importance Analysis

Ranking both the scenario and the elements of the system with respect to their risk or safety significance is one of the most important steps of a PRA. Ranking is simply arranging the elements of the system based on their increasing or decreasing contribution to the final risk outcomes (e.g., the consequences or frequency of an undesirable event). Importance measures rank hazard barriers, subsystems, or more basic system elements usually based on their contribution to the total risk of the system. Ranking should be performed with much care. During the interpretation of the results, since formal importance measures are context dependent and their meaning varies depending on the intended application of the risk results, the choice of the importance ranking measure and method is critical.

There are several unique importance measures used in PRAs. For example, the Fussell-Vesely, risk reduction worth, and risk achievement worth importance measures are all appropriate for use in PRAs, and are all representatives of the level of contribution of various elements of the system as modeled in the PRA. Similarly, the Birnbaum importance measure represents changes in the total risk of the system as a function of changes in the basic event probability of one component at a time. See Chapter 8 for a detailed discussion of the most commonly used importance measures.

Importance measures can be broadly categorized as either absolute or relative measures. Absolute measures express the fixed importance of one element of the system, independent of the importance of other elements, whereas relative importance measures express the significance of one element with respect to the importance of other elements. Absolute importance can be used to estimate the impact of component performance on the system regardless of how important other elements are, whereas relative importance estimates the significance of the risk impact of the component in comparison to the effect or contribution of others.

Absolute measures are useful when we speculate on improving actions, since they directly show the impact on the total risk of the system. Relative measures are preferred when resources or actions to improve or prevent failures are taken in a global and distributed manner. For additional discussions on the risk ranking methods and their implications in failure and success domains, see Azarkhail and Modarres (2004). The applications of importance measures may be categorized into these areas:

1. *(Re)design*: Supports decisions of the system design or redesign by adding or removing elements (barriers, subsystems, human interactions, etc.).
2. *Test and maintenance*: Addresses questions related to system or facility performance by changing the test and maintenance strategy for a given design.
3. *Configuration and control*: Measures the significance or the effect of the failure of a component on risk or safety, or the significance of the temporary removal of a component from service.
4. *Uncertainty*: Reduces uncertainties in the input variables of the PRAs.

If simultaneous changes in the basic event probabilities are being considered, a more complex representation would be needed.

An alternative to importance measures focuses on ranking the elements of the system with the highest contribution to the total uncertainty of the final risk results. This process is called *uncertainty ranking* and differs from component, subsystem, and barrier ranking. In this importance ranking, the analyst is interested to know which of the system elements drive the final risk uncertainties, so that resources can be focused on reducing important uncertainties.

The following are the major steps of risk ranking and importance analysis:

1. Determine the purpose of the ranking and select the appropriate importance measures or ranking measures that provide consistent interpretation for the results.
2. Perform risk ranking and uncertainty ranking.
3. Identify the most critical and important elements of the system with respect to the total risk values and total uncertainty associated with the calculated risk values.

9.3.12 Interpretation of Results

Once the total risk values are determined, they must be interpreted to decide whether any revisions are necessary to refine the risk results. A few main steps are involved in the interpretation process. The first is to determine whether the final values and details of the scenarios are logically and quantitatively meaningful. This step verifies the adequacy of the PRA model and the scope of analysis. The second is to characterize the role and contribution of each element of the system in the final results. This step highlights additional analyses, data and information gathering that would be necessary. It is also necessary to assess the extent to which appropriate input data have been used, which affects the trustworthiness of the results, and the quality of the documentation, which affects the traceability and repeatability of the results.

The interpretation process relies heavily on the examination of the details of the risk analysis to see whether the scenarios are logically meaningful (e.g., by examining the minimal cut sets of the scenarios), whether certain assumptions are significant and greatly control the risk results (using the sensitivity analysis results), and whether the absolute risk values are consistent with any historical data or expert opinions available. Based on the results of the interpretation, the details of the PRA logic or the assumptions and scope of the PRA may be modified to update the results into more realistic and dependable values.

The ranking and sensitivity analysis results may also be used to identify areas where gathering more information and performing better analysis (e.g., by using more accurate models) are warranted. The primary aim of the process is to reduce uncertainties in the risk results.

The interpretation step is a continuous process of receiving information from the quantification, sensitivity, uncertainty, and importance analysis activities of the PRA. The process continues until the final results can be best interpreted and used in the subsequent risk management steps.

The basic steps of the PRA results interpretation are:

1. Determine the accuracy of the logic models and scenario structures, assumptions, and scope of the PRA.

2. Identify those system elements for which better information is needed to reduce uncertainties in the failure probabilities and models used to calculate performance.
3. Revise the PRA and reinterpret the results until stable and accurate results are obtained.

9.4 STRENGTHS OF PRA

PRA is the most rigorous formal approach to QRA for engineering systems. The most important strengths that PRA approaches provide include:

1. An integrated and systematic examination of a broad set of design and operational features of an engineered system.
2. Insight into the root causes of failures and the effectiveness of multiple barriers.
3. Incorporation of the influence of causal factors, system interactions, and human-system interfaces that contribute to failure.
4. A process for incorporating operating experience and relevant data into the risk assessment, and a mechanism for updating risk estimates with new data.
5. A process for the explicit consideration of uncertainties.
6. The ability to analyze competing risks and compare risks (e.g., of one system vs. another, or of possible modifications to an existing system).
7. A formal evaluation of assumptions and data via sensitivity studies.
8. A measure of the absolute and relative importance of systems and components to the calculated risk value.
9. A quantitative measure of the overall level of health and safety of the engineered system.
10. A consistent and transparent framework for integrating information from multiple sources.
11. A documented process for exploring priorities, encouraging discourse, and bringing a variety of information to bear on system safety decisions.

Technical challenges to PRA come from weak or absent models or associated data for potentially important factors in the risk of the system, examples of which include:

1. Lack of objective data (e.g., due to extremely infrequent events).
2. Difficulty modeling human performance and human-system interactions and/or high uncertainties associated with the models used.
3. Failures occurring from complex dependent failures such as common cause failures, and extreme operating environments that are difficult to identify and model.
4. Combinatorial explosion and computational demands associated with models of real complex systems, if truncation is not used.

9.5 EXAMPLE: SIMPLE FIRE PROTECTION PRA

In this section we present a simplified risk analysis of a fire protection system at a generic power plant. The most critical steps of the risk analysis process for this system consist of the steps explained below.

9.5.1 Objectives and Methodology

The objective of this PRA is to assess the potential risk and financial losses from the possible failure scenarios of a fire protection system at a generic power plant. The methodology applied will follow the steps outlined in Section 9.3.

9.5.2 System Familiarization and Information Assembly

Figure 9.4 shows a schematic of the fire protection system. This system is designed to extinguish all possible fires in a power plant building. Two physically independent water-extinguishing nozzles are designed such that each can control all types of fires in the plant. Extinguishing nozzle-1 is the primary method of injection. Upon receiving a signal from the detector/alarm actuator device (DAA), pump-1 starts automatically, drawing water from the reservoir tank and injecting it into the fire area in the plant. If this pump's injection path is not actuated, plant operator OP_1 can start a second injection path manually. If the second path is not available, the operator OP_2 will call for help from the local fire department, although the detector also automatically sends a signal directly to the fire department. However, due to the delay in the arrival of the local fire department, the magnitude of damage would be higher than it would be if the local fire extinguishing nozzles were available to extinguish the fire.

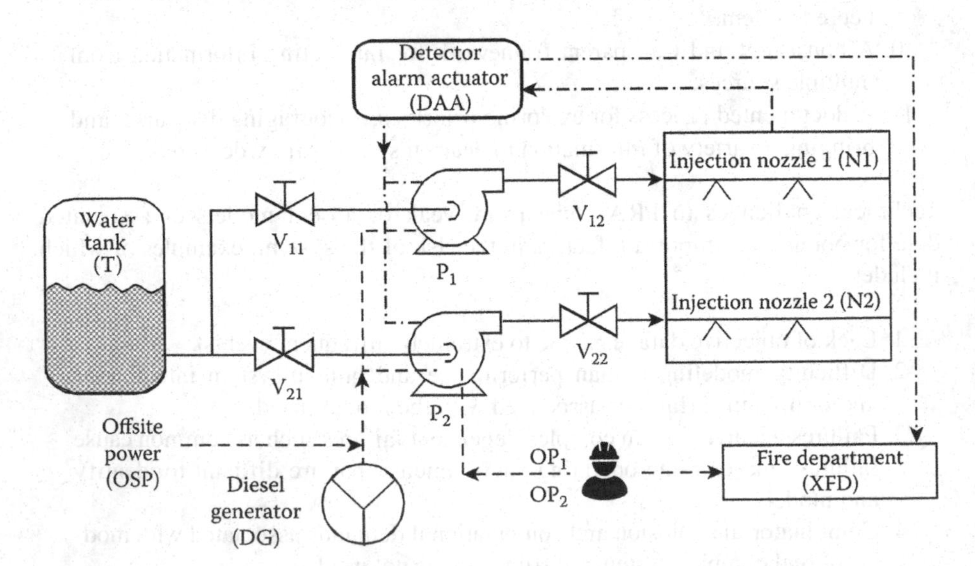

FIGURE 9.4 Fire protection system.

Under all conditions, if the normal off-site power (OSP) was not available due to the fire or other reasons, an on-site diesel generator (DG) would provide emergency electric power to the pumps. The power to the detector/alarm/actuator system is provided through the batteries that are constantly charged by the OSP. Even if the OSP is not available, the power provided by the battery is expected to be always available. The manual valves on the two sides of pump-1 and pump-2 are normally open but are manually closed when the pumps are being repaired. The entire fire protection system and generator are located outside of the main power plant building and are therefore not affected by an internal fire.

9.5.3 Identification of Initiating Events

In this step, all events that potentially cause a sustained fire (and thus thermal and chemical hazards) in the main power plant building must be identified. These include equipment malfunctions, human errors, and facility conditions. The frequency of each event should be estimated. If all events would lead to the same magnitude of fire, the ultimate initiating event is a fire, the frequency of which is the sum of the frequencies of the individual fire-causing events. Assume for this example that the frequency of fire is estimated as 1×10^{-6} per year and that fire is the only challenge to the plant in this example. Therefore, we end up with only one initiating event. However, in more complex situations, it is common to have a large set of initiating events, each posing a different challenge to the plant.

9.5.4 Scenario Development

In this step, we model the cause-and-effect relationship between the fire and the progression of events following the fire. We will use the event tree method to depict this relationship. Generally, this is done inductively, and the level of detail considered in the event tree depends on the analyst. Two types of protective measures are described in the system description: On-site protective measures (on-site pumps, tanks, etc.) and off-site protective measures (local fire department). If the on-site protective measures work correctly, the damage will be minimized, and the off-site fire department is not needed to put out the fire. There are different plant damage states depending on which protective measures are used. Figure 9.5 delineates the scenarios in the form of an event tree model.

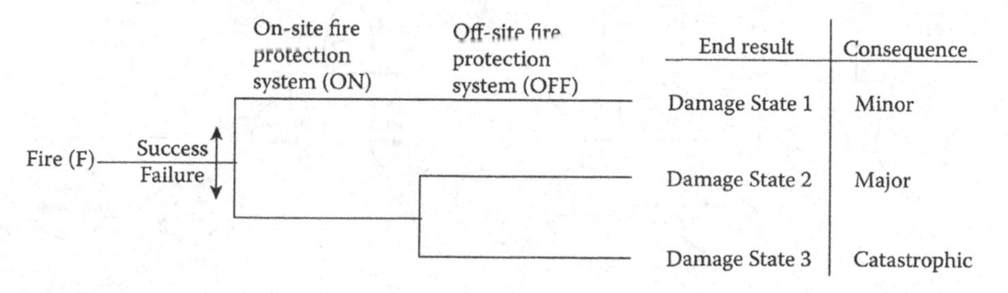

FIGURE 9.5 Scenario of events following a fire using the event tree method.

9.5.5 Causal Logic Model Development

In this step, we identify all failures (equipment or human) that lead to failure of the event tree pivotal events (i.e., on-site or off-site protective subsystems). Figure 9.6 shows the fault tree developed for the on-site fire protection system. In this fault tree, all the basic events that lead to the failure of the two independent water delivery paths are described. Note that the detector/alarm actuator, electric power to the pumps, and the water tank are shared by the two paths. Clearly these are physical dependencies. The effects of these dependencies will be accounted for in the quantification step of the risk analysis. In this fault tree, all external event failures and passive failures (such as pipe or tank leakage) are neglected.

Figure 9.7 shows the fault tree for the off-site fire protection system. This tree is simple since it includes only those failures that would cause a lack of on-time response from the local fire department.

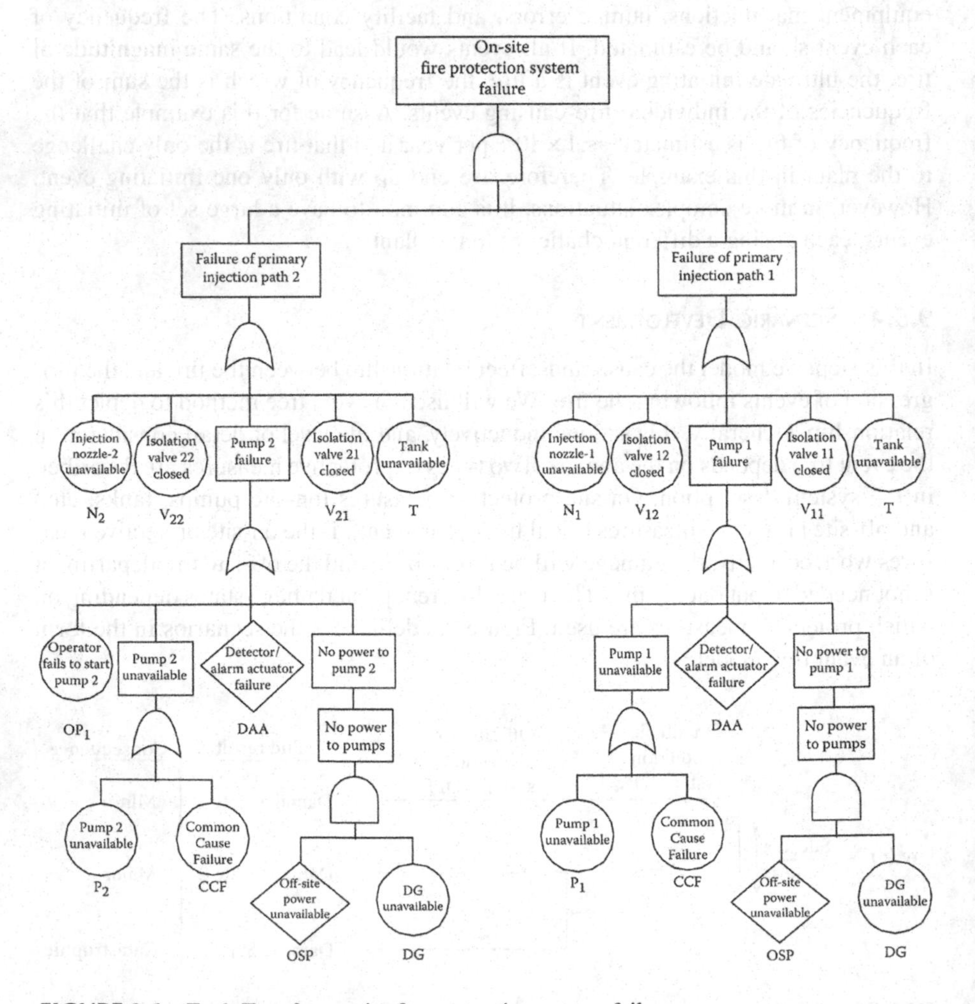

FIGURE 9.6 Fault Tree for on-site fire protection system failure.

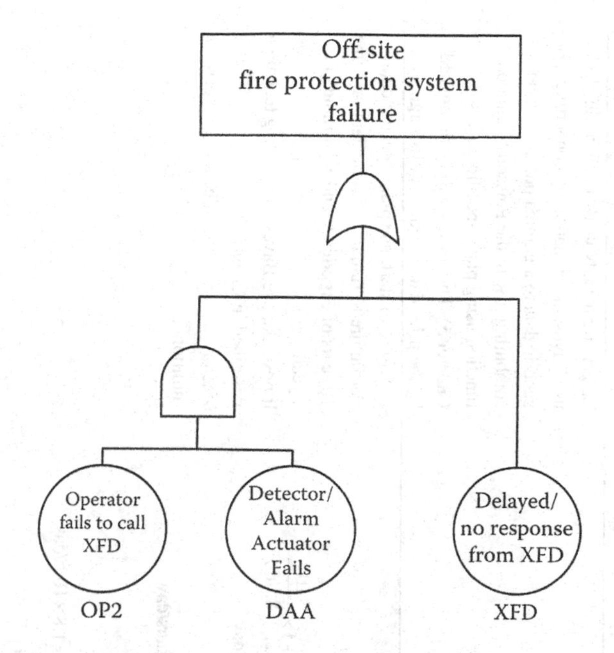

FIGURE 9.7 Fault Tree for off-site fire protection system failure.

9.5.6 Failure Data Analysis

At this point we calculate the frequency of the initiating event and the probabilities of the basic events described in the fault trees. Doing this entails using plant-specific data, generic data, or expert judgment. Table 9.3 describes the raw data used and their sources, the frequency of the initiating event, and the failure probability or unavailability of each event in the model. Furthermore, we assume that CCF can occur between pumps, but no significant CCF exists between valves and nozzles. It is assumed that at least 10 hours of operation is needed for the fire to be completely extinguished.

9.5.7 Quantification and Interpretation

To calculate the frequency of each scenario defined in the event tree of Figure 9.5, we must first determine the cut sets of the two fault trees shown in Figures 9.6 and 9.7. From this, the cut sets of each scenario are determined, followed by the calculation of the probabilities of each scenario based on the occurrence of one of its cut sets. These steps are described below.

First, the cut sets of the on-site fire protection system failure are obtained using the technique described in Chapter 6 and using the rare event approximation. These cut sets are listed in Table 9.4. Only cut set number 22, which is the failure of both pumps, is subject to a CCF. This is shown by adding a new cut set (cut set number 24), which represents this CCF.

The cut sets of the off-site fire protection system are similarly obtained and listed in Table 9.5.

TABLE 9.3
Sources of Data and Failure Probabilities for the Example

Failure Event	Plant-Specific Experience	Generic Data	Probability Used (Per Year or per Demand as Indicated by Units)	Comments
Fire initiation (and ignition) frequency	No such experience in 10 years of operation	Five fires in similar plants in 7,050 plant-years of experience	$F_i = \frac{5}{7,050} = 7.1 \times 10^{-4}$ /year	Use generic data. Note that one could also use Bayesian estimation by converting the generic data to a gamma pdf as prior and combining it with the Poisson likelihood function using plant-specific data (see Chapter 5). The results will be dominated by the prior data, because of its large size.
Pumps 1 and 2 failures	Monthly tests are performed (the test takes negligible time). Four failures to start have been observed for the two pumps in 10 years of tests. Repair time is about 10 hours at a frequency of 1 per year. No experience of failure to run. No common cause failures observed.	Failure to run $= 1 \times 10^{-5}$ /hour. Use the β-factor method, $\beta = 0.1$ for failure of pumps	Failure to start: $\frac{4}{2 \times 12 \times 10} = 1.7 \times 10^{-2}$ /day Unavailability $= 1.7 \times 10^{-2} + 1 \times \frac{10}{8,760}$ $= 1.8 \times 10^{-2}$ /day Failure to run: $= 1.0 \times 10^{-5}$ /hour $\times$ 10 hours/day $= 1.0 \times 10^{-4}$ /day Total for each pump: $= 1.8 \times 10^{-2} + 1.0 \times 10^{-4} \approx 1.8 \times 10^{-2}$ /day Independent failure of each pump: $P_1 = P_2 = 0.9 \times (1.8 \times 10^{-2}$ /day$)$ $= 1.6 \times 10^{-2}$ /day CCF of the pumps: $CCF = 0.1 \times 1.8 \times 10^{-2} = 1.8 \times 10^{-3}$ /day	For failure to start, use plant-specific data. For failure to run, use generic data. Assume 10 years of experience and 8,760 hours in 1 year. (If possible, use Bayesian updating technique described in Chapter 5). β-factor method used; value aligns expert intuition.

(Continued)

TABLE 9.3 (*Continued*)
Sources of Data and Failure Probabilities for the Example

Failure Event	Plant-Specific Experience	Generic Data	Probability Used (Per Year or per Demand as Indicated by Units)	Comments
Failure of isolation valves	One failure to reopen the valve following the monthly test of two pumps.	Not used	$V_{11} = V_{12} = V_{21} = V_{22} = \frac{1}{2 \times 10 \times 12}$ $= 4.2 \times 10^{-3}$ /day	Plant-specific data used.
Failure of nozzles	No such experience	1×10^{-5}/day	$N_1 = N_2 = 1.0 \times 10^{-5}$ /day	Generic data used.
Diesel generator failure	Three failures in the monthly tests conducted for 10 years.	Failure on demand: 3.0×10^{-2}/day and failure to run: 3.0×10^{-3}/hour	Failure on demand: $= \frac{3}{12 \times 10} = 2.5 \times 10^{-2}$ /day Failure to run: $= 3.0 \times 10^{-3}$ /hour × 10 hours/day $= 3.0 \times 10^{-2}$ /day Total failure of DG: $DG = 2.5 \times 10^{-2} + 3.0 \times 10^{-2} = 5.5 \times 10^{-2}$ /day	Plant-specific data used for both failure modes. The results are similar to the generic experience.
Loss of off-site power	No experience	0.1 year	$OSP = 0.1 \times \frac{10 \text{ hour/day}}{8,760 \text{ hour/year}}$ $OSP = 1.1 \times 10^{-4}$ /day	
Failure of DAA	No experience	No data available	$DAA = 1 \times 10^{-4}$ /day	This estimate is based on expert judgment.
Failure of operator to start pump 2	No such experience	Using SPAR-H, the NHEP for action tasks is 1×10^{-3}/day	Assume that time available ≈ time required (PIF multiplier 10). Assuming all other PIFs are nominal. $OP_1 = 1 \times 10^{-3} \times 10 = 1.0 \times 10^{-2}$ /day	Using SPAR-H.
Failure of operator to call the fire department	No such experience	1×10^{-3}/day	$OP_2 = 1 \times 10^{-3}$ /day	This is based on experience from no response to similar situations. Generic probability is used.
No or delayed response from fire department	No such experience	1×10^{-4}/day	$XFD = 1 \times 10^{-4}$ /day	This is based on the response to similar cases from the fire department. Delayed/no arrival is due to accidents, traffic, communication problems, etc.
Tank failure	No such experience	1×10^{-5}/day	$T = 1 \times 10^{-5}$ /day	This is based on data obtained from rupture of the tank or insufficient water content.

TABLE 9.4
Cut Sets of the On-Site Fire Protection System Failure

Cut Set No.	Cut Set	Probability/ (% of Total)	Cut Set No.	Cut Set	Probability/ (% of Total)
1	T	1.0×10^{-5} (0.36%)	13	$V_{21}{\cdot}V_{12}$	1.8×10^{-5} (0.65%)
2	DAA	1.0×10^{-4} (3.62%)	14	$V_{21}{\cdot}P_1$	6.7×10^{-5} (2.43%)
3	$OSP{\cdot}DG$	6.1×10^{-6} (0.22%)	15	$V_{21}{\cdot}V_{11}$	1.8×10^{-5} (0.65%)
4	$N_2{\cdot}N_1$	1.0×10^{-10} (~0%)	16	$OP_I{\cdot}N_1$	1.0×10^{-7} (~0%)
5	$N_2{\cdot}V_{12}$	4.2×10^{-8} (~0%)	17	$OP_I{\cdot}V_{12}$	4.2×10^{-5} (1.52%)
6	$N_2{\cdot}P_1$	1.6×10^{-7} (0.01%)	18	$OP_I{\cdot}P_1$	1.6×10^{-4} (5.80%)
7	$N_2{\cdot}V_{11}$	4.2×10^{-8} (~0%)	19	$OP_I{\cdot}V_{11}$	4.2×10^{-5} (1.52%)
8	$V_{22}{\cdot}N_1$	4.2×10^{-8} (~0%)	20	$P_2{\cdot}N_1$	1.6×10^{-7} (0.01%)
9	$V_{22}{\cdot}V_{12}$	1.8×10^{-5} (0.65%)	21	$P_2{\cdot}V_{12}$	6.7×10^{-5} (2.43%)
10	$V_{22}{\cdot}P_1$	6.7×10^{-5} (2.43%)	22	$P_2{\cdot}P_1$	2.6×10^{-4} (9.42%)
11	$V_{22}{\cdot}V_{11}$	1.8×10^{-5} (0.65%)	23	$P_2{\cdot}V_{11}$	6.7×10^{-5} (2.43%)
12	$V_{21}{\cdot}N_1$	4.2×10^{-8} (~0%)	24	CCF	1.8×10^{-3} (65.2%)

$$Pr(ON) = \sum_i C_i \approx 2.8\times10^{-3}$$

TABLE 9.5
Cut Sets of the Off-Site Fire Protection System

Cut Set No.	Cut Set	Probability
1	XFD	1×10^{-4}
2	$OP_2{\cdot}DAA$	1×10^{-7}

$$\text{Total } Pr(\text{OFF}) = \sum_i C_i \approx 1\times10^{-4}$$

Next, the cut sets of the three scenarios are obtained using the following Boolean equations representing each scenario:

$$\text{Scenario } 1 = F\cdot\overline{ON},$$

$$\text{Scenario } 2 = F\cdot ON\cdot\overline{OFF},$$

$$\text{Scenario } 3 = F\cdot ON\cdot OFF.$$

The cut sets from Tables 9.4 and 9.5 are integrated into these scenarios. Finally, the frequency of each scenario is obtained using the data and cut sets. These frequencies are shown in Table 9.6. The total frequency of each scenario is calculated using the rare event approximation. These are also shown in Table 9.6.

TABLE 9.6
Cut Sets of the Scenarios

Scenario No.	Cut Sets	Frequency	Comment
1	$F \cdot \overline{ON}$	7.1×10^{-4} $(1-2.8\times10^{-3})$ $\approx 7.1\times10^{-4}$	Since the probability can be directly evaluated for $\overline{ON}$ without the need to generate cut sets, only the probability is calculated.
2	$F \cdot DAA \cdot \overline{XFD} \cdot \overline{OP_2}$	7.1×10^{-8}	1. Only cut sets from Table 9.4 that have a contribution greater than 1% are shown. 2. Cut set $F \cdot DAA \cdot \overline{XFD} \cdot \overline{DAA}$ is eliminated since $DAA \cdot \overline{DAA} = \phi$.
	$F \cdot V_{22} \cdot P_1 \cdot \overline{XFD} \cdot \overline{OP_2}$	4.8×10^{-8}	
	$F \cdot V_{21} \cdot P_1 \cdot \overline{XFD} \cdot \overline{OP_2}$	4.8×10^{-8}	
	$F \cdot V_{22} \cdot P_1 \cdot \overline{XFD} \cdot \overline{DAA}$	4.8×10^{-8}	
	$F \cdot V_{21} \cdot P_1 \cdot \overline{XFD} \cdot \overline{DAA}$	4.8×10^{-8}	
	$F \cdot OP_1 \cdot V_{12} \cdot \overline{XFD} \cdot \overline{OP_2}$	3.0×10^{-9}	
	$F \cdot OP_1 \cdot V_{12} \cdot \overline{XFD} \cdot \overline{DAA}$	3.0×10^{-9}	
	$F \cdot OP_1 \cdot P_1 \cdot \overline{XFD} \cdot \overline{OP_2}$	1.1×10^{-7}	
	$F \cdot OP_1 \cdot P_1 \cdot \overline{XFD} \cdot \overline{DAA}$	1.1×10^{-7}	
	$F \cdot OP_1 \cdot V_{11} \cdot \overline{XFD} \cdot \overline{OP_2}$	3.0×10^{-9}	
	$F \cdot OP_1 \cdot V_{11} \cdot \overline{XFD} \cdot \overline{DAA}$	3.0×10^{-9}	
	$F \cdot P_2 \cdot V_{12} \cdot \overline{XFD} \cdot \overline{OP_2}$	4.8×10^{-8}	
	$F \cdot P_2 \cdot V_{12} \cdot \overline{XFD} \cdot \overline{DAA}$	4.8×10^{-8}	
	$F \cdot P_2 \cdot P_1 \cdot \overline{XFD} \cdot \overline{OP_2}$	1.8×10^{-7}	
	$F \cdot P_2 \cdot P_1 \cdot \overline{XFD} \cdot \overline{DAA}$	1.8×10^{-7}	
	$F \cdot P_2 \cdot V_{11} \cdot \overline{XFD} \cdot \overline{OP_2}$	4.8×10^{-8}	
	$F \cdot P_2 \cdot V_{11} \cdot \overline{XFD} \cdot \overline{DAA}$	4.8×10^{-8}	
	$F \cdot CCF \cdot \overline{XFD} \cdot \overline{OP_2}$	1.3×10^{-6}	
	$F \cdot CCF \cdot \overline{XFD} \cdot \overline{DAA}$	1.3×10^{-6}	
2 Total:		$\Sigma C_i = 3.7\times10^{-6}$	
3	$F \cdot DAA \cdot OP_2$	7.1×10^{-11}	1. Only cut sets from Tables 9.4 and 9.5 that have a contribution greater than 1% are shown.
	$F \cdot DAA \cdot XFD$	7.1×10^{-12}	
	$F \cdot V_{22} \cdot P_1 \cdot XFD$	4.8×10^{-12}	
	$F \cdot V_{21} \cdot P_1 \cdot XFD$	4.8×10^{-12}	
	$F \cdot OP_1 \cdot V_{12} \cdot XFD$	3.0×10^{-12}	
	$F \cdot OP_1 \cdot P_1 \cdot XFD$	1.1×10^{-11}	
	$F \cdot OP_1 \cdot V_{11} \cdot XFD$	3.0×10^{-12}	
	$F \cdot P_2 \cdot V_{12} \cdot XFD$	4.8×10^{-12}	
	$F \cdot P_2 \cdot P_1 \cdot XFD$	1.8×10^{-11}	
	$F \cdot P_2 \cdot V_{11} \cdot XFD$	4.8×10^{-12}	
	$F \cdot CCF \cdot XFD$	1.3×10^{-10}	
3 Total:		$\Sigma C_i = 2.6\times10^{-10}$	

9.5.8 Consequences Evaluation

In the scenario development and quantification tasks, we identified three distinct scenarios of interest, each with different outcomes and frequencies. The consequences associated with each scenario should be specified in terms of both economic and/or human losses. This part of the analysis is one of the most difficult for two reasons:

1. Each scenario poses different hazards and methods of hazard exposure and requires careful monitoring. Here, the model should include the ways the fire can spread through the plant, the ways people can be exposed, evacuation procedures, the availability of protective clothing, and so on.
2. The outcome of the scenario can be measured in terms of human losses. It can also be measured in terms of financial losses, that is, the total cost associated with the scenario. This involves monetizing human life or fatalities, which sometimes is a source of controversy.

Suppose a careful analysis of the spread of fire and fire exposure is performed, with consideration of the above issues, and ultimately results in damages measured only in terms of economic losses. These results are shown in Table 9.7.

9.5.9 Risk Calculation and Interpretation of Results

Using values from Table 9.7, we can calculate the total risk associated with each scenario. These risks are shown in Table 9.8.

It can be seen in Table 9.8 that the risk for the plant is very low: the value at risk (expected losses in dollars) indicates that fire risk is not important for this plant. However, Scenarios 1 and 2 are significantly more important than Scenario 3, even though they result in lower consequences because both of these scenarios are several

TABLE 9.7
Economic Consequences of Fire Scenarios

Scenario No.	Economic Consequence
1	\$1,000,000
2	\$92,000,000
3	\$210,000,000

TABLE 9.8
Risk Associated with Each Scenario

Scenario No.	Economic Consequence
1	(7.1×10^{-4}) (\$1,000,000) = \$710.00
2	(3.7×10^{-6}) (\$92,000,000) = \$340.40
3	(2.6×10^{-10}) (\$210,000,000) = \$0.05

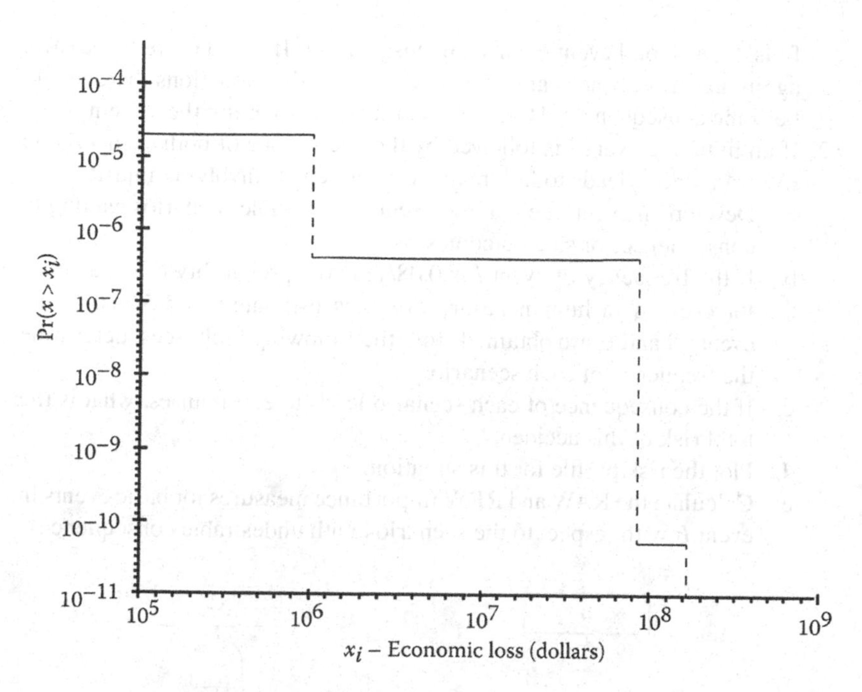

FIGURE 9.8 Risk profile for the fire example.

orders of magnitude more likely than Scenario 3. Therefore, if the risk were high, one should improve those components that are major contributors to Scenarios 1 and 2. The consequence of Scenario 1 is primarily influenced by CCFs between pumps P_1 and P_2, so reducing this failure is a potential source of improvement.

Since this analysis shows that the risk due to fire is rather low, uncertainty analysis is not very important. However, one of the methods described in Section 8.3 could be used to estimate the uncertainty associated with each component and the fire-initiating event if necessary. The uncertainties should be propagated through the cut sets of each scenario to obtain the uncertainty associated with the frequency estimation of each scenario. The uncertainty associated with the consequence estimates can also be obtained. When the uncertainty associated with the consequence values is combined with the scenario frequencies and their uncertainty, the uncertainty associated with the estimated consequences can be calculated. Although this is not a necessary step in this risk analysis example, it is critical to make estimates of the uncertainties when risk values are high.

Figure 9.8 shows the risk profile based on the values in Table 9.8.

9.6 EXERCISES

9.1 Three subsystems A, B, and C function so that, if successful, the system will pose no risks. The frequency of initiating event, I, is 0.1 events per year. Subsequent to the initiating event, pivotal event A should succeed, in which case pivotal event B is bypassed, after which pivotal event C must succeed. In this case no consequences are expected. If pivotal event A

fails, then pivotal event B can compensate for it. If B and C are successful, again no consequences are expected. For all other situations there would be some consequences. Draw an event tree representing the system.

9.2 If an initiating event I is followed by the occurrence of both events B and C, or A alone, it leads to failure and consequences involving injuries.

a. Develop an event tree that represents all possible scenarios leading to consequences or safe outcomes.

b. If the frequency of event I is 0.08/year the probability of occurrence for event A (a human failure event) is estimated as 5.5×10^{-3}, and events B and C are obtained from the following fault trees, determine the frequency of each scenario.

c. If the consequence of each scenario leads to 255 injuries, what is the total risk of this accident?

d. Plot the risk profile for this situation.

e. Calculate the RAW and RRW importance measures for basic events in event B with respect to the scenarios with undesirable consequences.

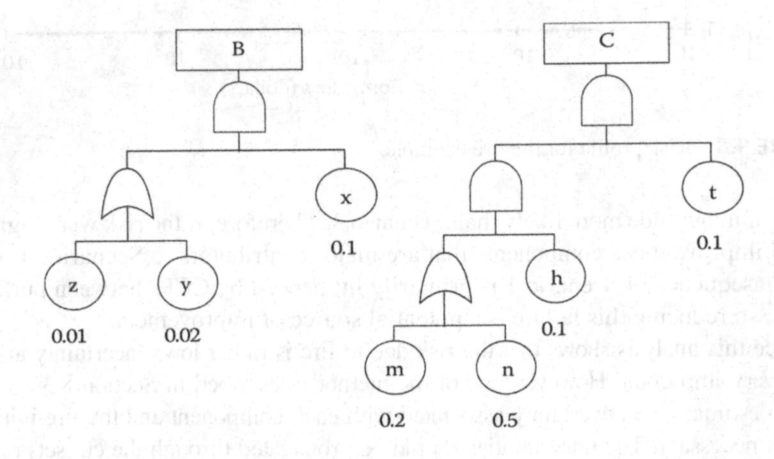

9.3 A PRA grouped risk scenarios into five categories, each associated with a different total frequency and consequence. The scenarios, frequencies, and consequences are shown below.

Scenario	Frequency of Exposure (Per Year)	Expected Number of Fatalities Given the Exposure
Release of substantial amount of chemical hazard and exposure due to failure of control	1×10^{-7}	500
Release and exposure due to external event factors (flooding, earthquake, etc.)	2.3×10^{-6}	350
Release and exposure due to internal flooding	4.0×10^{-6}	200
Release and exposure due to internal fire and delays in response	9.3×10^{-4}	800
All other forms of exposure	1.9×10^{-8}	900

a. Plot the risk profile in terms of "frequency per year of exceeding the given number of fatalities" versus "number of fatalities."
b. Calculate the total risk metric and discuss its tolerability.
c. Discuss uncertainties associated with the analysis.

9.4 Consider these two risk curves associated with two proposed design versions for a system.

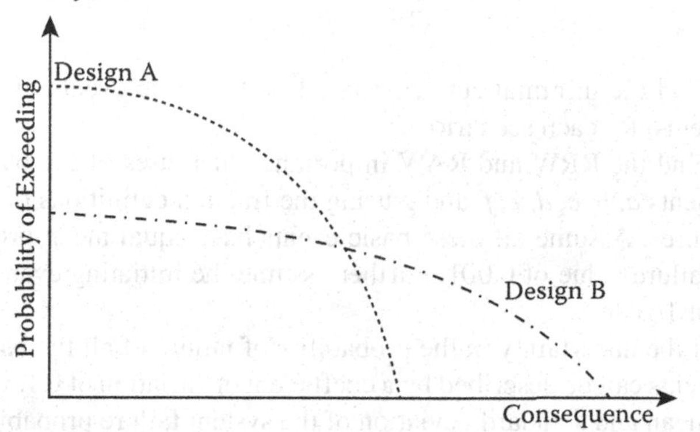

a. Explain which design is riskier and under what conditions.
b. How would you advise a rational organization faced with choosing between the two?

9.5 Consider a driver at night facing a deer that jumps in front of the car. Statistics show that over a 1 year period there have been 125 reports of such incidents in this area; assume that all incidents are counted. We want to know the possible outcomes. The important questions to address are: First, does the driver see the deer? Assume that the probability of "Yes" is 0.95. If the driver sees the deer, do they apply the brakes in time? Assume the probability of "Yes" is 0.75. If the brakes are applied, do they work as expected? Assume the probability of "Yes" is 0.9995 (mechanical systems are typically much more reliable than humans). Finally, would the deer evade the car? Assume the probability of "Yes" is 0.5.
a. Develop an event tree including the logic representation of each scenario and calculate the frequency of each scenario.
b. What is the total risk of hitting the deer?

9.6 Analyze the following event tree.

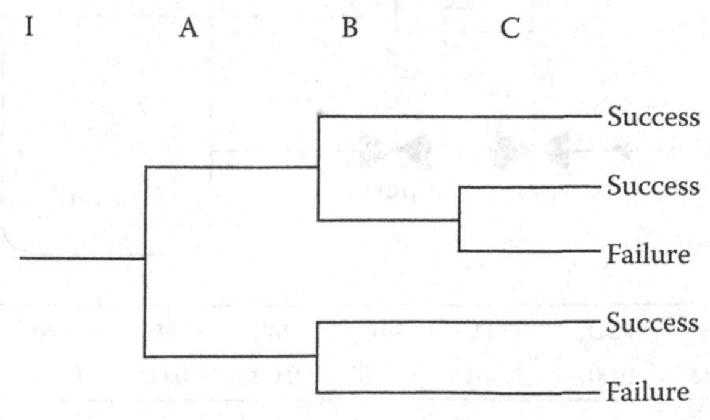

Assume the cut sets of the subsystems *A*, *B*, and *C* as a function of the basic events are:

$$A = ab + bc$$

$$B = ad$$

$$C = e + g + cf$$

a. Find the minimal cut sets (based on the basic events of the subsystems) for each scenario.
b. Find the RRW and RAW importance measures of the basic components *a*, *b*, *c*, *d*, *e*, *f*, and *g* using the fraction definitions of these measures. Assume all these basic events have equal mean probability of failure value of 0.001. Further assume the initiating event frequency of 1/year.
c. If the uncertainty in the probability of failure of all the basic components can be described by a coefficient of variation of 0.1, what are the mean and standard deviation of the system failure probability?

9.7 Consider the system below. Gas enters the tank through two identical process shutdown (PSD) valves, PSD_1 and PSD_2, arranged in series. The valves are fail-safe close and are held open by hydraulic pressure. If the valves are not closed it is possible that an operator can manually close either valve. When the pressure is too high, the pressure sensors send signals to PLC to close the process shutdown valves. Either valve can stop the flow. A pressure safety valve PSV can reduce the overpressure passively. Assume overpressure is the only potential cause of a tank failure. Assume the connecting pipes and wiring are perfect.

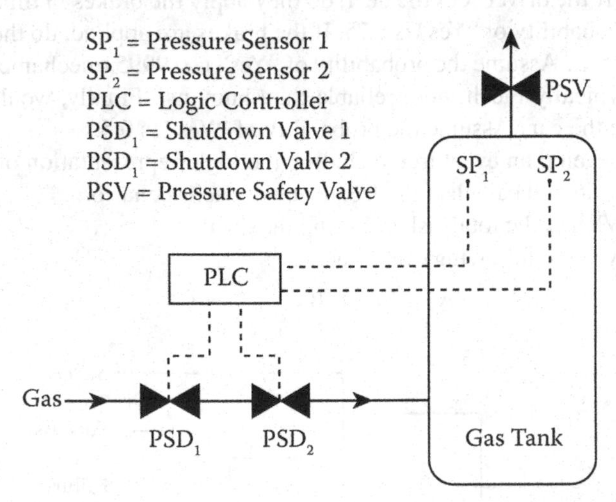

Name	PSD_1	PSD_2	PLC	SP_1	SP_2	PSV	Human Error
Probability of failure	0.001	0.001	0.0001	0.002	0.002	0.0001	0.08

a. Construct a fault tree for the top event "Overpressure in the tank."
b. Find the minimal cut sets of the top event.
c. If there is a common cause failure between the two sensors (SP_1 and SP_2) with $\beta = 0.05$ and considering the failure data given, find the probability of occurrence of the top event.

REFERENCES

Aldemir, T., "A Survey of Dynamic Methodologies for Probabilistic Safety Assessment of Nuclear Power Plants." *Annals of Nuclear Energy*, 52, 113–124, 2013.

ASME/ANS-RA-SA, *Standard for Level 1/Large Early Release Frequency PRA for NPP Applications*. ASME/ANS RA-SA-2009, 2009.

Aven, T. et al., "Society for Risk Analysis Glossary." Society for Risk Analysis, Aug, 2018. Available: https://www.sra.org/risk-analysis-introduction/risk-analysis-glossary/.

Azarkhail, M. and M. Modarres, "A Study of Implications of Using Importance Measures in Risk-Informed Decisions." In *PSAM-7, ESREL 04 Joint Conference*. Berlin, Germany, June 2004.

Center for Chemical Process Safety, *Guidelines for Chemical Process Quantitative Risk Analysis*, 2nd edition. Wiley Interscience, New York, 1999.

Dezfuli, H. and M. Modarres, "A Truncation Methodology for Evaluation of Large Fault Trees." *IEEE Transactions on Reliability*, 33, 325–328, 1984.

Farmer, F. R., "Reactor Safety and Siting: A Proposed Risk Criterion." *Nuclear Safety*, 539, 23–32, 1961.

Groth, K. M., and E. S. Hecht, "HyRAM: A Methodology and Toolkit for Quantitative Risk Assessment of Hydrogen Systems." *International Journal of Hydrogen Energy*, 42, 7485–7493, 2017.

Groth, K., C. Wang, and A. Mosleh, "Hybrid Causal Methodology and Software Platform for Probabilistic Risk Assessment and Safety Monitoring of Socio-Technical Systems." *Reliability Engineering & System Safety*, 95, 1276–1285, 2010.

Kaplan, S., and J. Garrick, "On the Quantitative Definition of Risk." *Risk Analysis*, 1(1), 11–28, 1981.

Kumamoto, H., and E. J. Henley, *Probabilistic Risk Assessment for Engineers and Scientists*. IEEE Press, New York, 1996.

LaChance J., A. Tchouvelev, and A. Engebø, "Development of Uniform Harm Criteria for Use in Quantitative Risk Analysis of the Hydrogen Infrastructure." *International Journal of Hydrogen Energy*, 36(3), 2381–2388, 2011.

Mannan, S. ed., *Lees' Loss Prevention in the Process Industries*, 4th edition. Elsevier, London, 2012.

Modarres, M., *Risk Analysis in Engineering, Techniques, Tools and Trends*. CRC Press, Boca Raton, FL. 2006.

Moradi, R., and K. M. Groth, "Modernizing Risk Assessment: A Systematic Integration of PRA and PHM Techniques." *Reliability Engineering & System Safety*, 204, 107194, 2020.

Parhizkar, T., J. E. Vinnem, I. B. Utne, and A. Mosleh, "Supervised Dynamic Probabilistic Risk Assessment of Complex Systems, Part 1: General Overview." *Reliability Engineering & System Safety*, 208, 107406, 2021.

Sattison, M. B., and K. W. Hall, *Analysis of Core Damage Frequency: Zion, Unit 1 Internal Events*, NUREG/CR-4550. US Nuclear Regulatory Commission. Washington, DC. 1990.

Smidts, C., "Software Reliability." In J. C. Whitaker, ed., *The Electronics Handbook*. CRC Press and IEEE Press, Boca Raton, FL, 1996.

Stamatelatos, M., and H. Dezfuli, eds., *Probabilistic Risk Assessment Procedures Guide for NASA Managers and Practitioners, Version 2*. National Aeronautics and Space Administration, Washington, DC, 2011.

Swaminathan, S., and C. Smidts, "The Event Sequence Diagram Framework for Dynamic Probabilistic Risk Assessment." *Reliability Engineering & System Safety*, 63 (1), 73–90, 1999.

U.S. NRC, WASH-1400, *Reactor Safety Study—An Assessment of Accident Risks in US Commercial Nuclear Power Plants*. U.S. Nuclear Regulatory Commission, Washington, DC. 1975.

Villa V., N. Paltrinieri, F. Khan, and V. Cozzani, "Towards Dynamic Risk Analysis: A Review of the Risk Assessment Approach and Its Limitations in the Chemical Process Industry." *Safety Science*, 89, 77–93, 2016.

Vinnem, J. E., *Offshore Risk Assessment*, 2nd edition. Springer, London, 2007.

Appendix A
Statistical Tables

TABLE A.1
The Standard Normal Cumulative Distribution Function

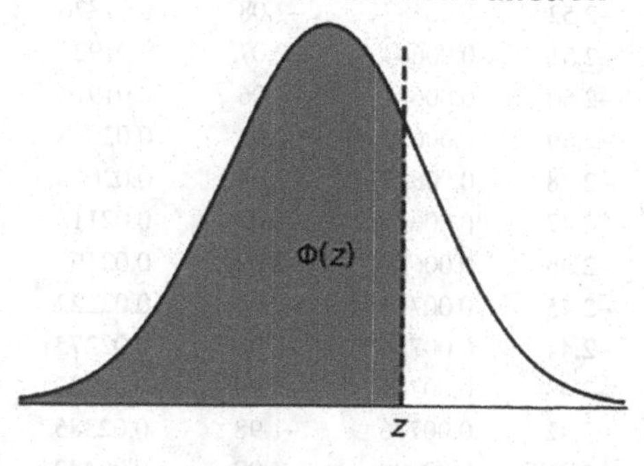

z	$\Phi(z)$	z	$\Phi(z)$	z	$\Phi(z)$	z	$\Phi(z)$
−4.00	0.00003	−3.75	0.00009	−3.51	0.00022	−3.26	0.00056
−3.99	0.00003	−3.74	0.00009	−3.50	0.00023	−3.25	0.00058
−3.98	0.00003	−3.73	0.00010	−3.49	0.00024	−3.24	0.00060
−3.97	0.00004	−3.72	0.00010	−3.48	0.00025	−3.23	0.00062
−3.96	0.00004	−3.71	0.00010	−3.47	0.00026	−3.22	0.00064
−3.95	0.00004	−3.70	0.00011	−3.46	0.00027	−3.21	0.00066
−3.94	0.00004	−3.69	0.00011	−3.45	0.00028	−3.20	0.00069
−3.93	0.00004	−3.68	0.00012	−3.44	0.00029	−3.19	0.00071
−3.92	0.00004	−3.67	0.00012	−3.43	0.00030	−3.18	0.00074
−3.91	0.00005	−3.57	0.00018	−3.42	0.00031	−3.17	0.00076
−3.90	0.00005	−3.66	0.00013	−3.41	0.00032	−3.16	0.00079
−3.89	0.00005	−3.65	0.00013	−3.40	0.00034	−3.15	0.00082
−3.88	0.00005	−3.64	0.00014	−3.39	0.00035	−3.14	0.00084
−3.87	0.00005	−3.63	0.00014	−3.38	0.00036	−3.13	0.00087
−3.86	0.00006	−3.62	0.00015	−3.37	0.00038	−3.12	0.00090
−3.85	0.00006	−3.61	0.00015	−3.36	0.00039	−3.11	0.00094
−3.84	0.00006	−3.60	0.00016	−3.35	0.00040	−3.10	0.00097
−3.83	0.00006	−3.59	0.00017	−3.34	0.00042	−3.09	0.00100
−3.82	0.00007	−3.58	0.00017	−3.33	0.00043	−3.08	0.00104
−3.81	0.00007	−3.57	0.00018	−3.32	0.00045	−3.07	0.00107
−3.80	0.00007	−3.56	0.00019	−3.31	0.00047	−3.06	0.00111
−3.79	0.00008	−3.55	0.00019	−3.30	0.00048	−3.05	0.00114
−3.78	0.00008	−3.54	0.00020	−3.29	0.00050	−3.04	0.00118
−3.77	0.00008	−3.53	0.00021	−3.28	0.00052	−3.03	0.00122
−3.76	0.00008	−3.52	0.00022	−3.27	0.00054	−3.02	0.00126

(Continued)

TABLE A.1 (*Continued*)
The Standard Normal Cumulative Distribution Function

z	$\Phi(z)$	z	$\Phi(z)$	z	$\Phi(z)$	z	$\Phi(z)$
−3.01	0.00131	−2.57	0.00508	−2.13	0.01659	−1.69	0.04551
−3.00	0.00135	−2.56	0.00523	−2.12	0.01700	−1.68	0.04648
−2.99	0.00139	−2.55	0.00539	−2.11	0.01743	−1.67	0.04746
−2.98	0.00144	−2.54	0.00554	−2.10	0.01786	−1.66	0.04846
−2.97	0.00149	−2.53	0.00570	−2.09	0.01831	−1.65	0.04947
−2.96	0.00154	−2.52	0.00587	−2.08	0.01876	−1.64	0.05050
−2.95	0.00159	−2.51	0.00604	−2.07	0.01923	−1.63	0.05155
−2.94	0.00164	−2.50	0.00621	−2.06	0.01970	−1.62	0.05262
−2.93	0.00169	−2.49	0.00639	−2.05	0.02018	−1.61	0.05370
−2.92	0.00175	−2.48	0.00657	−2.04	0.02068	−1.60	0.05480
−2.91	0.00181	−2.47	0.00676	−2.03	0.02118	−1.59	0.05592
−2.90	0.00187	−2.46	0.00695	−2.02	0.02169	−1.58	0.05705
−2.89	0.00193	−2.45	0.00714	−2.01	0.02222	−1.57	0.05821
−2.88	0.00199	−2.44	0.00734	−2.00	0.02275	−1.56	0.05938
−2.87	0.00205	−2.43	0.00755	−1.99	0.02330	−1.55	0.06057
−2.86	0.00212	−2.42	0.00776	−1.98	0.02385	−1.54	0.06178
−2.85	0.00219	−2.41	0.00798	−1.97	0.02442	−1.53	0.06301
−2.84	0.00226	−2.40	0.00820	−1.96	0.02500	−1.52	0.06426
−2.83	0.00233	−2.39	0.00842	−1.95	0.02559	−1.51	0.06552
−2.82	0.00240	−2.38	0.00866	−1.94	0.02619	−1.50	0.06681
−2.81	0.00248	−2.37	0.00820	−1.93	0.02680	−1.49	0.06811
−2.80	0.00256	−2.36	0.00842	−1.92	0.02743	−1.48	0.06944
−2.79	0.00264	−2.35	0.00866	−1.91	0.02807	−1.47	0.07078
−2.78	0.00272	−2.34	0.00889	−1.90	0.02872	−1.46	0.07215
−2.77	0.00280	−2.33	0.00914	−1.89	0.02938	−1.45	0.07353
−2.76	0.00289	−2.32	0.00939	−1.88	0.03005	−1.44	0.07493
−2.75	0.00298	−2.31	0.00964	−1.87	0.03074	−1.43	0.07636
−2.74	0.00307	−2.30	0.00990	−1.86	0.03144	−1.42	0.07780
−2.73	0.00317	−2.29	0.01017	−1.85	0.03216	−1.41	0.07927
−2.72	0.00326	−2.28	0.01044	−1.84	0.03288	−1.40	0.08076
−2.71	0.00336	−2.27	0.01072	−1.83	0.03362	−1.39	0.08226
−2.70	0.00347	−2.26	0.01191	−1.82	0.03438	−1.38	0.08379
−2.69	0.00357	−2.25	0.01222	−1.81	0.03515	−1.37	0.08534
−2.68	0.00368	−2.24	0.01255	−1.80	0.03593	−1.36	0.08691
−2.67	0.00379	−2.23	0.01287	−1.79	0.03673	−1.35	0.08851
−2.66	0.00391	−2.22	0.01321	−1.78	0.03754	−1.34	0.09012
−2.65	0.00402	−2.21	0.01355	−1.77	0.03836	−1.33	0.09176
−2.64	0.00415	−2.20	0.01390	−1.76	0.03920	−1.32	0.09342
−2.63	0.00427	−2.19	0.01426	−1.75	0.04006	−1.31	0.09510
−2.62	0.00440	−2.18	0.01463	−1.74	0.04093	−1.30	0.09680
−2.61	0.00453	−2.17	0.01500	−1.73	0.04182	−1.29	0.09853
−2.60	0.00466	−2.16	0.01539	−1.72	0.04272	−1.28	0.10027
−2.59	0.00480	−2.15	0.01578	−1.71	0.04363	−1.27	0.10204
−2.58	0.00494	−2.14	0.01618	−1.70	0.04457	−1.26	0.10383

(*Continued*)

TABLE A.1 ***(Continued)***
The Standard Normal Cumulative Distribution Function

z	$\Phi(z)$	z	$\Phi(z)$	z	$\Phi(z)$	z	$\Phi(z)$
−1.25	0.10565	−0.81	0.20897	−0.29	0.38591	+0.15	0.55962
−1.24	0.10749	−0.80	0.21186	−0.28	0.38974	+0.16	0.56356
−1.23	0.10935	−0.79	0.21476	−0.27	0.39358	+0.17	0.56749
−1.22	0.11123	−0.78	0.21770	−0.26	0.39743	+0.18	0.57142
−1.21	0.11314	−0.77	0.22065	−0.25	0.40129	+0.19	0.57535
−1.20	0.11507	−0.76	0.22363	−0.24	0.40517	+0.20	0.57926
−1.19	0.11702	−0.75	0.22663	−0.23	0.40905	+0.21	0.58317
−1.18	0.11900	−0.74	0.22965	−0.22	0.41294	+0.22	0.58706
−1.17	0.12100	−0.73	0.23270	−0.21	0.41683	+0.23	0.59095
−1.16	0.12302	−0.72	0.23576	−0.20	0.42074	+0.24	0.59483
−1.15	0.12507	−0.71	0.23885	−0.19	0.42465	+0.25	0.59871
−1.14	0.12714	−0.70	0.24196	−0.18	0.42858	+0.26	0.60257
−1.13	0.12924	−0.69	0.24510	−0.17	0.43251	+0.27	0.60642
−1.12	0.13136	−0.68	0.24825	−0.16	0.43644	+0.28	0.61026
−1.11	0.13350	−0.67	0.25143	−0.15	0.44038	+0.29	0.61409
−1.10	0.13567	−0.66	0.25463	−0.14	0.44433	+0.30	0.61791
−1.09	0.13786	−0.65	0.25785	−0.13	0.44828	+0.31	0.62172
−1.08	0.14007	−0.64	0.26109	−0.12	0.45224	+0.32	0.62552
−1.07	0.14231	−0.63	0.26435	−0.11	0.45620	+0.33	0.62930
−1.06	0.14457	−0.62	0.26763	−0.10	0.46017	+0.34	0.63307
−1.05	0.14686	−0.61	0.27093	−0.09	0.46414	+0.35	0.63683
−1.04	0.14917	−0.60	0.27425	−0.08	0.46812	+0.36	0.64058
−1.03	0.15151	−0.59	0.27760	−0.07	0.47210	+0.37	0.64431
−1.02	0.15386	−0.58	0.28096	−0.06	0.47608	+0.38	0.64803
−1.01	0.15625	−0.57	0.28434	−0.05	0.48006	+0.39	0.65173
−1.00	0.15866	−0.56	0.28774	−0.04	0.48405	+0.40	0.65542
−0.99	0.16109	−0.55	0.29116	−0.03	0.48803	+0.41	0.65910
−0.98	0.16354	−0.54	0.29460	−0.02	0.49202	+0.42	0.66276
−0.97	0.16602	−0.53	0.29806	−0.01	0.49601	+0.43	0.66640
−0.96	0.16853	−0.46	0.32276	+0.00	0.50000	+0.44	0.67003
−0.95	0.17106	−0.45	0.32636	+0.01	0.50399	+0.45	0.67364
−0.94	0.17361	−0.44	0.32997	+0.02	0.50798	+0.46	0.67724
−0.93	0.17619	−0.43	0.33360	+0.03	0.51197	+0.47	0.68082
−0.92	0.17879	−0.42	0.33724	+0.04	0.51595	+0.48	0.68439
−0.91	0.18141	−0.41	0.34090	+0.05	0.51994	+0.49	0.68793
−0.90	0.18406	−0.40	0.34458	+0.06	0.52392	+0.50	0.69146
−0.89	0.18673	−0.39	0.34827	+0.07	0.52790	+0.51	0.69497
−0.88	0.18943	−0.36	0.35942	+0.08	0.53188	+0.52	0.69847
−0.87	0.19215	−0.35	0.36317	+0.09	0.53586	+0.53	0.70194
−0.86	0.19489	−0.34	0.36693	+0.10	0.53983	+0.54	0.70540
−0.85	0.19766	−0.33	0.37070	+0.11	0.54380	+0.55	0.70884
−0.84	0.20045	−0.32	0.37448	+0.12	0.54776	+0.56	0.71226
−0.83	0.20327	−0.31	0.37828	+0.13	0.55172	+0.57	0.71566
−0.82	0.20611	−0.30	0.38209	+0.14	0.55567	+0.58	0.71904

(Continued)

TABLE A.1 ***(Continued)***
The Standard Normal Cumulative Distribution Function

z	$\Phi(z)$	z	$\Phi(z)$	z	$\Phi(z)$	z	$\Phi(z)$
+0.59	0.72240	+1.11	0.86650	+1.55	0.93943	+1.98	0.97615
+0.60	0.72575	+1.12	0.86864	+1.56	0.94062	+1.99	0.97670
+0.61	0.72907	+1.13	0.87076	+1.57	0.94179	+2.00	0.97725
+0.62	0.73237	+1.14	0.87286	+1.58	0.94295	+2.01	0.97778
+0.63	0.73565	+1.15	0.87493	+1.59	0.94408	+2.02	0.97831
+0.64	0.73891	+1.16	0.87698	+1.60	0.94520	+2.03	0.97882
+0.65	0.74215	+1.17	0.87900	+1.61	0.94630	+2.04	0.97932
+0.66	0.74537	+1.18	0.88100	+1.62	0.94738	+2.05	0.97982
+0.67	0.74857	+1.19	0.88298	+1.63	0.94845	+2.06	0.98030
+0.68	0.75175	+1.20	0.88493	+1.64	0.94950	+2.07	0.98077
+0.69	0.75490	+1.21	0.88686	+1.65	0.95053	+2.08	0.98124
+0.70	0.75804	+1.22	0.88877	+1.66	0.95154	+2.09	0.98169
+0.71	0.76115	+1.23	0.89065	+1.67	0.95254	+2.10	0.98214
+0.72	0.76424	+1.24	0.89251	+1.68	0.95352	+2.11	0.98257
+0.73	0.76730	+1.25	0.89435	+1.69	0.95449	+2.12	0.98300
+0.74	0.77035	+1.26	0.89617	+1.70	0.95543	+2.13	0.98341
+0.75	0.77337	+1.27	0.89796	+1.71	0.95637	+2.14	0.98382
+0.76	0.77637	+1.28	0.89973	+1.72	0.95728	+2.15	0.98422
+0.77	0.77935	+1.29	0.90147	+1.73	0.95818	+2.16	0.98461
+0.78	0.78230	+1.30	0.90320	+1.74	0.95907	+2.17	0.98500
+0.79	0.78524	+1.31	0.90490	+1.75	0.95994	+2.18	0.98537
+0.80	0.78814	+1.32	0.90658	+1.76	0.96080	+2.19	0.98574
+0.81	0.79103	+1.33	0.90824	+1.77	0.96164	+2.20	0.98610
+0.82	0.79389	+1.34	0.90988	+1.78	0.96246	+2.21	0.98645
+0.83	0.79673	+1.35	0.91149	+1.79	0.96327	+2.22	0.98679
+0.84	0.79955	+1.36	0.91309	+1.80	0.96407	+2.23	0.98713
+0.85	0.80234	+1.37	0.91466	+1.81	0.96485	+2.24	0.98745
+0.86	0.80511	+1.38	0.91621	+1.82	0.96562	+2.25	0.98778
+0.87	0.80785	+1.39	0.91774	+1.83	0.96638	+2.26	0.98809
+0.96	0.83147	+1.40	0.91924	+1.84	0.96712	+2.27	0.98840
+0.97	0.83398	+1.41	0.92073	+1.85	0.96784	+2.28	0.98870
+0.98	0.83646	+1.42	0.92220	+1.86	0.96856	+2.29	0.98899
+0.99	0.83891	+1.43	0.92364	+1.87	0.96926	+2.30	0.98928
+1.00	0.84134	+1.44	0.92507	+1.88	0.96995	+2.31	0.98956
+1.01	0.84375	+1.45	0.92647	+1.89	0.97062	+2.32	0.98983
+1.02	0.84614	+1.46	0.92785	+1.80	0.96407	+2.33	0.99010
+1.03	0.84849	+1.47	0.92922	+1.90	0.97128	+2.34	0.99036
+1.04	0.85083	+1.48	0.93056	+1.91	0.97193	+2.35	0.99061
+1.05	0.85314	+1.49	0.93189	+1.92	0.97257	+2.36	0.99086
+1.06	0.85543	+1.50	0.93319	+1.93	0.97320	+2.37	0.99111
+1.07	0.85769	+1.51	0.93448	+1.94	0.97381	+2.38	0.99134
+1.08	0.85993	+1.52	0.93574	+1.95	0.97441	+2.39	0.99158
+1.09	0.86214	+1.53	0.93699	+1.96	0.97500	+2.40	0.99180
+1.10	0.86433	+1.54	0.93822	+1.97	0.97558	+2.41	0.99202

(Continued)

TABLE A.1 (*Continued*)
The Standard Normal Cumulative Distribution Function

z	$\Phi(z)$	z	$\Phi(z)$	z	$\Phi(z)$	z	$\Phi(z)$
+2.42	0.99224	+2.85	0.99781	+3.28	0.99948	+3.73	0.99990
+2.43	0.99245	+2.86	0.99788	+3.29	0.99950	+3.74	0.99991
+2.44	0.99266	+2.87	0.99795	+3.30	0.99952	+3.75	0.99991
+2.45	0.99286	+2.88	0.99801	+3.31	0.99953	+3.76	0.99992
+2.46	0.99305	+2.89	0.99807	+3.32	0.99955	+3.77	0.99992
+2.47	0.99324	+2.90	0.99813	+3.33	0.99957	+3.78	0.99992
+2.48	0.99343	+2.91	0.99819	+3.34	0.99958	+3.79	0.99992
+2.49	0.99361	+2.92	0.99825	+3.35	0.99960	+3.80	0.99993
+2.50	0.99379	+2.93	0.99831	+3.36	0.99961	+3.81	0.99993
+2.51	0.99396	+2.94	0.99836	+3.37	0.99962	+3.82	0.99993
+2.52	0.99413	+2.95	0.99841	+3.38	0.99964	+3.83	0.99994
+2.53	0.99430	+2.96	0.99846	+3.39	0.99965	+3.84	0.99994
+2.54	0.99446	+2.97	0.99851	+3.40	0.99966	+3.85	0.99994
+2.55	0.99461	+2.98	0.99856	+3.41	0.99968	+3.86	0.99994
+2.56	0.99477	+2.99	0.99861	+3.42	0.99969	+3.87	0.99995
+2.57	0.99492	+3.00	0.99865	+3.43	0.99970	+3.88	0.99995
+2.58	0.99506	+3.01	0.99869	+3.44	0.99971	+3.89	0.99995
+2.59	0.99520	+3.02	0.99874	+3.45	0.99972	+3.90	0.99995
+2.60	0.99534	+3.03	0.99878	+3.46	0.99973	+3.91	0.99995
+2.61	0.99547	+3.04	0.99882	+3.47	0.99974	+3.92	0.99996
+2.62	0.99560	+3.05	0.99886	+3.48	0.99975	+3.93	0.99996
+2.63	0.99573	+3.06	0.99889	+3.49	0.99976	+3.94	0.99996
+2.64	0.99585	+3.07	0.99893	+3.50	0.99977	+3.95	0.99996
+2.65	0.99598	+3.08	0.99896	+3.51	0.99978	+3.96	0.99996
+2.66	0.99609	+3.09	0.99900	+3.52	0.99978	+3.97	0.99996
+2.67	0.99621	+3.10	0.99903	+3.53	0.99979	+3.98	0.99997
+2.68	0.99632	+3.11	0.99906	+3.54	0.99980	+3.99	0.99997
+2.69	0.99643	+3.12	0.99910	+3.55	0.99981		
+2.70	0.99653	+3.13	0.99913	+3.56	0.99981		
+2.71	0.99664	+3.14	0.99916	+3.57	0.99982		
+2.72	0.99674	+3.15	0.99918	+3.58	0.99983		
+2.73	0.99683	+3.16	0.99921	+3.59	0.99983		
+2.74	0.99693	+3.17	0.99924	+3.60	0.99984		
+2.75	0.99702	+3.18	0.99926	+3.61	0.99985		
+2.76	0.99711	+3.19	0.99929	+3.62	0.99985		
+2.77	0.99720	+3.20	0.99931	+3.63	0.99986		
+2.78	0.99728	+3.21	0.99934	+3.66	0.99987		
+2.79	0.99736	+3.22	0.99936	+3.67	0.99988		
+2.80	0.99744	+3.23	0.99938	+3.68	0.99988		
+2.81	0.99752	+3.24	0.99940	+3.69	0.99989		
+2.82	0.99760	+3.25	0.99942	+3.70	0.99989		
+2.83	0.99767	+3.26	0.99944	+3.71	0.99990		
+2.84	0.99774	+3.27	0.99946	+3.72	0.99990		

TABLE A.2
Critical Values of Student's *t*-Distribution

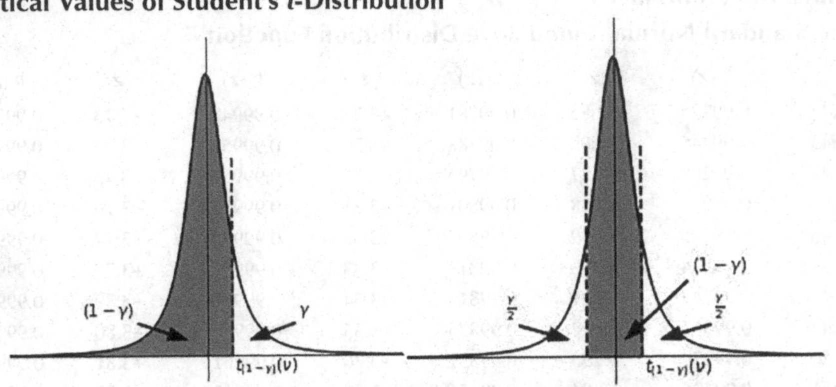

One-Sided *t*-distribution (Read Down)

$\gamma \rightarrow$	0.2	0.1	0.05	0.025	0.01	0.005	0.001	0.0005
$\nu \downarrow$	$t_{0.80}$	$t_{0.90}$	$t_{0.95}$	$t_{0.975}$	$t_{0.99}$	$t_{0.995}$	$t_{0.999}$	$t_{0.9995}$
1	1.3764	3.0777	6.3138	12.7062	31.8205	63.6567	318.3088	636.6192
2	1.0607	1.8856	2.9200	4.3027	6.9646	9.9248	22.3271	31.5991
3	0.9785	1.6377	2.3534	3.1824	4.5407	5.8409	10.2145	12.9240
4	0.9410	1.5332	2.1318	2.7764	3.7469	4.6041	7.1732	8.6103
5	0.9195	1.4759	2.0150	2.5706	3.3649	4.0321	5.8934	6.8688
6	0.9057	1.4398	1.9432	2.4469	3.1427	3.7074	5.2076	5.9588
7	0.8960	1.4149	1.8946	2.3646	2.9980	3.4995	4.7853	5.4079
8	0.8889	1.3968	1.8595	2.3060	2.8965	3.3554	4.5008	5.0413
9	0.8834	1.3830	1.8331	2.2622	2.8214	3.2498	4.2968	4.7809
10	0.8791	1.3722	1.8125	2.2281	2.7638	3.1693	4.1437	4.5869
11	0.8755	1.3634	1.7959	2.2010	2.7181	3.1058	4.0247	4.4370
12	0.8726	1.3562	1.7823	2.1788	2.6810	3.0545	3.9296	4.3178
13	0.8702	1.3502	1.7709	2.1604	2.6503	3.0123	3.8520	4.2208
14	0.8681	1.3450	1.7613	2.1448	2.6245	2.9768	3.7874	4.1405
15	0.8662	1.3406	1.7531	2.1314	2.6025	2.9467	3.7328	4.0728
16	0.8647	1.3368	1.7459	2.1199	2.5835	2.9208	3.6862	4.0150
17	0.8633	1.3334	1.7396	2.1098	2.5669	2.8982	3.6458	3.9651
18	0.8620	1.3304	1.7341	2.1009	2.5524	2.8784	3.6105	3.9216
19	0.8610	1.3277	1.7291	2.0930	2.5395	2.8609	3.5794	3.8834
20	0.8600	1.3253	1.7247	2.0860	2.5280	2.8453	3.5518	3.8495
21	0.8591	1.3232	1.7207	2.0796	2.5176	2.8314	3.5272	3.8193
22	0.8583	1.3212	1.7171	2.0739	2.5083	2.8188	3.5050	3.7921
23	0.8575	1.3195	1.7139	2.0687	2.4999	2.8073	3.4850	3.7676
24	0.8569	1.3178	1.7109	2.0639	2.4922	2.7969	3.4668	3.7454
25	0.8562	1.3163	1.7081	2.0595	2.4851	2.7874	3.4502	3.7251
26	0.8557	1.3150	1.7056	2.0555	2.4786	2.7787	3.4350	3.7066
27	0.8551	1.3137	1.7033	2.0518	2.4727	2.7707	3.4210	3.6896
	$t_{0.60}$	$t_{0.80}$	$t_{0.90}$	$t_{0.95}$	$t_{0.98}$	$t_{0.99}$	$t_{0.998}$	$t_{0.999}$
$\gamma \rightarrow$	0.4	0.2	0.1	0.05	0.02	0.01	0.002	0.001

Two-Sided *t*-distribution (Read Up)

(Continued)

TABLE A.2 (*Continued*)
Critical Values of Student's *t*-Distribution

	One-Sided *t*-distribution (Read Down)							
$\gamma\rightarrow$	0.2	0.1	0.05	0.025	0.01	0.005	0.001	0.0005
$\nu\downarrow$	$t_{0.80}$	$t_{0.90}$	$t_{0.95}$	$t_{0.975}$	$t_{0.99}$	$t_{0.995}$	$t_{0.999}$	$t_{0.9995}$
28	0.8546	1.3125	1.7011	2.0484	2.4671	2.7633	3.4082	3.6739
29	0.8542	1.3114	1.6991	2.0452	2.4620	2.7564	3.3962	3.6594
30	0.8538	1.3104	1.6973	2.0423	2.4573	2.7500	3.3852	3.6460
31	0.8534	1.3095	1.6955	2.0395	2.4528	2.7440	3.3749	3.6335
32	0.8530	1.3086	1.6939	2.0369	2.4487	2.7385	3.3653	3.6218
33	0.8526	1.3077	1.6924	2.0345	2.4448	2.7333	3.3563	3.6109
34	0.8523	1.3070	1.6909	2.0322	2.4411	2.7284	3.3479	3.6007
35	0.8520	1.3062	1.6896	2.0301	2.4377	2.7238	3.3400	3.5911
36	0.8517	1.3055	1.6883	2.0281	2.4345	2.7195	3.3326	3.5821
37	0.8514	1.3049	1.6871	2.0262	2.4314	2.7154	3.3256	3.5737
38	0.8512	1.3042	1.6860	2.0244	2.4286	2.7116	3.3190	3.5657
39	0.8509	1.3036	1.6849	2.0227	2.4258	2.7079	3.3128	3.5581
40	0.8507	1.3031	1.6839	2.0211	2.4233	2.7045	3.3069	3.5510
41	0.8505	1.3025	1.6829	2.0195	2.4208	2.7012	3.3013	3.5442
42	0.8503	1.3020	1.6820	2.0181	2.4185	2.6981	3.2960	3.5377
43	0.8501	1.3016	1.6811	2.0167	2.4163	2.6951	3.2909	3.5316
44	0.8499	1.3011	1.6802	2.0154	2.4141	2.6923	3.2861	3.5258
45	0.8497	1.3006	1.6794	2.0141	2.4121	2.6896	3.2815	3.5203
50	0.8489	1.2987	1.6759	2.0086	2.4033	2.6778	3.2614	3.4960
55	0.8482	1.2971	1.6730	2.0040	2.3961	2.6682	3.2451	3.4764
60	0.8477	1.2958	1.6706	2.0003	2.3901	2.6603	3.2317	3.4602
65	0.8472	1.2947	1.6686	1.9971	2.3851	2.6536	3.2204	3.4466
70	0.8468	1.2938	1.6669	1.9944	2.3808	2.6479	3.2108	3.4350
75	0.8464	1.2929	1.6654	1.9921	2.3771	2.6430	3.2025	3.4250
80	0.8461	1.2922	1.6641	1.9901	2.3739	2.6387	3.1953	3.4163
85	0.8459	1.2916	1.6630	1.9883	2.3710	2.6349	3.1889	3.4087
90	0.8456	1.2910	1.6620	1.9867	2.3685	2.6316	3.1833	3.4019
95	0.8454	1.2905	1.6611	1.9853	2.3662	2.6286	3.1782	3.3959
100	0.8452	1.2901	1.6602	1.9840	2.3642	2.6259	3.1737	3.3905
110	0.8449	1.2893	1.6588	1.9818	2.3607	2.6213	3.1660	3.3812
120	0.8446	1.2886	1.6577	1.9799	2.3578	2.6174	3.1595	3.3735
130	0.8444	1.2881	1.6567	1.9784	2.3554	2.6142	3.1541	3.3669
140	0.8442	1.2876	1.6558	1.9771	2.3533	2.6114	3.1495	3.3614
150	0.8440	1.2872	1.6551	1.9759	2.3515	2.6090	3.1455	3.3566
∞	0.8416	1.2816	1.6449	1.9600	2.3263	2.5758	3.0902	3.2905
	$t_{0.60}$	$t_{0.80}$	$t_{0.90}$	$t_{0.95}$	$t_{0.98}$	$t_{0.99}$	$t_{0.998}$	$t_{0.999}$
$\gamma\rightarrow$	0.4	0.2	0.1	0.05	0.02	0.01	0.002	0.001
	Two-Sided *t*-distribution (Read Up)							

For $t<0$ values, $F(t)$ can be obtained from $F(-t)=1-F(t)$.

TABLE A.3
Critical Values of Chi-Square $\chi^2_{1-\gamma}$ Distribution Function

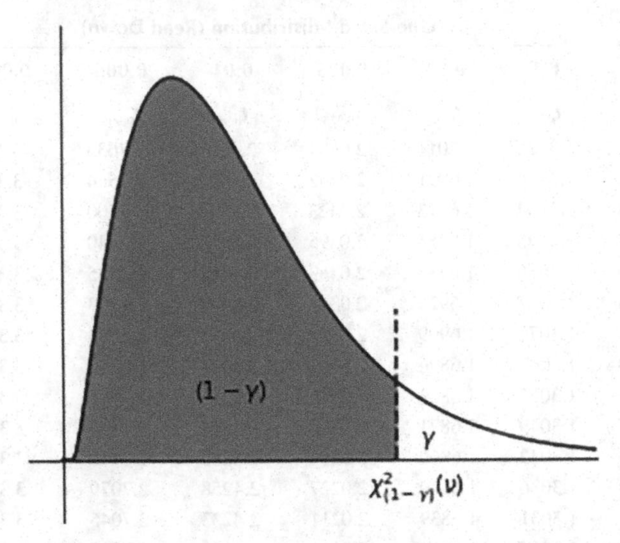

$\gamma\rightarrow$	0.9995	0.999	0.995	0.99	0.975	0.95	0.9	0.75
$df\downarrow$	$\chi^2_{0.0005}$	$\chi^2_{0.001}$	$\chi^2_{0.005}$	$\chi^2_{0.01}$	$\chi^2_{0.025}$	$\chi^2_{0.05}$	$\chi^2_{0.1}$	$\chi^2_{0.25}$
1	3.927E-7	1.571E-6	3.927E-5	1.571E-4	9.821E-4	3.932E-3	0.0158	0.1015
2	0.0010	0.0020	0.0100	0.0201	0.0506	0.1026	0.2107	0.5754
3	0.0153	0.0243	0.0717	0.1148	0.2158	0.3518	0.5844	1.2125
4	0.0639	0.0908	0.2070	0.2971	0.4844	0.7107	1.0636	1.9226
5	0.1581	0.2102	0.4117	0.5543	0.8312	1.1455	1.6103	2.6746
6	0.2994	0.3811	0.6757	0.8721	1.2373	1.6354	2.2041	3.4546
7	0.4849	0.5985	0.9893	1.2390	1.6899	2.1673	2.8331	4.2549
8	0.7104	0.8571	1.3444	1.6465	2.1797	2.7326	3.4895	5.0706
9	0.9717	1.1519	1.7349	2.0879	2.7004	3.3251	4.1682	5.8988
10	1.2650	1.4787	2.1559	2.5582	3.2470	3.9403	4.8652	6.7372
11	1.5868	1.8339	2.6032	3.0535	3.8157	4.5748	5.5778	7.5841
12	1.9344	2.2142	3.0738	3.5706	4.4038	5.2260	6.3038	8.4384
13	2.3051	2.6172	3.5650	4.1069	5.0088	5.8919	7.0415	9.2991
14	2.6967	3.0407	4.0747	4.6604	5.6287	6.5706	7.7895	10.1653
15	3.1075	3.4827	4.6009	5.2293	6.2621	7.2609	8.5468	11.0365
16	3.5358	3.9416	5.1422	5.8122	6.9077	7.9616	9.3122	11.9122
17	3.9802	4.4161	5.6972	6.4078	7.5642	8.6718	10.0852	12.7919
18	4.4394	4.9048	6.2648	7.0149	8.2307	9.3905	10.8649	13.6753
19	4.9123	5.4068	6.8440	7.6327	8.9065	10.1170	11.6509	14.5620
20	5.3981	5.9210	7.4338	8.2604	9.5908	10.8508	12.4426	15.4518
21	5.8957	6.4467	8.0337	8.8972	10.2829	11.5913	13.2396	16.3444

(Continued)

TABLE A.3 (*Continued*)
Critical Values of Chi-Square $\chi^2_{1-\gamma}$ Distribution Function

$\gamma \rightarrow$	0.9995	0.999	0.995	0.99	0.975	0.95	0.9	0.75
$df \downarrow$	$\chi^2_{0.0005}$	$\chi^2_{0.001}$	$\chi^2_{0.005}$	$\chi^2_{0.01}$	$\chi^2_{0.025}$	$\chi^2_{0.05}$	$\chi^2_{0.1}$	$\chi^2_{0.25}$
22	6.4045	6.9830	8.6427	9.5425	10.9823	12.3380	14.0415	17.2396
23	6.9237	7.5292	9.2604	10.1957	11.6886	13.0905	14.8480	18.1373
24	7.4527	8.0849	9.8862	10.8564	12.4012	13.8484	15.6587	19.0373
25	7.9910	8.6493	10.5197	11.5240	13.1197	14.6114	16.4734	19.9393
26	8.5379	9.2221	11.1602	12.1981	13.8439	15.3792	17.2919	20.8434
27	9.0932	9.8028	11.8076	12.8785	14.5734	16.1514	18.1139	21.7494
28	9.6563	10.3909	12.4613	13.5647	15.3079	16.9279	18.9392	22.6572
29	10.2268	10.9861	13.1211	14.2565	16.0471	17.7084	19.7677	23.5666
30	10.8044	11.5880	13.7867	14.9535	16.7908	18.4927	20.5992	24.4776
35	13.7875	14.6878	17.1918	18.5089	20.5694	22.4650	24.7967	29.0540
40	16.9062	17.9164	20.7065	22.1643	24.4330	26.5093	29.0505	33.6603
45	20.1366	21.2507	24.3110	25.9013	28.3662	30.6123	33.3504	38.2910
50	23.4610	24.6739	27.9907	29.7067	32.3574	34.7643	37.6886	42.9421
60	30.3405	31.7383	35.5345	37.4849	40.4817	43.1880	46.4589	52.2938
70	37.4674	39.0364	43.2752	45.4417	48.7576	51.7393	55.3289	61.6983
80	44.7910	46.5199	51.1719	53.5401	57.1532	60.3915	64.2778	71.1445
90	52.2758	54.1552	59.1963	61.7541	65.6466	69.1260	73.2911	80.6247
100	59.8957	61.9179	67.3276	70.0649	74.2219	77.9295	82.3581	90.1332
120	75.4665	77.7551	83.8516	86.9233	91.5726	95.7046	100.6236	109.2197
>df	$\frac{1}{2}[A-3.29]^2$	$\frac{1}{2}[A-3.09]^2$	$\frac{1}{2}[A-2.58]^2$	$\frac{1}{2}[A-2.33]^2$	$\frac{1}{2}[A-1.96]^2$	$\frac{1}{2}[A-1.64]^2$	$\frac{1}{2}[A-1.28]^2$	$\frac{1}{2}[A-0.67]^2$

Where $A = (2df - 1)^{\frac{1}{2}}$

$\gamma \rightarrow$	0.25	0.1	0.05	0.025	0.01	0.005	0.001	0.0005
$df \downarrow$	$\chi^2_{0.75}$	$\chi^2_{0.90}$	$\chi^2_{0.95}$	$\chi^2_{0.975}$	$\chi^2_{0.99}$	$\chi^2_{0.995}$	$\chi^2_{0.999}$	$\chi^2_{0.9995}$
1	1.3233	2.7055	3.8415	5.0239	6.6349	7.8794	10.8276	12.1157
2	2.7726	4.6052	5.9915	7.3778	9.2103	10.5966	13.8155	15.2018
3	4.1083	6.2514	7.8147	9.3484	11.3449	12.8382	16.2662	17.7300
4	5.3853	7.7794	9.4877	11.1433	13.2767	14.8603	18.4668	19.9974
5	6.6257	9.2364	11.0705	12.8325	15.0863	16.7496	20.5150	22.1053
6	7.8408	10.6446	12.5916	14.4494	16.8119	18.5476	22.4577	24.1028
7	9.0371	12.0170	14.0671	16.0128	18.4753	20.2777	24.3219	26.0178
8	10.2189	13.3616	15.5073	17.5345	20.0902	21.9550	26.1245	27.8680
9	11.3888	14.6837	16.9190	19.0228	21.6660	23.5894	27.8772	29.6658
10	12.5489	15.9872	18.3070	20.4832	23.2093	25.1882	29.5883	31.4198
11	13.7007	17.2750	19.6751	21.9200	24.7250	26.7568	31.2641	33.1366
12	14.8454	18.5493	21.0261	23.3367	26.2170	28.2995	32.9095	34.8213
13	15.9839	19.8119	22.3620	24.7356	27.6882	29.8195	34.5282	36.4778

(*Continued*)

TABLE A.3 (*Continued*)

Critical Values of Chi-Square $\chi^2_{1-\gamma}$ Distribution Function

$\gamma \rightarrow$	0.25	0.1	0.05	0.025	0.01	0.005	0.001	0.0005
$df \downarrow$	$\chi^2_{0.75}$	$\chi^2_{0.90}$	$\chi^2_{0.95}$	$\chi^2_{0.975}$	$\chi^2_{0.99}$	$\chi^2_{0.995}$	$\chi^2_{0.999}$	$\chi^2_{0.9995}$
14	17.1169	21.0641	23.6848	26.1189	29.1412	31.3193	36.1233	38.1094
15	18.2451	22.3071	24.9958	27.4884	30.5779	32.8013	37.6973	39.7188
16	19.3689	23.5418	26.2962	28.8454	31.9999	34.2672	39.2524	41.3081
17	20.4887	24.7690	27.5871	30.1910	33.4087	35.7185	40.7902	42.8792
18	21.6049	25.9894	28.8693	31.5264	34.8053	37.1565	42.3124	44.4338
19	22.7178	27.2036	30.1435	32.8523	36.1909	38.5823	43.8202	45.9731
20	23.8277	28.4120	31.4104	34.1696	37.5662	39.9968	45.3147	47.4985
21	24.9348	29.6151	32.6706	35.4789	38.9322	41.4011	46.7970	49.0108
22	26.0393	30.8133	33.9244	36.7807	40.2894	42.7957	48.2679	50.5111
23	27.1413	32.0069	35.1725	38.0756	41.6384	44.1813	49.7282	52.0002
24	28.2412	33.1962	36.4150	39.3641	42.9798	45.5585	51.1786	53.4788
25	29.3389	34.3816	37.6525	40.6465	44.3141	46.9279	52.6197	54.9475
26	30.4346	35.5632	38.8851	41.9232	45.6417	48.2899	54.0520	56.4069
27	31.5284	36.7412	40.1133	43.1945	46.9629	49.6449	55.4760	57.8576
28	32.6205	37.9159	41.3371	44.4608	48.2782	50.9934	56.8923	59.3000
29	33.7109	39.0875	42.5570	45.7223	49.5879	52.3356	58.3012	60.7346
30	34.7997	40.2560	43.7730	46.9792	50.8922	53.6720	59.7031	62.1619
35	40.2228	46.0588	49.8018	53.2033	57.3421	60.2748	66.6188	69.1986
40	45.6160	51.8051	55.7585	59.3417	63.6907	66.7660	73.4020	76.0946
45	50.9849	57.5053	61.6562	65.4102	69.9568	73.1661	80.0767	82.8757
50	56.3336	63.1671	67.5048	71.4202	76.1539	79.4900	86.6608	89.5605
60	66.9815	74.3970	79.0819	83.2977	88.3794	91.9517	99.6072	102.6948
70	77.5767	85.5270	90.5312	95.0232	100.4252	104.2149	112.3169	115.5776
80	88.1303	96.5782	101.8795	106.6286	112.3288	116.3211	124.8392	128.2613
90	98.6499	107.5650	113.1453	118.1359	124.1163	128.2989	137.2084	140.7823
100	109.1412	118.4980	124.3421	129.5612	135.8067	140.1695	149.4493	153.1670
120	130.0546	140.2326	146.5674	152.2114	158.9502	163.6482	173.6174	177.6029
>df	$\frac{1}{2}[A+0.67]^2$	$\frac{1}{2}[A+1.28]^2$	$\frac{1}{2}[A+1.64]^2$	$\frac{1}{2}[A+1.96]^2$	$\frac{1}{2}[A+2.33]^2$	$\frac{1}{2}[A+2.58]^2$	$\frac{1}{2}[A+3.09]^2$	$\frac{1}{2}[A+3.29]^2$

Where $A = (2df - 1)^{\frac{1}{2}}$

Notes: df is the number of degrees of freedom.

TABLE A.4
Critical Values of the Kolmogorov-Smirnov Statistic $D_n(\gamma)$

$\gamma \rightarrow$ $n \downarrow$	0.20	0.15	0.10	0.05	0.01
1	0.900	0.925	0.950	0.975	0.995
2	0.684	0.725	0.776	0.842	0.929
3	0.565	0.597	0.636	0.708	0.829
4	0.493	0.525	0.565	0.624	0.734
5	0.447	0.474	0.510	0.563	0.669
6	0.410	0.436	0.468	0.519	0.617
7	0.381	0.405	0.436	0.483	0.576
8	0.358	0.381	0.410	0.454	0.542
9	0.339	0.360	0.387	0.430	0.513
10	0.323	0.342	0.369	0.409	0.489
11	0.308	0.326	0.352	0.391	0.468
12	0.296	0.313	0.338	0.375	0.449
13	0.285	0.302	0.325	0.361	0.432
14	0.275	0.292	0.314	0.349	0.418
15	0.266	0.283	0.304	0.338	0.404
16	0.258	0.274	0.295	0.327	0.392
17	0.250	0.266	0.286	0.318	0.381
18	0.244	0.259	0.279	0.309	0.371
19	0.237	0.252	0.271	0.301	0.361
20	0.232	0.246	0.265	0.294	0.352
21	0.226	0.249	0.259	0.287	0.344
22	0.221	0.243	0.253	0.281	0.337
23	0.216	0.238	0.247	0.275	0.330
24	0.212	0.233	0.242	0.269	0.323
25	0.208	0.228	0.238	0.264	0.317
26	0.204	0.224	0.233	0.259	0.311
27	0.200	0.219	0.229	0.254	0.305
28	0.197	0.215	0.225	0.250	0.300
29	0.193	0.212	0.221	0.246	0.295
30	0.190	0.208	0.218	0.242	0.290
31	0.187	0.205	0.214	0.238	0.285
32	0.184	0.201	0.211	0.234	0.281
33	0.182	0.198	0.208	0.231	0.277
34	0.179	0.195	0.205	0.227	0.273
35	0.177	0.193	0.202	0.224	0.269
36	0.174	0.190	0.199	0.221	0.265
37	0.172	0.187	0.196	0.218	0.262
38	0.170	0.185	0.194	0.215	0.258
39	0.168	0.182	0.191	0.213	0.255
40	0.165	0.180	0.189	0.210	0.252
>40	**$1.07/(n)^{1/2}$**	**$1.14/(n)^{1/2}$**	**$1.22/(n)^{1/2}$**	**$1.36/(n)^{1/2}$**	**$1.63/(n)^{1/2}$**

n is the number of trials.

TABLE A.5A

F-Distribution Cumulative Distribution Function, Upper 10% ($\gamma = 0.1$)

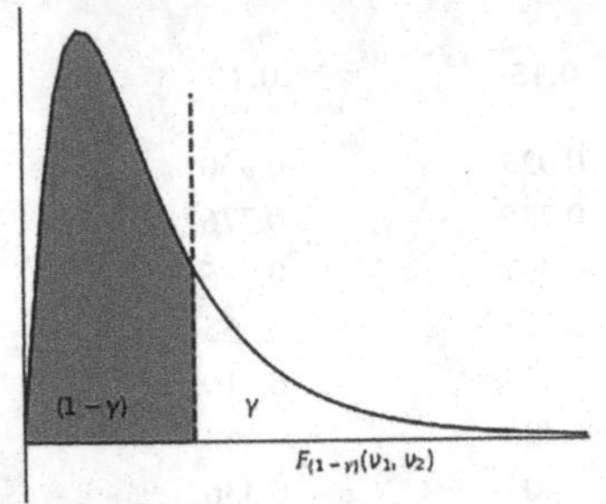

$\nu_1 \rightarrow$ / $\nu_2 \downarrow$	1	2	3	4	5	6	7	8	9	10	15	20	30	40	60	120	∞
1	39.9	49.5	53.6	55.8	57.2	58.2	58.9	59.4	59.9	60.2	61.2	61.7	62.3	62.5	62.8	63.1	63.3
2	8.53	9.00	9.16	9.24	9.29	9.33	9.35	9.37	9.38	9.39	9.42	9.44	9.46	9.47	9.47	9.48	9.49
3	5.54	5.46	5.39	5.34	5.31	5.28	5.27	5.25	5.24	5.23	5.20	5.18	5.17	5.16	5.15	5.14	5.13
4	4.54	4.32	4.19	4.11	4.05	4.01	3.98	3.95	3.94	3.92	3.87	3.84	3.82	3.80	3.79	3.78	3.76
5	4.06	3.78	3.62	3.52	3.45	3.40	3.37	3.34	3.32	3.30	3.24	3.21	3.17	3.16	3.14	3.12	3.11
6	3.78	3.46	3.29	3.18	3.11	3.05	3.01	2.98	2.96	2.94	2.87	2.84	2.80	2.78	2.76	2.74	2.72
7	3.59	3.26	3.07	2.96	2.88	2.83	2.78	2.75	2.72	2.70	2.63	2.59	2.56	2.54	2.51	2.49	2.47
8	3.46	3.11	2.92	2.81	2.73	2.67	2.62	2.59	2.56	2.54	2.46	2.42	2.38	2.36	2.34	2.32	2.29
9	3.36	3.01	2.81	2.69	2.61	2.55	2.51	2.47	2.44	2.42	2.34	2.30	2.25	2.23	2.21	2.18	2.16
10	3.29	2.92	2.73	2.61	2.52	2.46	2.41	2.38	2.35	2.32	2.24	2.20	2.16	2.13	2.11	2.08	2.06
11	3.23	2.86	2.66	2.54	2.45	2.39	2.34	2.30	2.27	2.25	2.17	2.12	2.08	2.05	2.03	2.00	1.97
12	3.18	2.81	2.61	2.48	2.39	2.33	2.28	2.24	2.21	2.19	2.10	2.06	2.01	1.99	1.96	1.93	1.90
13	3.14	2.76	2.56	2.43	2.35	2.28	2.23	2.20	2.16	2.14	2.05	2.01	1.96	1.93	1.90	1.88	1.85
14	3.10	2.73	2.52	2.39	2.31	2.24	2.19	2.15	2.12	2.10	2.01	1.96	1.91	1.89	1.86	1.83	1.80
15	3.07	2.70	2.49	2.36	2.27	2.21	2.16	2.12	2.09	2.06	1.97	1.92	1.87	1.85	1.82	1.79	1.76
16	3.05	2.67	2.46	2.33	2.24	2.18	2.13	2.09	2.06	2.03	1.94	1.89	1.84	1.81	1.78	1.75	1.72
17	3.03	2.64	2.44	2.31	2.22	2.15	2.10	2.06	2.03	2.00	1.91	1.86	1.81	1.78	1.75	1.72	1.69
18	3.01	2.62	2.42	2.29	2.20	2.13	2.08	2.04	2.00	1.98	1.89	1.84	1.78	1.75	1.72	1.69	1.66
19	2.99	2.61	2.40	2.27	2.18	2.11	2.06	2.02	1.98	1.96	1.86	1.81	1.76	1.73	1.70	1.67	1.63
20	2.97	2.59	2.38	2.25	2.16	2.09	2.04	2.00	1.96	1.94	1.84	1.79	1.74	1.71	1.68	1.64	1.61
21	2.96	2.57	2.36	2.23	2.14	2.08	2.02	1.98	1.95	1.92	1.83	1.78	1.72	1.69	1.66	1.62	1.59
22	2.95	2.56	2.35	2.22	2.13	2.06	2.01	1.97	1.93	1.90	1.81	1.76	1.70	1.67	1.64	1.60	1.57
23	2.94	2.55	2.34	2.21	2.11	2.05	1.99	1.95	1.92	1.89	1.80	1.74	1.69	1.66	1.62	1.59	1.55
24	2.93	2.54	2.33	2.19	2.10	2.04	1.98	1.94	1.91	1.88	1.78	1.73	1.67	1.64	1.61	1.57	1.53
25	2.92	2.53	2.32	2.18	2.09	2.02	1.97	1.93	1.89	1.87	1.77	1.72	1.66	1.63	1.59	1.56	1.52
26	2.91	2.52	2.31	2.17	2.08	2.01	1.96	1.92	1.88	1.86	1.76	1.71	1.65	1.61	1.58	1.54	1.50
27	2.90	2.51	2.30	2.17	2.07	2.00	1.95	1.91	1.87	1.85	1.75	1.70	1.64	1.60	1.57	1.53	1.49
28	2.89	2.50	2.29	2.16	2.06	2.00	1.94	1.90	1.87	1.84	1.74	1.69	1.63	1.59	1.56	1.52	1.48
29	2.89	2.50	2.28	2.15	2.06	1.99	1.93	1.89	1.86	1.83	1.73	1.68	1.62	1.58	1.55	1.51	1.47
30	2.88	2.49	2.28	2.14	2.05	1.98	1.93	1.88	1.85	1.82	1.72	1.67	1.61	1.57	1.54	1.50	1.46
40	2.84	2.44	2.23	2.09	2.00	1.93	1.87	1.83	1.79	1.76	1.66	1.61	1.54	1.51	1.47	1.42	1.38
60	2.79	2.39	2.18	2.04	1.95	1.87	1.82	1.77	1.74	1.71	1.60	1.54	1.48	1.44	1.40	1.35	1.29
120	2.75	2.35	2.13	1.99	1.90	1.82	1.77	1.72	1.68	1.65	1.55	1.48	1.41	1.37	1.32	1.26	1.19
∞	2.71	2.30	2.08	1.94	1.85	1.77	1.72	1.67	1.63	1.60	1.49	1.42	1.34	1.30	1.24	1.17	1.00

TABLE A.5B
F-Distribution Cumulative Distribution Function, Upper 5% ($\gamma = 0.05$)

$\nu_1 \rightarrow$ $\nu_2 \downarrow$	1	2	3	4	5	6	7	8	9	10	15	20	30	40	60	120	∞
1	161	200	216	225	230	234	237	239	241	242	246	248	250	251	252	253	254
2	18.51	19.00	19.16	19.25	19.30	19.33	19.35	19.37	19.38	19.40	19.43	19.45	19.46	19.47	19.48	19.49	19.50
3	10.13	9.55	9.28	9.12	9.01	8.94	8.89	8.85	8.81	8.79	8.70	8.66	8.62	8.59	8.57	8.55	8.53
4	7.71	6.94	6.59	6.39	6.26	6.16	6.09	6.04	6.00	5.96	5.86	5.80	5.75	5.72	5.69	5.66	5.63
5	6.61	5.79	5.41	5.19	5.05	4.95	4.88	4.82	4.77	4.74	4.62	4.56	4.50	4.46	4.43	4.40	4.37
6	5.99	5.14	4.76	4.53	4.39	4.28	4.21	4.15	4.10	4.06	3.94	3.87	3.81	3.77	3.74	3.70	3.67
7	5.59	4.74	4.35	4.12	3.97	3.87	3.79	3.73	3.68	3.64	3.51	3.44	3.38	3.34	3.30	3.27	3.23
8	5.32	4.46	4.07	3.84	3.69	3.58	3.50	3.44	3.39	3.35	3.22	3.15	3.08	3.04	3.01	2.97	2.93
9	5.12	4.26	3.86	3.63	3.48	3.37	3.29	3.23	3.18	3.14	3.01	2.94	2.86	2.83	2.79	2.75	2.71
10	4.96	4.10	3.71	3.48	3.33	3.22	3.14	3.07	3.02	2.98	2.85	2.77	2.70	2.66	2.62	2.58	2.54
11	4.84	3.98	3.59	3.36	3.20	3.09	3.01	2.95	2.90	2.85	2.72	2.65	2.57	2.53	2.49	2.45	2.40
12	4.75	3.89	3.49	3.26	3.11	3.00	2.91	2.85	2.80	2.75	2.62	2.54	2.47	2.43	2.38	2.34	2.30
13	4.67	3.81	3.41	3.18	3.03	2.92	2.83	2.77	2.71	2.67	2.53	2.46	2.38	2.34	2.30	2.25	2.21
14	4.60	3.74	3.34	3.11	2.96	2.85	2.76	2.70	2.65	2.60	2.46	2.39	2.31	2.27	2.22	2.18	2.13
15	4.54	3.68	3.29	3.06	2.90	2.79	2.71	2.64	2.59	2.54	2.40	2.33	2.25	2.20	2.16	2.11	2.07
16	4.49	3.63	3.24	3.01	2.85	2.74	2.66	2.59	2.54	2.49	2.35	2.28	2.19	2.15	2.11	2.06	2.01
17	4.45	3.59	3.20	2.96	2.81	2.70	2.61	2.55	2.49	2.45	2.31	2.23	2.15	2.10	2.06	2.01	1.96
18	4.41	3.55	3.16	2.93	2.77	2.66	2.58	2.51	2.46	2.41	2.27	2.19	2.11	2.06	2.02	1.97	1.92
19	4.38	3.52	3.13	2.90	2.74	2.63	2.54	2.48	2.42	2.38	2.23	2.16	2.07	2.03	1.98	1.93	1.88
20	4.35	3.49	3.10	2.87	2.71	2.60	2.51	2.45	2.39	2.35	2.20	2.12	2.04	1.99	1.95	1.90	1.84
21	4.32	3.47	3.07	2.84	2.68	2.57	2.49	2.42	2.37	2.32	2.18	2.10	2.01	1.96	1.92	1.87	1.81
22	4.30	3.44	3.05	2.82	2.66	2.55	2.46	2.40	2.34	2.30	2.15	2.07	1.98	1.94	1.89	1.84	1.78
23	4.28	3.42	3.03	2.80	2.64	2.53	2.44	2.37	2.32	2.27	2.13	2.05	1.96	1.91	1.86	1.81	1.76
24	4.26	3.40	3.01	2.78	2.62	2.51	2.42	2.36	2.30	2.25	2.11	2.03	1.94	1.89	1.84	1.79	1.73
25	4.24	3.39	2.99	2.76	2.60	2.49	2.40	2.34	2.28	2.24	2.09	2.01	1.92	1.87	1.82	1.77	1.71
26	4.23	3.37	2.98	2.74	2.59	2.47	2.39	2.32	2.27	2.22	2.07	1.99	1.90	1.85	1.80	1.75	1.69
27	4.21	3.35	2.96	2.73	2.57	2.46	2.37	2.31	2.25	2.20	2.06	1.97	1.88	1.84	1.79	1.73	1.67
28	4.20	3.34	2.95	2.71	2.56	2.45	2.36	2.29	2.24	2.19	2.04	1.96	1.87	1.82	1.77	1.71	1.65
29	4.18	3.33	2.93	2.70	2.55	2.43	2.35	2.28	2.22	2.18	2.03	1.94	1.85	1.81	1.75	1.70	1.64
30	4.17	3.32	2.92	2.69	2.53	2.42	2.33	2.27	2.21	2.16	2.01	1.93	1.84	1.79	1.74	1.68	1.62
40	4.08	3.23	2.84	2.61	2.45	2.34	2.25	2.18	2.12	2.08	1.92	1.84	1.74	1.69	1.64	1.58	1.51
60	4.00	3.15	2.76	2.53	2.37	2.25	2.17	2.10	2.04	1.99	1.84	1.75	1.65	1.59	1.53	1.47	1.39
120	3.92	3.07	2.68	2.45	2.29	2.18	2.09	2.02	1.96	1.91	1.75	1.66	1.55	1.50	1.43	1.35	1.25
∞	3.84	3.00	2.60	2.37	2.21	2.10	2.01	1.94	1.88	1.83	1.67	1.57	1.46	1.39	1.32	1.22	1.00

TABLE A.5C

F-Distribution Cumulative Distribution Function, Upper 2.5% ($\gamma = 0.025$)

$\nu_1 \rightarrow$ / $\nu_2 \downarrow$	1	2	3	4	5	6	7	8	9	10	15	20	30	40	60	120	∞
1	648	800	864	900	922	937	948	957	963	969	985	993	1001	1006	1010	1014	1018
2	38.51	39.00	39.17	39.25	39.30	39.33	39.36	39.37	39.39	39.40	39.43	39.45	39.46	39.47	39.48	39.49	39.50
3	17.44	16.04	15.44	15.10	14.88	14.73	14.62	14.54	14.47	14.42	14.25	14.17	14.08	14.04	13.99	13.95	13.90
4	12.22	10.65	9.98	9.60	9.36	9.20	9.07	8.98	8.90	8.84	8.66	8.56	8.46	8.41	8.36	8.31	8.26
5	10.01	8.43	7.76	7.39	7.15	6.98	6.85	6.76	6.68	6.62	6.43	6.33	6.23	6.18	6.12	6.07	6.02
6	8.81	7.26	6.60	6.23	5.99	5.82	5.70	5.60	5.52	5.46	5.27	5.17	5.07	5.01	4.96	4.90	4.85
7	8.07	6.54	5.89	5.52	5.29	5.12	4.99	4.90	4.82	4.76	4.57	4.47	4.36	4.31	4.25	4.20	4.14
8	7.57	6.06	5.42	5.05	4.82	4.65	4.53	4.43	4.36	4.30	4.10	4.00	3.89	3.84	3.78	3.73	3.67
9	7.21	5.71	5.08	4.72	4.48	4.32	4.20	4.10	4.03	3.96	3.77	3.67	3.56	3.51	3.45	3.39	3.33
10	6.94	5.46	4.83	4.47	4.24	4.07	3.95	3.85	3.78	3.72	3.52	3.42	3.31	3.26	3.20	3.14	3.08
11	6.72	5.26	4.63	4.28	4.04	3.88	3.76	3.66	3.59	3.53	3.33	3.23	3.12	3.06	3.00	2.94	2.88
12	6.55	5.10	4.47	4.12	3.89	3.73	3.61	3.51	3.44	3.37	3.18	3.07	2.96	2.91	2.85	2.79	2.72
13	6.41	4.97	4.35	4.00	3.77	3.60	3.48	3.39	3.31	3.25	3.05	2.95	2.84	2.78	2.72	2.66	2.60
14	6.30	4.86	4.24	3.89	3.66	3.50	3.38	3.29	3.21	3.15	2.95	2.84	2.73	2.67	2.61	2.55	2.49
15	6.20	4.77	4.15	3.80	3.58	3.41	3.29	3.20	3.12	3.06	2.86	2.76	2.64	2.59	2.52	2.46	2.40
16	6.12	4.69	4.08	3.73	3.50	3.34	3.22	3.12	3.05	2.99	2.79	2.68	2.57	2.51	2.45	2.38	2.32
17	6.04	4.62	4.01	3.66	3.44	3.28	3.16	3.06	2.98	2.92	2.72	2.62	2.50	2.44	2.38	2.32	2.25
18	5.98	4.56	3.95	3.61	3.38	3.22	3.10	3.01	2.93	2.87	2.67	2.56	2.45	2.38	2.32	2.26	2.19
19	5.92	4.51	3.90	3.56	3.33	3.17	3.05	2.96	2.88	2.82	2.62	2.51	2.39	2.33	2.27	2.20	2.13
20	5.87	4.46	3.86	3.51	3.29	3.13	3.01	2.91	2.84	2.77	2.57	2.46	2.35	2.29	2.22	2.16	2.09
21	5.83	4.42	3.82	3.48	3.25	3.09	2.97	2.87	2.80	2.73	2.53	2.42	2.31	2.25	2.18	2.11	2.04
22	5.79	4.38	3.78	3.44	3.22	3.05	2.93	2.84	2.76	2.70	2.50	2.39	2.27	2.21	2.14	2.08	2.00
23	5.75	4.35	3.75	3.41	3.18	3.02	2.90	2.81	2.73	2.67	2.47	2.36	2.24	2.18	2.11	2.04	1.97
24	5.72	4.32	3.72	3.38	3.15	2.99	2.87	2.78	2.70	2.64	2.44	2.33	2.21	2.15	2.08	2.01	1.94
25	5.69	4.29	3.69	3.35	3.13	2.97	2.85	2.75	2.68	2.61	2.41	2.30	2.18	2.12	2.05	1.98	1.91
26	5.66	4.27	3.67	3.33	3.10	2.94	2.82	2.73	2.65	2.59	2.39	2.28	2.16	2.09	2.03	1.95	1.88
27	5.63	4.24	3.65	3.31	3.08	2.92	2.80	2.71	2.63	2.57	2.36	2.25	2.13	2.07	2.00	1.93	1.85
28	5.61	4.22	3.63	3.29	3.06	2.90	2.78	2.69	2.61	2.55	2.34	2.23	2.11	2.05	1.98	1.91	1.83
29	5.59	4.20	3.61	3.27	3.04	2.88	2.76	2.67	2.59	2.53	2.32	2.21	2.09	2.03	1.96	1.89	1.81
30	5.57	4.18	3.59	3.25	3.03	2.87	2.75	2.65	2.57	2.51	2.31	2.20	2.07	2.01	1.94	1.87	1.79
40	5.42	4.05	3.46	3.13	2.90	2.74	2.62	2.53	2.45	2.39	2.18	2.07	1.94	1.88	1.80	1.72	1.64
60	5.29	3.93	3.34	3.01	2.79	2.63	2.51	2.41	2.33	2.27	2.06	1.94	1.82	1.74	1.67	1.58	1.48
120	5.15	3.80	3.23	2.89	2.67	2.52	2.39	2.30	2.22	2.16	1.94	1.82	1.69	1.61	1.53	1.43	1.31
∞	5.02	3.69	3.12	2.79	2.57	2.41	2.29	2.19	2.11	2.05	1.83	1.71	1.57	1.48	1.39	1.27	1.00

TABLE A.5D

F-Distribution Cumulative Distribution Function, Upper 1% ($\gamma = 0.01$)

$\nu_1 \rightarrow$ / $\nu_2 \downarrow$	1	2	3	4	5	6	7	8	9	10	15	20	30	40	60	120	∞
1	4052	5000	5403	5625	5764	5859	5928	5981	6022	6056	6157	6209	6261	6287	6313	6339	6366
2	98.50	99.00	99.17	99.25	99.30	99.33	99.36	99.37	99.39	99.40	99.43	99.45	99.47	99.47	99.48	99.49	99.50
3	34.12	30.82	29.46	28.71	28.24	27.91	27.67	27.49	27.35	27.23	26.87	26.69	26.50	26.41	26.32	26.22	26.13
4	21.20	18.00	16.69	15.98	15.52	15.21	14.98	14.80	14.66	14.55	14.20	14.02	13.84	13.75	13.65	13.56	13.46
5	16.26	13.27	12.06	11.39	10.97	10.67	10.46	10.29	10.16	10.05	9.72	9.55	9.38	9.29	9.20	9.11	9.02
6	13.75	10.92	9.78	9.15	8.75	8.47	8.26	8.10	7.98	7.87	7.56	7.40	7.23	7.14	7.06	6.97	6.88
7	12.25	9.55	8.45	7.85	7.46	7.19	6.99	6.84	6.72	6.62	6.31	6.16	5.99	5.91	5.82	5.74	5.65
8	11.26	8.65	7.59	7.01	6.63	6.37	6.18	6.03	5.91	5.81	5.52	5.36	5.20	5.12	5.03	4.95	4.86
9	10.56	8.02	6.99	6.42	6.06	5.80	5.61	5.47	5.35	5.26	4.96	4.81	4.65	4.57	4.48	4.40	4.31
10	10.04	7.56	6.55	5.99	5.64	5.39	5.20	5.06	4.94	4.85	4.56	4.41	4.25	4.17	4.08	4.00	3.91
11	9.65	7.21	6.22	5.67	5.32	5.07	4.89	4.74	4.63	4.54	4.25	4.10	3.94	3.86	3.78	3.69	3.60
12	9.33	6.93	5.95	5.41	5.06	4.82	4.64	4.50	4.39	4.30	4.01	3.86	3.70	3.62	3.54	3.45	3.36
13	9.07	6.70	5.74	5.21	4.86	4.62	4.44	4.30	4.19	4.10	3.82	3.66	3.51	3.43	3.34	3.25	3.17
14	8.86	6.51	5.56	5.04	4.69	4.46	4.28	4.14	4.03	3.94	3.66	3.51	3.35	3.27	3.18	3.09	3.00
15	8.68	6.36	5.42	4.89	4.56	4.32	4.14	4.00	3.90	3.80	3.52	3.37	3.21	3.13	3.05	2.96	2.87
16	8.53	6.23	5.29	4.77	4.44	4.20	4.03	3.89	3.78	3.69	3.41	3.26	3.10	3.02	2.93	2.84	2.75
17	8.40	6.11	5.18	4.67	4.34	4.10	3.93	3.79	3.68	3.59	3.31	3.16	3.00	2.92	2.83	2.75	2.65
18	8.29	6.01	5.09	4.58	4.25	4.01	3.84	3.71	3.60	3.51	3.23	3.08	2.92	2.84	2.75	2.66	2.57
19	8.18	5.93	5.01	4.50	4.17	3.94	3.77	3.63	3.52	3.43	3.15	3.00	2.84	2.76	2.67	2.58	2.49
20	8.10	5.85	4.94	4.43	4.10	3.87	3.70	3.56	3.46	3.37	3.09	2.94	2.78	2.69	2.61	2.52	2.42
21	8.02	5.78	4.87	4.37	4.04	3.81	3.64	3.51	3.40	3.31	3.03	2.88	2.72	2.64	2.55	2.46	2.36
22	7.95	5.72	4.82	4.31	3.99	3.76	3.59	3.45	3.35	3.26	2.98	2.83	2.67	2.58	2.50	2.40	2.31
23	7.88	5.66	4.76	4.26	3.94	3.71	3.54	3.41	3.30	3.21	2.93	2.78	2.62	2.54	2.45	2.35	2.26
24	7.82	5.61	4.72	4.22	3.90	3.67	3.50	3.36	3.26	3.17	2.89	2.74	2.58	2.49	2.40	2.31	2.21
25	7.77	5.57	4.68	4.18	3.85	3.63	3.46	3.32	3.22	3.13	2.85	2.70	2.54	2.45	2.36	2.27	2.17
26	7.72	5.53	4.64	4.14	3.82	3.59	3.42	3.29	3.18	3.09	2.81	2.66	2.50	2.42	2.33	2.23	2.13
27	7.68	5.49	4.60	4.11	3.78	3.56	3.39	3.26	3.15	3.06	2.78	2.63	2.47	2.38	2.29	2.20	2.10
28	7.64	5.45	4.57	4.07	3.75	3.53	3.36	3.23	3.12	3.03	2.75	2.60	2.44	2.35	2.26	2.17	2.06
29	7.60	5.42	4.54	4.04	3.73	3.50	3.33	3.20	3.09	3.00	2.73	2.57	2.41	2.33	2.23	2.14	2.03
30	7.56	5.39	4.51	4.02	3.70	3.47	3.30	3.17	3.07	2.98	2.70	2.55	2.39	2.30	2.21	2.11	2.01
40	7.31	5.18	4.31	3.83	3.51	3.29	3.12	2.99	2.89	2.80	2.52	2.37	2.20	2.11	2.02	1.92	1.80
60	7.08	4.98	4.13	3.65	3.34	3.12	2.95	2.82	2.72	2.63	2.35	2.20	2.03	1.94	1.84	1.73	1.60
120	6.85	4.79	3.95	3.48	3.17	2.96	2.79	2.66	2.56	2.47	2.19	2.03	1.86	1.76	1.66	1.53	1.38
∞	6.63	4.61	3.78	3.32	3.02	2.80	2.64	2.51	2.41	2.32	2.04	1.88	1.70	1.59	1.47	1.32	1.00

Appendix B
Generic Failure Data

Data in this section are adapted from:

- Ma, Z., T. E. Wierman and K. J. Kvarfordt. *Industry-Average Performance for Components and Initiating Events at U.S. Commercial Nuclear Power Plants: 2020 Update.* INL/EXT-21–65055, Idaho National Laboratory, Nov. 2021.
- Eide, S. A. et al., *Industry-Average Performance for Components and Initiating Events at U.S. Commercial Nuclear Power Plants*, NUREG/CR-6928, U.S. Nuclear Regulatory Commission, Washington, DC, Feb. 2007.

Raw data and updates are available at: https://nrcoe.inl.gov/AvgPerf/.

Ma et al. caution that if these distributions are to be used as priors in Bayesian updates using plant-specific data, then a check for consistency between the prior and the data should be performed first, as outlined in Section 6.2.3.5 of NUREG/CR-6823. The error factor provided is the 95th percentile divided by the median. The error factor is from an empirical Bayes analysis at the plant level, with Kass-Steffey adjustment.

Note that in the book we use scientific notation (e.g., 1.0×10^{-2}, but that for length in this table we use the alternative notation (e.g., 1.0E-02).

TABLE B.1
Generic Failure Rate Data for Mechanical Components

Component and Failure Mode		Data				Dist. for Failure Prob. (p) or Failure Rate (λ)				
		No. of Failures	No. of Demands (d) or Hours (hr)	d or hr	No. of Comps	Dist.	Mean	α	β	EF
Air-Operated Valve Fails To Open	AOV-FTO	50	165,942	d	1,755	Beta	3.04E-04	50.500	1.660E+05	1.3
Air-Operated Valve Fails To Close	AOV-FTC	27	165,942	d	1,755	Beta	1.89E-04	0.638	3.380E+03	6.4
Air-Operated Valve Fails To Open/Close	AOV-FTOC	83	165,942	d	1,755	Beta	5.58E-04	0.832	1.490E+03	5.0
Air-Operated Valve Fails To Control	AOV-FC	167	1,109,287,000	hr	8,788	Gamma	1.75E-07	1.260	7.170E+06	3.7
Air-Operated Valve Spurious Operation	AOV-SOP	61	1,109,287,000	hr	8,788	Gamma	5.83E-08	0.859	1.470E+07	4.9
Air-Operated Valve Internal Leakage (Small)	AOV-ILS	35	1,109,287,000	hr	8,788	Gamma	3.20E-08	35.500	1.110E+09	1.3
Air-Operated Valve Internal Leakage (Rupture)	AOV-ILL	(Note 1)	–	hr	8,788	Gamma	6.40E-10	0.300	4.688E+08	18.8
Air-Operated Valve External Leakage (Small)	AOV-ELS	35	1,109,287,000	hr	8,788	Gamma	3.43E-08	0.575	1.680E+07	7.2
Air-Operated Valve External Leakage (Rupture)	AOV-ELL	(Note 1)	–	hr	8,788	Gamma	2.40E-09	0.300	1.249E+08	18.8
Component Cooling Water Air-Operated Valve Spurious Operation	AOV-SOP-CCW	10	144,615,200	hr	1,164	Gamma	7.26E-08	10.500	1.450E+08	1.6
Instrument Air System Air-Operated Valve Spurious Operation	AOV-SOP-IAS	0	6,218,450	hr	50	Gamma	8.04E-08	0.500	6.220E+06	8.4
Motor-Operated Valve Fails To Open	MOV-FTO	190	593,626	d	7,120	Beta	3.43E-04	2.480	7.220E+03	2.6
Motor-Operated Valve Fails To Close	MOV-FTC	123	593,626	d	7,120	Beta	2.28E-04	0.972	4.260E+03	4.4
Motor-Operated Valve Fails To Open/Close	MOV-FTOC	346	593,626	d	7,120	Beta	6.40E-04	2.430	3.800E+03	2.6
Motor-Operated Feed Control Valve Fails To Control	MOV-FC	59	1,634,537,000	hr	13,344	Gamma	3.47E-08	0.798	2.300E+07	5.2
Motor-Operated Valve Spurious Operation	MOV-SOP	41	1,634,537,000	hr	13,344	Gamma	2.54E-08	41.500	1.630E+09	1.3
Motor-Operated Valve Internal Leakage (Small)	MOV-ILS	55	1,634,537,000	hr	13,344	Gamma	3.61E-08	0.451	1.250E+07	9.7
Motor-Operated Valve Internal Leakage (Rupture)	MOV-ILL	(Note 1)	–	hr	13,344	Gamma	7.22E-10	0.300	4.155E+08	18.8

(Continued)

TABLE B.1 ***(Continued)***
Generic Failure Rate Data for Mechanical Components

Component and Failure Mode		Data				Dist. for Failure Prob. (p) or Failure Rate (λ)				
		No. of Failures	No. of Demands (d) or Hours (hr)	d or hr	No. of Comps	Dist.	Mean	α	β	EF
Motor-Operated Valve External Leakage (Small)	MOV-ELS	29	1,634,537,000	hr	13,344	Gamma	1.88E-08	0.463	2.460E+07	9.3
Motor-Operated Valve External Leakage (Rupture)	MOV-ELL	(Note 1)	–	hr	13,344	Gamma	1.32E-09	0.300	2.280E+08	18.8
Butterfly Valve Fails To Open	MOV-FTO-BFV	24	89,399	d	983	Beta	2.74E-04	24.500	8.940E+04	1.4
Butterfly Valve Fails To Close	MOV-FTC-BFV	24	89,399	d	983	Beta	2.89E-04	1.270	4.390E+03	3.7
Butterfly Valve Fails To Open/Close	MOV-FTOC-BFV	54	89,399	d	983	Beta	7.69E-04	0.602	7.830E+02	6.8
Component Cooling Water Motor-Operated Valve Spurious Operation	MOV-SOP-CCW	4	183,661,900	hr	1,472	Gamma	2.45E-08	4.500	1.840E+08	2.0
Standby Service Water Motor-Operated Valve Spurious Operation	MOV-SOP-SWS	0	64,725,970	hr	566	Gamma	7.72E-09	0.500	6.470E+07	8.4
Component Cooling Water Butterfly Valve Spurious Operation	MOV-BFV-SOP-CCW	2	86,552,190	hr	738	Gamma	2.89E-08	2.500	8.660E+07	2.5
Hydraulic-Operated Valve Fails To Open	HOV-FTOC	17	16,401	d	219	Beta	1.23E-03	0.436	3.530E+02	10.1
Hydraulic-Operated Valve Fails To Control	HOV-FC	21	76,176,020	hr	603	Gamma	2.82E-07	21.500	7.620E+07	1.4
Hydraulic-Operated Valve Spurious Operation	HOV-SOP	10	76,176,020	hr	603	Gamma	1.23E-07	0.526	4.280E+06	7.9
Hydraulic-Operated Valve Internal Leakage (Small)	HOV-ILS	2	76,176,020	hr	603	Gamma	3.28E-08	2.500	7.620E+07	2.5
Hydraulic-Operated Valve Internal Leakage (Rupture)	HOV-ILL	(Note 1)	–	hr	603	Gamma	6.56E-10	0.300	4.573E+08	18.8
Hydraulic-Operated Valve External Leakage (Small)	HOV-ELS	7	76,176,020	hr	603	Gamma	9.66E-08	0.449	4.650E+06	9.7
Hydraulic-Operated Valve External Leakage (Rupture)	HOV-ELL	(Note 1)	–	hr	603	Gamma	6.76E-09	0.300	4.437E+07	18.8

(Continued)

TABLE B.1 (*Continued*)
Generic Failure Rate Data for Mechanical Components

Component and Failure Mode		Data				Dist. for Failure Prob. (p) or Failure Rate (λ)				
		No. of Failures	No. of Demands (d) or Hours (hr)	d or hr	No. of Comps	Dist.	Mean	α	β	EF
Solenoid-Operated Valve Fails To Open	SOV-FTOC	13	27,937	d	555	Beta	4.83E-04	13.500	2.790E+04	1.5
Solenoid-Operated Valve Fails To Control	SOV-FC	15	115,760,700	hr	921	Gamma	1.52E-07	0.609	4.010E+06	6.7
Solenoid-Operated Valve Spurious Operation	SOV-SOP	9	115,760,700	hr	921	Gamma	8.21E-08	9.500	1.160E+08	1.6
Solenoid-Operated Valve Internal Leakage (Small)	SOV-ILS	8	115,760,700	hr	921	Gamma	7.34E-08	8.500	1.160E+08	1.7
Solenoid-Operated Valve Internal Leakage (Rupture)	SOV-ILL	(Note 1)	–	hr	921	Gamma	1.47E-09	0.300	2.044E+08	18.8
Solenoid-Operated Valve External Leakage (Small)	SOV-ELS	2	115,760,700	hr	921	Gamma	2.16E-08	2.500	1.160E+08	2.5
Solenoid-Operated Valve External Leakage (Rupture)	SOV-ELL	(Note 1)	–	hr	921	Gamma	1.51E-09	0.300	1.984E+08	18.8
Explosive-Operated Valve Fails To Open	EOV-FTO	3	674	d	59	Beta	4.62E-03	1.010	2.170E+02	4.3
Vacuum Breaker Valve Fails To Open	VBV-FTO	1	23,202	d	167	Beta	6.46E-05	1.500	2.320E+04	3.3
Vacuum Breaker Valve Fails To Close	VBV-FTC	1	23,202	d	167	Beta	6.46E-05	1.500	2.320E+04	3.3
Vacuum Breaker Valve Fails To Open/Close	VBV-FTOC	2	23,202	d	167	Beta	1.08E-04	2.500	2.320E+04	2.5
Vacuum Breaker Valve Spurious Operation	VBV-SOP	0	43,685,040	hr	343	Gamma	1.14E-08	0.500	4.370E+07	8.4
Vacuum Breaker Valve Internal Leakage (Small)	VBV-ILS	2	43,685,040	hr	343	Gamma	5.72E-08	2.500	4.370E+07	2.5
Vacuum Breaker Valve Internal Leakage (Rupture)	VBV-ILL	(Note 1)	–	hr	343	Gamma	1.14E-09	0.300	2.622E+08	18.8
Turbine Bypass Valve Fails To Open	TBV-FTO	1	2,367	d	73	Beta	6.33E-04	1.500	2.370E+03	3.3
Turbine Bypass Valve Fails To Close	TBV-FTC	0	2,367	d	73	Beta	2.11E-04	0.500	2.370E+03	8.4
Turbine Bypass Valve Fails To Open/Close	TBV-FTOC	1	2,367	d	73	Beta	6.33E-04	1.500	2.370E+03	3.3

(*Continued*)

TABLE B.1 (*Continued*)
Generic Failure Rate Data for Mechanical Components

Component and Failure Mode		Data				Dist. for Failure Prob. (p) or Failure Rate (λ)				
		No. of Failures	No. of Demands (d) or Hours (hr)	d or hr	No. of Comps	Dist.	Mean	α	β	EF
Turbine Bypass Valve Fails To Control	TBV-FC	6	19,263,540	hr	153	Gamma	3.57E-07	0.492	1.380E+06	8.6
Main Steam Isolation Valve Fails To Open/ Close	MSV-FTOC	24	32,199	d	425	Beta	7.61E-04	24.500	3.220E+04	1.4
Main Steam Isolation Valve Spurious Operation	MSV-SOP	16	65,768,320	hr	520	Gamma	2.34E-07	0.501	2.140E+06	8.4
Main Steam Isolation Valve Internal Leakage (Small)	MSV-ILS	23	65,768,320	hr	520	Gamma	3.57E-07	23.500	6.580E+07	1.4
Main Steam Isolation Valve Internal Leakage (Rupture)	MSV-ILL	(Note 1)	–	hr	520	Gamma	7.14E-09	0.300	4.202E+07	18.8
Main Steam Isolation Valve External Leakage (Small)	MSV-ELS	1	65,768,320	hr	520	Gamma	2.28E-08	1.500	6.580E+07	3.3
Main Steam Isolation Valve External Leakage (Rupture)	MSV-ELL	(Note 1)	–	hr	520	Gamma	1.60E-09	0.300	1.880E+08	18.8
Check Valve Fails To Open	CKV-FTO	0	44,791	d	489	Beta	1.12E-05	0.500	4.480E+04	8.4
Check Valve Fails To Close	CKV-FTC	5	44,791	d	489	Beta	1.23E-04	5.500	4.480E+04	1.9
Check Valve Spurious Operation	CKV-SOP	0	806,744,700	hr	6,379	Gamma	6.20E-10	0.500	8.070E+08	8.4
Check Valve Internal Leakage (Small)	CKV-ILS	58	806,744,700	hr	6,379	Gamma	7.25E-08	58.500	8.070E+08	1.2
Check Valve Internal Leakage (Rupture)	CKV-ILL	(Note 1)	–	hr	6,379	Gamma	1.45E-09	0.300	2.069E+08	18.8
Check Valve External Leakage (Small)	CKV-ELS	3	806,744,700	hr	6,379	Gamma	4.34E-09	3.500	8.070E+08	2.2
Check Valve External Leakage (Rupture)	CKV-ELL	(Note 1)	–	hr	6,379	Gamma	3.04E-10	0.300	9.875E+08	18.8
Manual Valve Fails To Open	XVM-FTOC	1	2,875	d	66	Beta	5.22E-04	1.500	2.870E+03	3.3
Manual Valve Spurious Operation	XVM-SOP	2	132,674,000	hr	1,035	Gamma	1.88E-08	2.500	1.330E+08	2.5
Manual Valve Internal Leakage (Small)	XVM-ILS	3	132,674,000	hr	1,035	Gamma	2.64E-08	3.500	1.330E+08	2.2
Manual Valve Internal Leakage (Rupture)	XVM-ILL	(Note 1)	–	hr	1,035	Gamma	5.28E-10	0.300	5.682E+08	18.8

(*Continued*)

TABLE B.1 (*Continued*)
Generic Failure Rate Data for Mechanical Components

Component and Failure Mode		Data				Dist. for Failure Prob. (p) or Failure Rate (λ)				
		No. of Failures	No. of Demands (d) or Hours (hr)	d or hr	No. of Comps	Dist.	Mean	α	β	EF
Manual Valve External Leakage (Small)	XVM-ELS	11	132,674,000	hr	1,035	Gamma	8.67E-08	11.500	1.330E+08	1.6
Manual Valve External Leakage (Rupture)	XVM-ELL	(Note 1)	–	hr	1,035	Gamma	6.07E-09	0.300	4.943E+07	18.8
Standby Service Water Manual Valve Spuriously Transfers	XVM-SOP-SWS	0	18,055,700	hr	140	Gamma	2.77E-08	0.500	1.810E+07	8.4
Flow Control Valve Fails To Open/Close	FCV-FTOC	0	11,345	d	105	Beta	4.41E-05	0.500	1.130E+04	8.4
Flow Control Valve Fails To Control	FCV-FC	8	73,637,280	hr	595	Gamma	1.15E-07	8.500	7.360E+07	1.7
Flow Control Valve Spurious Operation	FCV-SOP	2	73,637,280	hr	595	Gamma	3.40E-08	2.500	7.360E+07	2.5
Feedwater Regulating Valve Fails To Operate	FRV-FTOP	49	27,637,200	hr	221	Gamma	1.88E-06	0.666	3.540E+05	6.1
Motor-Driven Pump Fails To Start, Normally Standby	MDP-FTS-NS	227	410,593	d	1,311	Beta	5.88E-04	2.070	3.520E+03	2.8
Motor-Driven Pump Fails To Run <1H	MDP-FTR<1H	31	378,369	hr	1,305	Gamma	9.13E-05	0.579	6.340E+03	7.1
Motor-Driven Pump Fails To Run >1H	MDP-FTR>1H	92	19,248,030	hr	1,311	Gamma	8.12E-06	0.511	6.290E+04	8.2
Motor-Driven Pump External Leakage (Small)	MDP-ELS	59	288,839,600	hr	2,351	Gamma	1.98E-07	0.684	3.450E+06	6.0
Motor-Driven Pump External Leakage (Rupture)	MDP-ELL	(Note 1)	–	hr	2,351	Gamma	1.39E-08	0.300	2.165E+07	18.8
Motor-Driven Pump Fails To Start, Normally Running	MDP-FTS-NR	89	125,005	d	649	Beta	7.86E-04	1.080	1.370E+03	4.1
Motor-Driven Pump Fails To Run, Normally Running	MDP-FTR-NR	129	56,750,330	hr	650	Gamma	2.26E-06	1.970	8.720E+05	2.8
Component Cooling Water Motor-Driven Pump Fails To Start	MDP-FTS-CCW	31	80,067	hr	288	Beta	4.57E-04	0.796	1.740E+03	5.2
Component Cooling Water Motor-Driven Pump Fails To Run	MDP-FTR-CCW	31	17,527,790	hr	288	Gamma	1.77E-06	1.850	1.040E+06	2.9

(Continued)

TABLE B.1 (*Continued*)
Generic Failure Rate Data for Mechanical Components

Component and Failure Mode		Data				Dist. for Failure Prob. (p) or Failure Rate (λ)				
		No. of Failures	No. of Demands (d) or Hours (hr)	d or hr	No. of Comps	Dist.	Mean	α	β	EF
Service Water Motor-Driven Pump Fails To Start	MDP-FTS-SWS	132	225,636	d	529	Beta	7.43E-04	0.848	1.140E+03	4.9
Service Water Motor-Driven Pump Fails To Run	MDP-FTR-SWS	100	25,635,460	hr	529	Gamma	4.20E-06	1.170	2.790E+05	3.9
Circulating Water Motor-Driven Pump Fails To Run	MDP-FTR-CWS	15	3,116,679	hr	31	Gamma	4.86E-06	4.570	9.410E+05	2.0
Turbine-Driven Pump Fails To Start (Pooled Systems), Normally Standby	TDP-FTS-NS	105	22,512	d	133	Beta	5.32E-03	1.260	2.350E+02	3.7
Turbine-Driven Pump Fails To Run (Pooled Systems), Early Term	TDP-FTR<1H	34	15,530	hr	133	Gamma	2.56E-03	0.444	1.730E+02	9.9
Turbine-Driven Pump Fails To Run (Pooled Systems), Late Term	TDP-FTR>1H	17	4,454	hr	133	Gamma	6.35E-03	0.441	6.950E+01	10.0
Turbine-Driven Pump External Leakage (Small)	TDP-ELS	10	24,190,380	hr	191	Gamma	4.13E-07	2.020	4.900E+06	2.8
Turbine Bypass Valve External Leakage (Rupture)	TDP-ELL	(Note 1)	–	hr	191	Gamma	2.89E-08	0.300	1.038E+07	18.8
Auxiliary Feedwater Turbine-Driven Pump Fails To Start, Normally Standby	TDP-FTS-NS-AFW	52	15,672	d	74	Beta	3.79E-03	0.831	2.180E+02	5.0
Auxiliary Feedwater Turbine-Driven Pump Fails To Run <1H	TDP-FTR<1H-AFW	18	10,670	hr	74	Gamma	1.73E-03	18.500	1.070E+04	1.4
Auxiliary Feedwater Turbine-Driven Pump Fails To Run >1H	TDP-FTR>1H-AFW	8	3,295	hr	74	Gamma	2.58E-03	8.500	3.300E+03	1.7
HCI-RCI Turbine-Driven Pump Fails To Start, Normally Standby	TDP-FTS-NS-HCI-RCI	25	4,026	d	31	Beta	6.68E-03	1.290	1.920E+02	3.6
HCI Turbine-Driven Pump Fails To Run <1H	TDP-FTR<1H-HCI-RCI	16	4,860	hr	59	Gamma	3.35E-03	2.220	6.640E+02	2.7

(*Continued*)

TABLE B.1 (*Continued*)
Generic Failure Rate Data for Mechanical Components

Component and Failure Mode		Data				Dist. for Failure Prob. (p) or Failure Rate (λ)				
		No. of Failures	No. of Demands (d) or Hours (hr)	d or hr	No. of Comps	Dist.	Mean	α	β	EF
HCI-RCI Turbine-Driven Pump Fails To Run >1H	TDP-FTR>1H-HCI-RCI	9	1,159	hr	59	Gamma	8.20E-03	9.500	1.160E+03	1.6
Main Feedwater Turbine-Driven Pump Fails To Start, Normally Running	TDP-FTS-NR-MFW	5	1,147	d	42	Beta	4.60E-03	0.633	1.370E+02	6.4
Main Feedwater Turbine-Driven Pump Fails To Run, Normally Running	TDP-FTR-NR-MFW	39	4,938,575	hr	42	Gamma	8.45E-06	0.824	9.760E+04	5.0
Engine-Driven Pump Fails To Start, Normally Standby	EDP-FTS-NS	13	17,773	d	44	Beta	7.60E-04	13.500	1.780E+04	1.5
Engine-Driven Pump Fails To Run <1H, Normally Standby	EDP-FTR<1H	6	9,888	hr	39	Gamma	6.57E-04	6.500	9.890E+03	1.8
Engine-Driven Pump Fails To Run >1H, Normally Standby	EDP-FTR>1H	15	4,754	hr	44	Gamma	3.26E-03	15.500	4.750E+03	1.5
Engine-Driven Pump External Leakage (Small)	EDP-ELS	6	7,690,189	hr	69	Gamma	8.45E-07	6.500	7.690E+06	1.8
Engine-Driven Pump External Leakage (Rupture)	EDP-ELL	(Note 1)	–	hr	69	Gamma	5.92E-08	0.300	5.072E+06	18.8
Auxiliary Feedwater Engine-driven Pump Fails To Start	EDP-FTS-AFW	1	1,163	d	5	Beta	1.29E-03	1.500	1.160E+03	3.3
Auxiliary Feedwater Engine-driven Pump Fails To Run <1H	EDP-FTR<1H-AFW	2	759	hr	5	Gamma	3.29E-03	2.500	7.590E+02	2.5
Auxiliary Feedwater Engine-driven Pump Fails To Run >1H	EDP-FTR>1H-AFW	2	234	hr	5	Gamma	1.07E-02	2.500	2.340E+02	2.5
Positive Displacement Pump Fails To Start, Normally Running	PDP-FTS-NR	53	28,865	d	57	Beta	2.47E-03	0.825	3.330E+02	5.0
Positive Displacement Pump Fails To Run, Normally Running	PDP-FTR-NR	40	2,353,162	hr	54	Gamma	1.91E-05	1.330	6.980E+04	3.6

(*Continued*)

TABLE B.1 (*Continued*)

Generic Failure Rate Data for Mechanical Components

Component and Failure Mode		Data				Dist. for Failure Prob. (p) or Failure Rate (λ)				
		No. of Failures	No. of Demands (d) or Hours (hr)	d or hr	No. of Comps	Dist.	Mean	α	β	EF
Positive Displacement Pump Fails To Start, Normally Standby	PDP-FTS-NS	10	9,064	d	72	Beta	1.16E-03	10.500	9.050E+03	1.6
Positive Displacement Pump Fails To Run <1H	PDP-FTR<1H	1	4,045	hr	72	Gamma	3.71E-04	1.500	4.050E+03	3.3
Positive Displacement Pump Fails To Run >1H	PDP-FTR>1H	0	1,505	hr	72	Gamma	3.32E-04	0.500	1.500E+03	8.4
Positive Displacement Pump External Leakage (Small)	PDP-ELS	15	21,211,980	hr	171	Gamma	7.31E-07	15.500	2.120E+07	1.5
Positive Displacement Pump External Leakage (Rupture)	PDP-ELL	(Note 1)	–	hr	171	Gamma	5.12E-08	0.300	5.863E+06	18.8
Pump Volute Fails To Run (Driver Independent Centrifugal Pumps)	PMP-Volute	16	133,247	hr	208	Gamma	1.24E-04	16.500	1.330E+05	1.5
Diesel Generator Fails To Start, Normally Standby	EDG-FTS	136	61,363	d	234	Beta	2.22E-03	23.800	1.070E+04	1.4
Diesel Generator Fails To Load And Run, Early	EDG-FTLR	172	53,343	hr	234	Gamma	3.31E-03	3.610	1.090E+03	2.2
Diesel Generator Fails To Run, Late Term	EDG-FTR	155	137,584	hr	234	Gamma	1.18E-03	3.830	3.250E+03	2.1
Hydro Turbine Generator Fails To Start	HTG-FTS	6	6,362	d	2	Beta	1.02E-03	6.500	6.360E+03	1.8
Hydro Turbine Generator Fails To Load And Run, Early	HTG-FTLR	2	4,582	hr	2	Gamma	5.46E-04	2.500	4.580E+03	2.5
Hydro Turbine Generator Fails To Run, Late Term	HTG-FTR	1	13,874	hr	2	Gamma	1.08E-04	1.500	1.390E+04	3.3
Combustion Turbine Generator Fails To Start, Normally Standby	CTG-FTS	21	419	d	3	Beta	7.03E-02	1.200	1.590E+01	3.5
Combustion Turbine Generator Fails To Load And Run, Early Term	CTG-FTLR	2	360	d	2	Gamma	6.94E-03	2.500	3.600E+02	2.5

(Continued)

TABLE B.1 (*Continued*)

Generic Failure Rate Data for Mechanical Components

		Data				Dist. for Failure Prob. (p) or Failure Rate (λ)				
Component and Failure Mode		No. of Failures	No. of Demands (d) or Hours (hr)	d or hr	No. of Comps	Dist.	Mean	α	β	EF
Combustion Turbine Generator Fails To Run, Late Term	CTG-FTR	4	959	hr	3	Gamma	4.69E-03	4.500	9.590E+02	2.0
High-Pressure Core Spray Generator Fails To Start	EDG-FTS-HCS	4	2,114	d	8	Beta	2.13E-03	4.500	2.110E+03	2.0
High-Pressure Core Spray Generator Fails To Run	EDG-FTR-HCS	3	4,196	hr	8	Gamma	8.34E-04	3.500	4.200E+03	2.2
Station Blackout Generator Fails To Start	EDG-FTS-SBO	14	625	d	5	Beta	2.94E-02	0.975	3.220E+01	4.3
Station Blackout Generator Fails To Run	EDG-FTR-SBO	2	2,204	hr	5	Gamma	1.13E-03	2.500	2.200E+03	2.5
Safety Relief Valve Fails To Open	SRV-FTO	7	3,548	d	–	Beta	2.11E-03	7.500	3.542E+03	1.7
BWR Automatic Depressurization System/ Safety Relief Valve Fails To Reclose	SRV-FTC	0	3,548	d	–	Beta	1.41E-04	0.500	3.547E+03	8.4
Safety Relief Valve (BWR Only) Fails To Control	SRV-FC	0	61,005,550	hr	519	Gamma	8.20E-09	0.500	6.100E+07	8.4
Safety Relief Valve Spurious Operation	SRV-SOP	4	61,005,550	hr	519	Gamma	7.38E-08	4.500	6.100E+07	2.0
Safety Relief Valve (BWR Only) Internal Leakage (Small)	SRV-ILS	23	61,005,550	hr	519	Gamma	3.85E-07	23.500	6.100E+07	1.4
Safety Relief Valve (BWR Only) Internal Leakage (Rupture)	SRV-ILL	(Note 1)	–	hr	519	Gamma	7.70E-09	0.300	3.896E+07	18.8
Safety Relief Valve (BWR Only) External Leakage (Small)	SRV-ELS	0	61,005,550	hr	519	Gamma	8.20E-09	0.500	6.100E+07	8.4
Safety Relief Valve (BWR Only) External Leakage (Rupture)	SRV-ELL	(Note 1)	–	hr	519	Gamma	5.74E-10	0.300	5.226E+08	18.8
Code Safety Valve Spurious Operation	SVV-SOP	1	171,647,800	hr	1,380	Gamma	8.74E-09	1.500	1.720E+08	3.3
Code Safety Valve Internal Leakage (Small)	SVV-ILS	5	171,647,800	hr	1,380	Gamma	3.20E-08	5.500	1.720E+08	1.9

(Continued)

TABLE B.1 (*Continued*)

Generic Failure Rate Data for Mechanical Components

Component and Failure Mode		Data				Dist. for Failure Prob. (p) or Failure Rate (λ)				
		No. of Failures	No. of Demands (d) or Hours (hr)	d or hr	No. of Comps	Dist.	Mean	α	β	EF
Code Safety Valve Internal Leakage (Rupture)	SVV-ILL	(Note 1)	–	hr	1,380	Gamma	6.40E-10	0.300	4.688E+08	18.8
Code Safety Valve External Leakage (Small)	SVV-ELS	1	171,647,800	hr	1,380	Gamma	8.74E-09	1.500	1.720E+08	3.3
Code Safety Valve External Leakage (Rupture)	SVV-ELL	(Note 1)	–	hr	1,380	Gamma	6.12E-10	0.300	4.904E+08	18.8
Safety Valve Fails To Open (PWRs)	SVV-FTO-PWR-MSS	0	745	d	–	Beta	6.70E-04	0.499	7.440E+02	8.5
Safety Valve Fails To Close (Main Steam System, PWRs)	SVV-FTC-PWR-MSS	4	745	d	–	Beta	6.03E-03	4.500	7.415E+02	2.0
Safety Valve Spurious Operation (Main Steam System, PWRs)	SVV-SOP-PWR-MSS	0	140,068,800	hr	1,109	Gamma	3.57E-09	0.500	1.400E+08	8.4
Safety Valve Fails To Open (Reactor Coolant System, PWRs)	SVV-FTO-PWR-RCS	0	4	d	–	Beta	6.63E-04	0.499	7.520E+02	8.5
Safety Valve Fails To Close (Reactor Coolant System, PWRs)	SVV-FTC-PWR-RCS	2	4	d	–	Beta	4.13E-02	2.487	5.769E+01	2.5
Safety Valve Spurious Operation (Reactor Coolant System, PWRs)	SVV-SOP-PWR-RCS	1	23,893,310	hr	207	Gamma	6.28E-08	1.500	2.390E+07	3.3
Power-Operated Relief Valve Fails To Open (Reactor Coolant System, PWRs)	PORV-FTO-RCS	4	377	d	–	Beta	1.19E-02	4.500	3.735E+02	2.0
Power-Operated Relief Valve Fails To Close (Reactor Coolant System, PWRs)	PORV-FTC-RCS	1	377	d	–	Beta	3.97E-03	0.494	1.240E+02	8.5
Power-Operated Relief Valve Fails To Open (Main Steam System, PWRs)	PORV-FTO-MSS	25	1,580	d	–	Beta	1.61E-02	25.500	1.556E+03	1.4
Power-Operated Relief Fails To Close (Main Steam System, PWRs)	PORV-FTC-MSS	7	1,580	d	–	Beta	4.35E-03	1.053	2.412E+02	4.1
Power-Operated Relief Fails To Control (Cooldown) (Main Steam System, PWRs)	PORV-FC-MSS	7	278	d	–	Beta	2.69E-02	7.500	2.715E+02	1.7

(*Continued*)

TABLE B.1 (*Continued*)
Generic Failure Rate Data for Mechanical Components

Component and Failure Mode		Data				Dist. for Failure Prob. (p) or Failure Rate (λ)				
		No. of Failures	No. of Demands (d) or Hours (hr)	d or hr	No. of Comps	Dist.	Mean	α	β	EF
Power Operated Relief Spurious Operation	PORV-SOP	13	57,223,460	hr	454	Gamma	2.36E-07	13.500	5.720E+07	1.5
Power-Operated Relief Valve Internal Leakage (Small)	PORV-ILS	3	57,223,460	hr	454	Gamma	6.12E-08	3.500	5.720E+07	2.2
Power-Operated Relief Valve Internal Leakage (Rupture)	PORV-ILL	(Note 1)	–	hr	454	Gamma	1.22E-09	0.300	2.451E+08	18.8
Power-Operated Relief Valve External Leakage (Small)	PORV-ELS	0	57,223,460	hr	454	Gamma	8.74E-09	0.500	5.720E+07	8.4
Power-Operated Relief Valve External Leakage (Rupture)	PORV-ELL	(Note 1)	–	hr	454	Gamma	6.12E-10	0.300	4.904E+08	18.8
Power-Operated Relief Valves Open During LOOP (Reactor Coolant System, PWRs)	PORV-LOOP	–	–	d	–	Point Estimate	9.23E-02	–	–	–
Power-Operated Relief Valves Open During Transient (Reactor Coolant System, PWRs)	PORV-Transient	–	–	d	–	Point Estimate	2.28E-02	–	–	–
Low Capacity Relief Valve Fails To Open	RVL-FTO	0	65	d	12	Beta	7.59E-03	0.500	6.540E+01	8.3
Low Capacity Relief Valve Fails To Close	RVL-FTC	0	65	d	12	Beta	7.59E-03	0.500	6.540E+01	8.3
Low Capacity Relief Valve Spurious Operation	RVL-SOP	0	9,165,162	hr	79	Gamma	5.46E-08	0.500	9.170E+06	8.4
Low Capacity Relief Valve Internal Leakage (Small)	RVL-ILS	3	9,165,162	hr	79	Gamma	3.82E-07	3.500	9.170E+06	2.2
Low Capacity Relief Valve Internal Leakage (Rupture)	RVL-ILL	(Note 1)	–	hr	79	Gamma	7.64E-09	0.300	3.927E+07	18.8
Low Capacity Relief Valve External Leakage (Small)	RVL-ELS	3	9,165,162	hr	79	Gamma	3.82E-07	3.500	9.170E+06	2.2
Low Capacity Relief Valve External Leakage (Rupture)	RVL-ELL	(Note 1)	–	hr	79	Gamma	2.67E-08	0.300	1.122E+07	18.8

(Continued)

TABLE B.1 (*Continued*)
Generic Failure Rate Data for Mechanical Components

Component and Failure Mode		Data				Dist. for Failure Prob. (p) or Failure Rate (λ)				
		No. of Failures	No. of Demands (d) or Hours (hr)	d or hr	No. of Comps	Dist.	Mean	α	β	EF
Battery Charger Fails To Operate	BCH-FTOP	161	99,754,050	hr	781	Gamma	1.76E-06	1.080	6.120E+05	4.1
BatteryFails To Operate	BAT-FTOP	21	52,018,730	hr	412	Gamma	4.05E-07	0.634	1.570E+06	6.5
Automatic Power Transfer Switch Fails To Transfer	ABT-FF	4	3,377	d	27	Beta	1.33E-03	4.500	3.370E+03	2.0
Automatic Power Transfer Switch Spurious Operation	ABT-SOP	0	4,010,342	hr	32	Gamma	1.25E-07	0.500	4.010E+06	8.4
Circuit Breaker Fails To Open/Close	CRB-FTOC	102	119,027	d	3,461	Beta	1.59E-03	0.793	4.990E+02	5.2
Circuit Breaker (All Voltages) Spurious Operation	CRB-SOP	57	552,883,300	hr	4,620	Gamma	1.73E-07	0.465	2.680E+06	9.3
High Voltage (13.8 and 16 kV) Circuit Breaker Fails To Open/Close	CRBHV-FTOC	17	9,198	d	244	Beta	1.90E-03	17.500	9.180E+03	1.4
High Voltage (13.8 and 16 Kv) Circuit Breaker Spurious Operation	CRBHV-SOP	14	37,600,840	hr	300	Gamma	3.86E-07	14.500	3.760E+07	1.5
Medium Voltage (4,160 V and 6.9 kV) Circuit Breaker Fails To Open/Close	CRBMV-FTOC	57	50,897	d	1,080	Beta	2.64E-03	0.466	1.760E+02	9.2
Medium Voltage (4,160 V and 6.9 kV) Circuit Breaker Spurious Operation	CRBMV-SOP	15	149,457,800	hr	1,240	Gamma	1.04E-07	15.500	1.490E+08	1.5
Low Voltage (480 V) Circuit Breaker Fails To Open/Close	CRB-FTOC-480	25	46,176	d	1,752	Beta	8.57E-04	0.497	5.790E+02	8.5
Low Voltage (480 V) Circuit Breaker Spurious Operation	CRB-SOP-480	27	310,690,800	hr	2,630	Gamma	8.85E-08	27.500	3.110E+08	1.3
DC Circuit Breaker Fails To Open/Close	CRBDC-FTOC	5	17,566	d	602	Beta	3.13E-04	5.500	1.760E+04	1.9
DC Circuit Breaker Spurious Operation	CRBDC-SOP	0	34,938,600	hr	270	Gamma	1.43E-08	0.500	3.490E+07	8.4
Inverter Fails To Operate	INV-FTOP	52	24,269,470	hr	199	Gamma	3.49E-06	0.986	2.820E+05	4.4

(*Continued*)

TABLE B.1 (*Continued*)
Generic Failure Rate Data for Mechanical Components

Component and Failure Mode		Data				Dist. for Failure Prob. (p) or Failure Rate (λ)				
		No. of Failures	No. of Demands (d) or Hours (hr)	d or hr	No. of Comps	Dist.	Mean	α	β	EF
AC Bus Fails To Operate	BUS-FTOP-AC	76	160,545,900	hr	1,296	Gamma	5.88E-07	0.986	1.680E+06	4.4
DC Bus Fails To Operate	BUS-FTOP-DC	1	2,103,936	hr	16	Gamma	7.13E-07	1.500	2.100E+06	3.3
Motor Control Center Fails To Operate	MCC-FTOP	7	28,535,130	hr	217	Gamma	2.43E-07	1.020	4.190E+06	4.3
Transformer Fails To Operate	TFM-FTOP	110	60,181,620	hr	512	Gamma	1.93E-06	1.630	8.470E+05	3.1
Sequencer Fails To Operate (as a Subcomponent of the Emergency Diesel Generator)	SEQ-FTOP	6	61,363	d	234	Beta	1.06E-04	6.500	6.140E+04	1.8
Fuse Spurious Operation	FUS-SOP	1	169,366,800	hr	1,288	Gamma	8.86E-09	1.500	1.690E+08	3.3
Strainer Plugging (Dirty Water Systems)	STR-FLT-PG-RAW	6	7,922,615	hr	62	Gamma	8.20E-07	6.500	7.920E+06	1.8
Filter External Leakage (Small) All Systems	STR-FLT-ELS	1	28,097,240	hr	223	Gamma	5.34E-08	1.500	2.810E+07	3.3
Filter External Leakage (Small) All Systems	STR-FLT-ELL	(Note 1)	–	hr	223	Gamma	3.74E-09	0.300	8.026E+07	18.8
Filter Plugging, Clean Systems	STR-FLT-PG-CLEAN	1	8,161,140	hr	68	Gamma	1.84E-07	1.500	8.160E+06	3.3
Clean Systems Passive Filter Bypass	STR-FLT-BYP-CLEAN	0	8,161,140	hr	68	Gamma	6.13E-08	0.500	8.160E+06	8.4
Instrument Air System Filter Plugs	STR-FLT-PG-IAS	0	210,384	hr	2	Gamma	2.38E-06	0.500	2.100E+05	8.4
Self Cleaning Filter Plugging	STR-FLTSC-PG	32	21,560,060	hr	167	Gamma	1.51E-06	32.500	2.160E+07	1.3
Self Cleaning Filter Bypass	STR-FLTSC-BYP	0	21,560,060	hr	167	Gamma	2.32E-08	0.500	2.160E+07	8.4
Self Cleaning Filter Fails To Operate	STR-FLTSC-FTOP	53	21,560,060	hr	167	Gamma	2.48E-06	53.500	2.160E+07	1.2
Self Cleaning Filter External Leakage (Small)	STR-FLTSC-ELS	2	21,560,060	hr	167	Gamma	1.16E-07	2.500	2.160E+07	2.5
Self Cleaning Filter External Leakage (Rupture)	STR-FLTSC-ELL	(Note 1)	–	hr	167	Gamma	8.12E-09	0.300	3.695E+07	18.8
Normally Running Service Water Strainer Plugs	STR-FLTSC-PG-SWN	19	13,235,010	hr	103	Gamma	1.47E-06	19.500	1.320E+07	1.4

(*Continued*)

TABLE B.1 (*Continued*)
Generic Failure Rate Data for Mechanical Components

Component and Failure Mode		Data				Dist. for Failure Prob. (p) or Failure Rate (λ)				
		No. of Failures	No. of Demands (d) or Hours (hr)	d or hr	No. of Comps	Dist.	Mean	α	β	EF
Standby Service Water Strainer Plugs	STR-FLTSC-PG-SSW	13	7,799,060	hr	60	Gamma	1.73E-06	13.500	7.800E+06	1.5
Standby Service Water Strainer Plugging, Environmental	STR-FLTSC-PG-EE-SSW	1	7,799,060	hr	60	Gamma	1.92E-07	1.500	7.800E+06	3.3
Containment Sump Plugging (BWRs, Suppression Pool Strainers)	STR-PG-SUMP-BWR	0	5,522,832	hr	42	Gamma	9.05E-08	0.500	5.520E+06	8.4
Containment Sump Plugging (PWRs)	STR-PG-SUMP-PWR	1	3,528,454	hr	29	Gamma	4.25E-07	1.500	3.530E+06	3.3
Traveling Screen Plugging	TSA-PG	37	25,155,920	hr	205	Gamma	1.49E-06	37.500	2.520E+07	1.3
Traveling Screen Bypass	TSA-BYP	2	25,155,920	hr	205	Gamma	9.94E-08	2.500	2.520E+07	2.5
Traveling Screen Fails To Operate	TSA-FTOP	45	25,155,920	hr	205	Gamma	2.12E-06	0.547	2.590E+05	7.6
Standby Service Water Traveling Screen Plugs	TSA-PG-SSW	0	1,972,440	hr	15	Gamma	2.53E-07	0.500	1.970E+06	8.4
Trash Rack Plugging	TRK-PG	0	1,314,960	hr	10	Gamma	3.80E-07	0.500	1.310E+06	8.4
Bistable Fails To Operate	BIS-FTOP	55	102,094	d	–	Beta	5.44E-04	0.500	9.193E+02	8.4
Process Logic (Flow) Fails To Operate	PLF-FTOP	(Note 2)	6,075	d	–	Beta	6.25E-04	0.500	7.990E+02	8.4
Process Logic (Level) Fails To Operate	PLL-FTOP	3	6,075	d	–	Beta	6.25E-04	0.500	7.990E+02	8.4
Process Logic (Pressure) Fails To Operate	PLP-FTOP	6	38,115	d	–	Beta	1.60E-04	0.500	3.124E+03	8.4
Process Logic (Delta Temperature) Fails To Operate	PLDT-FTOP	24	4,887	d	–	Beta	5.07E-03	0.500	9.805E+01	8.4
Sensor/Transmitter (Flow) Fails To Operate on Demand	STF-FTOP-D	(Note 2)	6,750	d	–	Beta	8.15E-04	0.500	6.132E+02	8.4
Sensor/Transmitter (Flow) Fails To Operate per Hour	STF-FTOP-R	(Note 2)	9,831,970	hr	–	Gamma	1.02E-07	0.500	4.916E+06	8.4

(*Continued*)

TABLE B.1 (*Continued*)
Generic Failure Rate Data for Mechanical Components

		Data				Dist. for Failure Prob. (p) or Failure Rate (λ)				
Component and Failure Mode		**No. of Failures**	**No. of Demands (d) or Hours (hr)**	**d or hr**	**No. of Comps**	**Dist.**	**Mean**	**α**	**β**	**EF**
Sensor/Transmitter (Level) Fails To Operate on Demand	STL-FTOP-D	5	6,750	d	–	Beta	8.15E-04	0.500	6.132E+02	8.4
Sensor/Transmitter (Level) Fails To Operate per Hour	STL-FTOP-R	0	9,831,970	hr	–	Gamma	1.02E-07	0.500	4.916E+06	8.4
Sensor/Transmitter (Pressure) Fails To Operate on Demand	STP-FTOP-D	2	23,960	d	–	Beta	1.17E-04	0.500	4.278E+03	8.4
Sensor/Transmitter (Pressure) Fails To Operate per Hour	STP-FTOP-R	35	43,430,500	hr	–	Gamma	8.22E-07	0.500	6.083E+05	8.4
Sensor/Transmitter (Temperature) Fails To Operate on Demand	STT-FTOP-D	17	40,759	d	–	Beta	4.32E-04	0.500	1.157E+03	8.4
Sensor/Transmitter (Temperature) Fails To Operate per Hour	STT-FTOP-R	29	35,107,400	hr	–	Gamma	8.40E-07	0.500	5.950E+05	8.4
RPS Breaker (Mechanical) Fails To Open/Close	RTB-FTOC-BME	1	97,359	d	–	Beta	1.54E-05	0.500	3.245E+04	8.4
RPS Breaker (Shunt Trip) Fails To Operate	RTB-FTOP-BSN	14	44,104	d	–	Beta	3.29E-04	0.500	1.520E+03	8.4
RPS Breaker (Undervoltage Trip) Fails To Operate	RTB-FTOP-BUV	23	57,199	d	–	Beta	4.13E-04	0.500	1.211E+03	8.4
RPS Breaker (Combined) Fails To Open/Close	RTB-FTOC	–	–	d	–	Beta	1.55E-05	0.500	3.217E+04	8.4
Manual Switch Fails To Open/Close	MSW-FTOC	2	19,789	d	–	Beta	1.26E-04	0.500	3.958E+03	8.4
Relay Fails To Operate	RLY-FTOP	24	974,417	d	–	Beta	2.48E-05	0.500	2.013E+04	8.4
Control Rod Drive Fails To Insert Rod	CRD-FTOP	19	145,016,900	d	1,198	Gamma	1.68E-07	0.560	3.340E+06	7.4
Control Rod Drive Spurious Operation	CRD-SOP	23	145,016,900	hr	1,198	Gamma	1.62E-07	23.500	1.450E+08	1.4
Control Rod Fails To Operate/ Insert Rod	ROD-FTOP	10	110,389,200	d	844	Gamma	9.51E-08	10.500	1.100E+08	1.6

(*Continued*)

TABLE B.1 (*Continued*)
Generic Failure Rate Data for Mechanical Components

Component and Failure Mode		Data				Dist. for Failure Prob. (p) or Failure Rate (λ)				
		No. of Failures	No. of Demands (d) or Hours (hr)	d or hr	No. of Comps	Dist.	Mean	α	β	EF
Control Rod Spurious Operation	ROD-SOP	11	110,389,200	hr	844	Gamma	1.04E-07	11.500	1.100E+08	1.6
Hydraulic Control Unit Components Fail	HCU-FTI	–	–	d	–	Lognormal	1.10E-07	20.000	--	20.0
Hydraulic Control Unit Fails To Operate	HCU-FTOP	19	1,347,114,000	hr	10,425	Gamma	1.45E-08	19.500	1.350E+09	1.4
Hydraulic Control Unit Spurious Operation	HCU-SOP	27	1,347,114,000	hr	10,425	Gamma	1.99E-08	4.300	2.160E+08	2.1
Air-Operated Damper Fails To Open/Close	AOD-FTOC	0	6,602	d	50	Beta	7.57E-05	0.500	6.600E+03	8.4
Air-Operated Damper Spurious Operation	AOD-SOP	4	24,287,000	hr	207	Gamma	1.61E-07	0.579	3.600E+06	7.1
Air-Operated Damper Internal Leakage (Small)	AOD-ILS	3	24,287,000	hr	207	Gamma	1.44E-07	3.500	2.430E+07	2.2
Air-Operated Damper Internal Leakage (Rupture)	AOD-ILL	(Note 1)	–	hr	207	Gamma	2.88E-09	0.300	1.042E+08	18.8
Hydraulic-Operated Damper Fails To Open/Close	HOD-FTOC	4	6,113	d	42	Beta	7.36E-04	4.500	6.110E+03	2.0
Hydraulic-Operated Damper Spurious Operation	HOD-SOP	2	16,454,520	hr	126	Gamma	1.52E-07	2.500	1.650E+07	2.5
Hydraulic-Operated Damper Internal Leakage (Small)	HOD-ILS	0	16,454,520	hr	126	Gamma	3.04E-08	0.500	1.650E+07	8.4
Hydraulic-Operated Damper Internal Leakage (Rupture)	HOD-ILL	(Note 1)	–	hr	126	Gamma	6.08E-10	0.300	4.934E+08	18.8
Motor-Operated DamperFails To Open	MOD-FTOC	11	28,949	d	52	Beta	3.56E-04	0.981	2.760E+03	4.4
Motor-Operated Damper Spurious Operation	MOD-SOP	0	14,134,270	hr	109	Gamma	3.54E-08	0.500	1.410E+07	8.4
Motor-Operated Damper Internal Leakage (Small)	MOD-ILS	0	14,134,270	d	109	Gamma	3.54E-08	0.500	1.410E+07	8.4
Motor-Operated Damper Internal Leakage (Rupture)	MOD-ILL	(Note 1)	–	hr	109	Gamma	7.08E-10	0.300	4.237E+08	18.8

(*Continued*)

TABLE B.1 (*Continued*)

Generic Failure Rate Data for Mechanical Components

		Data				Dist. for Failure Prob. (p) or Failure Rate (λ)				
Component and Failure Mode		No. of Failures	No. of Demands (d) or Hours (hr)	d or hr	No. of Comps	Dist.	Mean	α	β	EF
Air Handling Unit Fails To Start, Normally Running	AHU-FTS-NR	23	15,981	d	145	Beta	1.47E-03	23.500	1.600E+04	1.4
Air Handling Unit Fails To Run, Normally Running	AHU-FTR-NR	39	15,131,330	hr	145	Gamma	2.61E-06	39.500	1.510E+07	1.3
Air Handling Unit Fails To Start, Normally Standby	AHU-FTS-NS	33	158,866	d	403	Beta	2.11E-04	33.500	1.590E+05	1.3
Air Handling Unit Fails To Run <1H, Normally Standby	AHU-FTR<1H	0	147,963	hr	395	Gamma	3.38E-06	0.500	1.480E+05	8.4
Air Handling Unit Fails To Run >1H, Normally Standby	AHU-FTR>1H	27	9,928,068	hr	403	Gamma	2.77E-06	27.500	9.930E+06	1.3
Chiller Unit Fails To Start, Normally Running	CHL-FTS-NR	66	21,137	d	92	Beta	5.09E-03	0.438	8.560E+01	10.0
Chiller Unit Fails To Run, Normally Running	CHL-FTR-NR	179	7,250,769	hr	92	Gamma	3.87E-05	0.524	1.350E+04	8.0
Chiller Unit Fails To Start, Normally Standby	CHL-FTS-NS	0	18,006	d	64	Beta	2.78E-05	0.500	1.800E+04	8.4
Chiller Unit Fails To Run <1H, Normally Standby	CHL-FTR<1H	34	233,781	hr	64	Gamma	1.48E-04	34.500	2.340E+05	1.3
Chiller Unit Fails To Run >1H, Normally Standby	CHL-FTR>1H	34	233,781	hr	64	Gamma	1.48E-04	34.500	2.340E+05	1.3
HVC Fan Fails To Start, Normally Standby	FAN-FTS-NS	17	63,511	d	154	Beta	2.76E-04	17.500	6.350E+04	1.5
HVC Fan Fails To Run <1H, Normally Standby	FAN-FTR<1H	17	39,405	hr	133	Gamma	4.44E-04	17.500	3.940E+04	1.5
HVC Fan Fails To Run >1H, Normally Standby	FAN-FTR>1H	3	120,200	hr	154	Gamma	2.91E-05	3.500	1.200E+05	2.2
HVC Fan Fails To Start, Normally Running	FAN-FTS-NR	28	87,323	d	233	Beta	7.15E-04	0.456	6.360E+02	9.5
HVC Fan Fails To Run, Normally Running	FAN-FTR-NR	50	16,050,850	hr	233	Gamma	3.23E-06	0.674	2.090E+05	6.1
Motor-Driven Compressor Fails To Start, Normally Running	MDC-FTS-NR	52	7,855	d	65	Beta	1.36E-02	0.456	3.310E+01	9.3

(Continued)

TABLE B.1 (*Continued*)
Generic Failure Rate Data for Mechanical Components

		Data				Dist. for Failure Prob. (p) or Failure Rate (λ)				
Component and Failure Mode		No. of Failures	No. of Demands (d) or Hours (hr)	d or hr	No. of Comps	Dist.	Mean	α	β	EF
Motor-Driven Compressor Fails To Run	MDC-FTR-NR	173	4,802,083	hr	65	Gamma	4.03E-05	2.690	6.680E+04	2.5
Motor-Driven Compressor Fails To Start, Normally Standby	MDC-FTS-NS	34	21,074	d	57	Beta	2.93E-03	0.847	2.890E+02	4.9
Motor-Driven Compressor Fails To Run <1H	MDC-FTR<1H	1	20,248	hr	54	Gamma	7.41E-05	1.500	2.020E+04	3.3
Motor-Driven Compressor Fails To Run >1H	MDC-FTR>1H	90	1,573,366	hr	57	Gamma	5.75E-05	90.500	1.570E+06	1.2
Engine-Driven Compressor Fails To Start, Normally Standby	EDC-FTS-NS	14	1,459	d	4	Beta	9.93E-03	14.500	1.450E+03	1.5
Engine-Driven Compressor Fails To Run <1H, Normally Standby	EDC-FTR<1H	1	1,459	hr	4	Gamma	1.03E-03	1.500	1.460E+03	3.3
Engine-Driven Compressor Fails To Run >1H, Normally Standby	EDC-FTR>1H	12	1,609	hr	4	Gamma	7.77E-03	12.500	1.610E+03	1.5
Engine-Driven Compressor Fails To Run, Normally Running	EDC-FTR-NR	10	163,321	d	3	Gamma	6.43E-05	10.500	1.630E+05	1.6
Instrument Air System Motor-Driven Compressor Fails To Run	MDC-FTR-IAS	117	2,376,803	hr	36	Gamma	4.93E-05	7.620	1.540E+05	1.7
Containment Instrument Air Motor-Driven Compressor Fails To Run	MDC-FTR-CIA	0	98,561	hr	2	Gamma	5.07E-06	0.500	9.860E+04	8.4
Air Dryer Unit Fails To Operate	ADU-FTOP	–	–	hr	0	Gamma	5.00E-06	0.300	6.000E+04	18.8
Accumulator Fails To Operate	ACC-FTOP	11	79,315,180	hr	617	Gamma	1.45E-07	11.500	7.930E+07	1.6
Accumulator External Leakage (Small)	ACC-ELS	8	79,315,180	hr	617	Gamma	1.07E-07	8.500	7.930E+07	1.7
Accumulator External Leakage (Rupture)	ACC-ELL	(Note 1)	–	hr	617	Gamma	7.49E-09	0.300	4.005E+07	18.8
Cooling Tower Fan Fails To Start (Standby)	CTF-FTS-NS	14	37,307	d	55	Beta	3.89E-04	14.500	3.730E+04	1.5
Cooling Tower Fan Fails To Run <1H (Standby)	CTF-FTR<1H	0	37,231	hr	54	Gamma	1.34E-05	0.500	3.720E+04	8.4

(*Continued*)

TABLE B.1 (*Continued*)
Generic Failure Rate Data for Mechanical Components

Component and Failure Mode		Data				Dist. for Failure Prob. (p) or Failure Rate (λ)				
		No. of Failures	No. of Demands (d) or Hours (hr)	d or hr	No. of Comps	Dist.	Mean	α	β	EF
Cooling Tower Fan Fails To Run >1H (Standby)	CTF--FTR>1H	0	895,323	hr	55	Gamma	5.58E-07	0.500	8.950E+05	8.4
Cooling Tower Fan Fails To Start	CTF-FTS-NR	1	2,239	d	20	Beta	6.70E-04	1.500	2.240E+03	3.3
Cooling Tower Fan Fails To Run	CTF-FTR-NR	6	1,253,930	hr	20	Gamma	5.18E-06	6.500	1.250E+06	1.8
Tank Rupture	TNK-FC	16	46,469,300	hr	383	Gamma	4.18E-07	0.420	1.000E+06	10.7
Pressurized Liquid Tank External Leakage (Small)	TNK-PRESS-LIQ-ELS	5	19,535,510	hr	156	Gamma	2.51E-07	0.489	1.950E+06	8.7
Pressurized Liquid Tank External Leakage (Rupture)	TNK-PRESS-LIQ-ELL	(Note 1)	–	hr	156	Gamma	1.76E-08	0.300	1.707E+07	18.8
Unpressurized Liquid Tank External Leakage (Small)	TNK-UNPRESS-LIQ-ELS	4	22,725,910	hr	195	Gamma	1.98E-07	4.500	2.270E+07	2.0
Unpressurized Liquid Tank External Leakage (Rupture)	TNK-UNPRESS-LIQ-ELL	(Note 1)	–	hr	195	Gamma	1.39E-08	0.300	2.165E+07	18.8
Instrument Air System Tank Fails To Control	TNK-FC-IAS	0	3,287,400	hr	25	Gamma	1.52E-07	0.500	3.290E+06	8.4
Standby Service Water Tank Fails To Control	TNK-FC-SSW	0	880,966	hr	7	Gamma	5.68E-07	0.500	8.810E+05	8.4
Gas Tank External Leakage (Small)	TNK-GAS-ELS	0	4,207,872	hr	32	Gamma	1.19E-07	0.500	4.210E+06	8.4
Gas Tank External Leakage (Rupture)	TNK-GAS-ELL	(Note 1)	–	hr	32	Gamma	8.33E-09	0.300	3.601E+07	18.8
Orifice Plugging	ORF-PG	–	–	hr	0	Gamma	1.00E-06	0.300	3.000E+05	18.8
Piping Non-Service Water System External Leak Small	PIPE-OTHER-ELS	5	15,830,000,000	hr-ft	0	Gamma	2.53E-10	0.500	1.979E+09	8.4
Piping Non-Service Water System External Leak Large	PIPE-OTHER-ELL	(Note 1)	15,830,000,000	hr-ft	0	Gamma	2.53E-11	0.300	1.187E+10	18.8
Piping Service Water System External Leak Small	PIPE-SWS-ELS	9	13,060,000,000	hr-ft	0	Gamma	6.89E-10	0.500	7.256E+08	8.4

(*Continued*)

TABLE B.1 (*Continued*)
Generic Failure Rate Data for Mechanical Components

Component and Failure Mode		Data: No. of Failures	Data: No. of Demands (d) or Hours (hr)	Data: d or hr	Data: No. of Comps	Dist. for Failure Prob. (p) or Failure Rate (λ): Dist.	Mean	α	β	EF
Piping Service Water System External Leak Large	PIPE-SWS-ELL	(Note 1)	13,060,000,000	hr-ft	0	Gamma	1.38E-10	0.300	2.177E+09	18.8
Heat Exchanger Plugging/Loss of Heat Transfer	HTX-LOHT	67	222,831,700	hr	1,750	Gamma	3.39E-07	0.483	1.420E+06	8.8
Heat Exchanger Internal Leakage (Small)	HTX-ILS	61	222,831,700	hr	1,750	Gamma	2.76E-07	61.500	2.230E+08	1.2
Heat Exchanger Internal Leakage (Rupture)	HTX-ILL	(Note 1)	–	hr	1,750	Gamma	5.52E-09	0.300	5.435E+07	18.8
Heat Exchanger External Leakage (Small)	HTX-ELS	38	222,831,700	hr	1,750	Gamma	1.90E-07	0.825	4.350E+06	5.0
Heat Exchanger External Leakage (Rupture)	HTX-ELL	(Note 1)	–	hr	1,750	Gamma	2.85E-08	0.300	1.053E+07	18.8
Heat Exchanger Plugging Non-Standby	HTX-PG-CCW	8	28,273,230	hr	223	Gamma	3.01E-07	8.500	2.830E+07	1.7
Component Cooling Water Heat Exchanger Plugging Non-ExEE (hr^{-1})	HTX-PG-NE-CCW	3	28,273,230	hr	223	Gamma	1.24E-07	3.500	2.830E+07	2.2
Instrumentation and Control Circuit Fails	ICC-FC	0	–	d	0	Lognorm	1.70E-03	2.200	–	2.2
Failure of Instrument and Control for a Turbine Trip	ICC-FA	0	–	d	0	Lognorm	1.70E-03	2.200	–	2.2
Automatic Actuation and Control Circuit Fails	ACT-FC	0	–	d	0	Lognorm	1.70E-03	2.200	–	2.2

Notes:

- Note 1: External and internal large leakage (ELL and ILL) events are defined as greater than 50 gpm. Because ELL and ILL events are rare, good estimates for ELL and ILL cannot be obtained using data from only one component. The NUREG/CR-6928 study (Table A.1.2-1 of NUREG/CR-6928) shows the mean of ELL is the ELS mean multiplied by 0.07 for pump, valves, tanks, and heat exchanger shells, multiplied by 0.2 for Emergency Service Water (ESW) pipe, multiplied by 0.1 for non-ESW pipe, and multiplied by 0.15 for heat exchanger tubes. The ILL mean is the ILS mean multiplied by 0.02.
- Note 2: The flow process logic (PLF) reliability was estimated by using the level process logic (PLL) data. The flow sensor/transmitter (STF) reliability was estimated by using the level sensor/transmitter (STL) data.

Index

Printed in the United States
by Baker & Taylor Publisher Services

Printed in the United States
by Baker & Taylor Publisher Services